Android 核心编程

——Activity、BroadcastReceiver、Service与ContentProvider实战

温淑鸿　田沛　主编

朱兵　孙象然　编著

清华大学出版社

北　京

内容简介

本书是一本以 Android 应用开发为主题的基础教材，读者对象为已经具有 Java 基础的高等院校学生、开发人员。本书配有大量的实验案例，实验步骤特别详细，非常适合初学者入门。

本书的读者对象是高等院校计算机类、电子类、电气类、控制类等专业本科生和初学 Android 应用程序开发的技术人员。

图书在版编目(CIP)数据

Android 核心编程：Activity、BroadcastReceiver、Service 与 ContentProvider 实战/温淑鸿等主编. —北京：清华大学出版社，2019

ISBN 978-7-302-53113-5

Ⅰ. ①A…　Ⅱ. ①温…　Ⅲ. ①移动终端－应用程序－程序设计　Ⅳ. ①TN929.53

中国版本图书馆 CIP 数据核字(2019)第 101133 号

责任编辑：赵　凯
封面设计：何凤霞
责任校对：焦丽丽
责任印制：杨　艳

出版发行：清华大学出版社
　　网　　址：http://www.tup.com.cn，http://www.wqbook.com
　　地　　址：北京清华大学学研大厦 A 座　　**邮　　编**：100084
　　社 总 机：010-62770175　　**邮　　购**：010-62786544
　　投稿与读者服务：010-62776969，c-service@tup.tsinghua.edu.cn
　　质量反馈：010-62772015，zhiliang@tup.tsinghua.edu.cn
　　课件下载：http://www.tup.com.cn，010-62795954
印 装 者：清华大学印刷厂
经　　销：全国新华书店
开　　本：185mm×260mm　　**印　　张**：35.25　　**字　　数**：855 千字
版　　次：2019 年 9 月第 1 版　　**印　　次**：2019 年 9 月第 1 次印刷
定　　价：99.00 元

产品编号：081943-01

前言

PREFACE

Android 是一个流行、开源的移动终端平台，一直保持高速的增长率，众多开发人员已把 Android 应用开发列为重点选择。

本书是一本以 Android 的应用开发为主题的基础教材，面向已经具有 Java 基础的高等院校学生和开发人员。通过对 Android 平台基础知识以及应用程序开发基本技术的讲解，帮助读者迅速掌握 Android 应用开发技能，为今后从事基于 Android 的应用软件开发打下坚实的基础。

本书介绍了 Android 四类应用组件，每一章尽可能把理论部分简明扼要地讲清楚，又在每一章后面都附有实验，实验步骤特别详细，便于初学者学习，以便加深对理论知识的理解，达到 Android 应用开发入门的目的。

第 1 章：搭建 Android 开发环境，主要包括 Android Studio、SDK 的安装，以及插件 Android Code Generator、Android Parcelable Code Generator 和 Android Layout ID Converter for IntelliJ IDEA 的安装。

第 2 章：Android 清单文件，主要介绍了清单文件的作用和语法。

第 3 章：Android resource 介绍，主要介绍了字符串资源、菜单资源、Drawable 资源中的 Shape Drawable、StateListDrawable 和 LayerDrawable。

第 4 章：Gradle 的 Android 插件，主要介绍了 Gardle 初始化脚本的文件位置、工程级别和模块级别的 build 脚本的配置、尤其是依赖配置。

第 5 章：Activity 与 Fragment，主要介绍了怎样创建和协调 Activity、在清单文件中注册组件、使用< intent-filter >元素声明组件功能、Activity 生命周期、任务与回退栈的管理，以及 Android 结构组件 LiveDate 和 ViewModel。

第 6 章：Intent 和 IntentFilter，主要介绍了 Intent 的主要成员变量，Intent 和 IntentFilter 的匹配规则。

第 7 章：线性、表格、栅格、相对布局与帧布局，主要介绍了这几种布局以及 CardView 的基本使用方法。

第 8 章：ConstraintLayout，主要介绍了用于控制视图在水平方向或者垂直方向位置的两种约束：沿水平或者垂直方向的单条边约束、双条边约束，以及三种链条的特点，Guideline 的使用。

第 9 章：TextView，主要介绍了 TextView、EditText 和 Button 的属性及使用方法。

第 10 章：Android 的双向数据绑定，主要介绍了实现双向数据绑定的步骤。

第 11 章：滚动与翻页，介绍了垂直方向的滚动与嵌套滚动、水平方向的滚动与翻页，主要包括 RecyclerView、ScrollView、HorizontalScrollView、CoordinatorLayout、NestedScrollView、

Toolbar、AppBarLayout、ViewPager、DrawerLayout 和 NavigationView。

第 12 章：BroadcastReceiver，主要介绍了普通广播、顺序广播和用于同一进程内广播的 LocalBroadcastManager。

第 13 章：Handler 与 Service，主要介绍了 Handler 发送和处理 Message 的机制，系统通知的发送，Service 的两种工作方式：启动的 Service 和绑定的 Service，以及创建绑定的 Service 的三种方法：继承 Binder 类进行同一进程内通信，通过 Messenger 进行跨进程通信、使用 AIDL 语言定义跨进程的通信接口。

第 14 章：数据存储，主要介绍了用于保存和读取少量基本数据类型的 SharedPreferences(用于应用私有的数据，形式为键值对)，以文件的形式保存应用私有的数据到内存存储或者外部存储，保存所有应用可以共享的数据到外部存储的公共区域、以 sqlite 保存结构化的数据。还介绍了 SQLite 语句的语法，SQLiteDatabase 类的用法以及 Room 持久库的使用。

第 15 章：ContentProvider，主要介绍了 ContentProvider 和它的子类 FileProvider 的使用方法。

第 16 章：访问互联网，主要介绍了使用 HttpURLConnection 类访问互联网的使用方法，JSON 的数据格式，以及 JSON 和 XML 文件的解析。

由于作者水平有限，编写时间仓促，书中难免存在疏漏和不足。敬请读者不吝赐教，对本书给予建议和批评指正，以便我们在改版或再版的时候及时纠正补充。

作　者

2019 年 3 月

目录

CONTENTS

第 1 章 搭建 Android 开发环境

CHAPTER 1

Android 是 Google 于 2007 年 11 月 5 日发布的基于 Linux 内核的开源手机操作系统的名称，该平台由操作系统、中间件、用户界面和应用软件组成，Android 系统架构如图 1-1 所示。

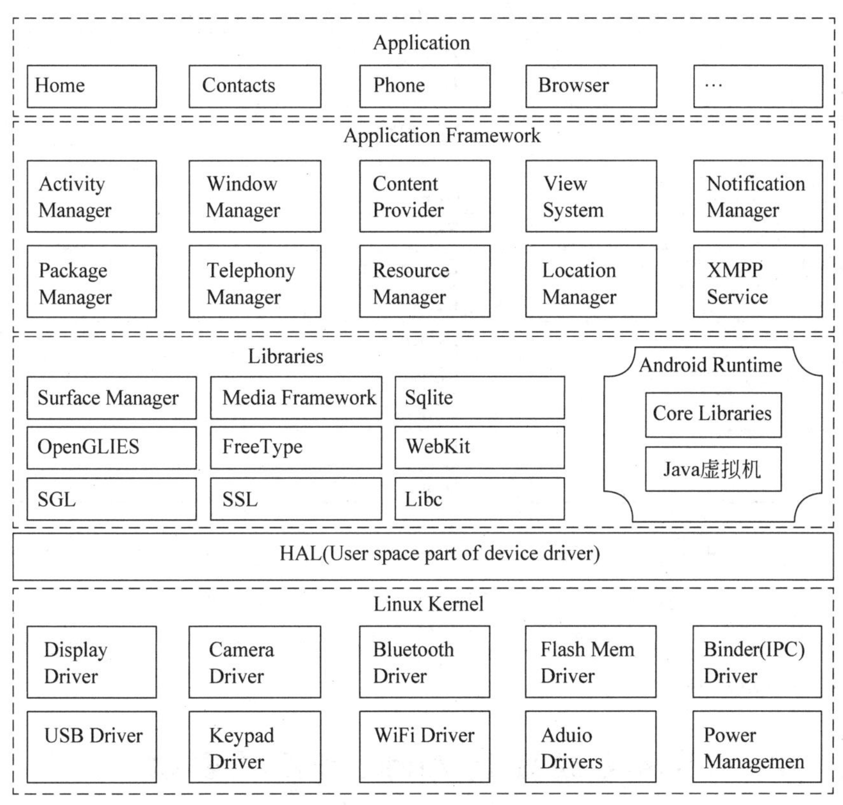

图 1-1　Android 系统架构

Android API level 与版本对应关系如表 1-1 所示。

表 1-1　Android API level 与版本对应关系

平台版本	版本编码	API level
Android 9.0	Pie	28
Android 8.1	Oreo	27

续表

平台版本	版本编码	API level
Android 8.0	Oreo	26
Android 7.1.1	Nougat	25
Android 7.0	Nougat	24
Android 6.0	Marshmallow	23
Android 5.1.1	Lollipop_Mr1	22
Android 5.0.1	Lollipop	21
Android 4.4W	Kitkat_Watch	20
Android 4.4.2	Kitkat	19
Android 4.3	Jelly_Bean_Mr2	18
Android 4.2，4.2.2	Jelly_Bean_Mr1	17
Android 4.1，4.1.1	Jelly_Bean	16
Android 4.0.3，4.0.4	Ice_Cream_Sandwich_Mr1	15
Android 4.0，4.0.1，4.0.2	Ice_Cream_Sandwich	14
Android 3.2	Honeycomb_Mr2	13
Android 3.1.x	Honeycomb_Mr1	12
Android 3.0.x	Honeycomb	11
Android 2.3.3,2.3.4	Gingerbread_Mr1	10
Android 2.3，2.3.1，2.3.2	Gingerbread	9
Android 2.2.x	Froyo	8
Android 2.1.x	Eclare_Mr1	7
Android 2.0.1	Eclare_0_1	6
Android 2.0	Eclare	5
Android 1.6	Donut	4
Android 1.5	Cupcake	3
Android 1.1	Base_1_1	2
Android 1.0	Base	1

本章将介绍 Android 开发环境的安装和配置，首先，安装 Java 开发工具包(JDK)，然后安装 Android Studio 以及 Android 软件开发工具包(SDK)，这些都是开发 Android 应用程序所必需的工具，还可以选择安装一些常用的插件。如果已经安装好了 Android 开发环境，阅读时可以跳过这一章。

1.1 Windows 系统安装 Android 系统要求

在 Windows 系统下安装 Android Studio，需要 Microsoft Windows 7/8/10(64 位系统)，最低 3GB 内存、10GB 可用磁盘空间，最低屏幕分辨率 1280×800。对于加速模拟器，需要 64 位操作系统、支持 Intel VT-x、Intel EM64T(Intel 64)和禁止执行(XD)位功能的 Intel 处理器。Intel Hardware Accelerated Execution Manager 在 Windows XP 系统中不能正常安装，即不能正常运行 intelhaxm-android.exe。

Android Studio 使用 Java 编译环境构建，安装 Android Studio 就同时在安装目录的 jre

目录里安装了 Java 开发工具包(JDK),所以单独安装 JDK 不是必需的。当然也可以从 Oracle 网站上下载 JDK,例如 jdk-8u91-windows-x64. exe 并安装。

安装 JDK 之前,在计算机 C 盘的根目录下新建一个名为 android 的文件夹,将 Android Studio 相关的软件全部安装在 C:\android 文件夹内,这使得开发环境更有条理。为了使 Java、Android Studio、SDK 安装在 C:\android 目录,可以在 C:\android 文件夹内再创建一文件夹名为 Java,再在 C:\android\Java 文件夹内创建文件夹 jdk1. 8. 0_91 和 jre1. 8. 0_91,以便把 jdk-8u91-windows-x64. exe 安装到此文件夹。找到下载的安装文件 jdk-8u91-windows-x64. exe,然后双击运行。安装开始后,可以看到安装向导。单击"下一步"按钮,可以看见 Java 定制安装界面,单击"更改"按钮,更改安装目录的位置为 C:\android\Java\jdk1. 8. 0_91 目录,单击"确定"按钮,单击弹出界面中的"下一步"按钮,直到安装完成。JDK 安装完成后,继续安装 Java 运行环境(JRE),单击"更改"按钮,把 JRE 安装目录设置在 C:\android\Java\jre1. 8. 0_91,单击"确定"按钮,单击弹出界面中的"下一步"按钮,单击"关闭"按钮。这时,jdk-8u91-windows-x64. exe 就安装好了。

在安装 JDK 后,需要配置 JDK 环境变量。配置过程如下,按住 Windows 键和 Pause 键打开"控制面板|系统和安全|系统"设置界面,单击界面左边的"高级系统设置"选项,单击"环境变量"按钮,出现环境变量设置界面,在"系统变量"列表中找到 JAVA_HOME 变量。如果 JAVA_HOME 不存在,单击"新建"按钮来创建,否则,单击"编辑"按钮,出现系统变量设置对话框,如图 1-2 所示,在变量名中输入 JAVA_HOME,然后在变量值中输入刚才 JDK 的安装位置 C:\android\Java\jdk1. 8. 0_91(去掉最后面斜杠),然后单击"确定"按钮。

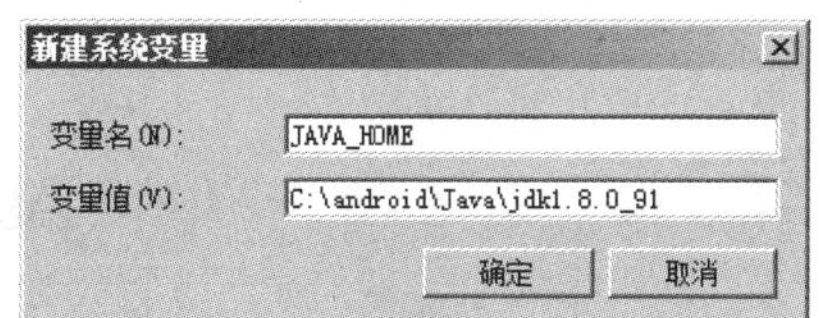

图 1-2 Window 系统变量设置界面

同样,像编辑 JAVA_HOME 环境变量一样,把 java. exe 执行路径加入 Path 环境变量。将光标移动到 Path 变量值的最前面,然后输入路径:

```
%JAVA_HOME%\bin;
```

其中分号为路径分隔符。然后单击"确定"按钮若干次,保存这些更改并退出系统属性。

1.2 安装 Android Studio 和 SDK

首先在 C:\android 文件夹内新建文件夹 androidstudio 和 sdk,分别用于安装 Android Studio 和 SDK。从 https://developer. android. google. cn/studio/index. html 查看 Android Studio 安装说明。下载最新版本的 Android Studio,例如 android-studio-ide-181. 5014246-windows. exe,双击下载的文件,安装向导开始之后,一路单击 Next 按钮,直到出现 Configuation Settings Install Locations 界面,单击 Browse 按钮,选择 C:\android\androidstudio 文件夹作为 Android Studio 安装位置,然后单击 Next 按钮,单击 Install 按钮,就开始安装 Android Studio 了,安装完成后,启动 Android Studio。启动后,单击 Do not import settings 按钮,单击 OK 按钮,若弹出 unable to access Android SDK add-on list 对话框,原因是计算机没有安装 SDK 而且下载的 Android Studio 又是不带 SDK 的,单击 Cancel

按钮,单击下一界面中的 Next 按钮,出现 Install Type 界面,选择 Custom 选项,单击 Next 按钮,根据自己喜好选择 UI Theme,例如选择 IntelliJ 选项,单击 Next 按钮,在出现 SDK Components Setup 界面时,勾选 Android Virtual Device 选项,并更改 Android SDK Location 为 C:\android\sdk,单击 Next 按钮,Emulator Settings 使用默认值,单击 Next 按钮,弹出 Verify Settings 界面,单击 Finish 按钮,就开始从网上下载 SDK 了。

1.3 配置 Android Studio

启动 Android Studio 后,便可进入 Welcome to Android Studio 界面,如图 1-3 所示,单击图中右下角的 Configure 按钮,在弹出窗口中依次单击 Project Defaults|Project Structure 选项,最后弹出 Project Structure 配置界面,如图 1-4 所示。可以设置 SDK 位置、JDK 位置和 NDK 位置,JDK 可以选择 Use embedded JDK 选项,也可以选择使用自己单独安装的 JDK。单击 OK 按钮,关闭 Project Structure 界面。

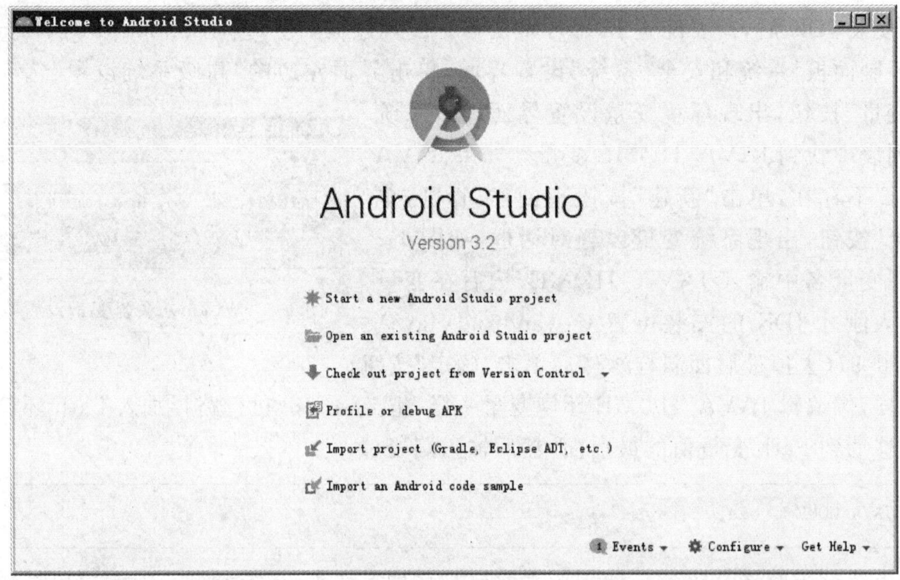

图 1-3 Welcome to Android Studio 界面

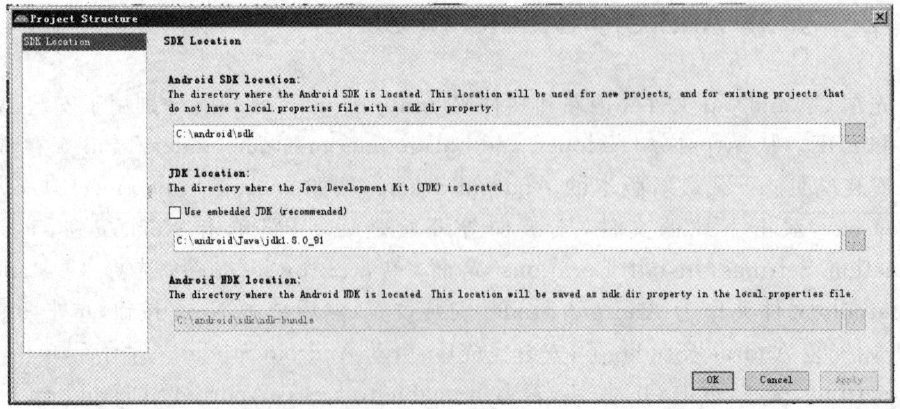

图 1-4 Android Studio 的 Project Structure 配置界面

1.4　安装 ndk-bundle，Cmake 和 LLDB

如果需要使用 C/C++开发 Android 应用程序，需要安装 Android Native Development Kit(NDK)和 Cmake。CMake 是一个跨平台的工具，可以用简单的语句来描述所有平台的安装或者编译过程，它能够输出各种各样的 makefile 或者 project 文件，能测试编译器所支持的 C++特性，类似 UNIX 系统中的 automake。LLDB 是个开源的具有 REPL(read-eval-print-loop)特征的 Debugger，可以调试 C，Objective-C 和 C++。

单击图 1-3 Welcome to Android Studio 界面中的 Configure 按钮，在弹出窗口中依次单击 SDK manager | Android SDK | SDK Tools 选项，出现 SDK manager 界面，如图 1-5 所示。

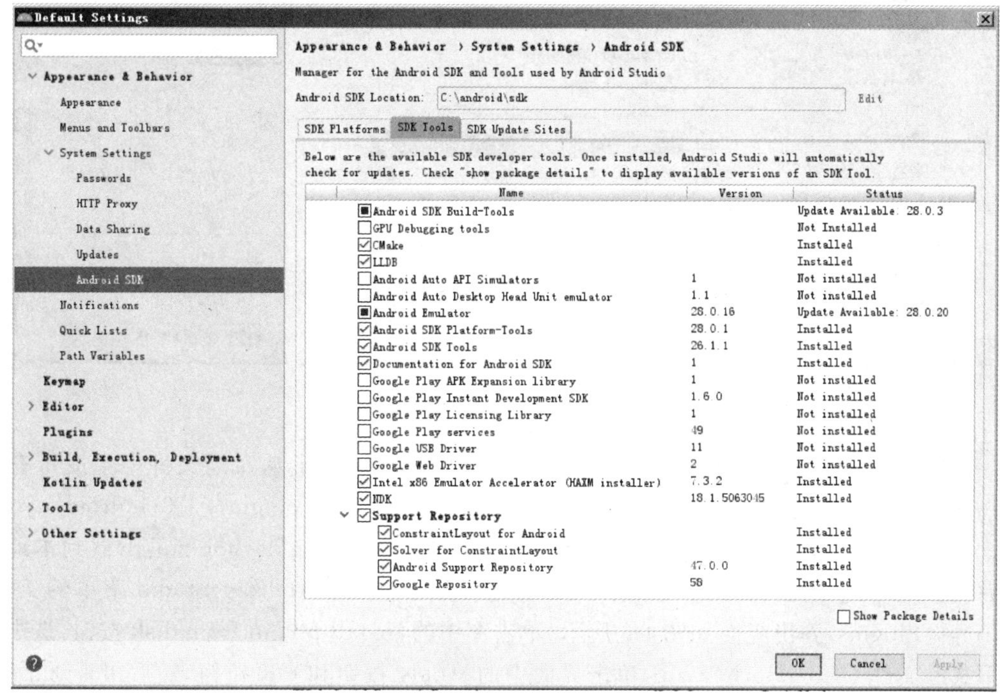

图 1-5　SDK manager 界面

勾选 Cmake、LLDB 和 NDK，单击 Apply 按钮，弹出确认对话框，单击 Confirm Change 对话框中的 OK 按钮，单击 Accept 按钮接受 Licence Agreement，单击 Next 按钮继续安装，单击 Finish 按钮，单击 OK 按钮，关闭 Default Settings 界面。

1.5　Android Studio 常用插件的安装

android-codegenerator-plugin-intellij 插件的网址为 https://github.com/tmorcinek.github，用于快速生成一些 Android 代码。插件 Android parcelable code generator 可以帮助类快速实现 Parcelable 接口序列化。

安装过程：单击 Android Studio 菜单的 File | Settings | Plugins，弹出 Plugins 界面，如图 1-6 所示，单击 Browse repositories 按钮，在弹出窗口的左上角的搜索框里，输入 Android

Code Generator，选中列表框中搜索到的 Android Code Generator 项，单击 Install 按钮，弹出 Third-party PluginsPrivacy Note 对话框，单击 Accept 按钮，就可以安装插件了，安装完成后单击 close 按钮关闭界面。使用同样的方法安装插件 Android Parcelable Code Generator：在图 1-6 的 Plugins 界面中，单击 Browse repositories 按钮，在弹出窗口的左上角的搜索框里，输入 Android Parcelable code generator，选中列表框中搜索到的 Android Parcelable code generator 项，单击 Install 按钮，就可以安装插件了，安装完成后单击 Close 按钮关闭界面。

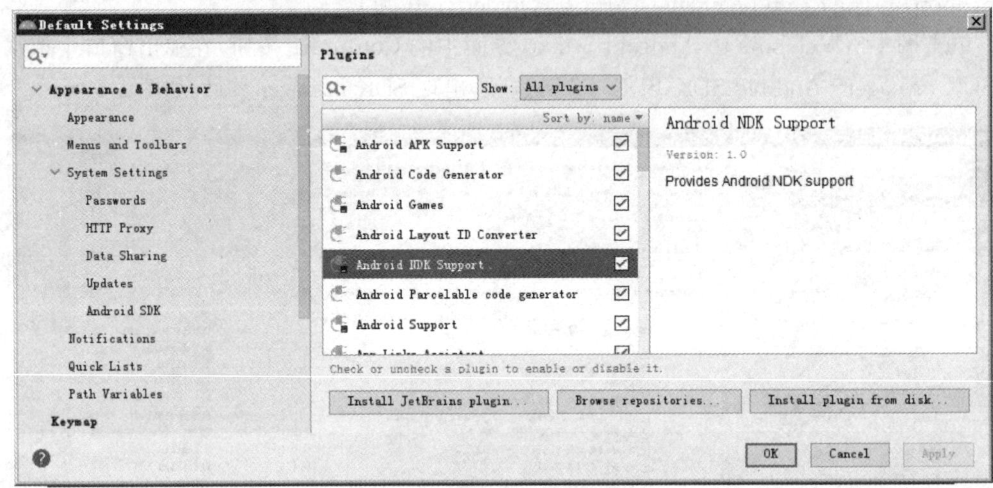

图 1-6 Android Studio 的 Plugins 界面

插件 Android Layout ID Converter for IntelliJ IDEA，用于从布局文件中生成带有 id 的视图的代码。首先从 https://plugins. jetbrains. com/idea/plugin/7373-android-layout-id-converter 下载最新的版本 1. 6 到本地磁盘，文件名为 OffingHarbor. zip，下载到本地磁盘后，就可以从本地磁盘开始安装了。安装过程如下，单击 Android Studio 菜单的 File|Settings|Plugins，弹出界面如图 1-6 所示，单击图中的 Install plugin from disk 按钮，选中刚刚下载的文件 OffingHarbor. zip，单击界面中的 Apply 按钮和 OK 按钮，然后出现对话框，提示“Restart Android Studio to activate changes in plugins”，单击对话框中的 Restart 按钮，重新启动 Android Studio。

1.6 Android Studio 界面介绍

1.6.1 主菜单栏

主菜单栏位于 Android Studio 的最上面，用户利用主菜单及其子菜单可以执行任何操作，但更多的则是通过快捷键和相应的上下文菜单实现。不像 Android Studio 中其他的一些菜单，主菜单不能被隐藏。

1.6.2 ToolBar 工具栏

通过主菜单中的 View 菜单可以显示或者隐藏 TooBar。工具栏中还包含 SDK 管理器和 Android 虚拟设备管理器。工具栏中还有设置和帮助按钮以及运行和调试应用程序按钮。工具栏中所有的按钮都有相应的菜单项和快捷键。用户可以通过取消勾选 View 菜单下的 ToolBar 菜单项来释放屏幕空间。

1.6.3 Navigation Bar 导航栏

通过主菜单中的 View 菜单可以显示或者隐藏 Navagation Bar。导航栏是以水平箭头的链状结构方式来显示从项目根目录(左边)依次到编辑器中选中的选项卡(右边)。

1.6.4 Status Bar 状态栏

通过主菜单中的 View 菜单可以显示或者隐藏 Status Bar。状态栏显示一些相关的上下文敏感的反馈信息,比如正在运行中的进程或者项目中 Git 版本库状态的信息。

1.6.5 Tool Button

通过主菜单中的 View 菜单可以显示或者隐藏 Tool Button。Tool Button 主要位于界面的最左边、最下边、最右边。

Project:提供了多种视图模式来查看项目结构,在该窗口中可以对项目中的文件和目录进行各种操作。

Structure:结构工具窗口会以树状形式展现文件中元素的层次结构,单击元素可以跳转到编辑器中对应代码的位置。

Captures:快照工具窗口中存放 Logcat 中 dump 出的 heap 文件和 allocation 文件,在这里可以导出 hprof 文件,并且支持一键转成 Java Heap Dump。

Build Variants:构建变体工具窗口用来选择当前多渠道版本的构建,执行 Build APK 会构建选中的版本。BuildVariants 其实就是 buildTypes + productFlavors。在 build.gradle 文件中定义 buildTypes 和 productFlavors。默认情况下,Android plugin 会自动的构建 release 和 debug 两种 buildType。而 productFlavors 是指打包不同版本 app 的主要方式。ProductFlavors 可以是单维度的或者多维度的。下面是单维度的例子:

```
productFlavors{
    flavor1 { minSdkVersion 10
        versionCode 1
    }
    flavor2 {
        minSdkVersion 14
        versionCode 2
    }
}
```

下面是多维度的例子:

```
    flavorDimensions "manufacturer", "isFree"
productFlavors {
    huawei {
```

```
        dimension "manufacturer"
    }
    lenovo {
        dimension "manufacturer"
    }
    xiaomi {
        dimension "manufacturer"
    }
    free {
        dimension "isFree"
    }
    paid {
        dimension "isFree"
    }
}
```

创建多维度的 ProductFlavor 的步骤为：首先使用 flavorDimensions 定义维度和顺序，然后在 productFlavors 下添加一个自定义名称的元素，必须指定维度。Build Variants 是 Gradle 按照特定规则集合并 buildTypes 和 productFlavors 的结果。上面的多维度的例子默认可以生成 3×2×2＝12 种 build variants，即 manufacturer 维度的数量乘以 isFree 维度的数量乘以 buildTypes 的数量。因此不能直接配置构建变体，而是配置组成变体的构建类型和产品风格。

Logcat：Logcat 窗口主要包含以下区域：连接的 Android 虚拟设备或者硬件设备、设备上运行的应用进程、日志级别、日志过滤关键字搜索框、日志过滤信息配置、日志信息、日志操作按钮。日志操作按钮包括：清除日志、滚动到日志的最后、向上查看堆栈信息、向下查看堆栈信息、开启日志自动换行、打印、重启、Logcat 设置、Logcat 帮助。

Profiler：Android Studio 3.0 中的 Profiler 窗口替代早期版本中 Android Monitor 工具。Profiler 工具可为应用程序的 CPU、内存和网络活动提供实时数据。可以通过 Android Studio 菜单 View | Tool Windows | Profiler 打开 Profiler 窗口，也可以点击工具栏中的 Profiler 图标打开 Profiler 窗口。在 Profiler 窗口的顶部，可以选择设备和应用程序。当启动一个可调试的应用程序时，该进程是默认选择的。Profiler 初始显示的是一个共享的时间线视图，其中包括一个时间轴，用于 CPU、内存和网络使用的实时显示。窗口还包括时间线缩放按钮和跳转到实时更新的按钮。

TODO：待处理任务工具按钮。在 Android Studio 中待处理的任务可以加上 TODO 注释，它表明了这个地方是待处理状态。

Terminal：终端工具窗口。在终端工具窗口中可以执行终端命令，使用起来非常方便。

Event Log：事件日志窗口。

Device File Explorer：从版本 3.0 开始，Android Studio 包括设备文件资源管理器，用于查看、复制和删除 Android 设备上的文件。当检查由应用程序创建的文件或要将文件传输到设备或从设备传输文件时，这非常有用。可以通过 Android Studio 菜单 View | Tool Windows | Device File Explorer 或者直接在右侧的工具栏窗口单击 Device File Explorer 按钮就可以打开设备文件资源管理器。

Gradle：Gradle 工具窗口。Gradle 工具窗口列出了当前项目和模块中支持的所有 Gradle 任务和运行配置，以方便快速操作。可以通过 Android Studio 菜单 View|Tool Windows|Gradle 或者直接在右侧的工具栏窗口单击 Gradle 按钮就可以打开 Gradle 窗口。

1.6.6 上下文菜单 Context Menus

大多数的上下文菜单都可以在 IDE 中通过鼠标右击来激活。Android Studio 中的大多数的窗格、图标和边栏，如果被鼠标右击将会出现一个上下文菜单。

1.6.7 设置 Auto Import 自动导入包

设置自动导入包的过程如下：单击 Android Studio 集成环境的 File|Settings|Editor|General，|Auto Import，勾选右边的全部选项，再单击 Apply 按钮使更改生效，单击 OK 按钮关闭界面。

1.7 本章主要参考文献

1. https://developer.android.google.cn/studio/profile/android-profiler
2. https://developer.android.google.cn/studio/profile/cpu-profiler.html
3. https://developer.android.google.cn/studio/build/build-variants

第 2 章 Android 清单文件

CHAPTER 2

每个 Android 应用的根目录中都必须包含一个 AndroidManifest.xml 文件(且文件名精确无误),清单文件向 Android 系统提供如下的应用的必要信息:

(1) 为应用的 Java 软件包命名。软件包名称是应用的唯一标识符。

(2) 描述 Application 的各个组件,包括构成应用的 Activity、Service、BroadcastReceiver 和 ContentProvider,并发布其功能,例如它们可以处理的 Intent 消息。这些声明向 Android 系统告知有关组件以及可以启动这些组件的条件信息。

(3) 确定托管应用组件的进程。

(4) 声明应用必须具备哪些权限才能访问 API 中受保护的部分并与其他应用交互。还声明其他应用与该应用组件交互所需具备的权限。

(5) 列出 Instrumentation 类,这些类可在应用运行时提供分析和其他信息。这些声明只会在应用处于开发阶段时出现在清单中,在应用发布之前将移除。

(6) 声明应用所需的最低 Android API 级别。

(7) 列出应用必须链接到的库。

第一次阅读本书时,可以跳过本章节下面部分,等到阅读后面的章节需要详细了解清单文件时,再来查阅。

2.1 AndroidManifest.xml 文件结构

下面的代码段显示了 AndroidManifest.xml 文件的通用结构及其可包含的每个元素。

```
<?xml version = "1.0" encoding = "utf - 8"?>
<manifest>
    <uses - permission />
    <permission />
    <permission - tree />
    <permission - group />
    <instrumentation />
    <uses - sdk />
    <uses - configuration />
    <uses - feature />
    <supports - screens />
    <compatible - screens />
```

```
    <supports-gl-texture />
    <application>
        <activity>
            <intent-filter>
                <action />
                <category />
                <data />
            </intent-filter>
            <meta-data />
        </activity>
        <activity-alias>
            <intent-filter> . . . </intent-filter>
            <meta-data />
        </activity-alias>
        <service>
            <intent-filter> . . . </intent-filter>
            <meta-data/>
        </service>
        <receiver>
            <intent-filter> . . . </intent-filter>
            <meta-data />
        </receiver>
        <provider>
            <grant-uri-permission />
            <meta-data />
            <path-permission />
        </provider>
        <uses-library />
    </application>
</manifest>
```

关于XML文件的基础知识，请参阅"16.3 XML文件解析"章节。一个XML元素包含一个开始标记、一个结束标记以及标记之间的数据内容。

XML元素的格式如下：

```
<标记名称  属性名1="属性值1"  属性名2="属性值2"…>内容</标记名称>
```

2.2 元素

只有< manifest >和< application >元素是必需的，它们都必须存在并且只能出现一次。其他大部分元素可以出现多次或者根本不出现。一个元素可以包含其他内容，即包含其他元素，元素所有的值均通过属性进行设置，而不是通过在元素的开始标签和结束标签之间的字符数据进行设置。也就是说，元素的开始标签和结束标签之间的内容不可能有直接的字符数据，如果有，只可能有其他元素。

同一级别的元素通常不分先后顺序。例如，< activity >、< receiver >、< provider >和< service >元素可以按任何顺序混合在一起，作为< application >的子元素。这条规则有下面两个主要例外：

< activity-alias >元素必须跟在别名所指的< activity >之后。

< application >元素必须是< manifest >元素内最后一个元素。换言之,</manifest >结束标记必须紧接在</application >结束标记后。

从某种意义上说,元素的所有属性都是可选的。但是,必须指定某些属性,元素才可实现其目的。对于真正可选的属性,它将指定默认值或声明缺乏规范时将执行何种操作。

除了根元素< manifest >的一些属性外,所有属性名称均以 android: 前缀开头,例如, android: alwaysRetainTaskState。由于该前缀是通用的,因此在按名称引用属性时,本文档通常会将其忽略。

大量元素具有 icon 和 label 属性,用作显示给用户的小图标和文本标签。某些元素还具有 description 属性,用作显示在屏幕上的较长说明文本。例如,< permission >元素具有 icon、label、和 description 这三个属性。因此,当系统询问用户是否授权给请求获得权限的应用时,代表该权限的图标、权限名称以及所需信息的说明均会呈现给用户。

无论何种情况下,在容器元素(containing element)中设置的 icon 图标和 label 标签都将成为容器的所有子元素的默认 icon 和 label。因此,在< application >元素中设置的 icon 和 label 是每个应用组件的默认图标和标签。同样,为组件(例如< activity >元素)设置的 icon 和 label 是组件每个< intent-filter >元素的默认设置。如果< application >元素设置 label,但是 Activity 及其 intent-filter 没有设置 label,则< application >元素的 label 将被视为 Activity 和 intent-filter 的标签。

在实现 intent-filter 公布的功能时,只要向用户呈现组件,系统便会使用为 intent-filter 设置的 icon 和 label 表示该组件。例如,具有 android. intent. action. MAIN 和 android. intent. category. LAUNCHER 设置的过滤器将 Activity 公布为应显示在应用启动器中的功能,在过滤器中设置的图标和标签显示在启动器中。

2.3 声明类名

许多元素对应于 Java 对象,包括 application 元素本身及其主要组件: Activity、Service、BroadcastReceiver 和 ContentProvider,这些组件类的子类通过 name 属性来声明其类名,该名称必须包含完整的软件包名称。例如,Service 子类可能会声明如下:

```
<manifest … >
  <application … >
    <service android:name = "com.example.project.SecretService" … >
      …
      </service>
    </application>
</manifest>
```

但是,如果定义 name 属性的字符串的第一个字符是句点,则由 manifest 元素的 package 属性指定的应用的包名称将附加到该字符串的前面,作为类的名字。以下赋值与上述方法相同。

```
<manifest package = "com.example.project" … >
  <application … >
      <service android:name = ".SecretService" …>
          …
      </service>
  </application>
</manifest>
```

当启动组件时,Android 系统会创建已命名子类的实例。如果未指定子类,则会创建基类的实例。

2.4 多个值

如果可以指定多个值,则几乎总是在重复此元素,而不是列出单个元素内的多个值。例如,intent 过滤器可以列出多个操作:

```
<intent-filter … >
    <action android:name = "android.intent.action.EDIT" />
    <action android:name = "android.intent.action.INSERT" />
    <action android:name = "android.intent.action.DELETE" />
    …
</intent-filter>
```

2.4.1 资源值

某些属性的值可以显示给用户,例如,Activity 的标签和图标。这些属性的值应该本地化,并通过资源或主题进行设置。资源值用以下格式表示:

```
@[<i>package</i>:]<i>type</i>/<i>name</i>
```

格式说明:

成对的中括号内的内容是可选的,中括号本身并不出现在资源值中。

从<i>到</i>结束,表示是必需的,并将被<i>和</i>之间的内容所替换,实际使用时,<i>name</i>应该替换成资源的名字,<i>type</i>应该替换成资源的类型,类型和名称是必需的。[<i>package</i>:]是可选的,即[<i>package</i>:可以不出现,若出现,里面的<i>package</i>应该替换成包名。格式中的@字符,<i>type</i>与<i>name</i>之间的/都是必需的,即资源名必须以@开头,必须包含资源类型和资源名字,资源类型和资源名字之间必须以斜线分隔,资源定义的包名不是必需的,若没有指定包名,资源就定义在<manifest>的 package 属性指定的包里;若需要指定包名,包名和资源类型之间必须用冒号分隔。

如果资源与应用在同一个软件包中,可以省略软件包名称。下面是示例:

```
<activity android:icon = "@drawable/smallPic" … >
```

引用 style 中的<item>定义的值用类似的方法表示，但是以？开头，而不是以@开头。

```
?[<i>package</i>:]<i>type</i>/<i>name</i>
```

例如：资源文件 styles.xml 定义如下：

```
<resources>
    <!-- Base application theme. -->
    <style name="AppTheme" parent="Theme.AppCompat.Light.DarkActionBar">
        <!-- Customize your theme here. -->
        <item name="colorPrimary">@color/colorPrimary</item>
        <item name="colorPrimaryDark">@color/colorPrimaryDark</item>
        <item name="colorAccent">@color/colorAccent</item>
    </style>
    <style name="AppTheme.NoActionBar">
        <item name="windowActionBar">false</item>
        <item name="windowNoTitle">true</item>
    </style>
    <style name="AppTheme.AppBarOverlay" parent="ThemeOverlay.AppCompat.Dark.ActionBar" />

    <style name="AppTheme.PopupOverlay" parent="ThemeOverlay.AppCompat.Light" />
</resources>
```

下面的布局文件中引用了<item name="colorPrimary">定义的资源：

```
<android.support.design.widget.CoordinatorLayout
xmlns:android="http://schemas.android.com/apk/res/android"
    xmlns:app="http://schemas.android.com/apk/res-auto"
    xmlns:tools="http://schemas.android.com/tools"
    android:layout_width="match_parent"
    android:layout_height="match_parent"
    tools:context="cuc.toolbarasactionbar1.MainActivity">
    <android.support.design.widget.AppBarLayout
        android:layout_width="match_parent"
        android:layout_height="wrap_content"
        android:theme="@style/AppTheme.AppBarOverlay">
        <android.support.v7.widget.Toolbar
            app:layout_scrollFlags="scroll|enterAlways"
            android:id="@+id/toolbar"
            android:layout_width="match_parent"
            android:layout_height="?attr/actionBarSize"
            android:background="?attr/colorPrimary"
            app:popupTheme="@style/AppTheme.PopupOverlay" />
        <TextView
            android:layout_width="match_parent"
            android:layout_height="wrap_content"
            android:text="donot scroll"
            android:textSize="30sp"/>
    </android.support.design.widget.AppBarLayout>
</android.support.design.widget.CoordinatorLayout>
```

字符串值

如果属性值为字符串,则必须使用双反斜杠(\\)转义字符,例如,使用\\n 表示换行符或使用\\uxxxx 表示 Unicode 字符。

2.5 theme 属性

为了让一个 application 中的所有 activity 元素共用一个 theme,打开 AndroidManifest.xml 文件,编辑< application >标签,设置< application >标签的 android: theme 属性。例如:

```
<application android:theme = "@style/CustomTheme">
```

如果一个 theme 仅仅适用于应用中的某一个 activity,把 android: theme 属性添加到这个< activity >标签内部。

```
<activity android:theme = "@android:style/Theme.Dialog">
```

如果< activity >元素没有定义 android: theme 属性,则它将继承< application >元素中定义的 android: theme 属性,如果 application 元素中也没有定义 android: theme 属性,则< activity>元素将使用系统默认的 theme 属性。

2.6 权限

权限是一种限制,施加限制是为了保护可能被误用以致破坏的关键数据和代码。对于 Android 6.0 以下的权限,在安装的时候,根据权限声明产生一个权限列表,用户只有在同意之后才能完成 App 的安装,造成了用户想要使用 App 的部分功能,就必须授予一些不必要的权限。而在 Android 6.0 以后,App 可以直接安装,App 在运行时需要授予某些权限的时候,系统会弹出一个对话框让用户选择是否授予某个权限给 App(这个 Dialog 不能由开发者定制),用户可以拒绝授权。< permission >元素用于定义一种新的权限;应用组件中的 permission 属性指明这个组件受什么权限保护,< uses-permission >元素用于请求获取权限。

2.6.1 permission 元素

如果要定义一个新的 permission,可以用< permission >元素来定义。表 2-1 解释了 permission 元素的属性的用途。

```
<permission android:description = "string resource"
            android:icon = "drawable resource"
            android:label = "string resource"
            android:name = "string"
            android:permissionGroup = "string"
            android:protectionLevel = ["normal" | "dangerous" |
                                       "signature" | "signatureOrSystem"] />
```

表 2-1　permission 元素属性的解释

permission 元素属性	解　释
description	对权限的描述,比 label 更长,包含的信息也更多。一般包括这个权限所针对的操作和用户授予 App 这个权限会带来的后果。这个属性只能设置为另一字符串的引用
icon	对代表权限的 drawable 资源的引用
label	对权限的一个简短描述,用来显示给用户
name	权限的唯一标识,在代码中使用该名字引用权限。name 属性是必需的,其他的可选。在 uses-permission 元素中和应用组件的 permission attributes 属性中使用这个名字。系统不允许多个包声明同名的权限
permissionGroup	权限所属权限组的名称。如果没有设置此属性,权限不属于任何一个组
protectionLevel	权限的等级：normal、dangerous、signature 和 signatureOrSystem

从表 2-1 可以看出,permission 的 protectionLevel 分为 4 种：normal、dangerous、signature 和 signatureOrSystem,其中 signature 和 signatureOrSystem 使用较少。

normal 权限不涉及用户隐私,是最低的等级,只需要在 Manifest 中使用 uses-permission 元素请求获取权限即可,比如网络、蓝牙、NFC 等,安装即授权,不需要每次使用时都检查权限,而且用户不能取消以上授权,除非用户卸载 App。声明此权限的 App,系统会默认授予此权限,不会提示用户显式地同意。

Dangerous 权限涉及到用户隐私信息,用户必须明确授予应用使用这些权限,比如 SD 卡读写、联系人、短信读写等。Android 6.0 以前系统在安装 App 时会提示用户确认；从 Android 6.0 开始,App 运行时必须确认授权。所有的 Dangerous Permissions 被分入到 9 个权限组,见表 2-2。无论获取哪种保护级别的权限,都必须在 Manifest 中使用< uses-permission >元素请求获取权限,包括危险权限。如果危险权限没有使用< uses-permission >元素请求获取权限,申请时不提示用户授权。

表 2-2　危险权限的分组

Permission Group(元素属性)	Permissions(描述)
android. permission-group. CALENDAR	android. permission. READ_CALENDAR android. permission. WRITE_CALENDAR
android. permission-group. CAMERA	android. permission. CAMERA
android. permission-group. CONTACTS	android. permission. READ_CONTACTS android. permission. WRITE_CONTACTS android. permission. GET_ACCOUNTS
android. permission-group. LOCATION	android. permission. ACCESS_FINE_LOCATION android. permission. ACCESS_COARSE_LOCATION
android. permission-group. MICROPHONE	android. permission. RECORD_AUDIO
android. permission-group. PHONE	android. permission. READ_PHONE_STATE android. permission. CALL_PHONE android. permission. READ_CALL_LOG android. permission. WRITE_CALL_LOG com. android. voicemail. permission. ADD_VOICEMAIL android. permission. USE_SIP android. permission. PROCESS_OUTGOING_CALLS

续表

Permission Group(元素属性)	Permissions(描述)
android. permission-group. SENSORS	android. permission. BODY_SENSORS
android. permission-group. SMS	android. permission. SEND_SMS android. permission. RECEIVE_SMS android. permission. READ_SMS android. permission. RECEIVE_WAP_PUSH android. permission. RECEIVE_MMS android. permission. READ_CELL_BROADCASTS
android. permission-group. STORAGE	android. permission. READ_EXTERNAL_STORAGE android. permission. WRITE_EXTERNAL_STORAGE

Dangerous Permissions 是按照权限组分配使用的。申请某一个危险权限的时候，系统弹出的对话框是对整个权限组的说明，而不是单个权限。当应用申请某个危险权限的时候，如果这个 App 早已被用户授权了同一组的某个危险权限，那么系统会立即授权同一权限组的所有权限，而不需要用户去单独授权。即同一个权限组的任何一个危险权限被授权了，这个权限组的其他权限也自动被授权。比如某个 App 已经获得了 READ_CONTACTS 授权，当这个 App 申请在同一权限组的 WRITE_CONTACTS 时，系统会直接授权通过。

如果 App 运行在 Android 6.0(API level 23)或者更高级别的设备中，而且 targetSdkVersion>=23 时，系统将会自动采用动态权限管理策略，危险权限都需要在运行的时候去申请，如果在涉及到危险权限操作时没有申请权限而直接调用了相关代码，App 可能就崩溃了。

如果 App 运行在 Android 5.1(API level 22)或者更低级别的设备时，或者 targetSdkVersion<=22 时(此时设备可以是 Android 6.0 或者更高)，系统仍将采用旧的权限管理策略，要求用户在安装的时候授予权限。

signature 权限：仅仅当声明权限的应用和请求授予权限的应用使用相同的证书签名时才应用。signatureOrSystem 与 signature 类似，只是增加了 rom 中自带的 App 的声明，一般不推荐使用。

2.6.2 permission 属性

一个应用组件可以使用 permissions 属性保护自己，它能使用 Android 系统定义的 permission(在 android. Manifest. permission 中列出)，也可以使用应用内定义的或者其他应用定义的 permission。Android 应用的 4 大组件(Activity、BroadcastReceiver、Service 和 ContentProvider)都可以使用 permission 属性来声明该组件受权限保护，即启动该组件需要权限。如果某个应用组件没有设置 permission 属性，则该组件使用< application >元素的 permission 属性设置的权限保护自己。如果该组件和< application >元素都没有设置 permission 属性，则 Activity 不受权限保护。以 Activity 为例，如果系统未向 startActivity() 或 startActivityForResult()的调用者授予指定权限，其 Intent 将不会传递给受权限保护的 Activity。

2.6.3 uses-permission 元素

如果应用需要访问受 permission 保护的组件，包括访问在自身应用内受 permission 保护的组件，也必须在 AndroidManifest.xml 文件中使用< uses-permission >元素请求获取相应的权限。无论 permission 的保护级别是哪一种，为了获取权限，都必须使用< uses-permission >元素请求获取权限。当应用安装到设备上之后，安装程序会通过检查签署应用证书的颁发机构，并在某些情况下询问用户，确定是否授予请求的权限。如果权限被授予了，则应用才能够使用受保护的功能。否则，其访问这些功能的尝试将会失败，并且不会向用户发送任何通知。

例如，在下面的清单文件中，首先使用< permission >元素定义了一个名叫 com.example.project.DEBIT_ACCT 的权限，组件 com.example.project.FreneticActivity 使用 permission 属性声明自己受名字为 com.example.project.DEBIT_ACCT 的权限保护，要让应用的其他组件能够启动这个受保护的 Activity，必须使用< uses-permission >元素请求用户授予其使用权限，即便保护是由应用本身施加的。

```
<manifest … >
<permission android:name = "com.example.project.DEBIT_ACCT" />
  <uses - permission android:name = "com.example.project.DEBIT_ACCT" />
  <application … >
    <activity android:name = "com.example.project.FreneticActivity"
           android:permission = "com.example.project.DEBIT_ACCT" … >
           …
        </activity>
  </application>
</manifest>
```

2.6.4 动态权限请求的实现步骤

因为权限动态检查相关的 API 是 Android 6.0 才加入的，所以 minSdkVersion 不是 23 时，推荐使用 SupportLibrary 来实现，好处是：程序里不必加 if 来判断当前设备的版本。为了保持兼容性，建议使用 v4 包的兼容方法：ContextCompat.checkSelfPermission()、ActivityCompat.requestPermissions()、ActivityCompat.OnRequestPermissionsResultCallback、ActivityCompat.shouldShowRequestPermissionRationale()。

(1) 在 AndroidManifest 文件中使用< uses-permission >元素请求获取权限。这个步骤和低版本的开发并没有什么变化，试图去申请一个没有声明的权限可能会导致程序崩溃。

(2) 检查权限状态。如果执行的操作需要一个 dangerous permission，那么每次在执行操作的地方都必须检查是否有这个 permission，因为用户可以在"设置"里随意地更改授权情况，所以必须每次在使用前都检查是否有权限。可以调用 checkSelfPermission (@NonNull Context context, @NonNullString permission)检查权限，该方法的返回值是 PERMISSION_GRANTED 或者 PERMISSION_DENIED。

(3) 有时候可能需要解释为什么需要这个权限。如果上面权限检查的结果是 PERMISSION_DENIED，那么就需要显式地向用户请求这个权限了。用户需要使用这个功能，但是拒绝了相应的权限，那么很可能是用户不理解 App 为什么需要这个权限，这时候可能会需要跟用户解释权限的用途。不是每条权限都需要解释，太多的解释会降低用户体验。为了发现这种用户可能需要解释的情形，Android 提供了一个工具类方法：shouldShowRequestPermissionRationale()。第一次请求权限时，被用户拒绝了，下一次 shouldShowRequestPermissionRationale()返回 true，应该显示一些为什么需要这个权限的说明；第二次请求同一权限时，用户拒绝了，并选择了“Don't ask again”的选项时，shouldShowRequestPermissionRationale()返回 false；设备的策略禁止当前应用获取这个权限的授权时，shouldShowRequestPermissionRationale()返回 false。具体解释原因的 dialog 系统没有提供，需要自己实现。

(4) 请求权限。请求权限的方法是：requestPermissions(Activity activity, String[] permissions, int requestCode)，这个方法是异步的，它会立即返回。当用户和对话框交互完成之后，系统会调用回调方法 onRequestPermissionsResult(int requestCode, String[] permissions, int[] grantResults)，传回用户的选择结果和对应的 request code。

当调用 requestPermissions()时，系统会显示一个获取权限的提示对话框，应用不能配置和修改这个对话框。第一次弹出请求权限对话框中没有 Don't ask again 选项，第二次请求权限时，才会增加一个 Don't ask again 选项，如果用户一直拒绝，并没有选择该选项，下次请求权限时，会继续有该选项。如果选择了 Don't ask again 选项，程序就不会再询问是否授予权限了。如果需要提示用户这个权限相关的解释信息，需要在调用 requestPermissions()之前处理。

如果用户已经允许了 permission group 中的 A 权限，那么当下次调用 requestPermissions()方法请求同一个 group 中的 B 权限时，系统会直接调用 onRequestPermissionsResult()回调方法，并传回 PERMISSION_GRANTED 的结果。如果用户选择了不再询问此权限，那么 App 再次调用 requestPermissions()方法来请求同一权限的时候，系统会直接调用 onRequestPermissionsResult()回调，返回 PERMISSION_DENIED。

一个例子

```
if (PackageManager.PERMISSION_GRANTED == ContextCompat.checkSelfPermission(MainActivity.
this, Manifest.permission.READ_CONTACTS)) {      //has permission, do operation directly
    ContactsUtils.readPhoneContacts(this);
    Log.i(DEBUG_TAG, "user has the permission already!");
} else {                                          //do not have permission
    Log.i(DEBUG_TAG, "user do not have this permission!");
    // Should we show an explanation?
    if (ActivityCompat.shouldShowRequestPermissionRationale(MainActivity.this,
            Manifest.permission.READ_CONTACTS)) {
        // Show an explanation to the user *asynchronously* -- don't block
        // this thread waiting for the user's response! After the user
        // sees the explanation, try again to request the permission.
        Log.i(DEBUG_TAG, "we should explain why we need this permission!");
    } else {
        // No explanation needed, we can request the permission.
```

```
            Log.i(DEBUG_TAG, " == request the permission == ");

            ActivityCompat.requestPermissions(MainActivity.this,
                    new String[]{Manifest.permission.READ_CONTACTS},
                    MY_REQUEST_CODE_READ_CONTACTS);

            // MY_PERMISSIONS_REQUEST_READ_CONTACTS is an
            // app-defined int constant. The callback method gets the
            // result of the request.
        }
    }
```

(5) 处理请求权限的响应。当用户对请求权限的对话框做出响应之后，系统会调用onRequestPermissionsResult(int requestCode, String[] permissions, int[] grantResults)方法，这个回调中request code即为调用requestPermissions()时传入的参数，是App自定义的一个整型值。

一个实例

```
@Override
public void onRequestPermissionsResult(int requestCode, String permissions[], int[]
grantResults) {
    switch (requestCode) {
        case MY_REQUEST_CODE_READ_CONTACTS: {
            // If request is cancelled, the result arrays are empty.
            if (grantResults.length > 0
                    && grantResults[0] == PackageManager.PERMISSION_GRANTED) {
            // permission granted, yay! Do the task you need to do.
                ContactsUtils.readPhoneContacts(this);
                Log.i(DEBUG_TAG, "user granted the permission!");
            } else {
//permission denied,Disable the functionality that depends on this permission.
                Log.i(DEBUG_TAG, "user denied the permission!");
            }
            return;
        }
        // other 'case' lines to check for other
        // permissions this app might request
    }
}
```

2.7 使用uses-feature元素声明应用要求

基于Android系统的设备多种多样，并非所有设备都提供相同的特性和功能。为防止将应用安装在缺少应用所需特性的设备上，必须在清单文件中使用uses-feature元素声明支持的设备类型和软件要求。其中的大多数声明只是为了提供信息，系统不会读取它们，但Google Play等外部服务会根据uses-feature过滤设备不支持的应用，以便当用户在其设备中搜索应用时为用户提供过滤功能。Android 用户可能会注意到一些高版本的应用没有在手机上的 Android Market 中显示，这必定是应用使用了 uses-feature 的结果。例如，如果应用需要相机，并使用 Android 2.1(API 级别 7)中引入的 API，可以在清单文件中这样声明：

```
<manifest … >
    <uses - feature android:name = "android.hardware.camera.any"
                  android:required = "true" />
    <uses - sdk android:minSdkVersion = "7" android:targetSdkVersion = "19" />
</manifest>
```

现在，没有相机或者 Android 版本低于 2.1 的设备将无法从 Google Play 安装该应用。不过，也可以声明：应用使用相机，但并不要求必须使用。在这种情况下，应用必须将 required 属性设置为 false，并在运行时检查设备是否具有相机，然后根据需要停用任何相机功能。

这里必须说明一点，要区别 uses-feature 和 uses-permisstion，声明了一个< uses-feature android: name = " android. hardware. camera"/>并不代表可以不写< uses-permission android: name="android. permission. CAMERA"/>。uses-feature 是供 GooglePlay 用的，而 uses-permission 是供 Android 系统使用的，应用需要使用某个硬件设备或者软件功能就必须采用< uses-permisstion >申请权限。

2.8　intent-filter

应用组件(Activity、Service、BroadcastReceiver)将通过< intent-filter >公布它们可响应的 Intent 类型。intent-filter 元素申明了它的父组件的能力：一个 Activity 或者 Service 能做什么，一个 BroadcastReceiver 能响应什么类型的广播。一个组件可有任意数量的 intent-filter 元素，其中每个 intent-filter 描述一种不同的功能。

显式指定目标组件名字的 Intent 将激活该组件，因此< intent-filter >对显式 Intent 不起作用。没有指定目标组件名字的 Intent 只有在能够匹配组件的一个< intent-filter >时才可激活该组件。如需了解< intent-filter >如何匹配 Intent 对象的信息，请参阅“第 6 章 Intent 和 IntentFilter”。本节只介绍< intent-filter >元素的语法。intent-filter 语法如下：

```
<intent - filter android:icon = "drawable resource"
              android:label = "string resource"
              android:priority = "integer" >
<action android:name = "string">
<category android:name = "string">
<data …>
</intent - filter>
```

intent-filter 元素包含在 activity、activity-alias、service 或者 receiver 元素中。intent-filter 必须包含一个 action 元素，也可以包含多个 action 元素，还可选择包含(即可以包含，也可以不包含)category 元素和 data 元素。intent-filter 元素的属性包括：android: icon，android: label，android: priority。

intent-filter 的属性 icon：当父组件因为有这个 intent-filter 中描述的能力而呈现给用户时，图标用来代表父组件（Activity、Service 或者 BroadcastReceiver）。这个属性必须设置为包含图像定义的 drawable 资源的引用，该属性的默认值是父组件的 android：icon 属性。如果父组件没有指定 android：icon 属性，默认值是< application >元素指定的 android：icon 属性。

intent-filter 的属性 label：当父组件因为有这个 intent-filter 中描述的能力而呈现给用户时，这个 label（而不是父组件中设定的 label）用来代表父组件（Activity、Service 或者 BroadcastReceiver）。该属性的默认值是父组件的 android：label 属性。如果父组件没有指定 android：label 属性，默认值是< application >元素指定的 android：label 属性。这个 label 属性必须设置为字符串资源的引用，以便能像用户接口中的其他字符串一样能本地化。在开发应用时，为了方便，该属性也可以设置成原始的字符串。

intent-filter 的属性 priority：这个属性值对 Activity 和 BroadcastReceiver 都有意义。它提供了父组件 Activity 响应能够匹配该过滤器的 Intent 的优先权。当一个 Intent 可以被几个具有不同优先权的 Activity 处理时，Android 仅仅考虑具有最高优先权的 Activity 作为目标。priority 也控制广播接收器被执行接收广播消息的顺序，那些具有高优先权的接收器优先被调用。这个顺序仅仅应用于有序广播，处理普通广播时将忽略。只有当真正需要控制广播接收的特定顺序，或者强制 Android 优先使用某些 Activity（相对于其他 Acivity）时，才使用 priority 这个属性。这个属性的值必须是一个整数，取值范围是大于-1000 且小于 1000，较大的数具有较高的优先权，默认值是 0。

2.8.1 action 元素

action 元素包含在 intent-filter 元素中。一个 intent-filter 元素必须包含一个或者多个 action 元素。如果 intent-filter 不包含 action 元素，任何 Intent 对象都不可能通过 intent-filter。action 的 name 属性值必须是文本字符串值，而不是类常量。

一些标准的 action 定义在 Intent 类中，形式为 ACTION_string 常数，给这些常数分配值时，将 android. intent. action. 放置在 string 的前面。例如常数 ACTION_MAIN 的值是 android. intent. action. MAIN，常数 ACTION_WEB_SEARCH 的值是 android. intent. action. WEB_SEARCH。对于自己定义的 action，最好使用应用包名作为前缀，以保证唯一性。例如 TRANSMOGRIFY action 可以如下定义：

```
<action android:name = "com.example.project.TRANSMOGRIFY" />
```

2.8.2 category 元素

category 元素包含在 intent-filter 元素中。category 元素的 name 属性是 category 的名字，该值必须是操作的文本字符串值，而不是类常量。标准的 category 以 CATEGORY_name 常数的形式定义在 Intent 类中。实际的名字可以从 CATEGORY_name 常数导出来，即 android. intent. category. 再加上 CATEGORY_name 中 CATEGORY_后面的 name 部分。例如，CATEGORY_LAUNCHER 代表的字符串就是 android. intent. category. LAUNCHER。对于自定义的 category，最好使用应用包名作为前缀，以保证唯一性。

为了使用隐式的 Intent 启动 Activity，activity 的 intent-filter 元素中必须包含 Intent.

CATEGORY_DEFAULT。因为 startActivity()和 startActivityForResult()的 Intent 参数，好像申明了 Intent.CATEGORY_DEFAULT 一样。如果 activity 的 intent-filter 元素中没有包含 Intent.CATEGORY_DEFAULT，将没有任何隐式的 Intent 匹配该 Activity。

2.8.3 data 元素

语法：

```
<data android:scheme = "string"
          android:host = "string"
          android:port = "string"
          android:path = "string"
          android:pathPattern = "string"
          android:pathPrefix = "string"
          android:mimeType = "string" />
```

注意：跟 RFC(Request For Comments)文件不一样，Android 框架的 android：scheme、android：host、android：mimeType 的匹配都是大小写敏感的，应该使用小写字母指定 scheme、host 以及 mimeType。

data 元素包含在 intent-filter 内。一个 intent-filter 元素可以声明 0 个或者多个 data 元素，每个 data 元素均可指定 URI 和数据 MIME 类型。data 元素可以仅仅包含 mimeType(数据类型)、仅仅包含 URI，或者包含这两者。

1. URI

URI(Universal Resource Identifier，RFC1630)统一资源标示符，可以唯一标识一个资源。URI 一般由三部分组成：访问资源的命名机制、存放资源的主机名、资源自身的名称(路径)。URL(Uniform Resource Locator，RFC1738)统一资源定位符，可以提供找到该资源的路径，是 Internet 上用来描述信息资源的字符串，主要用在各种 WWW 客户程序和服务器程序上。URL 的格式由下列三部分组成：协议(或称为服务方式)、存有该资源的主机 IP 地址(有时也包括端口号)、资源的具体地址(如目录和文件名等)。

URI 由以下几个属性指定：

```
<scheme>://<host>:<port>[<path>|<pathPrefix>|<pathPattern>][#fragment]
```

上面格式中的尖括号表示里面的元素(参数、值或信息)是必需的，需要用相应的信息来替换尖括号里面的文本。格式中的方括号是可选项，表示里面的元素(参数、值或信息)是可选的，方括号内的内容可写可不写，不要输入方括号本身。格式中竖线的含义是“或者”，如果两个参数由竖线分隔开，可以选择分隔符左边的元素，也可选择分隔符右边的元素，但在一次命令使用中不能同时选择两个元素。在方括号中，这些选项是可选的。在尖括号中，至少需要一个选项。就 Android 平台而言，URI 主要分三个部分：scheme，authority 和 path。其中 authority 又分为 host 和 port 两个部分。格式如下：

```
scheme://host: port/path
```

举个例子：

```
content://com.example.project: 200/folder/subfolder/etc
```

在此 URI 中，scheme 是 content，host 是 com. example. project，port 是 200，path 是 /folder/subfolder/etc。而对于那些基于服务器的 URI 来说，例如 mailto：tom@sina. com，authority 结构为[userinfo@]host[：port]，authority 不是以//开头，这个 URI 就是不透明的，以//开头的 URI 称为透明的。有的 URI 指向一个资源的内部，这种 URI 以 # 号结束，并跟着一个 anchor 标志符。例如

```
https://developer.android.google.cn/guide/topics/manifest/manifest - intro.html#filestruct
```

其中 https 是协议，developer. android. google. cn/guide/topics/manifest/manifest-intro. html 是资源位置，filestruct 是 anchor 标识符，即资源。

组成 URI 的这些属性在 data 元素中都是可选的，但存在如下的依赖关系：

(1) 如果 data 元素没有包含 scheme 属性，那么 URI 所有参数会被忽略。

(2) 如果 data 元素没有包含 host 属性，那么 port 属性和所有的 path 属性会被忽略。

(3) 在同一个 intent-filter 内可以包含任意数量的 data 元素，所有 data 元素对同一个过滤器起作用，data 元素的所有属性都没有默认值。

```
<intent - filter … >
    <data android:scheme = "something" android:host = "project.example.com" />
</intent - filter>
```

等价于

```
<intent - filter … >
      <data android:scheme = "something" />
      <data android:host = "project.example.com" />
  </intent - filter>
```

2. scheme 属性

data 元素的 scheme 是指 URI 的 scheme，这是 URI 的最基本的属性。如果没有指定 scheme 属性，URI 的其他属性就没有意义。指定 scheme 不需要结尾的冒号，例如 http 而不是 http：。

3. host 属性

data 元素的 host 属性是指 URI authority 的主机部分。如果 scheme 属性没有指定，android：host 属性就没有意义。为了匹配多个子域，可以使用一个星号(*)匹配零个或者多个字符。例如 *. google. com 可以匹配 www. google. com 和 developer. google. com。星号必须是 android：host 属性的第一个字符。例如 google. co. * 是无效的，因为星号不是第一个字符。

4. port 与 path 属性

data 元素的 port 属性是指 URI authority 的端口部分。仅仅当指定了 scheme 属性和

host 属性后，port 属性才有意义。path、pathPrefix 和 pathPattern 这 3 个属性仅仅当 scheme 属性和 host 属性指定后才有意义。path 属性指定了完整的路径，android: path 必须用"/"开始，用于匹配 Intent 对象中的完整路径。pathPrefix 仅仅指定了部分路径，用于匹配 Intent 对象中的路径的初始部分。pathPattern 指定了完整的路径，用于匹配 Intent 对象中的完整路径，但它可以包含以下通配符：①一个星号('*')匹配它前面的字符的 0 次或者多次出现。②一个点紧随星号(".*")匹配 0 个或者多个字符。

5. mimeType 属性

MIME(Multipurpose Internet Mail Extensions)多用途互联网邮件扩展类型，设定某种扩展名文件的打开方式，当该扩展名文件被访问的时候，浏览器会自动使用指定应用程序来打开。例如，能够显示图像的 Activity 可能无法播放音频文件，因此，指定数据的 MIME 类型有助于 Android 系统找到接受 Intent 的最佳组件。但有时，MIME 类型可以从 URI 中推断得出，特别当数据是 content: URI 时尤其如此，这表明数据位于设备中，且由 ContentProvider 控制，这使得数据 MIME 类型对系统可见。android: mimeType 中的子类型可以采用通配符(*)匹配任意的子类型。

2.9 uses-library 元素

每个应用均链接到默认的 Android 库，该库中包括用于开发应用的基本软件包(如 Activity、Service、intent、View、Button、ContentProvider)。但是，某些软件包驻留在自己的库中。如果应用使用来自其中这些软件包的代码，则必须使用单独的< uses-library >元素来命名其中每个库。

2.10 本章主要参考文献

1. https://developer.android.google.cn/guide/topics/manifest/manifest-intro.html。
2. https://developer.android.google.cn/guide/topics/manifest/intent-filter-element.html
3. https://developer.android.google.cn/guide/topics/manifest/action-element.html
4. https://developer.android.google.cn/guide/topics/manifest/permission-element.html
5. https://developer.android.google.cn/guide/topics/manifest/data-element

第3章 Android resource 介绍

CHAPTER 3

Android 应用除了包含源代码，还包含与源代码分离的资源，如图像、音频文件以及任何与应用的视觉呈现有关的内容。例如，可以使用 XML 文件定义 Activity 用户界面的菜单和布局。使用资源的优点在于，资源可以单独维护、提供针对不同设备配置（如不同的屏幕尺寸）的替代资源，能够在不修改代码的情况下轻松地更新应用的各种特性。

对于任意类型的资源，都可以指定默认资源和多个替代资源。默认资源系指无论设备配置如何，或者在没有替代资源与当前配置相匹配时，均应使用的资源。替代资源是指用于特定配置的资源。要指明某组资源适用的特定配置，请将相应的配置限定符追加到目录名称中。限定符是一种加入到资源目录名称中、用来定义这些资源适用的设备配置的简短字符串。例如，尽管默认 UI 布局保存在 res/layout/目录中，但是可以指定在屏幕处于横向时要使用的不同布局，方法是将其保存在 res/layout-land/目录中。Android 系统可以通过将设备的当前配置与资源目录名称进行匹配，自动使用合适的资源。

res 目录中的资源按类型和配置组织成子目录。对于 Android 项目中的每一项资源，SDK 构建工具都会定义一个唯一的整型 ID，以便在代码中或 XML 文件定义的其他资源中引用该资源。

3.1 提供资源

本节介绍 Android 项目中的资源目录，以及如何为特定的设备配置提供备用资源。表 3-1 中列出的子目录下的资源是默认资源，即这些资源定义应用的默认设计和内容。各种资源文件应该放入项目 res 目录的特定子目录下，切勿将资源文件直接保存在 res 目录内，这将导致编译错误。

表 3-1 res 目录内支持的默认资源目录

资源目录	解释
animator/	用于定义属性动画的 XML 文件
anim/	定义渐变动画的 XML 文件。（属性动画也可以保存在此目录中，但是为了区分这两种类型，属性动画首选 animator/目录）
color/	用于定义颜色状态列表的 XML 文件
drawable/	位图文件（.png、.9.png、.jpg、.gif）或编译为以下可绘制对象资源子类型的 XML 文件：位图文件、九宫格（可调整大小的位图）、状态列表、形状、动画可绘制对象、其他可绘制对象

续表

资源目录	解释
mipmap/	适用于不同启动器图标密度的可绘制对象文件。如果 Launcher 使用 mipmap，那么 Launcher 会自动加载更合适密度的资源。而在 App 中，无论将图片放在 drawable 还是 mipmap 目录，系统只会加载对应 density 中的图片
layout/	用于定义用户界面布局的 XML 文件
menu/	用于定义应用菜单(如选项菜单、上下文菜单或子菜单)的 XML 文件
raw/	要以原始形式保存的任意文件。要使用原始 InputStream 打开这些资源，调用 Resources. openRawResource()以资源 ID(即 R. raw. filename)作为参数。但是，如需访问原始文件名和文件层次结构，则可以考虑将某些资源保存在 app\src\main\assets 目录下(而不是 app\src\main\res 目录)。assets 中的文件没有资源 ID，只能使用 AssetManager 读取这些文件
values/	包含字符串、整型数和颜色等简单值的 XML 文件。values 目录中的单个文件可描述多个资源。对于此目录中的文件，resources 元素的每个子元素均定义一个资源，例如，每个 string 元素创建一个 R. string 资源，每个 color 元素都创建一个 R. color 资源。由于每个资源均由其自己的 XML 元素定义，因此文件名可根据自己需要定义，并且可以把不同的资源类型放在一个文件中。但是，为了清晰起见，还是需要把不同的资源类型放在不同的文件中。例如，对于可在此目录中创建的资源，下面给出了相应的文件名约定：①arrays. xml 用于定义资源数组(类型化数组)；②colors. xml 用于定义颜色值；③dimens. xml 用于定义尺寸值；④strings. xml 用于定义字符串值；5styles. xml 用于定义样式。 除 values 目录外，res 其他子目录中的 XML 资源文件是根据 XML 文件名定义单个资源
xml/	可以在运行时通过调用 Resources. getXML()读取的任意 XML 文件。各种 XML 配置文件(如可搜索配置)都必须保存在此处

采用 Android 技术的不同设备类型可能需要不同类型的资源。例如，如果设备的屏幕尺寸大于标准屏幕，就应该提供替代的布局资源，以充分利用额外的屏幕空间。对于不同的屏幕密度，应该提供替代的可绘制对象资源。如果设备设置不同的语言，应该提供替代的字符串资源，以转换用户界面中的文本。几乎每个应用都应提供替代资源以支持特定的设备配置，在运行时，Android 系统会检测当前设备配置并为应用加载合适的资源。为一组资源指定特定于配置的替代资源的方法，就是在 res 目录中创建一个以< resources_name >-< config_qualifier >形式命名的新目录，可以追加多个< config_qualifier >，每个< config_qualifier >的前面用短划线与其他部分隔开。< resources_name >是相应默认资源的目录名称(表 3-1 中定义)。< config_qualifier >指定要使用这些资源的各个配置的名称，如表 3-2 所示。资源目录使用可以使用多个限定符，必须按照表 3-2 中列出的先后顺序将它们添加到目录名称中，如果限定符的顺序有错误，则该资源将被忽略。将相应的替代资源保存在新目录下，这些资源文件的名称必须与默认资源文件完全一样。

表 3-2 有效的配置限定符

配置	限定符值	解释
MCC 和 MNC	示例： mcc310 mcc310-mnc004 mcc208-mnc00 等等	移动国家代码(MCC)，(可选)后跟设备 SIM 卡中的移动网络代码(MNC)。例如，mcc310 是指美国任一运营商，mcc310-mnc004 是指美国的 Verizon 公司，mcc208-mnc00 是指法国的 Orange 公司 如果设备使用无线电连接(GSM 手机)，则 MCC 和 MNC 值来自 SIM 卡
语言和区域	示例： en fr en-rUS 等等	语言通过由两个字母组成的 ISO 639-1 语言代码定义，可以选择后跟两个字母组成的 ISO 3166-1-alpha-2 区域码(前带小写字母 r，r 前缀用于区分区域码)。这些代码不区分大小写；不能单独指定区域 如果用户更改系统设置中的语言，它有可能在应用生命周期中发生改变。另请参阅 locale 配置字段，该字段表示当前的语言区域

布局方向	ldrtl
	ldltr
		ldrtl 即 "layout-direction-right-to-left"。ldltr 即 "layout-direction-left-to-right"。ldltr 是默认值，它适用于布局、图片或值等任何资源。
		例如，若要针对阿拉伯语提供某种特定布局，并针对任何其他“从右到左”语言(如波斯语或希伯来语)提供某种通用布局，则可编码如下：
		res/
		    layout/
		        main.xml (Default layout)
		    layout - ar/
		        main.xml (Specific layout for Arabic)
		    layout - ldrtl/
		        main.xml (Any "right - to - left" language, except for Arabic, because the "ar" language qualifier has a higher precedence.)
		注：要为应用启用从右到左的布局功能，必须将 supportsRtl 设置为"true"，并将 targetSdkVersion 设置为 17 或更高版本。此项为 API 级别 17 中新增配置

smallestWidth	sw<N>dp 示例： sw320dp sw600dp sw720dp 等等	屏幕的尺寸特性，最小宽度是指可用屏幕区域的宽和高的较小值，它不会随屏幕方向的变化而改变。例如，如果布局要求屏幕区域的最小尺寸至少为 600dp，则可使用 res/layout-sw600dp 创建布局资源，仅当可用屏幕的最小尺寸至少为 600dp 时，系统才会使用这些资源，而不考虑 600dp 所代表的边是用户所认为的高度还是宽度。smallestWidth 是设备的固定屏幕尺寸特性；smallestWidth 将屏幕装饰元素和系统 UI 考虑在内。例如，如果设备的

续表

配　置	限定符值	解　释
smallestWidth	sw<N>dp 示例： sw320dp sw600dp sw720dp 等等	屏幕上有一些永久性UI元素占据沿smallestWidth轴的空间，则系统会声明smallestWidth小于实际屏幕尺寸，因为这些屏幕像素不适用于UI。因此，使用的值应该是布局所需要的实际最小尺寸。应用为多个资源目录提供不同的smallestWidth限定符值时，系统会使用最接近(但未超出)设备smallestWidth的值
可用宽度	w<N>dp 示例： w720dp w1024dp 等等	指定资源应该使用的最小可用屏幕宽度，以dp为单位，由<N>值定义。在横向和纵向之间切换时，为了匹配当前实际宽度，此配置值也会随之发生变化。 应用为多个资源目录提供不同的此配置值时，系统会使用最接近(但未超出)设备当前屏幕宽度的值。此处的值考虑到了屏幕装饰元素，因此如果设备显示屏的左边缘或右边缘上有一些永久性UI元素，考虑到这些UI元素，它会使用小于实际屏幕尺寸的宽度值，这样会减少应用的可用空间。此项为API级别13中新增配置
可用高度	h<N>dp 示例： h720dp h1024dp 等等	指定资源应该使用的最小可用屏幕高度，以dp为单位，由<N>值定义。在横向和纵向之间切换时，为了匹配当前实际高度，此配置值也会随之发生变化。 应用为多个资源目录提供不同的此配置值时，系统会使用最接近(但未超出)设备当前屏幕高度的值。此处的值考虑到了屏幕装饰元素，因此如果设备显示屏的上边缘或下边缘有一些永久性UI元素，考虑到这些UI元素减少了应用的可用空间，它会使用小于实际屏幕尺寸的高度值。非固定的屏幕装饰元素(例如，全屏时可隐藏的手机状态栏)并不在考虑范围内，标题栏或操作栏等窗口装饰也不在考虑范围内，因此应用必须准备好处理稍小于其所指定值的空间
屏幕尺寸	small normal large xlarge	①small屏幕的布局尺寸约为320×426dp单位，例如QVGA和VGA。②normal屏幕的布局尺寸约为320×470dp单位，例如WQVGA、HVGA、WVGA。③large屏幕的布局尺寸约为480×640dp单位，例如VGA和WVGA中等密度屏幕。④xlarge屏幕的布局尺寸约为720×960dp单位，最常见的是平板式设备。API级别9中的新增配置 使用尺寸限定符并不表示资源仅适用于该尺寸的屏幕。如果没有为备用资源提供最符合当前设备配置的限定符，则系统可能使用其中最匹配的资源。如果所有资源均使用大于当前屏幕的尺寸限定符，则系统不会使用这些资源，并且应用在运行时将会崩溃(例如，如果所有布局资源均用xlarge限定符标记，但设备是标准尺寸的屏幕)

续表

配　置	限定符值	解　释
屏幕纵横比	long notlong	long：宽屏，如 WQVGA、WVGA、FWVGA notlong：非宽屏，如 QVGA、HVGA 和 VGA 它完全基于屏幕的纵横比(宽屏较宽)，而与屏幕方向无关
圆形屏幕	round notround	round：圆形屏幕，例如圆形可穿戴式设备 notround：方形屏幕，例如手机或平板电脑
屏幕方向	port land	port：设备处于纵向(垂直)。land：设备处于横向(水平) 如果用户旋转屏幕，它有可能在应用生命周期中发生改变。orientation 配置字段指示当前的设备方向
UI 模式	car desk television appliance watch	car：设备正在车载手机座上显示 desk：设备正在桌面手机座上显示 television：显示电视图像，其 UI 位于远离用户的大屏幕上，主要面向方向键或其他非指针式交互 appliance：设备用于不带显示屏的装置 watch：设备配有显示屏，戴在手腕上
夜间模式	night notnight	如果夜间模式停留在自动模式(默认)，它有可能在应用生命周期中发生改变。在这种情况下，该模式会根据当天的时间进行调整。可以使用 UiModeManager 启用或禁用此模式。
屏幕像素密度(dpi)	ldpi mdpi hdpi xhdpi xxhdpi xxxhdpi nodpi tvdpi anydpi	①ldpi 约为 120dpi；②mdpi 约为 160dpi，例如传统 HVGA 屏幕；③hdpi 约为 240dpi；④xhdpi 约为 320dpi；⑤xxhdpi 约为 480dpi；⑥xxxhdpi 仅限启动器图标，约为 640dpi；⑦nodpi 用于不希望缩放以匹配设备密度的位图资源；⑧tvdpi 约为 213dpi，密度介于 mdpi 和 hdpi 之间的屏幕，主要用于电视机，它并不属于主要密度组；⑨anydpi：此限定符适合所有屏幕密度，其优先级高于其他限定符。这对于矢量可绘制对象很有用。六个主要密度之间的缩放比为 3∶4∶6∶8∶12∶16(忽略 tvdpi 密度)。使用密度限定符并不表示资源仅适用于该密度的屏幕。如果没有为备用资源提供最符合当前设备配置的限定符，则系统可能使用其中最匹配的资源
触摸屏类型	notouch finger	notouch：设备没有触摸屏 finger：设备有专供用户通过手指直接与其交互的触摸屏
键盘可用性	keysexposed keyshidden keyssoft	keysexposed：设备具有可用的键盘。如果设备启用了软键盘，那么即使没有硬键盘，也可以使用此限定符。如果没有提供或已经禁用软键盘，则只有在显示硬键盘时才会使用此限定符 keyshidden：设备具有可用的硬键盘，但它处于隐藏状态，且设备没有启用软键盘 keyssoft：设备已经启用软键盘(无论是否可见) 如果提供了 keysexposed 资源，但未提供 keyssoft 资源，那么只要系统已经启用软键盘，就会使用 keysexposed 资源，而不考虑键盘是否可见

续表

配置	限定符值	解释
主要文本输入法	nokeys qwerty 12key	nokeys：设备没有用于文本输入的硬按键 qwerty：设备具有标准硬键盘(无论是否对用户可见) 12key：设备具有 12 键硬键盘(无论是否对用户可见)
导航键可用性	navexposed navhidden	navexposed：导航键可供用户使用 navhidden：导航键不可用(例如，位于密封盖子后面)。如果用户显示导航键，它有可能在应用生命周期中发生改变。navigationHidden 配置字段，指示导航键是否处于隐藏状态
主要非触摸导航方法	nonav dpad trackball wheel	nonav：除了使用触摸屏以外，设备没有其他导航设施 dpad：设备具有用于导航的方向键 trackball：设备具有用于导航的轨迹球 wheel：设备具有用于导航的方向盘(不常见)
平台版本(API 级别)	示例：v4 等等	设备支持的 API 级别。v4 支持 API 级别 4 或更高版本系统的设备)

并非所有版本的 Android 系统都支持所有限定符，使用新限定符会隐式添加平台版本限定符，因此较旧版本系统的设备必然会忽略它。例如，使用 w600dp 限定符会自动包括 v13 限定符，因为可用宽度限定符是 API 级别 13 中的新增配置。为了避免出现任何问题，应该始终包含一组默认资源(不带限定符的资源)。

例如，以下是一些默认资源和替代资源：

```
res/
    drawable/
        icon.png
    drawable-hdpi/
        icon.png
```

hdpi 限定符表示该目录中的资源适用于屏幕密度较高的设备，其中每个可绘制对象目录中的图像已针对特定的屏幕密度调整了大小，但是文件名完全相同。这样一来，用于引用 icon.png 图像的资源 ID 始终相同，但是 Android 会通过将设备配置信息与资源目录名称中的限定符进行比较，选择最符合当前设备的资源版本。

单组资源可以指定多个限定符，必须遵循表 3-2 中列出的顺序，并使用短划线分隔。例如，drawable-en-rUS-land 适用于横排美国英语设备。例如：

错误：drawable-hdpi-port/

正确：drawable-port-hdpi/

不能嵌套备用资源目录。例如 res/drawable/drawable-en 是错误的。

对于每种限定符类型，仅支持一个值。例如，若要对西班牙语和法语使用相同的可绘制对象文件，肯定不能使用名为 drawable-rES-rFR/的目录，而是需要两个包含相应文件的资源目录，如 drawable-rES/和 drawable-rFR/。目录名称不区分大小写，资源编译器会将目录名称转换为小写，以避免不区分大小写的文件系统出现问题。名称中使用的任何大写字母只是为了便于认读。

将替代资源保存到以这些限定符命名的目录中之后，Android 会根据当前设备配置在应用中自动使用这些资源。每次请求资源时，Android 都会检查替代资源目录是否包含所请求的资源文件，然后查找最佳匹配资源。如果没有与特定设备配置匹配的备用资源，则 Android 会使用相应的默认资源（不含配置限定符）。

3.2 访问资源

在应用中提供资源后，可通过资源 ID 来使用该资源。编译应用时，aapt 会自动生成 R 类，R 类包含 res 目录中所有资源的 ID。每个资源类型都有对应的 R 子类（例如，R. drawable 对应于所有可绘制对象资源），而该类型的每个资源都有对应的静态整型数（例如，R. drawable. icon），这个整型数就是可用来检索资源的资源 ID。

R. java 文件是在编译项目时由 aapt 工具自动生成的，下次编译时所有更改都会被替代，因此切勿手动修改 R. java。尽管资源 ID 是在 R 类中指定的，但是编写代码时永远不需要在 R 类中查找资源 ID。资源 ID 始终由资源类型和资源名称组成。每个资源都被分到一个类型组中，例如 string、drawable 和 layout。资源名称是不包括扩展名的文件名，或者是 XML 文件中元素的 name 属性值（如果资源是简单值的话，例如字符串）。

3.2.1 在代码中访问资源

资源 ID 可以用作方法的参数，在代码中使用资源。在代码中引用资源的语法：

```
[<package_name>.]R.<resource_type>.<resource_name>
```

<package_name>是资源所在包的名称（如果引用的资源来自自己的资源包，则不需要指定包名）。<resource_type>是 R 类的子类。<resource_name>是不带扩展名的资源文件名，或 XML 元素中的 android：name 属性值（如果资源是简单值，例如字符串）。例如，R. string. hello，string 是资源类型，hello 是资源名称。

例如，ImageView 通过方法 setImageResource（）使用 res/drawable/myimage. png 资源：

```
setContentView(R.layout.main_screen);
TextView msgTextView = (TextView) findViewById(R.id.msg);
msgTextView.setText(R.string.hello_message);
ImageView imageView = (ImageView) findViewById(R.id.myimageview);
imageView.setImageResource(R.drawable.myimage);
```

可以通过 Context. getResources()获得 Resources 的实例，再利用 Resources 中的方法检索个别资源。

3.2.2 在 XML 中访问资源

在 XML 资源文件中引用资源的语法：

```
@[<package_name>:]<resource_type>/<resource_name>
```

<package_name>是资源所在包的名称(如果引用的资源来自相同的包,则不需要)。<resource_type>是R类的子类。<resource_name>是不带扩展名的资源文件名,或XML元素的name属性值(如果资源是简单值,例如字符串)。例如:@string/hello,string是资源类型,hello是资源名称。

某些情况下,必须使用资源作为XML中的值(例如,对widget应用可绘制图像),也可以在XML中任何接受简单值的地方使用资源。例如,以下资源文件中,其中包括一个颜色资源和一个字符串资源,都是简单值。

```
<?xml version = "1.0" encoding = "utf-8"?>
<resources>
    <color name = "opaque_red">#f00</color>
    <string name = "hello">Hello!</string>
</resources>
```

以下布局文件使用上面定义的资源来设置文本颜色和文本字符串,资源来自自己的资源包,因而无须在资源引用中指定包名称。如果要引用系统资源,必须加入包名称。

```
<?xml version = "1.0" encoding = "utf-8"?>
<EditText xmlns:android = "http://schemas.android.com/apk/res/android"
      android:layout_width = "fill_parent"
      android:layout_height = "fill_parent"
      android:textColor = "@color/opaque_red"
      android:text = "@string/hello" />
```

3.2.3 访问系统资源

Android包含许多标准资源,例如样式、风格主题和布局。要访问这些资源,请通过android包名称限定资源引用。例如,下面代码使用Android平台提供的布局simple_list_item_1,而不必自己创建列表项布局:

```
setListAdapter(new ArrayAdapter<String>(this, android.R.layout.simple_list_item_1,
myarray));
```

在xml文件使用系统资源,下面代码使用Android平台提供的颜色secondary_text_dark例如:

```
<?xml version = "1.0" encoding = "utf-8"?>
<EditText xmlns:android = "http://schemas.android.com/apk/res/android"
      android:layout_width = "fill_parent"
      android:layout_height = "fill_parent"
      android:textColor = "@android:color/secondary_text_dark"
      android:text = "@string/hello" />
```

3.2.4 引用 style 属性

一个应用应该保持一套统一的样式，包括 Button、EditText、ProgressBar、Toast、Checkbox 等各种控件的样式，还包括控件间隔、文字大小和颜色、阴影等等。样式主要通过 shape、selector、layer-list、level-list、style、theme 等组合实现。当然，也可以为个别视图定义 theme。XML 布局的元素可指定 theme 属性，而该属性将引用一个主题资源。可以在当前应用的 theme 中引用某个 style 属性资源的值，使用此属性定义的样式。引用 style 属性的语法为：

```
?[<package_name>:][<resource_type>/]<resource_name>
```

引用 style 属性的名称语法几乎与普通资源格式完全相同，引用普通资源采用@符号，引用 style 属性改为问号(?)，resource_type 部分为可选项。

例如，下面的代码将文本颜色设置为与系统 theme 的文本颜色匹配：

```
<EditText id = "text"
    android:layout_width = "fill_parent"
    android:layout_height = "wrap_content"
    android:textColor = "?android:textColorSecondary"
    android:text = "@string/hello_world" />
```

在以上代码中，textColorSecondary 是在当前风格主题中定义的样式属性的值。由于系统资源工具知道此环境中肯定存在某个属性资源，因此无须显式声明类型(类型应为 ? android：attr/textColorSecondary)。

Material design 提供的主题有很多，例如@android：style/Theme. Material(深色版本)、@android：style/Theme. Material. Light(浅色版本)。材料主题仅在 Android 5.0 (API 级别 21)及更高版本中提供。v7 支持内容库为一些小组件提供附带 Material Design 风格的主题，同时为配色工具定制提供支持。可以使用主题属性为操作栏和状态栏着色，为应用定制配色工具、触摸反馈动画以及操作行为转换。如果要为状态栏设置定制颜色，可以在扩展材料主题时使用 statusBarColor 属性，statusBarColor 默认将继承 colorPrimaryDark 的值。也可把 statusBarColor 属性设置为@android：color/transparent，以便以透明的方式显示状态栏，同时利用细微的深色渐变以确保白色状态图标仍保持可见，可以根据需要调整窗口标志，也可以使用 Window. setStatusBarColor()方法进行动画或淡出设置。定制导航栏和状态栏时，可以选择将导航栏和状态栏变透明或仅修改状态栏。在所有其他情况中，导航栏均应保持黑色。其中材料设计的颜色定义见图 3-1。如果要定制主题的基色，可以继承材料主题时使用主题属性定义颜色，例如

```
<resources>
    <!-- Base application theme. -->
    <style name = "AppTheme" parent = "@android:style/Theme.Material">
        <!-- Customize your theme here. -->
        <item name = "colorPrimary">@color/colorPrimary</item>
        <item name = "colorPrimaryDark">@color/colorPrimaryDark</item>
        <item name = "colorAccent">@color/colorAccent</item>
    </style>
</resources>
```

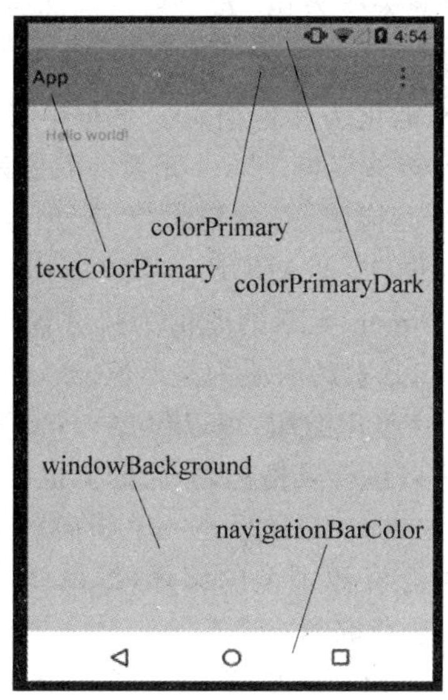

图 3-1 材料设计的颜色定义

3.3 字符串资源与其他简单值

string、string-array、plurals、bool、color、dimen、integer、string-array、integer-array、array 都是使用 name 属性的值作为资源 ID 进行引用的简单资源，不能采用 XML 文件的名称进行引用。因此，可以把各种简单资源放在一个 XML 文件中，作为< resources >元素的子元素。定义这些简单值的 XML 文件的根结点必须是 resources 元素，这是必需的，< resources >标签不需要任何属性。下面是定义这些简单值的语法：

```
<?xml version = "1.0" encoding = "utf - 8"?>
< resources >
< string name = "string_name"> text_string </string >
< bool name = "bool_name">[true | false]</bool >
< color name = "color_name"> hex_color </color >
< dimen name = "dimension_name"> dimension </dimen >
< integer name = "integer_name"> integer </integer >
< string - array name = "string_array_name">
        < item > text_string </item >
    </string - array >
< integer - array name = "integer_array_name">
        < item > integer </item >
    </integer - array >
< array name = "array_name">
        < item > resource </item >
    </array >
</resources >
```

字符串资源为应用提供了文本字符串，有三种类型的字符串资源：string 元素提供单个字符串，string-array 元素提供字符串数组，plurals 元素提供复数数量字符串。应该始终使用字符串资源，以便将应用本地化为其他语言。字符串可以包含风格标识，在风格标识中，撇号和引号标记前一定要加转移字符。bool 元素的取值为 true 或者 false，color 元素的值用 RGB 值和可选的 alpha 通道指定，必须以数字符号(#)开头，紧接着是 Alpha-Red-Green-Blue 信息，格式如下：#RGB，#ARGB，#RRGGBB，#AARRGGBB，其中 A、R、G 和 B 都用十六进制数值表示。dimen 元素的取值为尺寸值，即一个浮点数，后面加上单位(dp，sp，pt，px，mm，in)。integer 元素的取值为整型值。

string-array 元素定义一个字符串数组，可以包含一个或者多个< item >元素，每个 item 元素定义一个字符串，字符串可以包含风格标志，它的值也可以是对另一个字符串资源的引用，item 元素没有属性。integer-array 编译后是一个指向整型数组的指针，array 并不需要是同类的资源数组，也可以是混合资源类型的编译，编译后是指向 TypedArray 的指针，可以使用 TypedArray 的 get()方法获取每一项。

string、bool、color、dimen 和 integer 资源的定义和引用方式，如表 3-3 所示。string-array、integer-array、array 资源的定义与引用，如表 3-4 所示。

表 3-3 string、bool、color、dimen 和 integer 资源的定义与引用

文件位置	res/values/filename. xml 文件名是任意的
资源 ID	name 属性的值将用作资源 ID
在 Java 代码中引用资源	R. type. name 属性值，type 为 string、bool、color、dimen 和 integer
在 XML 文件中引用资源	@[package：]type/name 属性值

表 3-4 string-array、integer-array、array 资源的定义与引用

文件位置	res/values/filename. xml 文件名是任意的
资源 ID	name 属性的值将用作资源 ID
在 Java 代码中引用资源	R. array. name 属性值
在 XML 文件中引用资源	@[package：]array/name 属性值

一个例子，文件 res/values/strings. xml 内容如下：

```
<?xml version = "1.0" encoding = "utf - 8"?>
< resources >
    < string name = "hello"> Hello! </string >
< bool name = "screen_small"> true </bool >
    < bool name = "adjust_view_bounds"> true </bool >
< color name = "opaque_red"> # f00 </color >
     < color name = "translucent_red"> # 80ff0000 </color >
< dimen name = "textview_height"> 25dp </dimen >
     < dimen name = "textview_width"> 150dp </dimen >
     < dimen name = "ball_radius"> 30dp </dimen >
     < dimen name = "font_size"> 16sp </dimen >
< integer name = "max_speed"> 75 </integer >
     < integer name = "min_speed"> 5 </integer >
< string - array name = "planets_array">
```

```
            <item>Mercury</item>
            <item>Venus</item>
            <item>Earth</item>
        </string-array>
    <integer-array name="bits">
            <item>4</item>
            <item>8</item>
        </integer-array>
    <array name="icons">
            <item>@drawable/home</item>
            <item>@drawable/settings</item>
            <item>@drawable/logout</item>
        </array>
        <array name="colors">
            <item>#FFFF0000</item>
            <item>#FF00FF00</item>
            <item>#FF0000FF</item>
        </array>
    </resources>
```

在Java代码中可以引用上述资源。

```
Resources res = getResources();
String helloString = res.getString(R.string.hello);           //去除了样式
CharSequence helloText = res.getText(R.string.hello);         //包含样式
boolean screenIsSmall = res.getBoolean(R.bool.screen_small);
int color1 = res.getColor(R.color.opaque_red);
float fontSize = res.getDimension(R.dimen.font_size);
int maxSpeed = res.getInteger(R.integer.max_speed);
String[] planets = res.getStringArray(R.array.planets_array);
int[] bits = res.getIntArray(R.array.bits);
TypedArray icons = res.obtainTypedArray(R.array.icons);
Drawable drawable = icons.getDrawable(0);
TypedArray colors = res.obtainTypedArray(R.array.colors);
int color2 = colors.getColor(0,0);
```

在布局文件中可以引用上面定义的资源，例如：

```
<TextView
    android:layout_width="fill_parent"
    android:layout_height="wrap_content"
    android:text="@string/hello" />
<ImageView
    android:layout_height="fill_parent"
    android:layout_width="fill_parent"
    android:src="@drawable/logo"
    android:adjustViewBounds="@bool/adjust_view_bounds" />
<TextView
    android:layout_width="fill_parent"
    android:layout_height="wrap_content"
```

```
    android:textColor = "@color/translucent_red"
    android:text = "Hello"/>
<TextView
    android:layout_height = "@dimen/textview_height"
    android:layout_width = "@dimen/textview_width"
    android:textSize = "@dimen/font_size"/>
```

plurals 元素语法如下：

```
<?xml version = "1.0" encoding = "utf - 8"?>
<resources>
    <plurals
        name = "plural_name">
        <item
            quantity = ["zero" | "one" | "two" | "few" | "many" | "other"]
            >text_string</item>
    </plurals>
</resources>
```

定义 string-array 的 xml 文件的根结点必须是 resources 元素，这是必需的。resources 标签不需要属性。plurals 定义一个字符串集合，根据事物数量提供其中的一个字符串，包含一个或多个 item 元素。plurals 的属性 name 将用作资源 ID。item 元素必须是 plurals 元素的子项，用于定义一个字符串，其值可以是对另一字符串资源的引用。

3.3.1 关于字符串的值

如果字符串中有撇号(')，就必须在撇号前加上反斜线(\')，或者把整个字符串包含在双引号中。下面是一些能工作和不能工作的例子：

```
<string name = "good_example">This\'ll work</string>
<string name = "good_example_2">"This'll also work"</string>
<string name = "bad_example">This doesn't work,Cause compile error</string>
```

如果字符串中有双引号(")，必须在双引号前加上反斜线(\")。把整个字符串包含在单引号中，将不能工作。下面是一些能工作和不能工作的例子：

```
<string name = "good_example">This is a \"good string\".</string>
<string name = "bad_example">This is a "bad string".</string>
    <!-- Quotes are stripped; displays as: This is a bad string. -->
<string name = "bad_example_2">'This is another "bad string".'</string>
    <!-- Causes a compile error -->
```

3.3.2 设置字符串的格式

如果需要使用 String. format(String, Object,…) 设置字符串格式，可以通过在字符串资源中加入格式参数来实现。例如，对于以下资源：

```
<string name = "welcome_messages"> Hello, %1$s! You have %2$d new messages.</string>
```

在本例中，格式字符串有两个参数：%1$s 是一个字符串，其中%1 代表第一个变量，$s 代表 string。而%2$d 是一个十进制数字，其中%2 代表第二个变量，$d 代表 decimal。可以这样使用应用中的参数设置字符串格式：

```
Resources res = getResources();
String text = String.format(res.getString(R.string.welcome_messages), username,
mailCount);
```

也可以使用 HTML 标记为字符串添加样式设置。支持的 HTML 元素包括：<b>表示 bold，粗体；<i>表示 italic 斜体。<u>表示 underline(下画线)。例如：

```
<?xml version = "1.0" encoding = "utf-8"?>
<resources>
    <string name = "welcome"> Welcome to <b> Android</b>!</string>
</resources>
```

3.4 菜单资源

一个菜单资源定义一个应用程序菜单(选项菜单、上下文菜单或子菜单)。menu 语法如下：

```
<?xml version = "1.0" encoding = "utf-8"?>
<menu xmlns:android = "http://schemas.android.com/apk/res/android">
    <item android:id = "@[+][package:]id/resource_name"
          android:title = "string"
          android:titleCondensed = "string"
          android:icon = "@[package:]drawable/drawable_resource_name"
          android:onClick = "method name"
          app:showAsAction = ["ifRoom" | "never" | "withText" | "always" | "collapseActionView"]
          android:actionLayout = "@[package:]layout/layout_resource_name"
          android:actionViewClass = "class name"
          android:actionProviderClass = "class name"
          android:alphabeticShortcut = "string"
          android:numericShortcut = "string"
          android:checkable = ["true" | "false"]
          android:visible = ["true" | "false"]
          android:enabled = ["true" | "false"]
          android:menuCategory = ["container" | "system" | "secondary" | "alternative"]
          android:orderInCategory = "integer" />
    <group android:id = "@[+][package:]id/resource name"
           android:checkableBehavior = ["none" | "all" | "single"]
           android:visible = ["true" | "false"]
```

```
            android:enabled = ["true" | "false"]
            android:menuCategory = ["container" | "system" | "secondary" | "alternative"]
            android:orderInCategory = "integer" >
        <item />
    </group>
    <item >
        <menu >
            <item />
        </menu >
    </item >
</menu >
```

菜单资源文件必须使用 menu 标签作为根结点，menu 元素内部可以包含 item 和/或者 group。item 内部还可以包含 menu，表示子菜单。item 和 group 标签用于设置菜单项和分组。xmlns：android 定义了 XML 名称空间，是必需的，取值必须为 http://schemas.android.com/apk/res/android。

3.4.1 item 元素

item 元素内部可以包含一个 menu 用作子菜单。item 标签中不能再嵌入 item 标签。item 元素必须是 menu 元素或者 group 元素的子元素。item 元素的属性，如表 3-5 所示。

表 3-5 菜单 item 元素的属性

属　性	解　释
android：id	菜单项的资源 ID。形式为@＋id/name，其中的加号表示创建一个新的 ID
android：title	菜单项标题(菜单项显示的文本)，必须是字符串资源或者原始的字符串
android：titleCondensed	菜单项的短标题，当菜单项标题太长时会显示该属性值。必须是字符串资源或者原始的字符串
android：icon	用作菜单项图标，必须是 Drawable 资源
android：onClick	方法名，方法应该在 Activity 中声明为 public，并且接受 MenuItem 作为唯一的参数，它表示被单击的菜单项。当菜单项被单击时，调用该方法，这个方法优先于标准的回调函数 onOptionsItemSelected()
app：showAsAction	这个属性描述了在 Android 的高版本中，菜单项何时以何种方式加入到 AppBar 中。这个属性是设置菜单该怎么显示，取值有 5 种，主要应用的有 ifRoom、never、always 这三种，ifRoom 表示如果 Toolbar 上有显示空间就显示在 Toolbar 上，如果没有空间就展显在溢出菜单里。如果没有空间把所有标有 ifRoom 的菜单项显示为 action，含有最低的 orderInCategory 的菜单项将被显示为 action，其余的将显示在溢出菜单中 never 表示永远不显示在 Appbar 上，总是显示在溢出菜单里 always 表示总是显示在 Appbar 上，应当避免使用这一选项。把多个菜单项设置成 always 会导致它们和 Appbar 上的其他 UI 相互覆盖 withText 表示在 action 项上也包含由 android：title 定义的文本 collapseActionView：与 action 项相关联的视图(由 android：actionLayout 或者 android：actionViewClass 声明)是可以折叠的
android：actionLayout	布局资源。指定某个布局资源作为 action 视图

续表

属性	解释
android：actionViewClass	指定一个全限定的 View 类名作为 action 视图
android：actionProviderClass	指定一个全限定的 ActionProvider 类名代替 action 项，例如 android.widget.ShareActionProvider
android：alphabeticShortcut	字符。菜单项的字母快捷键
android：numericShortcut	整数。菜单项的数字快捷键
android：checkable	表示菜单项是否带复选框。取值为 true 或 false
android：checked	如果菜单项带复选框(checkable 属性为 true)，该属性表示复选框默认状态是否被选中。可设置的值为 true 或 false
android：visible	菜单项默认状态是否可见
android：enabled	菜单项默认状态是否被使能
android：menuCategory	该属性可取 4 个值：container、system、secondary 和 alternative，对应于 Menu 的 CATEGORY_常数。通过 menuCategroy 属性可以控制菜单项的位置
android：orderInCategory	该属性需要设置一个整数值，表示同一个组里菜单重要性的顺序，值小的显示在前面。例如 menuCategory 属性值都为 system 的 3 个菜单项(item1、item2 和 item3)。将这 3 个菜单项的 orderInCategory 属性值设为 3、2、1，那么 item3 会显示在最前面，而 item1 会显示在最后面

3.4.2 group 元素

菜单组创建了一组共享特点(例如是否可见、使能、可选择)的菜单项，对菜单项进行分组。group 必须是< menu >元素的子元素。group 可以包含一个或者多个 item 元素。group 对菜单进行分组，分组后的菜单显示效果并没有区别，唯一的区别在于可以针对菜单组进行操作，Menu 对象的方法 setGroupCheckable(int group，boolean checkable，boolean exclusive)用于菜单组内的菜单是否都可选，setGroupVisible(int group，boolean visible)设置是否隐藏菜单组的所有菜单，setGroupEnabled(int group，boolean enabled)设置菜单组的菜单是否可用。group 标签的属性，如表 3-6 所示。

表 3-6 菜单 group 元素的属性

属性	解释
id	菜单组的 ID。形式为@+id/name，其中的加号表示创建一个新的 ID
checkableBehavior	设置这个菜单组的 checkable 行为的类型，有效值为 none、all 和 single. 如果将该属性值设为 all，显示 CheckBox 组件，所有的 item 可以同时被 check；如果设为 single，显示 Radio Button 组件，只有一个 item 被 check；如果设为 none，显示正常的菜单项(不显示任何选择组件)
visible	取值是 true 或 false，设置当前组中所有菜单项是否显示
enabled	取值是 true 或 false，设置当前组中所有菜单项是否被激活
menuCategory	与< item >标签的同名属性含义相同，只是作用域为菜单组。关键字对应于 Menu 的 CATEGORY_常数。该属性可取 4 个值：container、system、secondary 和 alternative。通过 menuCategroy 属性可以控制菜单项的位置。例如将属性设为 system，表示该菜单项是系统菜单，应放在其他种类菜单项的后面
orderInCategory	与< item >标签的同名属性含义相同。只是作用域为菜单组

一个例子，文件 res/menu/ example_menu.xml 内容如下：

```
<menu xmlns:android="http://schemas.android.com/apk/res/android">
    <item android:id="@+id/item1"
          android:title="@string/item1"
          android:icon="@drawable/group_item1_icon"
          android:showAsAction="ifRoom|withText"/>
    <group android:id="@+id/group">
        <item android:id="@+id/group_item1"
              android:onClick="onClickMy"
              android:title="@string/group_item1"
              android:icon="@drawable/group_item1_icon" />
        <item android:id="@+id/group_item2"
              android:onClick="onClickMy"
              android:title="@string/group_item2"
              android:icon="@drawable/group_item2_icon" />
    </group>
    <item android:id="@+id/submenu"
          android:title="@string/submenu_title"
          android:showAsAction="ifRoom|withText" >
        <menu>
            <item android:id="@+id/submenu_item1"
                  android:title="@string/submenu_item1" />
        </menu>
    </item>
</menu>
```

当创建好一个 XML 菜单资源文件之后，可以使用 MenuInflater.inflate()方法填充菜单资源，使 XML 资源变成一个 Menu 类型的对象。菜单资源采用文件名作为资源 id。

```
public boolean onCreateOptionsMenu(Menu menu) {
    MenuInflater inflater = getMenuInflater();
    inflater.inflate(R.menu.example_menu, menu);
    return true;
}
public void onClickMy(MenuItem item) {
    // The item parameter passed here indicates which item it is
    // All other menu item clicks are handled by onOptionsItemSelected()
}
```

3.5 颜色状态列表资源 ColorStateList

ColorStateList 是可以在 XML 文件中定义用作颜色的对象，它可以根据它所应用的视图状态而自动改变颜色。如果仅仅想提供一个静态的颜色资源，应该使用简单的 Color 值。例如，一个按钮可以以几种不同的状态存在(pressed，focused，或者两者都不是)，使用颜色状态列表，可以为每种状态提供一种不同的颜色。ColorStateList 语法如下：

```
<?xml version = "1.0" encoding = "utf - 8"?>
< selector xmlns:android = "http://schemas.android.com/apk/res/android" >
    < item
        android:color = "hex_color"
        android:state_pressed = ["true" | "false"]
        android:state_focused = ["true" | "false"]
        android:state_selected = ["true" | "false"]
        android:state_checkable = ["true" | "false"]
        android:state_checked = ["true" | "false"]
        android:state_enabled = ["true" | "false"]
        android:state_window_focused = ["true" | "false"] />
</selector >
```

ColorStateList 的 Xml 文件的根元素必须是 selector，这是必需的。selector 元素可以包含一个或者多个 item 元素。item 定义了在某种状态(由属性值描述)时使用的一种颜色，item 元素必须是 selector 元素的子元素，每个 item 使用不同的属性描述它应该使用的状态，item 的属性如表 3-7 所示。

表 3-7　ColorStateList 的 item 元素的属性

属　　性	解　　释
android: color	十六进制颜色值，是 item 元素的必需的属性。颜色值用 RGB 值和可选的 alpha 通道指定，必须以数字符号(#)开头，紧接着是 Alpha-Red-Green-Blue 信息。格式如下：#RGB，#ARGB，#RRGGBB，#AARRGGBB
android: state_pressed	布尔量。true 表示这一项将在对象被按下时(例如当一个按钮被触摸或者被单击时)使用；false 表示这一项将在默认的、没有被按下的状态使用
android: state_focused	布尔量。true 表示这一项将在对象获得焦点时(例如当一个按钮通过轨迹球或者方向键高亮时)使用；false 表示这一项将在默认的、没有被聚焦的状态时使用
android: state_selected	布尔量。true 表示这一项将在对象被选择时(例如当一个选项卡被打开时)使用；false 表示这一项将在没有被选择的状态使用
android: state_checkable	布尔量。true 表示这一项可以被勾选；false 表示这一项不能被勾选
android: state_checked	布尔量。true 表示这一项在被勾选状态时使用；false 表示这一项在未被勾选状态时使用
android: state_enabled	布尔量。true 表示这一项在使能状态时(可以接受触摸或单击状态时使用)使用；false 表示这一项将在对象处于被禁止状态时使用
android: state_window_focused	布尔量。true 表示这一项将在应用窗口拥有焦点时(应用处于前台)使用；false 表示这一项将在应用窗口不具有焦点时(例如通知阴影被拉下时或者对话框出现时)使用

当状态发生改变的时候，就会从上到下遍历这个状态列表，第一个和对象的现在状态匹配的 item 将会被使用，不是去选择最适合的匹配，而是简单地选择符合现在状态的最小标准的第一个 item，因而默认值必须放在列表的最后面。

ColorStateList 文件位置为 res/color/filename.xml，文件名用作资源 id，编译后的资源数据类型为指向 ColorStateList 的资源指针。在 Java 文件中，采用 R.color.filename 引用

资源。在 XML 文件中，采用@[package：]color/filename 引用资源。

一个例子：res/color/button_text. xml 文件内容如下：

```
<?xml version = "1.0" encoding = "utf - 8"?>
<selector xmlns:android = "http://schemas.android.com/apk/res/android">
    <item android:state_pressed = "true"
          android:color = "#ffff0000"/> <!-- pressed -->
    <item android:state_focused = "true"
          android:color = "#ff0000ff"/> <!-- focused -->
    <item android:color = "#ff000000"/> <!-- default -->
</selector>
```

布局文件中的一个 Button 可以使用上面的颜色状态列表作为文字颜色。

```
<Button
    android:layout_width = "fill_parent"
    android:layout_height = "wrap_content"
    android:text = "@string/button_text"
    android:textColor = "@color/button_text" />
```

3.6 Drawable 资源

Drawable 资源是指可在屏幕上绘制的图形，可以使用 getDrawable(int) 等 API 检索，或者用作 android：drawable 和 android：icon 等属性的取值。Drawable 资源主要包括以下类型：Bitmap、Nine-Patch、Layer List、State List、Level List、Transition Drawable、Inset Drawable、Clip Drawable、Scale Drawable、Shape Drawable、Animation Resource。

Bitmap 文件(文件扩展名. png、. jpg，或者. gif)创建 BitmapDrawable。Nine-Patch 文件(. 9. png)创建 NinePatchDrawable，具有可拉伸区域，允许根据内容调整图像大小。StateList 文件创建 StateListDrawable，在不同状态引用不同位图图形。Level List 文件创建 LevelListDrawable，用于管理大量备选可绘制对象，每个可绘制对象都分配有最大的备选数量。Transition Drawable 文件创建 TransitionDrawable，可在两种可绘制对象资源之间交错淡出。Inset Drawable 文件用于以指定距离插入其他可绘制对象。当视图需要小于视图实际边界的背景时，此类可绘制对象很有用。Layer List 文件创建 LayerDrawable，列表中的每个可绘制对象按照列表的顺序绘制，列表中的最后一个可绘制对象绘于顶部，每个可绘制对象由 layer-list 元素内的< item >元素表示。Clip Drawable 文件创建 ClipDrawable，用于对其他可绘制对象进行裁剪(根据其当前级别值)。Scale Drawable 文件创建 ScaleDrawable，用于更改其他可绘制对象大小(根据其当前级别值)。Shape Drawable 文件创建 GradientDrawable，用于定义几何形状(包括颜色和渐变)。Animation Resource 创建 AnimationDrawable。

3.6.1 ShapeDrawable

android. graphics. drawable. ShapeDrawable 对应的 XML 的根元素是< shape >，用于

设置控件自身的形状，这些形状不会因为状态的改变而改变，比如圆形、圆角，边框效果等。一般定义 shape 的 XML 文件存放在 drawable 目录下，而不要将它放到 drawable-hdpi 等目录中。Shape Drawable 文件位置为 res/drawable/filename. xml，文件名用作资源 id，编译后的资源数据类型为指向 GradientDrawable 的资源指针。在 Java 文件中，采用 R. drawable. filename 引用资源。在 XML 文件中，采用@[package：]drawable/filename 引用资源。shape 语法如下：

```
<?xml version = "1.0" encoding = "utf - 8"?>
<shape
    xmlns:android = "http://schemas.android.com/apk/res/android"
    android:shape = ["rectangle" | "oval" | "line" | "ring"] >
    <corners
        android:radius = "integer"
        android:topLeftRadius = "integer"
        android:topRightRadius = "integer"
        android:bottomLeftRadius = "integer"
        android:bottomRightRadius = "integer" />
    <gradient
        android:angle = "integer"
        android:centerX = "float"
        android:centerY = "float"
        android:centerColor = "integer"
        android:endColor = "color"
        android:gradientRadius = "integer"
        android:startColor = "color"
        android:type = ["linear" | "radial" | "sweep"]
        android:useLevel = ["true" | "false"] />
    <padding
        android:left = "integer"
        android:top = "integer"
        android:right = "integer"
        android:bottom = "integer" />
    <size
        android:width = "integer"
        android:height = "integer" />
    <solid
        android:color = "color" />
    <stroke
        android:width = "integer"
        android:color = "color"
        android:dashWidth = "integer"
        android:dashGap = "integer" />
</shape>
```

shape 元素可以通过 android：shape 属性指定形状，shape 属性取值为 rectangle、oval、line 或者 ring。rectangle 是默认的形状，用于填充视图，可以画出直角矩形、圆角矩形、弧形等。oval 椭圆形，填充在视图内，用得比较多的是画圆。line 线，必须使用 stroke 元素定义线宽，可以画实线和虚线。ring 环，可以画环形进度条。当 shape 元素的 android：shape 属性取值为 ring 时，其他属性如表 3-8 所示。

表 3-8 shape 元素的 android：shape 属性取值为 ring 时的其他属性

属 性	解 释
innerRadius	尺寸，以尺寸值或尺寸资源表示环内部(中间的孔)的半径
innerRadiusRatio	浮点型。环内部的半径，以环宽度的比率表示。例如，如果 innerRadiusRatio="5"，则内半径等于环宽度除以 5。此值被 innerRadius 覆盖。默认值为 9
thickness	尺寸，以尺寸值或尺寸资源表示环的厚度
thicknessRatio	浮点型，表示为环宽度的比率。例如，如果 thicknessRatio="2"，则厚度等于环宽度除以 2。此值被 innerRadius 覆盖。默认值为 3
useLevel	布尔值。如果用作 LevelListDrawable，则此值为 true。这通常应为 false，否则形状不会显示

1. corners 元素

< corners >元素为形状产生圆角，仅当形状为矩形时适用。属性包括 radius、topLeftRadius、topRightRadius、bottomLeftRadius、bottomRightRadius，它们的取值为尺寸，以尺寸值或尺寸资源表示。radius 指定所有角的半径，会被以下属性覆盖：左上角会被 topLeftRadius 属性覆盖，右上角会被 topRightRadius 属性覆盖，左下角会被 bottomLeftRadius 属性覆盖，右下角会被 bottomRightRadius 属性覆盖。

2. gradient 元素

gradient 元素指定形状的渐变颜色，它的 angle 属性指定渐变的角度(度)，0 度方向为从左到右，90 度方向为从下到上，必须是 45 的倍数，默认值为 0。它的 centerX 和 centerY 属性指定渐变中心的相对位置(0～1.0)，浮点型。它的 startColor、endColor 和 centerColor 分别指定渐变的开始颜色、结束颜色、起始颜色与结束颜色之间的可选颜色，以十六进制值或颜色资源表示。它的 type 属性指定渐变的类型：linear、radial 或者 sweep。gradientRadius 浮点型，指定渐变的半径，仅在 android：type="radial"时适用。

3. padding 元素

padding 元素指定要应用到包含视图元素的内边距(这会填充视图内容的位置)。

4. size 元素

size 指定形状的大小，默认情况下，形状按照此处定义的尺寸按比例缩放至容器视图的大小。在 ImageView 中使用形状时，可通过将 android：scaleType 设置为 center 来限制缩放。

5. solid 元素

solid 用于指定填充 shape 的颜色，它的属性 color，以十六进制值或颜色资源表示。

6. stroke 元素

stroke 元素用于画 shape 的边框。属性 width，以尺寸值或尺寸资源指定线宽。属性 color，以十六进制值或颜色资源指定线的颜色。属性 dashGap 和 dashWidth，取值为尺寸值或尺寸资源，分别指定了短划线的间距和每个短划线的长度，两者必须同时设置才能有效，即 dashWidth 仅在设置了 dashGap 时有效，dashGap 仅在设置了 dashWidth 时有效。

3.6.2 StateListDrawable

StateListDrawable 类定义了不同状态值下与之对应的图片资源。根据对象不同的状态，用几张不同的图片来代表相同的图形。比如，一个按钮有多种状态，获取焦点、失去焦点、单击等等，使用 StateListDrawable 可以根据不同的状态提供不同的背景。StateListDrawable 可以在 XML 文件中用 selector 标签定义。在唯一的一个 selector 标签下，每个状态的图片使用嵌入在 selector 标签内的 item 标签来定义。StateListDrawable 语法如下：

```
<?xml version = "1.0" encoding = "utf - 8"?>
< selector xmlns:android = "http://schemas.android.com/apk/res/android"
    android:constantSize = ["true" | "false"]
    android:dither = ["true" | "false"]
    android:variablePadding = ["true" | "false"] >
    < item
        android:drawable = "@[package:]drawable/drawable_resource"
        android:state_pressed = ["true" | "false"]
        android:state_focused = ["true" | "false"]
        android:state_hovered = ["true" | "false"]
        android:state_selected = ["true" | "false"]
        android:state_checkable = ["true" | "false"]
        android:state_checked = ["true" | "false"]
        android:state_enabled = ["true" | "false"]
        android:state_activated = ["true" | "false"]
        android:state_window_focused = ["true" | "false"] />
</selector >
```

文件位置为 res/drawable/filename.xml，文件名用作资源 id，编译后的资源数据类型为指向 StateListDrawable 的资源指针。在 Java 文件中，采用 R.drawable.filename 引用资源。在 XML 文件中，采用@[package:]drawable/filename 引用资源。表 3-9 列出了 StateListDrawable 语法中属性的含义。

表 3-9　StateListDrawable 的 xml 属性

属　性	解　释
constantSize	布尔型，默认为 false。false 表示图像的大小将根据现在的状态变化；true 表示图像的大小不随状态改变，大小取所有状态的最大值
state_activated	布尔型，当一个 View 或者它的父视图被激活成持久的选择（例如在持久的导航视图中高亮之前选中的列表项）设置，意味着已经把它标记为感兴趣区。这是当状态需要沿着层级视图向下传播时 state_checked 的替代表示。
state_checkable	布尔型，若设置 true，表示对象可以被选中，显示一个打勾 check 的标记
state_checked	布尔型，若设置 true，表示对象现在已经被选中(checked)
state_enabled	布尔型，当对象被使能时（处于可单击状态，能够接收触摸/单击事件）设置为 true，可以调用 setEnable()方法改变其状态
state_focused	布尔型，当视图获得输入焦点时设置。一个窗口只能有一个视图获取输入焦点，而一个窗口可以同时有多个视图处于 selected 状态。一般由用户交互所致，不需要应用程序直接改变

续表

属　性	解　释
state_pressed	布尔型，当用户在这个视图里按下时设置，一般由用户交互所致
state_selected	布尔型，当视图或者父视图中的某个视图被选中时设置。应用程序可以主动调用 setSelected()改变其状态，例如 ListView 或者 GridView 会使某个 View 处于 selected 状态，并且获得该状态的 View 处于高亮状态
state_window_focused	布尔型，当视图的窗口有输入焦点时设置，视图所在窗口是否为当前交互窗口，该状态值由系统自动确定，应用程序不能改变其状态。
variablePadding	布尔型，若设置 false，表示图像的 padding 将不随状态变化(基于所有状态的最大 padding)，若设置 true，允许图像的 padding 跟随状态改变
visible	布尔型，指示视图的初始状态是否可见

当状态发生改变的时候，就会从上到下遍历这个状态列表，第一个和对象的现在状态匹配的 item 将会被使用，不是去选择最适合的匹配，而是简单地选择符合现在状态的最小标准的第一个 item，这就是默认值必须放在表中最后面的原因。

3.6.3 LayerDrawable

LayerDrawable 对应的 XML 的根元素是< layer-list >，表示可绘制对象阵列，可以包含多个 item 元素，每个 item 包含一个 Drawable，按照列表的顺序绘制，列表中的后一个可绘制对象绘于上层。layer-list 元素语法如下：

```
<?xml version = "1.0" encoding = "utf - 8"?>
< layer - list
    xmlns:android = "http://schemas.android.com/apk/res/android" >
    < item
        android:drawable = "@[package:]drawable/drawable_resource"
        android:id = "@[ + ][package:]id/resource_name"
        android:top = "dimension"
        android:right = "dimension"
        android:bottom = "dimension"
        android:left = "dimension" />
</layer - list >
```

layer-list 元素，必须是根元素，必须包含一个或多个 item 元素。item 元素定义要放在图层中的可绘制对象，item 里面可以放置< bitmap >或者< shape >子元素，后添加的 item 放置在上层。位置由以下属性定义：top、bottom、left 和 right，取值为尺寸值，默认值为 0，分别指定该 item 与上边、下边、左边和右边的距离。drawable 属性，引用可绘制对象资源，是必需的。

默认情况下，所有可绘制项都会缩放以适应包含视图的大小。为避免缩放列表中的项目，请在 item 元素内使用 bitmap 元素指定可绘制对象，并且对某些不缩放的项目定义 gravity(例如"center")。例如，以下< item >定义缩放以适应其容器视图的项目：

```
< item android:drawable = "@drawable/image" />
```

为避免缩放，以下示例使用 gravity 居中的 bitmap 元素：

```
<item>
  <bitmap android:src="@drawable/image"android:gravity="center" />
</item>
```

layer-list 文件位置为 res/drawable/*filename*. xml，文件名用作资源 ID。编译后的资源数据类型为指向 LayerDrawable 的资源指针。在 Java 文件中，采用 R. drawable. filename 引用资源。在 XML 文件中，采用@[package:]drawable/filename 引用资源。

采用 layer-list，可以实现简单的图层叠加、图像旋转叠加、阴影、只有一条边、两条边、三条边的矩形，同心圆环等效果。

实例 1：只有部分边线的矩形，文件 app\src\main\res\drawable\rectangle_side. xml 文件内容如下：

```
<?xml version="1.0" encoding="utf-8"?>
<layer-list xmlns:android="http://schemas.android.com/apk/res/android">
    <item>
        <shape android:shape="rectangle">
            <solid android:color="@android:color/holo_blue_light"/>
        </shape>
    </item>
    <item android:top="5dp">
        <shape android:shape="rectangle">
            <solid android:color="@android:color/white"/>
        </shape>
    </item>
</layer-list>
```

app\src\main\res\layout\activity_main. xml 布局文件使用了 rectangle_side，内容如下：

```
<?xml version="1.0" encoding="utf-8"?>
<android.support.constraint.ConstraintLayout xmlns:android="http://schemas.android.com/
apk/res/android"
    xmlns:app="http://schemas.android.com/apk/res-auto"
    xmlns:tools="http://schemas.android.com/tools"
    android:layout_width="match_parent"
    android:layout_height="match_parent"
    tools:context=".MainActivity">
    <Button
        android:id="@+id/btn_test"
        android:background="@drawable/rectangle_side"
        android:layout_width="200dp"
        android:layout_height="0dp"
        app:layout_constraintDimensionRatio="1:1"
        android:text="Hello World!"
        app:layout_constraintBottom_toBottomOf="parent"
        app:layout_constraintLeft_toLeftOf="parent"
```

```
        app:layout_constraintRight_toRightOf = "parent"
        app:layout_constraintTop_toTopOf = "parent" />

</android.support.constraint.ConstraintLayout>
```

上边的布局文件中，btn_test 按钮的 android：background 属性设置为@drawable/rectangle_side，btn_test 按钮的 layout_width 为 200dp，layout_constraintDimensionRatio 为 1∶1，所以 layout_height 也是 200dp，rectangle_side 有两个图层，下层的矩形没有包含 top、bottom、left、right 属性，因而这几个属性取默认值 0，所以占据 btn_test 按钮的整个矩形区域，该层为蓝色。上层的图层也是矩形，android：top＝"5dp"，该图层的顶部与 btn_test 按钮的矩形区域的顶部间距为 5dp，android：bottom、android：left、android：right 都为默认值 0dp，所以下面的图层只有顶部 5dp 没有被上层覆盖，没有被覆盖的区域为蓝色。如果希望显示的矩形的上边线和左边线，只需要给 layer-list 的第二个 item 元素再添加 android：left＝"5dp"，修改后第二个< item >元素的定义如下：

```
<item android:top = "5dp" android:left = "5dp">
    <shape android:shape = "rectangle">
        <solid android:color = "@android:color/white"/>
    </shape>
</item>
```

实例 2：同心圆，文件 app\src\main\res\drawable\ring.xml 文件内容如下：

```
<?xml version = "1.0" encoding = "utf - 8"?>
<layer - list xmlns:android = "http://schemas.android.com/apk/res/android">
    <item>
        <shape android:shape = "oval">
            <solid android:color = "@android:color/holo_red_light"/>
            <stroke android:color = "@android:color/holo_blue_light" android:width = "10dp"/>
        </shape>
    </item>
    <item android:top = "40dp" android:bottom = "40dp" android:left = "40dp" android:right =
"40dp">
        <shape android:shape = "oval">
            <solid android:color = "@android:color/holo_green_light"/>
            <stroke android:color = "@android:color/holo_blue_light" android:width =
"10dp"/>
        </shape>
    </item>
</layer - list>
```

然后把实例 1 的布局文件 activity_main.xml 中的 btn_test 按钮的 background 属性设置为@drawable/ring。ring.xml 有两个图层，下层的椭圆指定 stroke 的宽度为 5dp、颜色为蓝色，内部的填充色为红色。上面的图层设置了 top、bottom、left 和 right 属性，指定了该图层与 btn_test 按钮的矩形区域的上下左右四个方向的间距，这个图层也指定了 stroke 的宽度为 5dp、颜色为蓝色，内部的填充色为绿色。

3.7 本章主要参考文献

1. https://developer.android.google.cn/guide/topics/resources/providing-resources.html
2. https://developer.android.google.cn/guide/topics/resources/more-resources
3. https://developer.android.google.cn/guide/topics/resources/string-resource.html
4. https://developer.android.google.cn/guide/topics/resources/menu-resource.html
5. https://developer.android.google.cn/guide/topics/resources/color-list-resource.html
6. https://developer.android.google.cn/guide/topics/resources/drawable-resource.html
7. https://developer.android.google.cn/training/material/theme.html

第4章 Gradle的Android插件

CHAPTER 4

Gradle是一个基于JVM的构建工具，基于Groovy的特定领域语言(DSL)声明项目设置，抛弃了基于XML(pom.xml和ivy.xml配置文件)的各种繁琐配置。Gradle支持maven和Ivy仓库，支持传递性依赖管理。Gradle脚本使用Groovy编写，Gradle的实质是配置脚本，执行一种类型的配置脚本时就会创建一个关联的对象，譬如执行Build Script就会创建一个Project对象。

Gradle使用两个主要的目录执行和管理它的工作：Gradle用户的home目录和工程的根目录。Gradle用户的home目录(默认值为$USER_HOME/.gradle)用于存储全局的配置属性和初始化脚本以及cache和日志文件。Gradle用户的home目录里主要包括caches、daemon、init.d、wrapper子目录和文件gradle.properties。

Gradle初始化脚本(Init script)类似于build.gradle脚本，也是groovy语言脚本。这种脚本在构建开始之前运行(在运行build script之前运行)，主要的用途是为接下来的Build Script做一些准备工作，定制一些全局设置：例如设置仓库、JDK安装位置、设置企业级别的配置等。运行初始化脚本文件的方法有以下4种：

(1) 在命令行指定脚本文件，使用-I或者--init-script选项后接脚本文件的路径，这个选项可以在命令行上出现多次，每次添加另一个脚本文件。例如：gradle --init-script yourdir/init.gradle -q taskName。

(2) 把init.gradle文件放到USER_HOME/.gradle/目录下。

(3) 把以.gradle结尾的文件放到USER_HOME/.gradle/init.d/目录下。

(4) 把以.gradle结尾的文件放到GRADLE_HOME/init.d/目录下。

如果存在多个脚本文件，gradle会按上面的(1)～(4)顺序依次执行这些文件，如果给定目录下存在多个init脚本，会按字母a～z顺序执行这些脚本。执行Gradle命令时，就会创建一个Gradle对象，整个构建执行过程中只有这么一个对象，初始化脚本的任何属性引用和方法调用，都会委托给这个Gradle实例。每个init脚本都实现了Script接口。

工程的根目录包含源文件，也包含Gradle生成的临时文件和目录。工程的根目录主要包括：.gradle、gradle子目录和build.gradle、gradle.properties、gradlew、gradlew.bat和settings.gradle文件。gradle子目录中主要包含wrapper子目录。

对于多模块工程的构建，工程的根目录里必须包含settings.gradle和build.gradle文件。settings.gradle文件描述工程和子工程的结构，用于配置Settings对象，这个文件一般放置在工程的根目录中。对于单一模块的工程，工程的根目录里可以不包含settings.

gradle。子工程里可以包含自己的 build. gradle 文件(如果子工程仅仅用作其他子工程的容器,也可以不包含 build. gradle 文件)。另外工程的 build 脚本名不是必须为 build. gradle,也可以是 *. gradle,这可以在 settings. gradle 文件中设置。通常,多模块工程的目录结构要求将子模块放在父模块的根目录中,但是如果有特殊的目录结构,可以在 settings. gradle 文件中配置。默认情况下,Gradle 使用包含 settings. gradle 文件的名字作为根工程的名字。工程根目录里的 build. gradle 通常用来共享子工程的公用配置。

Gradle 配置过程如下:首先创建一个 Settings 实例;如果根目录下存在 settings. gradle 文件,就使用它来配置 Settings 对象;使用已经配置了的 Settings 对象来创建分级的 Project 实例;如果存在 build. gradle 文件,就执行 build. gradle 文件来解析该 project。每个 project 要先于它的子 project 被解析,这个解析顺序可以通过调用 evaluationDependsOnChildren()或者通过使用 evaluationDependsOn(String)显式地添加解析依赖性来覆盖,解析 build. gradle 后会建立一个有向图来描述 Task 之间的依赖关系。每个 build. gradle 文件会转换成一个 Project 对象,build. gradle 文件和 Project 对象是一对一的关系。在构建脚本中通过 project(':sub-project-name ')来引用子项目对应的 Project 对象。Gradle 脚本类型与关联对象,如表 4-1 所示。

表 4-1　Gradle 脚本类型与关联对象

脚本类型	关联文件名称	关联对象类型
Init script		Gradle
Settings script	settings. gradle	Settings
Build script	build. gradle	Project

Gradle 命令行运行格式如下,选项可以在任务名称的前面或者后面,选项的含义如表 4-2 所示。

```
gradle [option...] [task...]
```

表 4-2　Gradle 选项

选　　项	解　　释
-?,-h,--help	显示帮助信息
-a,--no-rebuild	不重新编译工程 dependencies
--all	显示任务列表中的所有附加细节
-b,--build-file	指定 build-file
-c,--settings-file	指定 settings-file
--console	plain 在 console 仅仅输出文字,不含颜色和 rich 输出。auto,当 build 进程依附于一个 console 时,输出颜色和 rich 输出。rich,不管 build 进程是否依附于一个 console 时,输出颜色和 rich 输出
--continue	某个任务失败后,继续执行
--configure-on-demand(incubating)	仅仅按需要配置有关系的工程
-D,--system-prop	设置 JVM 的系统属性,例如-Dmyprop=myvalue
-d,--debug	debug 模式

续表

选　项	解　释
-g,--gradle-user-home	指定 gradle-user-home
--gui	启动图形用户接口模式
-I,--init-script	指定 init-script
--offline	指定 build 不连接网络
-P,--project-prop	指定 root project 的工程属性。例如-Pmyprop=myvalue
-p,--project-dir	指定 gradle 的开始文件夹,默认是当前文件夹
--parallel(incubating)	并行地 build 工程
--project-cache-dir	指定工程的 cache 文件夹
-q,--quiet	仅仅输出错误信息
-v,--version	显示版本信息
-x,--exclude-task	执行时排除某个任务
--daemon	使用 Gradle daemon 运行 build
--foreground	在前台启动 Gradle daemon,主要用于调试
--no-daemon	不使用 Gradle daemon 运行 build
--stop	如果 daemon 正在运行,停止 daemon

为了看见一个工程的任务列表,运行 gradle < project-path >: tasks。每个子项目名字前面有一个冒号。为了看见某个任务的详细信息,运行 gradle help --task < task >。为了运行 build 任务,运行 gradle < task >。任务可以使用冒号分隔的完全限定名,冒号相当于 Windows 或者 Linux 操作系统里的路径分隔符'/'或者'\',最前面的冒号代表根工程自身。

示例:执行 gradle -q projects 输出如下:

```
> gradle - q projects
------------------------------------------------------------
Root project
------------------------------------------------------------
Root project 'multiproject'
 +--- Project ':api'
 +--- Project ':services'
 | +--- Project ':services:shared'
 | \--- Project ':services:webservice'
 \--- Project ':shared'
To see a list of the tasks of a project, run gradle <project - path>:tasks
For example, try running gradle :api:tasks
```

子工程名为文件夹的名字,例如 webservice 位于< root >/services/webservice。对于多模块的构建,需要明确指出需要执行哪个模块的任务。有两种方法:

(1) 切换到子工程对应的文件夹,像单模块工程一样执行 gradle < task >。例如 gradle test 命令将执行当前文件夹中所有子工程的 test 任务。如果从工程根目录执行此命令,将执行 api,shared,services:shared 和 services:webservice 中的 test 任务。如果从 service 工程文件夹执行此命令,将执行 services:shared 和 services:webservice 中的 test 任务。

(2) 在工程的任何文件夹,使用完全限定的任务名称,例如:gradle :services:webservice:

build 将执行 webservice 子工程以及它依赖的子工程的 build 任务。

4.1　Project 接口介绍

```
public interface Project extends Comparable<Project>, ExtensionAware, PluginAware
```

Project 的主要方法有很多，表 4-3 列出了开发 Android 应用时 build.gradle 中可能用到的部分方法。

表 4-3　Project 的方法（部分）

方　法	解　释
String absoluteProjectPath(String path)	把名字转换为绝对路径
void afterEvaluate(Action<? super Project> action)	添加一个在 project 被评估后立即执行的动作
void afterEvaluate(Closure closure)	添加一个在 project 被评估后立即被调用的闭包
void allprojects(Action<? super Project> action)	配置这个工程和每个子工程
void allprojects(Closure configureClosure)	配置这个工程和每个子工程
AntBuilder ant(Closure configureClosure)	使用 AntBuilder 指定闭包
void artifacts(Closure configureClosure)	配置这个工程的发布环境
void beforeEvaluate(Closure closure)	添加一个在 project 被评估前立即被调用的闭包
void beforeEvaluate(Action<? super Project> action)	添加一个在 project 被评估前立即执行的动作
void buildscript(Closure configureClosure)	配置这个工程的 build 脚本的 classpath
Object configure(Object object, Closure configureClosure)	通过闭包配置一个对象，闭包的代理设置为提供的对象
WorkResult copy(Action<? super CopySpec> action)	拷贝指定的文件
WorkResult copy(Closure closure)	拷贝指定的文件
CopySpec copySpec()	创建一个 copySpec 对象，以后可以用来拷贝文件或者创建 archive 文件
CopySpec copySpec(Closure closure)	创建一个 copySpec 对象，以后可以用来拷贝文件或者创建 archive 文件
void defaultTasks(String... defaultTasks)	设置这个工程的默认任务的名字
WorkResult delete(Action<? super DeleteSpec> action)	删除指定的文件
boolean delete(Object... paths)	删除指定的文件和文件夹
void dependencies(Closure configureClosure)	配置这个工程的依赖文件
FileTree tarTree(Object tarPath)	创建一个新的 FileTree，它包含了给定的 tar 文件的内容
Task task(String name)	用给定的任务名创建一个任务，并把这个任务添加到工程中
Task task(String name,Closure configureClosure)	用给定的任务名创建一个任务，并把这个任务添加到工程中

续表

方　法	解　释
URIuri(Object path)	把一个文件路径解析成相对于工程根目录的 URI
FileTreezipTree(Object zipPath)	创建一个新的 FileTree,它包含了给定的 zip 文件的内容

4.2 Gradle Android 插件

Android 的构建系统使用 Gradle 来构建应用。Gradle 是一种构建工具,它使用一种基于 Groovy 的特定领域语言来申明项目设置。Gradle 通过根据一个名为 build. gradle 配置文件对项目进行管理,这包含添加项目的依赖、打包、签名、发布等一系列操作。那么,如何为 Android 项目提供这个 Gradle 构建环境呢？这就需要通过安装 Gradle Android 插件来使系统能支持运行 Gradle。安装 Android Studio 就安装了 Gradle 插件。但 Gradle 插件是独立于 Android Studio 运行的,所以它的更新也是与 Android Studio 分开的。

指定 Android plugin for Gradle 版本的方式：可以单击 Android Studio 的菜单 File | Project Structure | Project,也可以编辑工根目录中的 build. gradle 文件。

Gradle 插件需要一个默认的文件夹结构。Gradle 遵循约定优先于配置的概念,在可能的情况尽可能提供合理的默认配置参数。

基本的 Java 工程包含 main 和 test 两个 source sets,Android 工程还包含另外一个 source sets,名字为 androidTest。这 3 个 source sets 的文件夹分别为：src/main/,src/test/,src/androidTest/。对于 Java plugin 和 Android plugin 来说,source sets 中包含 Java 的路径和 resources 的路径。但对于 Android plugin 来说,它还拥有以下特有的文件和文件夹结构,main 目录里的内容如下：

src/main/	
	AndroidManifest. xml
	java/
	res/
	aidl/
	cpp/
	assets/
	jni/

当默认的项目结构不适用的时候,可以使用以下方法为 Java 项目配置 sourceSets：

```
sourceSets {
    main {
        java {
            srcDir 'src/java'
        }
```

```
            resources {
                srcDir 'src/resources'
            }
        }
    }
```

4.3 setting.gradle 解析

当创建新工程时，Android Studio 会默认创建两个 build.gradle 和一个 settings.gradle，build.gradle 分别放在了根目录和根目录里的 app 目录下，下面是 Gradle 相关文件的位置：

root-project/		
	app/	
		src
		build.gradle
	build.gradle	
	settings.gradle	
	gradle.properties	

当工程只有一个模块的时候，setting.gradle 内容如下：

```
include ':app'
```

下面的 setting.gradle 文件中，描述工程包含三个模块：app、pcasserver 和 libpcasserver。每个子项目名字前面有一个冒号(：)，每个子项目的目录里都有一个 build.gradle 文件。

```
include ':app', ':pcasserver', ':libpcasserver'
```

4.4 Android 项目根目录里的 build.gradle

根目录下的 build.gradle 文件定义了这个工程下的所有模块的公共属性。module 是通过 settings.gradle 文件来配置。app 文件夹就是一个 module，如果在当前工程中添加了一个新的 module 例如 pcasserver，就需要在 settings.gralde 文件中包含这个新的 module。根目录下的 build.gradle 默认包含两个方法：buildscript 和 allprojects。

```
// Top - level build file where you can add configuration options common to all sub - projects/
modules.
buildscript {
    repositories {
        google()
```

```
        jcenter()
    }
    dependencies {
        classpath 'com.android.tools.build:gradle:3.0.1'
    // NOTE: Do not place your application dependencies here; they belong
    // in the individual module build.gradle files
    }
}
allprojects {
    repositories {
        google()
        jcenter()
    }
}
task clean(type: Delete) {
    delete rootProject.buildDir
}
```

buildscript：buildscript 中的声明是 Gradle 脚本自身运行需要使用的资源，可以声明的资源包括 maven 仓库地址、依赖项、第三方插件等。repositorie 就是存放代码的库，是一些文件的集合，这些文件通过 group、name 和 version 三个属性组织在一起，它能很好地进行版本控制和访问。应用广泛的仓库类型有两种：Ivy 和 Maven，Ivy 主要应用在使用 Ant 工具构建的系统中。Maven 仓库有 Jcenter 和 Maven Central，它们都是 Maven 仓库的实现。而这两个仓库在以 Gradle 为构建工具的 Android Studio 中都能使用。JCenter 由 bintray.com 维护，Maven Central 则是由 sonatype.org 维护，它们分别存储在不同的服务器，两者没有任何关系。与 Maven Central 相比，JCenter 仓库更大，对开发者更加友好，因而新版的 Android Studio 默认采用 jcenter 作为仓库。默认情况下，Gradle 并没有定义任何的仓库。在使用外部依赖之前需要至少定义一个仓库。

buildscript 里的 dependencies 中的 classpath 是配置运行 build script 所需要的 classpath。

allprojects：allprojects 是父 Project 的一个属性，该属性用于该 Project 对象以及其所有子项目。在父项目的 build.gradle 脚本里，可以通过给 allprojects 传一个包含配置信息的闭包，来配置所有项目（包括父项目）的共同设置，通常可以在这里配置 IDE 的插件，group 和 version 等信息。子项目可以配置自己的 repositories 以获取自己独需的依赖包。allprojects 块的 repositories 用于多项目构建，为所有项目提供共同所需依赖包。

4.5 Android 模块内的 build.gradle

模块内的 build.gradle 文件只对该模块起作用，而且其可以重写来自于根目录下的 build.gradle 文件中的任何参数。该模块文件主要包含以下内容：

```
apply plugin: 'com.android.application'
android {
    dataBinding.enabled = true
```

```
    compileSdkVersion 26
    defaultConfig {
        applicationId "cuc.viewpager1"
        minSdkVersion 19
        targetSdkVersion 26
        versionCode 1
        versionName "1.0"
        testInstrumentationRunner "android.support.test.runner.AndroidJUnitRunner"
    }
    buildTypes {
        release {
            minifyEnabled false
            proguardFiles getDefaultProguardFile('proguard-android.txt'), 'proguard-
rules.pro'
        }
    }
}

dependencies {
    implementation fileTree(include: ['*.jar'], dir: 'libs')
    implementation 'com.android.support:appcompat-v7:26.1.0'
    implementation 'com.android.support.constraint:constraint-layout:1.1.2'
    testImplementation 'junit:junit:4.12'
    androidTestImplementation 'com.android.support.test:runner:1.0.2'
    androidTestImplementation 'com.android.support.test.espresso:espresso-core:3.0.2'
    implementation 'com.android.support:design:26.1.0'
}
```

该文件中的 apply plugin：'com.android.application'声明构建的项目类型。注意，Groovy 支持函数调用的时候，通过 参数名 1：参数值 1，参数名 2：参数值 2 的方式来传递参数。

该文件中的 Android 配置用于设置 Android 属性，而必需的属性有 compileSdkVersion。compileSdkVersion 指定了编译该 app 时使用的 api 版本。defaultConfig 方法包含了该 app 的核心属性，该属性会重写在 AndroidManifest.xml 中的对应属性。minSdkVersion、targetSdkVersion、versionCode、versionName 等所有属性都是重写了 AndroidManifest 文件中的属性，所以没必要在 AndroidManifest 中重复定义这些属性。第一个属性 applicationId，将被用在设备和各大应用商店中作为唯一的标示，在 AndroidManifest 文件中定义的 package name 依然被用来作为包名和 R 文件的包名。

4.5.1 依赖配置

大部分项目并不是依靠项目本身独立完成的，需要加入其他文件或者库以便进行编译或测试。对于这些由项目依赖导入的文件，Gradle 可能从远程的 Maven 仓库下载这些依赖文件，或者从本地目录中，或者在一些多项目构建中由其他项目构建而来。只需要简单声明依赖文件的名称，就可以确定从哪里获取这些依赖文件。在 Gradle 中，依赖被组织到不同的配置项。每一个配置项中有一组的依赖，这些配置项被称作依赖配置项。可以用这些配置项声明项目的外部依赖，也可以用来声明项目的产物。可以使用特定的配置关键字告诉 Gradle 如何以及何时使用某个依赖项，下面介绍可以用来配置依赖项的一些关键字。

1. implementation

implementation 定义了编译工程的产品源代码所必需的依赖，这些依赖不是工程需要暴露的 api，即 implementation 依赖仅仅用于当前模块内部，仅对当前模块可见，对该模块的消费者不可见，简单地说 implementation 依赖不会传递。该模块的消费者指依赖于该模块的组件。对于大型多项目构建，使用 implementation 而不是 api/compile 可以显著缩短构建时间，因为它可以减少构建系统需要重新编译的项目量。大多数应用和测试模块都应使用此配置。implementation 支持 jar 文件和 aar 文件。testImplementation 定义了编译和运行工程的测试源代码所必需的依赖。

例如，有一个模块 testsdk 依赖于 gson：

```
implementation 'com.google.code.gson:gson:2.8.2'
```

这时候，在 testsdk 里边的 java 代码是可以访问 gson 的。假设另一个模块 app 依赖于 testsdk：

```
implementation project(':testsdk')
```

这时候，因为 testsdk 使用的是 implementation 指令来依赖 gson，所以 app 里边不能引用 gson。

2. api

api 定义了编译工程的产品源代码所必需的依赖，这些依赖也是工程需要暴露的 api。api 依赖将作为库 API 暴露给其他模块，在编译、运行期间对该模块和该模块的消费者都可见。此配置的行为类似于 compile（现在已弃用），一般情况下，api 应该仅在库模块中使用它。应用模块应该使用 implementation，除非想要将其 API 公开给单独的测试模块。api 支持 jar 文件和 aar 文件。

3. compileOnly

compileOnly 代替废弃的 provided。只参与编译，编译期间对该模块和该模块的消费者都可见。运行时不需要此依赖项，所以不打包到 apk，这将有助于缩减 APK 的大小。compileOnly 支持 jar 文件，不支持 aar 文件。

4. runtimeOnly

runtimeOnly 代替废弃的 apk。runtimeOnly 依赖项仅在运行时对该模块及该模块的消费者可见。编译期间不可见，直接打包到 apk。

Gradle 中有多种不同类型的依赖，其中一种就是外部依赖，即依赖于不在当前构建项目中的文件，并且这些文件被存储在某种类型的仓库中，或者一个本地文件系统的目录中。要定义一个外部依赖，需要将它添加到一个依赖配置项中。例如

```
implementation 'com.android.support:appcompat-v7:25.3.1'
```

一个外部依赖由 group、name 和 version 三个属性构成。可以使用快捷方式声明外部依赖，形式为 group：name：version。

4.6 配置 build 环境

Gradle 提供了多种措施配置 Gradle 自身的行为和特定工程的行为。下面列出的配置方法优先级从低到高(如果一个选项在多个位置被设置，最后的设置覆盖前面的设置)。

(1) 环境变量，可以在本地环境中通过环境变量 GRADLE_OPTS、JAVA _OPTS 进行配置。

(2) Gradle 属性，可以把某些设置(例如 JVM 存储器设置、Java home、daemon on/off 等)放入 gradle. properties 文件中，gradle. properties 文件的位置为工程的根文件夹或者环境变量 GRADLE_USER_HOME 指定的文件夹。在 Windows 系统，Administrator 用户的主文件夹通常为 C:\Users\Administrator\. gradle。工程根文件夹中 gradle. properties 文件里的配置优先级低于来自于 Gradle 用户主文件夹里的 gradle. properties 文件里的配置。使用文件 gradle. properties 比使用环境变量更容易维护，这样可以使得整个团队在一致的环境下工作。

(3) 系统属性，形式为 systemProp. http. proxyHost=somehost. org，它典型地存储在 gradle. properties 文件中。工程根文件夹中 gradle. properties 文件里的配置优先级低于来自于 Gradle 用户主文件夹里的 gradle. properties 文件里的配置。

(4) 命令行标志，例如--build-cache。命令行标志比所有 gradle. properties 文件中属性和环境变量有更高的优先权，在命令行上除了配置 Gradle 自身的行为外，还可以使用工程属性为给定的工程进行配置，例如-PreleaseType=final。

4.6.1 环境变量

Windows 系统可以使用 GRADLE_OPTS、GRADLE_USER_HOME 和 JAVA _HOME 环境变量来设置 Gradle 的 BUILD 环境。环境变量的命名习惯是相邻的单词用下画线隔开。

GRADLE_OPTS：指定启动 Gradle 客户端时使用的命令行参数。例如可以设置 GRADLE_OPTS="-Dorg. gradle. daemon=true"，来代替每次运行 gradle 命令时——daemon 选项。

GRADLE_USER_HOME：指定 Gradle 用户主文件夹。如果 GRADLE_USER_HOME 环境变量没有设置，它的默认值是 $USER_HOME/. gradle 文件夹。在 Windows 系统中，与 $USER_HOME 对应的环境变量为 HOMEPATH，例如 Administrator 用户的 HOMEPATH 为 C:\Users\Administrator，因而 Administrator 用户的主文件夹默认值为 C:\Users\Administrator\. gradle。

JAVA_HOME：指定要使用的 JDK 安装文件夹。

如果环境变量的形式为 ORG_GRADLE_PROJECT_prop=somevalue，Gradle 将为给定的 Project 对象设置属性 prop 的值为 somevalue。

JVM 选项：运行 Gradle 的 JVM 选项可以通过环境变量 GRADLE_OPTS 或者 JAVA _OPTS，或者两者设置，目前还不可能在 Gradle 命令行上设置。JAVA_OPTS 设置的环境变量被所有的 Java 应用程序所共享。典型的应用是在 GRADLE_OPTS 中设置存储器选项。

4.6.2 Gradle 属性

通过 gradle. properties 文件或者命令行选项都可以为 Gradle 设置环境。下面列出了 gradle. properties 文件中的 gradle 属性,可以用于配置 Gradle 环境。

org. gradle. caching=(true,false)

org. gradle. caching. debug=(true,false)

org. gradle. configureondemand=(true,false)

org. gradle. console=(auto,plain,rich,verbose)

org. gradle. daemon=(true,false)当设置为 true 时,Gradle daemon 用于运行 build。从 Gradle 3.0 开始,daemon 默认是使能的。对于本地的 build 任务,总是采用 daemon 运行 gradle 工作,以便取得更快的速度

org. gradle. daemon. idletimeout=(# of idle millis)

org. gradle. debug=(true,false)

org. gradle. java. home=(path to JDK home) 指定 Gradle build 使用的 Java home,这个值可以设置为 JDK 或者 JRE 的安装位置,通常设置 JDK 的安装位置更安全一些。如果没有设置,将使用合理的默认值

org. gradle. jvmargs=(JVM arguments)

org. gradle. logging. level=(quiet,warn,lifecycle,info,debug)

org. gradle. parallel=(true,false)

org. gradle. warning. mode=(all,none,summary)

org. gradle. workers. max=(max # of worker processes)

4.6.3 系统属性

可以在 Gradle 命令行选项上加上-D 直接为运行 Gradle 的 JVM 传递系统属性。系统属性实际是指的 JVM 的系统属性。在运行 Java 程序时,可以使用-D 命令行选项来设置 Java 的系统变量,在运行 Gradle 时,同样可以使用-D 命令行选项向运行 Gradle 的 JVM 传递系统属性,即 Gradle 命令的-D 命令行选项和 Java 命令的-D 命令行选项具有相同的效果。

类似地,也可以在 gradle. properties 文件中添加前缀 systemProp. 设置系统属性。如果 gradle. properties 文件中的属性名类似 systemProp. propName= somevalue,即属性名有一个前缀 systemProp. ,Gradle 将设置系统属性 propName 的值为 somevalue,系统属性没有前缀 systemProp. 。对于多工程的构建,只检查根工程的 gradle. properties 文件中以 systemProp. 为前缀的属性,忽略子工程文件夹里 gradle. properties 文件里的以 systemProp. 为前缀的属性。

例如根工程的 gradle. properties 文件内容如下:

```
systemProp.gradle.wrapperUser = myuser
systemProp.gradle.wrapperPassword = mypassword
systemProp.gradle.user.home = mypath
```

上述文件设置了系统属性 gradle. wrapperUser，指定用户名，使用基本的 http 认证从服务器下载 Gradle 发布版的；gradle. wrapperPassword 指定使用 Gradle wrapper 下载 Gradle 发布版的密码；gradle. user. home 指定 Gradle 用户的主目录。

4.6.4 工程属性

工程属性是 Gradle 专门为 Project 对象定义的属性。可以在命令行上使用-P 命令行选项为给定的工程对象添加属性，也可以通过特殊命名的环境变量或者 gradle. properties 文件中的特殊命名的属性向 Project 对象添加属性，这种情形对于没有管理员权限时特别有用。如果环境变量的形式为 ORG_GRADLE_PROJECT_prop＝somevalue，Gradle 将为给定的 Project 对象设置属性 prop 的值为 somevalue。如果 gradle. properties 文件中的属性形式为 org. gradle. project. prop＝ somevalue，Gradle 将为给定的 Project 对象设置属性 prop 的值为 somevalue。

gradle. properties 可以定义在 Gradle 用户的主文件夹里或者工程的文件夹里。Gradle 用户的主文件夹是由 GRADLE_USER_HOME 环境变量定义的。对于多工程的构建，可以把 gradle. properties 文件放在任何子工程的文件夹里。在 gradle. properties 文件里设置的属性只能通过 Project 对象访问。用户的主文件夹里的属性文件比工程文件夹里的属性文件有更高的优先权。

下面的两种方法都能把工程对象的 foo 属性设置为 bar。

方法 1：通过环境变量设置给定工程对象的属性。

```
ORG_GRADLE_PROJECT_foo = bar
```

方法 2：通过 gradle. properties 文件设置给定工程对象的属性。

```
org.gradle.project.foo = bar
```

4.6.5 Ext(ra)Properties

可以为 Project 或者 Task 定义 Ext 属性，必须使用关键字 ext(对应 ExtraPropertiesExtension 的实例)去定义 Ext 属性。对于 ext 定义的属性，并不能通过外部的方式修改它的值，只能在 build. gradle 中去设置或者修改它的值。

一个例子：在全局的 gradle 文件中定义一些属性，然后在模块中运用它们。比如在根目录下的 build. gradle 中定义：

```
ext {
        compileSdkVersion = 22
        buildToolsVersion = "22.0.1"
}
```

那么在子模块中就可以使用这些属性了。

```
android {
        compileSdkVersion rootProject.ext.compileSdkVersion
        buildToolsVersion rootProject.ext.buildToolsVersion
}
```

4.6.6 一个属性设置的实例

新建一个文件夹 rootProject，在 rootProject 文件夹里文件 gradle.properties 和 build.gradle。gradle.properties 文件内容如下：

```
gradlePropertiesProp = gradlePropertiesValue
envProjectProp = shouldBeOverWrittenByEnvProp
sysProp = shouldBeOverWrittenBySysProp
systemProp.systemABC = systemABCValue
```

build.gradle 内容如下：

```
task printProps {
    doLast {
        println commandLineProjectProp
        println systemProjectProp
        println gradlePropertiesProp
        println envProjectProp
        println sysProp
        println System.properties['systemABC']
    }
}
```

在命令行模式下，进入 rootProject 文件夹，运行：

```
gradle -q -PcommandLineProjectProp = commandLineProjectPropValue
-Dorg.gradle.project.systemProjectProp = systemPropertyValue printProps
```

执行 printProps，输出内容如下：

```
commandLineProjectPropValue
systemPropertyValue
gradlePropertiesValue
shouldBeOverWrittenByEnvProp
shouldBeOverWrittenBySysProp
systemABCValue
```

4.7 本章主要参考文献

1. https://docs.gradle.org/current/userguide/installation.html
2. https://docs.gradle.org/current/userguide/userguide_single.html
3. https://docs.gradle.org/current/userguide/java_library_plugin.html
4. https://developer.android.google.cn/studio/releases/gradle-plugin
5. https://developer.android.google.cn/studio/build/gradle-plugin-3-0-0-migration
6. https://docs.gradle.org/current/userguide/dependency_management_for_java_projects.html
7. https://developer.android.google.cn/studio/build/build-variants#dependencies

第5章 Activity 与 Fragment

CHAPTER 5

Android 操作系统是一种多用户 Linux 系统，其中的每个应用都是一个不同的用户。默认情况下，系统会为每个应用分配一个唯一的 Linux 用户 ID(该 ID 仅由系统使用，应用并不知晓)，每个应用都在其自己的 Linux 进程内运行。Android 会在需要执行任何应用组件时启动该进程，当不再需要该进程或系统必须为其他应用恢复内存时关闭该进程。每个进程都具有自己的虚拟机(VM)，因此应用代码是在与其他应用隔离的环境中运行。系统为应用中的所有文件设置权限，使得只有分配给该应用的用户 ID 才能访问这些文件。Android 系统可以通过这种方式实现最小权限原则。也就是说，默认情况下，每个应用都只能访问执行其工作所需的组件，而不能访问其他组件。不过，应用仍然可以通过一些途径与其他应用共享数据以及访问系统服务。例如安排两个应用共享同一 Linux 用户 ID，在这种情况下，它们能够相互访问彼此的文件。为了节省系统资源，可以安排具有相同用户 ID 的应用在同一 Linux 进程中运行，并共享同一虚拟机(应用还必须使用相同的证书签署)。

Android 具有 4 种应用组件：Activity、Service、BroadcastReceiver、ContentProvider。Android 系统设计的独特之处在于，任何应用都可以启动其他应用的组件。当系统启动某个组件时，会启动该应用的进程(如果尚未运行)，并实例化该组件所需的类。从视觉效果来看，一个 Activity 占据当前的窗口，具备控件、菜单等界面元素，响应所有窗口事件，用户可与其提供的 UI 进行交互，Activity 表示具有用户界面的单一屏幕。例如，电子邮件应用可能具有一个显示新电子邮件列表的 Activity、一个用于撰写电子邮件的 Activity 以及一个用于阅读电子邮件的 Activity。尽管这些 Activity 通过协作在电子邮件应用中形成了一种紧密结合的用户体验，但每一个 Activity 都独立于其他 Activity 而存在。因此，如果电子邮件应用允许，其他应用可以启动其中任何一个 Activity，例如，相机应用可以启动电子邮件应用内用于撰写新电子邮件的 Activity，以便用户共享图片。又例如，正在开发的应用需要用户使用设备的相机拍摄照片，如果相机应用可以执行该操作，就不需要开发一个 Activity 来拍摄照片，也不需要集成甚至链接该相机应用的代码，而是只需启动相机应用中拍摄照片的 Activity，则该 Activity 在属于相机应用的进程中运行，而不是在原始应用的进程中运行。拍摄完成后，系统可以把照片回传给原始应用。对用户而言，就好像相机是应用的组成部分。因此，与大多数其他系统上的应用不同，Android 应用并没有单一入口点(例如没有 main 函数)。

5.1 启动 Activity

Activity、Service 和 BroadcastReceiver 都可以通过名为 Intent 的异步消息进行启动，Intent 不能用于启动 ContentProvider，如图 5-1 所示。ContentProvider 会在成为 ContentResolver 的请求目标时启动，可以通过调用 ContentResolver 对象的 query()方法来对 ContentProvider 执行查询。

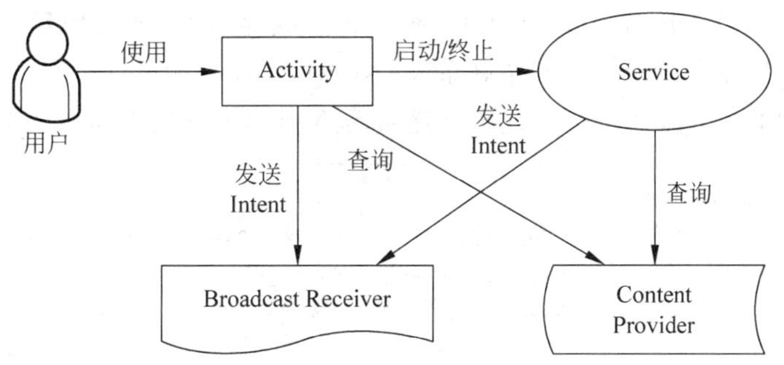

图 5-1 Android 组件关系

每种类型的组件有不同的启动方法，把 Intent 对象传递给 startActivity()或 startActivityForResult()来启动 Activity；把 Intent 对象传递给 startService()来启动 Service，或者把 Intent 对象传递给 bindService()来绑定到 Service；把 Intent 对象传递给 sendBroadcast()、sendOrderedBroadcast()等方法来发送广播。

Intent 对象定义的消息用于启动特定组件或特定类型的组件。Intent 可以是显式的，也可以是隐式的。显式 Intent 对象是指 Intent 指定了被启动组件的组件名。隐式 Intent 对象描述了想要执行的 action 类型(系统会选择合适的组件，甚至是来自其他应用的组件)。Intent 对象还可能携带少量的数据，供被启动的组件使用。

可以通过调用 startActivity(Intent intent)来启动另一个 Activity。在自己的应用内工作时，通常只需要启动某个已知 Activity，可以通过创建一个 Intent 对象，显式地指定要启动的 Activity 类名来实现此目的。例如，可以通过以下代码让一个 Activity 启动另一个名为 SignInActivity 的 Activity：

```
Intent intent = new Intent(this, SignInActivity.class);
startActivity(intent);
```

有时候，应用可能需要利用设备上其他应用提供的 Activity 来执行某些操作，可以创建一个 Intent 对象，并描述想要执行的操作，并且可以指定操作数据的 URI，以及正在启动的组件可能需要了解的信息，系统会从其他应用启动相应的 Activity。如果有多个 Activity 可以处理这个 Intent 对象，用户可以选择其中一个 Activity。例如，如果用户想发送电子邮件，可以创建以下 Intent：

```
Intent intent = new Intent(Intent.ACTION_SEND);
intent.putExtra(Intent.EXTRA_EMAIL, recipientArray);
startActivity(intent);
```

添加到 Intent 中的 EXTRA_EMAIL 是一个字符串数组，其中包含要发送的电子邮件的目的地址。当电子邮件应用响应此 Intent 时，它会读取 extra 中提供的字符串数组，并将它们放入电子邮件撰写窗体的“收件人”字段。

有时可能需要从被启动的 Activity 获得结果。例如 ActivityA 启动 ActivityB，并希望获得 ActivityB 返回的结果，可以在 ActivityA 中调用 startActivityForResult() 来启动 ActivityB，并实现 onActivityResult() 回调方法以便接收 ActivityB 返回的结果。当 ActivityB 完成时，它会通过 Intent 向 ActivityA 的 onActivityResult()方法传递结果。

例如，可以发出一个 Intent，让用户选取某位联系人并把选择结果返回给原始的 Activity，返回的 Intent 包括指向所选联系人的 URI，以便原始的 Activity 对该联系人中的信息执行某项操作。可以通过以下代码创建启动 Activity，并处理返回结果。

```
private void pickContact() {
//Create an intent to "pick" a contact, as defined by content provider URI
  Intent intent = new Intent(Intent.ACTION_PICK, Contacts.CONTENT_URI);
  startActivityForResult(intent, PICK_CONTACT_REQUEST);
}
@Override
protected void onActivityResult(int requestCode, int resultCode, Intent intent) {
// If the request went well (OK) and the request was PICK_CONTACT_REQUEST
if (resultCode == Activity.RESULT_OK && requestCode == PICK_CONTACT_REQUEST) {
 // query the contact's content provider for the contact's name
    Cursor cursor = getContentResolver().query(intent.getData(),
    new String[] {Contacts.DISPLAY_NAME}, null, null, null);
    if (cursor.moveToFirst()) { // True if the cursor is not empty
      int columnIndex = cursor.getColumnIndex(Contacts.DISPLAY_NAME);
      String name = cursor.getString(columnIndex);
      // Do something with the selected contact's name...
    }
  }
}
```

上例显示的是，在处理 Activity 结果时应该在 onActivityResult()方法中使用的基本逻辑。第一个条件检查请求是否成功(如果成功，则 resultCode 为 RESULT_OK)以及此结果响应的请求 requestCode 是否和 startActivityForResult()发送的第二个参数匹配，满足条件时根据 Intent 对象返回的数据进行查询，即 ContentResolver 对一个 ContentProvider 执行查询，后者返回一个 Cursor，让查询的数据能够被读取。

5.2 在 application 元素中声明组件

Activity、Service 和 ContentProvider 必须在清单文件中的 application 元素内注册后，系统才能使用。包括在源代码中、但没有在清单文件中声明的 Activity、Service 和 ContentProvider 对系统不可见，因此也永远不会运行。不过，BroadcastReceiver 可以在清单文件中注册，也可以在代码中创建 BroadcastReceiver 对象，并调用 registerReceiver()在系统中注册。

<activity>元素的语法如下：

```
<activity android:allowEmbedded = ["true" | "false"]
        android:allowTaskReparenting = ["true" | "false"]
        android:alwaysRetainTaskState = ["true" | "false"]
        android:autoRemoveFromRecents = ["true" | "false"]
        android:banner = "drawable resource"
        android:clearTaskOnLaunch = ["true" | "false"]
        android:configChanges = ["mcc", "mnc", "locale",
                   "touchscreen", "keyboard", "keyboardHidden","navigation", "screenLayout",
"fontScale","uiMode", "orientation", "screenSize", "smallestScreenSize"]
        android:documentLaunchMode = ["intoExisting" | "always" |"none" | "never"]
        android:enabled = ["true" | "false"]
        android:excludeFromRecents = ["true" | "false"]
        android:exported = ["true" | "false"]
        android:finishOnTaskLaunch = ["true" | "false"]
        android:hardwareAccelerated = ["true" | "false"]
        android:icon = "drawable resource"
        android:label = "string resource"
             android:launchMode = ["standard" | "singleTop" |"singleTask" | "singleInstance"]
        android:maxRecents = "integer"
        android:multiprocess = ["true" | "false"]
        android:name = "string"
        android:noHistory = ["true" | "false"]
        android:parentActivityName = "string"
        android:permission = "string"
        android:process = "string"
        android:relinquishTaskIdentity = ["true" | "false"]
        android:resizeableActivity = ["true" | "false"]
          android:screenOrientation = ["unspecified" | "behind" | "landscape" | "portrait" |
"reverseLandscape" |
              "reversePortrait" |"sensorLandscape" | "sensorPortrait" |
              "userLandscape" | "userPortrait" |"sensor" | "fullSensor" | "nosensor" |
                     "user" | "fullUser" | "locked"]
        android:stateNotNeeded = ["true" | "false"]
        android:supportsPictureInPicture = ["true" | "false"]
        android:taskAffinity = "string"
        android:theme = "resource or theme"
        android:uiOptions = ["none" | "splitActionBarWhenNarrow"]
              android: windowSoftInputMode = [ " stateUnspecified "," stateUnchanged ",
"stateHidden",
             "stateAlwaysHidden", "stateVisible", "stateAlwaysVisible", "adjustUnspecified",
                   "adjustResize", "adjustPan"] >
</activity>
```

name 属性是唯一必需的属性，它指定 Activity 的类名。应用一旦发布，就不应该更改此类名，否则，可能会破坏诸如应用快捷方式等一些功能。

5.3 使用 intent-filter 声明组件功能

如上所述，可以通过传递 Intent 对象来启动 Activity、Service 和 BroadcastReceiver。可以给 Intent 对象显式指定目标组件的类名来启动组件，不过，Intent 的真正强大之处在于隐式 Intent 概念。隐式 Intent 是指没有指定要启动的目标组件的类名，而是描述要执行的操作类型(还可选择描述要操作的数据)，让系统能够在设备上找到可执行某操作的组件、并启动该组件。系统通过接收到的 Intent 对象与系统中注册的组件 IntentFilter 进行匹配，来确定可以响应 Intent 的组件。如果有多个组件可以执行 Intent 所描述的操作，则由用户选择启动哪一个组件。

可以把 intent-filter 元素作为组件的子元素添加进清单文件，声明组件的功能。例如，在应用的清单文件中声明 Activity 时，可以选择性地加入声明 Activity 功能的 intent-filter 元素，以便响应隐式的 Intent 对象。

当使用 Android Studio 创建新应用时，系统自动创建的 Activity 就包含了一个 intent-filter 元素。这个 intent-filter 的内容如下：

```
<activity android:name = ".ExampleActivity" android:icon = "@drawable/app_icon">
    <intent - filter>
      <action android:name = "android.intent.action.MAIN" />
      <category android:name = "android.intent.category.LAUNCHER" />
   </intent - filter>
</activity>
```

名称为 android.intent.action.MAIN 的 action 元素指定组件是应用的主入口点。名称为 android.intent.category.LAUNCHER 的 category 元素指定此 Activity 应列入系统的应用启动器内，以便用户启动该 Activity。如果打算让应用成为独立应用，不允许其他应用激活其 Activity，就不再需要任何其他 intent-filter 了。不过，如果需要 Activity 对从其他应用(以及同一应用)传递过来的隐式 Intent 作出响应，则必须为 Activity 定义 intent-filter，每个 intent-filter 至少包括一个 action 元素，category 元素(至少包含名为 android.intent.category.DEFAULT 的 category)、零个或者多个 data 元素，这些元素指定了 Activity 可以响应的 Intent 类型。

如需了解有关 Activity 如何响应 Intent 的详细信息，请参阅“第 6 章 Intent 和 IntentFilter”。

5.4 Activity 生命周期

Activity 是一个应用组件，用户可与其提供的 UI 进行交互，以执行拨打电话、拍摄照片等操作。每个 Activity 都会获得一个用于绘制其用户界面的窗口，窗口通常会充满屏幕，但也可小于屏幕并浮动在其他窗口之上。图 5-2 说明了 Activity 在状态转变期间可能经过的路径。矩形表示回调方法，当 Activity 在不同状态之间转变时，可以实现这些方法来执行特定的操作。

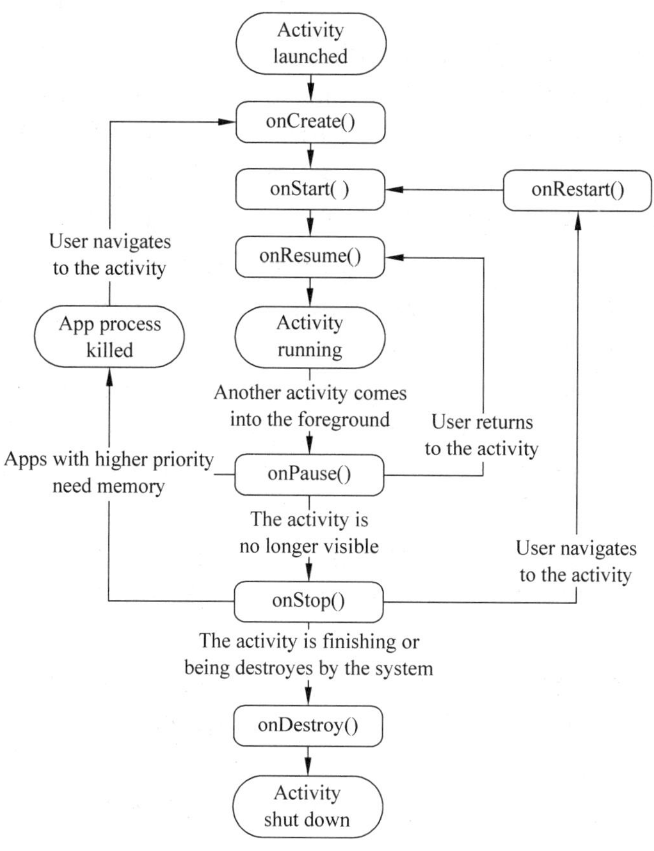

图 5-2 Activity 生命周期和运行状态

Activity 的整个生命周期发生在 onCreate()调用与 onDestroy()调用之间。首次创建 Activity 时调用 onCreate(),应该在此方法中执行所有正常的静态设置:创建视图、将数据绑定到列表,等等,系统向此方法传递一个 Bundle 对象,其中包含 Activity 的上一状态,不过前提是捕获了该状态(请参阅后文的保存 Activity 状态)。onCreate()始终后接 onStart()。

在 Activity 被销毁前调用 onDestroy(),这是 Activity 最后收到的系统调用。当 Activity 调用了 finish(),或系统为节省空间而暂时销毁该 Activity 实例时,onDestroy()方法被调用。例如,如果 Activity 有一个在后台运行的线程,用于从网络下载数据,可以在 onCreate()方法中创建该线程,然后在 onDestroy()方法中停止该线程。

Activity 的可见生命周期发生在 onStart()调用与 onStop()调用之间,在这期间,用户可以在屏幕上看到 Activity,但可能不在前台无法与用户进行交互。在 Activity 即将对用户可见之前调用 onStart(),如果 Activity 转入前台,则 onStart()后接 onResume();如果 Activity 转入隐藏状态,则 onStart()后接 onStop()。

当 Activity 对用户不再可见时调用 onStop()。如果 Activity 被销毁,或另一个 Activity(一个现有 Activity 或新 Activity)继续执行并将其覆盖,原来 Activity 就可能对用户不再可见。如果 Activity 恢复与用户的交互,则 onStop()后接 onRestart();如果 Activity 被销毁,则 onStop()后接 onDestroy()。在 Activity 的整个生命周期内,当 Activity 在对用户可见和隐藏两种状态中交替变化时,系统可能会多次调用 onStart()和

onStop()。

在 Activity 已停止并即将再次启动前调用 onRestart()。onRestart()始终后接 onStart()。

Activity 的前台生命周期发生在 onResume()调用与 onPause()调用之间，在这段时间，Activity 处于 Activity 堆栈的顶层，位于屏幕上的所有其他 Activity 之前，并具有用户输入焦点。在 Activity 即将开始与用户进行交互之前调用 onResume()。onResume()始终后接 onPause()。当另一个 Activity 即将调用 onResume()到前台时，当前的 Activity 调用 onPause()转到后台。onPause()方法通常用于确认对持久性数据的未保存更改、停止动画以及其他可能消耗 CPU 的内容，它应该非常迅速地执行所需操作，以便下一个 Activity 继续执行。如果 Activity 返回前台，则 onPause()后接 onResume()，如果 Activity 转入对用户不可见状态，则 onPause()后接 onStop()。Activity 可频繁进入前台或者退出前台。例如，当设备转入休眠状态或出现对话框时，系统会调用 onPause()。由于此状态可能经常发生转变，因此这两个方法中应采用适度轻量级的代码，以避免因转变速度慢而让用户等待。

一个应用通常由多个彼此松散联系的 Activity 组成，一般会指定应用中的某个 Activity 为主 Activity，即首次启动应用时呈现给用户的那个 Activity，而且在一个 Activity 中可以启动另一个 Activity，以便执行不同的操作。当一个 Activity 启动另一个 Activity 时，启动第二个 Activity 的过程与停止第一个 Activity 的过程存在重叠。当两个 Activity 位于同一进程，由 Activity A 启动 Activity B 时一系列操作的发生顺序如下：

Activity A 的 onPause()方法执行。

Activity B 的 onCreate()、onStart()和 onResume()方法依次执行，Activity B 最后获取用户焦点。

然后，如果 ActivityA 在屏幕上不再可见，则其 onStop()方法执行。

可以利用这种可预测的生命周期回调顺序管理从一个 Activity 到另一个 Activity 的信息转变。例如，如果必须在第一个 Activity 停止时向数据库写入数据，以便下一个 Activity 能够读取该数据，则应在 onPause()而不是 onStop()执行期间向数据库写入数据。

Activity 基本上有三种状态存在：resumed、paused 和 stopped。

Resumed 状态：也称为运行状态，Activity 位于屏幕前台并具有用户焦点。

Paused 状态：Activity 已经失去了焦点但仍然可见，也就是说，有另一个不是全屏大小的 Activity 显示在讨论的 Activity 的上面，处于 paused 状态的 Activity 处于完全活动状态(Activity 所有状态和成员信息保留在内存中，并与窗口管理器保持连接)，但在内存极度不足的情况下，可能会被系统终止。

Stopped 状态：Activity 被另一个 Activity 完全遮盖，对用户不再可见，已停止的 Activity 同样仍处于活动状态(Activity 对象保留在内存中，它保留了所有状态和成员信息，但未与窗口管理器连接)，当其他应用需要内存时可能会被系统终止。

每次启动新 Activity 时，前一 Activity 便会停止，但系统会在任务回退栈中保留该 Activity。当新 Activity 启动时，系统将其推送到回退栈的栈顶，并取得用户焦点。回退栈遵循基本的“后进先出”堆栈机制，因此，当用户完成当前 Activity 并按 Back 按钮时，系统会从堆栈中将其弹出并销毁，然后恢复前一 Activity。关于任务和回退栈的详细内容，请参阅 5.7 节。

5.5 创建 Activity

要创建 Activity,必须创建 Activity 的子类(或使用其现有子类),并在子类中实现 Activity 的各种回调方法,在实现这些生命周期方法时必须始终先调用父类实现,然后再执行其他操作。以下 Activity 包括主要的生命周期方法:

```
public class ExampleActivity extends Activity {
   @Override
   public void onCreate(Bundle savedInstanceState) {
      super.onCreate(savedInstanceState);
      // The activity is being created.
   }
   @Override
   protected void onStart() {
      super.onStart();
      // The activity is about to become visible.
   }
   @Override
   protected void onResume() {
      super.onResume();
      // The activity has become visible (it is now "resumed").
   }
   @Override
   protected void onPause() {
      super.onPause();
   // Another activity is taking focus (this activity is about to be "paused").
   }
   @Override
   protected void onStop() {
      super.onStop();
      // The activity is no longer visible (it is now "stopped")
   }
   @Override
   protected void onDestroy() {
      super.onDestroy();
      // The activity is about to be destroyed.
   }
}
@Override
public void onSaveInstanceState(Bundle outState, PersistableBundle outPersistentState) {
   super.onSaveInstanceState(outState, outPersistentState);
}
```

必须实现 onCreate()方法,系统会在创建 Activity 时调用此方法。在 onCreate()方法内初始化 Activity 的必需组件,最重要的是,必须在此方法内调用 setContentView()方法定义 Activity 用户界面的布局。Activity 的用户界面是由分级的 View(View 的派生类)提供的。每个视图都控制 Activity 窗口内的特定矩形空间,可对用户交互作出响应。

Android提供了很多现成的视图，例如Button、TextView、EditText、CheckBox、ImageView等。Layout是从ViewGroup派生出来的视图，为其子视图提供唯一布局模型，例如LinearLayout、GridLayout或RelativeLayout等。还可以为View类和ViewGroup类创建子类或者使用其现有子类来自行创建Widget和Layout，然后将它们应用到Activity的布局中。

定义布局的最常用方法是借助保存在应用资源内的布局资源文件。这样，定义用户界面的XML文件与定义Activity行为的Java源文件就可以分开维护。可以通过在Activity的onCreate()方法中调用setContentView()设置布局，布局的资源ID传递给setContentView()作为参数。不过，也可以在Activity代码中创建新View，并通过将新View插入ViewGroup来创建视图层次，然后通过将根ViewGroup传递到setContentView()作为参数来使用该布局。可以使用这些现成的视图来设计和组织布局。

onPause()方法作为用户离开Activity的第一个信号(但并不总是意味着Activity会被销毁)进行调用。通常在此方法内提交在当前用户会话结束后仍然有效需要持久化的任何数据(因为用户可能不会回退)。

还应该实现其他生命周期回调方法，以便处理导致Activity停止甚至被销毁的意外中断。还可以调用Activity的finish()方法来结束该Activity，或者调用finishActivity()结束之前启动的另一个Activity。在大多数情况下，不应该使用这些方法显式结束Activity。调用这些方法可能对预期的用户体验产生不良影响，因此只有在确实不想让用户回退到此Activity实例时才使用。

5.5.1 保存Activity状态

Activity生命周期部分简要提及，当Activity onPause或onStop时，有关成员和当前状态的所有信息仍处于活动状态。因此，用户在Activity内所做的任何更改都会得到保留，这样，当Activity onResume时，这些更改仍然存在。

不过，系统为了获取内存可能销毁某个Activity对象，在用户回退Activity时，系统必须重建Activity对象，但用户并不知道系统销毁Activity后又对其进行了重建，因而很可能认为Activity状态毫无变化。在这种情况下，可以实现Activity的另一个回调方法onSaveInstanceState()对有关Activity的状态信息进行保存，以确保有关Activity状态的重要信息得到保留。系统会先调用onSaveInstanceState()方法，然后再使Activity变得易于销毁。系统会向该方法传递一个Bundle对象，可以使用Bundle对象的putString()和putInt()等方法以名称-值对形式保存有关Activity状态的信息。然后，如果系统杀死了这个应用进程，随后用户回退到该Activity时，则系统会重建该Activity，并将Bundle对象同时传递给onCreate()方法和onRestoreInstanceState()方法，可以在上述两个方法的任何一个方法中，从Bundle对象中提取已经保存的状态并恢复该Activity状态。如果没有状态信息需要恢复，则传递的Bundle是空值(如果是首次创建该Activity，就会出现这种情况)。图5-3示意了Activity在销毁后重建和停止后继续执行两种情况下的状态保存情况。

注意：无法保证系统会在销毁Activity前调用onSaveInstanceState()，因为存在不需要保存状态的情况，例如用户使用Back按钮离开Activity时，用户的行为是显式关闭Activity。如果系统调用onSaveInstanceState()，它会在调用onStop()之前，并且可能会在调用 onPause()之前进行调用。由于无法保证系统会调用 onSaveInstanceState()，因此只能在 onSaveInstanceState()方法中记录 Activity 的瞬态(UI 的状态)，切勿使用它来存储持久性数据，而应在 onPause()方法中保存用户离开 Activity 后需要持久化的数据(例如保存到数据库的数据)。

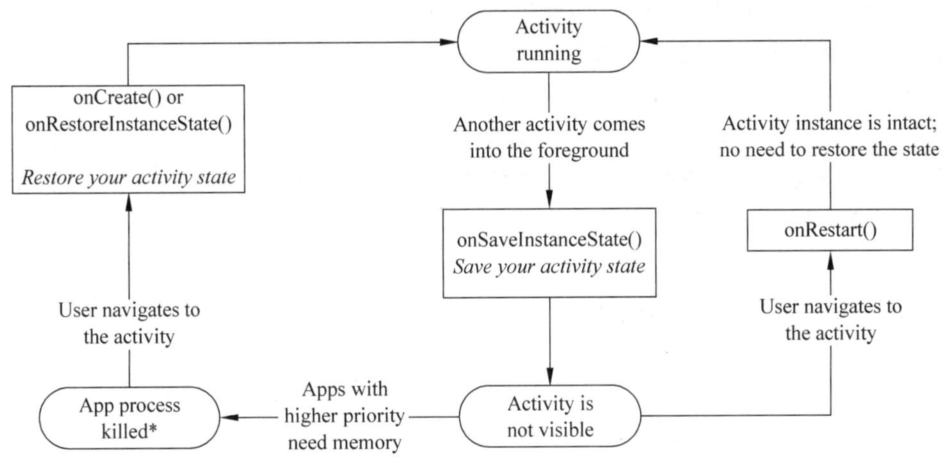

图 5-3　Activity 在销毁后重建和停止后继续执行两种情况下的状态保存

不过，即使不实现 onSaveInstanceState()，Activity 类的 onSaveInstanceState()默认实现也会恢复部分 Activity 状态。具体地讲，默认实现会为布局中的每个 View 调用相应的 onSaveInstanceState()方法，让每个 View 都能提供有关自身的应保存信息。Android 框架中几乎每个 widget 都会根据需要实现此方法，以便在重建 Activity 时自动保存和恢复对 UI 所做的任何可见的变化。例如，EditText 保存用户输入的任何文本，CheckBox 保存复选框的选中或未选中状态。只需通过 android：id 属性为想要保存其状态的每个 widget 提供一个唯一的 ID，系统就可以保存该 widget 的状态。如果 widget 没有设置 ID，则系统无法保存其状态。尽管 onSaveInstanceState()的默认实现会保存有关 Activity UI 的有用信息，可能仍需替换它以保存更多信息。由于 onSaveInstanceState()的默认实现有助于保存 UI 的状态，因此如果为了保存更多状态信息而替换该方法，应始终先调用 onSaveInstanceState()的父类实现，然后再执行任何操作。同样，如果替换 onRestoreInstanceState()方法，也应调用它的超类实现，以便默认实现能够恢复视图状态。例如，可能需要保存在 Activity 生命周期内发生了变化的成员值(它们可能与 UI 中恢复的值有关联，但默认情况下系统不会恢复储存这些 UI 值的成员)。

当然，也可以通过把 android：saveEnabled 属性设置为 false，或通过调用 setSaveEnabled()方法显式阻止布局内的视图保存其状态，通常该属性不应该被停用，但如果希望以不同方式恢复 ActivityUI 的状态，就应该停用该属性。

5.5.2　处理配置变更

有些设备配置可能会在运行时发生变化(例如屏幕方向、键盘可用性及语言)。发生此类变化时，Android 系统调用 onDestroy()，然后立即调用 onCreate() 重建运行中的 Activity。此行为旨在通过利用备用资源(例如适用于不同屏幕方向和屏幕尺寸的不同布局)、自动重新加载应用、来帮助设备适应新配置。

如果对 Activity 进行了适当设计，让它能够处理屏幕方向变化带来的重启并恢复 Activity 状态，那么在遭遇 Activity 生命周期中的其他意外事件时，这个 Activity 将具有更强的适应性。处理此类重启的最佳方法是利用 onSaveInstanceState() 和 onRestoreInstanceState()，也可以在 onCreate() 中恢复 Activity 的状态。可以旋转设备让屏幕方向发生变化，测试应用的状态恢复能力。当屏幕方向变化时，系统会销毁并重建 Activity，以便应用可供新屏幕配置使用的备用资源。用户在执行应用时可能会旋转屏幕，因而 Activity 在重建时能否完全恢复其状态就显得非常重要。

5.6 Android 结构组件

Android 框架管理 Activity 和 Fragment 的生命周期，决定销毁或重新创建 Activity。当因配置更改重新创建 Activity 时，新 Activity 必须重新获取以前的数据。对于简单数据，Activity 可以使用 onSaveInstanceState() 方法并从 onCreate() 中的数据包中恢复其数据，这仅仅适用于可以序列化然后反序列化的少量数据，而不适用于潜在的大量数据，例如用户列表或位图。另一个问题是 Activity/Fragment 中经常需要管理异步请求，确保在系统销毁自己时，清理这些请求以避免潜在的内存泄漏。这种管理需要大量的维护代码，并且在因配置更改而重新创建 UI 控制器时，可能不得不重新发出已经发出过的请求，这样会浪费许多资源。

Android 提供了架构组件 ViewModel 和 LiveData 解决上述问题。ViewModel 对象允许 UI 相关的数据经历配置变化之后还能幸存下来，这些数据可以立即提供给新的 Activity 或 Fragment 实例。android. arch. lifecycle. ViewModel 类用于以 Activity 生命周期感知的方式存储管理 UI 相关的数据，通常作为 LiveData 的容器，把 Activity/Fragment 中数据的所有权从 Activity/Fragment 中分离出来，让 Activity/Fragment 主要用于显示 UI 数据，对用户操作做出响应，或处理与操作系统之间的通信(如权限请求)。而获取 UI 数据的工作让 ViewModel 去做，而不是 Activity 或 Fragment。一个 Activity 中的多个 Fragment 相互通信是很常见的，这些 Fragment 可以使用它们的 Activity 共享 ViewModel 来实现数据共享。ViewModel 不可以持有 Activity Fragment 的引用，否则会导致内存泄漏。

ViewModel 是个抽象类，里面只定义了一个 onCleared() 方法，该方法在 ViewModel 不再被使用时调用。ViewModel 有一个子类 AndroidViewModel，这个类便于在 ViewModel 中使用 Context 对象，因为前面提到不能在 ViewModel 中持有 Activity 的引用。

android. arch. lifecycle. LiveData 类是能感知生命周期的可观察(observable)数据持有类，能够感知 Activity、Fragment 和 Service 的生命周期，这种感知确保 LiveData 仅仅更新处于活跃状态(STARTED 或者 RESUMED 状态)的应用组件观察者(observer)。android. arch. lifecycle. State 枚举类定义了 LifecycleOwner 的 5 种状态：DESTROYED、INITIALIZED、CREATED、STARTED 和 RESUMED，对于 Activity 而言，已经调用了构造方法还没有调用 onCreate 方法为 INITIALIZED 状态，在调用 onCreate 后和正要调用 onStop 之前为

CREATED 状态，在调用 onStart 后和正要调用 onPause 之前为 STARTED 状态，在调用 onResume 后为 RESUMED 状态，正要调用 onDestroy 之前为 DESTROYED 状态，在 Destroyed 后，Lifecycle 就不再发出任何事件，其中 STARTED 或者 RESUMED 状态称为活跃状态。

LiveData 是一个可以被观察的数据持有者，这意味着组件中的观察者能够观察 LiveData 对象的更改，而无须在它们之间创建明确的和严格的依赖关系，这属于被观察者/观察者(obserable/observer)模式。观察者是和 LifecycleOwner 对象绑定的，当 LifecycleOwner 被 onDestroy()时，对应的观察者就被移除。当 LifecycleOwner 处于非活跃状态时，观察者也处于 inactive 状态时，观察者不会收到 LiveData 中数据变化的通知。这意味着观察者会在 LifecycleOwner 的生命周期处于 started 或 resumed 时作出相应更新，而在 LifecycleOwner 被销毁时停止更新。通常情况下，LiveData 只在数据有变化时，给活跃的 Observer 进行通知。此行为的一个例外是，Observer 在从非活跃状态变为活跃状态时也会收到通知。并且，如果 Observer 第二次从非活跃状态变为活跃状态，则只有在自上一次变为活跃状态以来该数据发生变化时才会接收到更新。所以，当 Activity/Fragment 等生命周期变化或者 LiveData 对象持有的数据发生更改后，LiveData 通知观察者，观察者的 onChanged()会被触发调用，执行 UI 更新。

LiveData 是个抽象类，MutableLiveData 类继承于 LiveData，公开暴露 postValue、setValue 两个方法。MediatorLiveData 类继承于 MutableLiveData，可观察多个 LiveData 对象，响应来自所观察 LiveData 对象的 onChanged 事件。MediatorLiveData 用于合并多个 LiveData 源。例如，如果在 UI 中有一个从本地数据库或网络获取更新的 LiveData 对象，则可以把与数据库中的数据关联的 LiveData 对象和与从网络访问的数据关联的 LiveData 对象添加到 MediatorLiveData 对象，Activity 只需观察 MediatorLiveData 对象即可接收来自两个数据源的更新。

ViewModelProviders 是 ViewModel 工具类，该类提供了通过 Fragment 和 Activity 得到 ViewModel 的方法，而具体实现又是由 ViewModelProvider 实现的。ViewModelProviders 的静态方法 of 可以返回一个 ViewModelProvider。在 ViewModelProvider 中定义了一个创建 ViewModel 的接口类 Factory。ViewModelProvider 中有个 ViewModelStore 对象，用于存储 ViewModel 对象。ViewModelStore 是存储 ViewModel 的类，具体实现是通过 HashMap 来保存 ViewModel 对象，Fragment 和 Activity 作为 key 访问获取 ViewModel 对象。ViewModel 不做持久化数据存储。

图 5-4 是 Google 公司推荐的 Android 应用体系结构。Model 层表示用户程序的数据和业务逻辑，ViewModel 和 Model 进行交互，并且 ViewModel 可以被 View 观察，ViewModel 与视图是分离的，ViewModel 不能持有 Activity，Fragment 的引用，避免内存泄漏。在 View 层创建观察者，利用 LiveData 的变化更新 UI。LiveData 可以配合 Room 使用。Room 持久性库支持 Observable 查询返回 LiveData 对象。Observable 查询成为数据库访问对象(DAO)的一项功能。当更新数据库时，会生成所有必要的代码来更新 LiveData 对象。生成的代码在需要时在后台线程上异步运行查询。这种模式对于保持用户界面中显示的数据与存储在数据库中的数据同步很有用。

使用 LiveData 主要包括 4 步：添加依赖，创建 LiveData 实例，观察 LiveData 对象，更

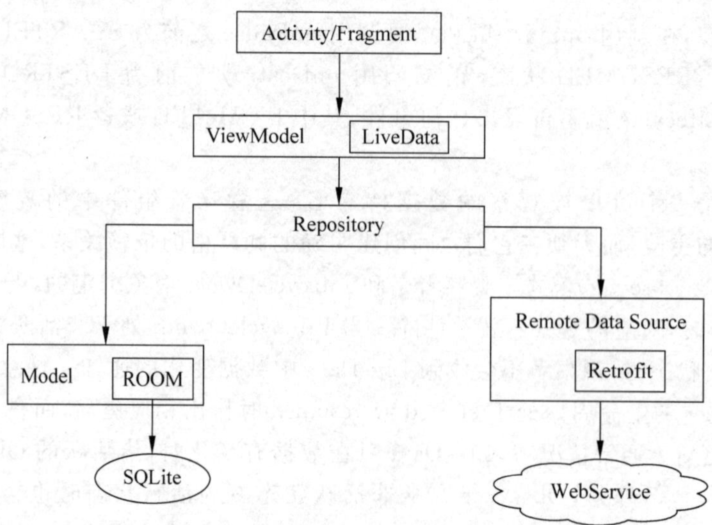

图 5-4　Google 公司推荐的 Android 应用体系结构

新 LiveData 对象。

第一步，在 build.gradle(module: app)中直接添加

```
implementation 'android.arch.lifecycle:extensions:1.1.1'
annotationProcessor 'android.arch.lifecycle:compiler:1.1.1'
```

第二步，创建 LiveData 实例。创建 LiveData 实例持有特定类型的数据，通常放在自定义的 ViewModel 类中。

```
public class NameViewModel extends ViewModel {
// Create a LiveData with a String
private MutableLiveData<String> mCurrentName;
    public MutableLiveData<String> getCurrentName() {
        if (mCurrentName == null) {
            mCurrentName = new MutableLiveData<String>();
        }
        return mCurrentName;
    }
// Rest of the ViewModel...
}
```

第三步，观察 LiveData 对象。这通常包括：初始化 ViewModel、从初始化后的 ViewModel 对象中获取 LiveData 对象、创建 Observer 对象实现 onChanged()方法、调用 LiveData 对象的 observe(@NonNull LifecycleOwner owner, @NonNull Observer<? super T> observer)方法把 Observer 对象附着到 LiveData 对象。

一般通过如下代码初始化 ViewModel：

```
viewModel = ViewModelProviders.of(this).get(NameViewModel.class);
```

this 参数一般为 Activity 或 Fragment，ViewModelProviders 的 of 方法返回一个

ViewModelProvider，因此 ViewModelProvider 可以获取组件的生命周期。

应用程序组件的 onCreate()方法通常是开始观察 LiveData 对象的最佳位置，这样可以保证当应用程序组件处于 STARTED 状态，就可以从 LiveData 对象中接收到最新的值，一旦变为活跃状态时，就有可展示的数据。也能避免在 onResume()方法中进行多余的调用。

```
public class NameActivity extends AppCompatActivity {
    private NameViewModel mModel;
    @Override
    protected void onCreate(Bundle savedInstanceState) {
        super.onCreate(savedInstanceState);
        // Other code to setup the activity…
        // Get the ViewModel.
        mModel = ViewModelProviders.of(this).get(NameViewModel.class);
        // Create the observer which updates the UI.
        final Observer<String> nameObserver = new Observer<String>() {
          @Override
          public void onChanged(@Nullable final String newName) {
              // Update the UI, in this case, a TextView.
              mNameTextView.setText(newName);
          }
        };
        /* Observe the LiveData, passing in this activity as the LifecycleOwner and the observer. */
        mModel.getCurrentName().observe(this, nameObserver);
    }
}
```

第四步，更新 LiveData 对象。

LiveData 没有公用的方法来更新存储的数据。MutableLiveData 类暴露公用的 setValue(T)和 postValue(T)方法，如果需要编辑存储在 LiveData 对象中的值，必须使用这两个方法。从主线程更新 LiveData 对象只能调用 setValue(T)方法，从工作线程更新 LiveData 对象只能使用 postValue(T)方法。通常在 ViewModel 中使用 MutableLiveData，然后 ViewModel 仅向 Observer 公开不可变的 LiveData 对象。

在建立被观察者/观察者关系之后，可以更新 LiveData 对象的值，如以下示例所示，当用户单击按钮时向所有观察者发出通知：

```
mButton.setOnClickListener(new OnClickListener() {
    @Override
    public void onClick(View v) {
        String anotherName = "John Doe";
        mModel.getCurrentName().setValue(anotherName);
    }
});
```

5.7 任务和回退栈

在 Android 应用开发中，界面的引导和跳转直接影响用户体验，与界面跳转联系比较紧密的概念是 Task(任务)和 Back Stack(回退栈)。任务是指在执行特定作业时与用户交互的一系列 Activity 的集合。这些 Activity 按照各自的打开顺序排列在回退栈中。回退栈的基本行为是，当用户从一个 Activity 启动另一个 Activity 时，把新创建的 Activity 压进堆栈，当用户按 Android 手机上的回退键时，把当前 Activity 弹出堆栈并销毁，显示同一任务栈的前一个 Activity，连续按 Back 键需要遍历完该任务栈中的所有 Activity 才能回退到调用者任务中或者回退到主屏幕。对 Task 进行整体调度主要包括以下操作：

按 Home 键，可将之前的任务切换到后台。

按多任务键，会显示出最近执行过的任务列表。

在 Launcher 主屏幕上单击 App 图标，开启一个新任务，或者将已经运行的任务调度到前台。

Activity 的启动模式会影响 Task 和回退栈的状态，进而影响用户体验。除了启动模式之外，Intent 类中定义的一些标志(以 FLAG_ACTIVITY_开头)也会影响 Task 和回退栈的状态。

应用通常包含多个 Activity，每个 Activity 都能够启动其他 Activity，甚至可以启动设备上其他应用中存在的 Activity。例如，如果应用想要发送电子邮件，则可将 Intent 定义为执行"发送"操作并加入一些数据，如电子邮件地址和电子邮件。然后，系统将打开其他应用中声明自己能处理此类 Intent 的 Activity。如果多个 Activity 支持相同 Intent，则系统会让用户选择要使用的 Activity。发送电子邮件时，看起来好像电子邮件 Activity 是应用的一部分，即使这两个 Activity 可能来自不同的应用，但是 Android 仍会将 Activity 保留在相同的任务中，以维护这种无缝的用户体验。

5.7.1 taskAffinity

Activity 的 taskAffinity 属性表示 Activity 优先属于哪个任务，指出了它希望进入的任务栈。每个 Activity 都有 taskAffinity 属性，如果一个 Activity 没有显式地声明 taskAffinity，那么它就继承了 Application 声明的 taskAffinity；如果 Application 也没有显式地声明 taskAffinity 属性，那么该 taskAffinity 的默认值就是 manifest 元素的 package 属性的值。Activity 的 taskAffinity 值必须不同于在 manifest 元素 package 属性的值，因为系统使用该名称作为应用的 taskAffinity 的默认值。

默认情况下，同一个应用中的所有 Activity 具有相同的 taskAffinity 值、优先位于相同任务中，可以给 activity 元素单独设置 taskAffinity 属性，即同一应用中的 Activity 可以分配不同的 taskAffinity，不同应用中的 Activity 也可以定义相同的 taskAffinity，以便把不同应用中的 Activity 放在同一个任务内。taskAffinity 属性的值是一个字符串，字符串中必须至少包含一个小圆点。为了指定一个 Activity 不属于任何任务，可以把 taskAffinity 属性设置为空的字符串。

从概念上讲，具有相同 taskAffinity 属性的 Activity 归属同一任务。任务的 taskAffinity 由其根 Activity 的 taskAffinity 确定。在以下两种情况下，taskAffinity 属性会起如下作用：

第一种：Activity 的启动模式为"singleTask"或者启动 Activity 的 Intent 包含 FLAG_ACTIVITY_NEW_TASK 标志。

如果 FLAG_ACTIVITY_NEW_TASK 标志导致 Activity 开始新任务，且用户单击 Home 键离开，则必须为用户提供导航回任务的方式。有些实体(如通知管理器)始终在外部任务中启动 Activity，而从不作为其自身的一部分启动 Activity，因此它们始终将 FLAG_ACTIVITY_NEW_TASK 放入传递给 startActivity()的 Intent 中。如果应用有一个 Activity 可以被外部组件使用 FLAG_ACTIVITY_NEW_TASK 标志启动，用户可以通过独立方式回退到启动的任务，例如，使用启动器图标(任务的根 Activity 具有 CATEGORY_LAUNCHER Intent 过滤器)。

第二种：Activity 将其 allowTaskReparenting 属性设置为 true。

在这种情况下，Activity 可以从其启动的任务移动到与其具有关联的任务(如果该任务出现在前台)。例如，假设将报告所选城市天气状况的 Activity 定义为旅行应用的一部分，它与同一应用中的其他 Activity 具有相同的关联(默认应用关联)，并允许利用此属性重定父级。当 Activity A 启动天气预报 Activity 时，天气预报 Activity 最初所属的任务与 Activity A 相同。但是，当旅行应用的任务出现在前台时，系统会将天气预报 Activity 重新分配给该任务并显示在其中。

5.7.2 管理任务

Android 默认的管理任务和回退栈的方式为将所有连续启动的 Activity 放入同一任务中，任务栈使用后进先出原则。大多数应用都不应该中断 Activity 和任务的默认行为，默认的方式适用于大多数应用，开发人员不必担心 Activity 如何与任务关联或者如何存在于回退栈中。但是，App 开发人员可能希望中断正常行为，例如把即将被启动的 Activity 放置在新任务中(而不是在当前任务中)；或者，希望把即将被启动的 Activity 的现有实例上移到栈顶部(而不是在回退栈的顶部创建新实例)；或者，希望在用户离开任务时，清除回退栈中除根 Activity 以外的所有其他 Activity。如果确定必须修改 Activity 的默认行为，则当使用 Back 键从其他 Activity 和任务导航回到该 Activity 时或者启动 Activity 期间，请务必要测试该 Activity 的可用性、测试导航行为是否有可能与用户的预期行为冲突。

清单文件中 Activity 元素的属性(包括 taskAffinity、launchMode、allowTaskReparenting、clearTaskOnLaunch、alwaysRetainTaskState、finishOnTaskLaunch)和传递给 startActivity()的 Intent 对象中的标志(包括 FLAG_ACTIVITY_NEW_TASK、FLAG_ACTIVITY_CLEAR_TOP、FLAG_ACTIVITY_SINGLE_TOP)，可以用来定义 Activity 与任务的关联方式以及 Activity 在回退栈中的行为方式。

1. 定义 Activity 启动模式

启动模式用于定义 Activity 的新实例如何与当前任务关联，可以通过清单文件和 Intent 对象标志两种方法来定义启动模式。在清单文件中声明 Activity 时，可以指定 Activity 在启动时应该如何与任务关联；在 Java 代码中调用 startActivity()时，可以在 Intent 对象中加入标志(例如 FLAG_ACTIVITY_NEW_TASK)，用于声明新 Activity 如何(或是否)与当前任务关联。Java 代码中 startActivity()的请求 Intent 的标志的优先级要高于清单文件中定义的启动模式。注意，某些适用于清单文件的启动模式不可用作 Intent 标志，同样，某些可用作 Intent 标志的启动模式无法在清单文件中定义。

设备主屏幕是大多数任务的起点。当用户触摸应用启动器中的图标(或主屏幕上的快捷方式)时，该应用中的任务将出现在前台，但如果应用还没有运行，就会创建一个新任务，并且该应用的主 Activity 将作为任务的根 Activity。在清单文件中，可以使用 activity 元素的 launchMode 属性指定 Activity 应该如何与任务关联。launchMode 属性指定了如何将 Activity 启动到任务中，launchMode 属性的取值有以下 4 种：

(1) standard 模式(默认值)

在 standard 模式中，系统总是在启动 Activity 的任务中创建 Activity 的新实例并向其传送 Intent，新启动的 Activity 会被放置到堆栈顶部并获取焦点，前一个 Activity 仍保留在堆栈中，但是处于停止状态。Activity 停止时，系统会保持其用户界面的状态。堆栈中的 Activity 永远不会重新排列，仅推入和弹出堆栈：用户按下 Back 键时，当前的 Activity 从堆栈顶部弹出并销毁(销毁 Activity 时，系统不会保留该 Activity 的状态)，堆栈中的前一个 Activity 恢复执行，以显示前一个 Activity，连续按 Back 键需要遍历完该任务栈中的所有 Activity 才能回退到调用者任务中或者回退到主屏幕。当该任务的所有 Activity 都从堆栈中移除后，任务就不再存在了。因此，回退栈以"后进先出"方式运行。图 5-5 通过时间线显示 Activity 之间的进度以及每个时间点的当前回退栈，直观呈现了这种行为。

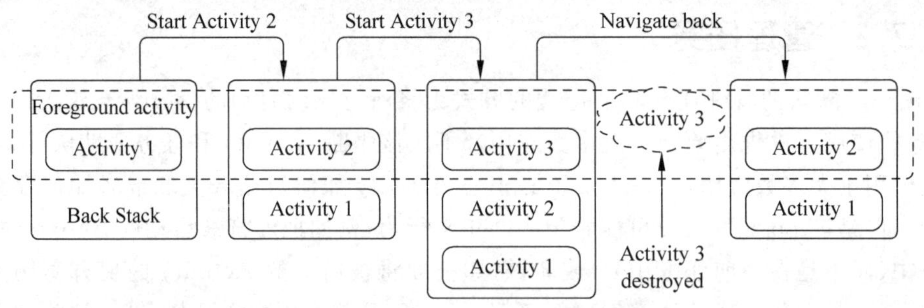

图 5-5 standard 模式的 Activity 任务栈

standard 模式下 Activity 可以多次实例化，而每个实例均可属于不同的任务，并且一个任务可以拥有多个实例。回退栈中的 Activity 永远不会重新排列，因此如果应用允许用户从多个 Activity 中启动某个特定 Activity，当 Activity 的启动模式为默认模式时，会创建该 Activity 的新实例并放置在堆栈顶部(而不是将 Activity 的任一先前实例置于堆栈顶部)。因此，应用中的一个 Activity 可能会多次实例化(即使 Activity 来自不同的任务)，如图 5-6 所示。如果用户使用 Back 键向后导航，则会按 Activity 每个实例的打开顺序显示这些实例(每个实例的 UI 状态各不相同)。但是，如果不希望 Activity 多次实例化，则可以通过设置为其他启动模式修改此行为。

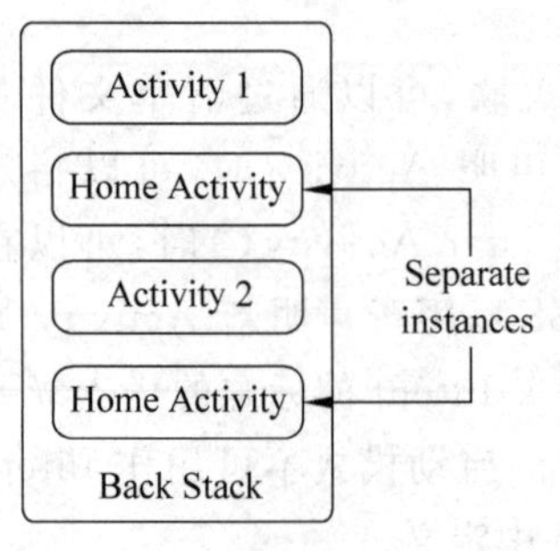

图 5-6 standard 模式下 Activity 可以多次实例化

一个任务是一个有机整体，当用户开始新任务或通过 Home 键转到 Home 屏幕时，当前 Activity 将停止且其所在的任务转移到后台，系统将保留任务中每个 Activity 的状态。尽管在后台时，该任务中的所有 Activity 全部停止，但是任务的回退栈仍旧不变，也就是说，当另一个任务发生时，原来的任务仅仅失去焦点而已。如果用户稍后通过选择那个任务的启动器图标来恢复任务，则任务将出现在前台并恢复执行堆栈顶部的 Activity。如图 5-7 所示，假设当前任务(任务 A)的堆栈中有两个 Activity，用户先按 Home 键显示 Home 屏幕时,任务A进入后台,然后从应用启动器启动新应用,新应用启动时,系统为该应用启动了任务B,任务B使用自己的Activity堆栈。与该应用交互之后,用户再次单击Home键回退到主屏幕并选择最初启动任务A的应用。现在,任务A出现在前台,其堆栈中的两个Activity没有改变,而位于堆栈顶部的Activity则会恢复执行。此时,用户还可以通过单击Home键转到主屏幕并选择启动任务B的应用图标(或者,通过从概览屏幕选择该应用的任务)切换回任务B。这是Android系统中的一个多任务实例。

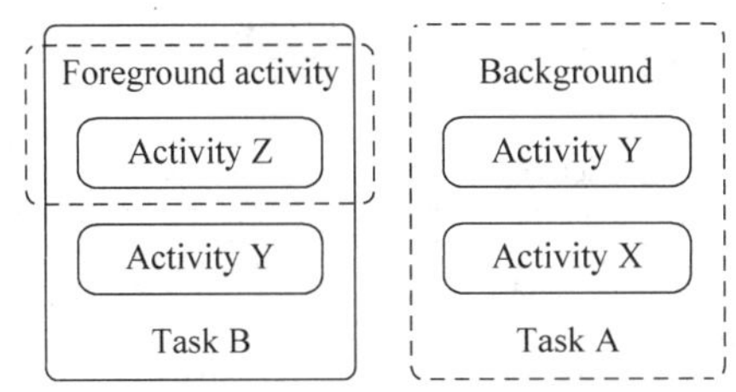

图5-7 任务B在前台接收用户交互,而任务A则在后台等待恢复

(2) singleTop模式

singleTop模式的表现几乎和standard模式一样,与standard模式唯一的区别是如果当前任务栈的栈顶就是要启动的Activity的一个实例,则系统会通过调用该实例的onNewIntent()方法向其传送Intent,而不是创建Activity的新实例。

例如,假设任务的回退栈包含A、B、C和D共四个Activity(堆栈是A-B-C-D,A位于栈底,D位于顶部),收到启动D类Activity的Intent。如果D具有默认的standard启动模式,则启动该类的新实例,且堆栈变成A-B-C-D-D。但是,如果D的启动模式是singleTop,则D的现有实例会通过onNewIntent()接收Intent,因为它位于堆栈的顶部;而堆栈仍为A-B-C-D。但是,如果收到针对B类Activity的Intent,且B的启动模式为singleTop,仍然会向堆栈添加B的新实例。

为某个Activity创建新实例时,用户可以按Back键回退到前一个Activity。但是,当Activity的现有实例处理新Intent时,用户无法按Back按钮回退到新Intent在onNewIntent()之前的Activity的状态。

(3) singleTask模式

当即将被启动的Activity的launchMode属性为singleTask时,该Activity将在它的taskAffinity属性指定的任务中启动。若指定的任务不存在,则会启动一个新的任务,在这个新任务中创建该Activity的一个实例;若指定的任务存在,则会将这个任务调度到前台,并检查这个任务中是否有即将被启动的Activity的实例:如果这个任务中已经存在一个即将被启动的Activity的实例,就清除这个实例之上的所有Activity,系统会调用该实例的onNewIntent()方法向其传送Intent,而不是创建新的Activity实例;如果这个任务中不存在即将被启动的Activity的实例,则在这个任务的顶端创建一个新实例。

启动了standard模式或者singleTop模式的Activity后,若用户再按Back键,将始终会回退到启动它的那个Activity,这是因为standard模式或者singleTop模式启动的Activity和调用者都在同一任务中。Android系统的任务调度以任务为单位。例如图5-8

中,任务 A 中的 Activity2 启动了任务 B 中的 ActivityY(Y 的 launchMode 属性为 singleTask),含有 Activit Y 实例的整个任务都会转移到前台。此时,回退栈包括上移到堆栈顶部的任务 B 中的所有 Activity。可以这么理解,按 Back 键回退时,总是先回退到同一任务栈下面的一个 Activity,这个任务栈的所有的 Activity 都回退完后,然后回退到启动该任务中的 Activity。

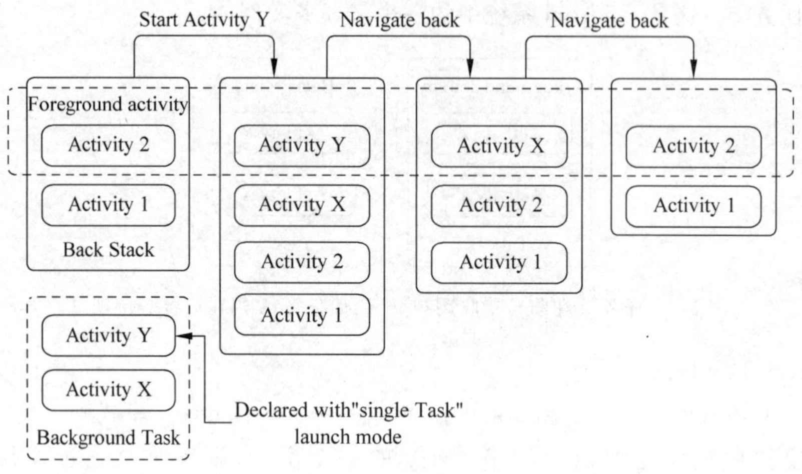

图 5-8 多个任务之间的 Activity 的回退(ActivityY 启动模式为“singleTask”)

(4) singleInstance 模式

这个模式非常接近于 singleTask,区别在于持有这个 Activity 的任务中只能有一个 Activity:即这个单例本身,由被该实例启动的其他 Activity 会自动运行于另一个任务中。以 singleInstance 模式启动的 Activity 具有任务独占性,即它会独自占用一个任务,被它开启的任何 activity 都会运行在其他任务中;当 singleInstance 模式的 Activity 启动其他 Activity 时,启动时系统也会为它加上 FLAG_ACTIVITY_NEW_TASK 标志。

启动 singleInstance 模式的 Activity 时,系统检查是否已经存在一个这个 Activity 的实例,如果存在,会将这个实例所在的任务调度到前台,并且会调用这个实例的 onNewIntent() 方法,将 Intent 实例传递到该实例中。如果不存在,会开启一个新任务,并在这个新任务中创建 Activity 的一个实例。因而,以 singleInstance 模式启动的 Activity 实例具有全局唯一性,即整个系统中只会存在一个这样的实例。

2. 使用 Intent 标志

当调用 startActivity()启动 Activity 时,可以给 startActivity()的参数 Intent 加入适当的标志,修改 Activity 与其任务的默认关联方式。可用于修改默认行为的标志包括:

(1) FLAG_ACTIVITY_NEW_TASK 标志

如果设置 startActivity()的参数 Intent 的标志为 FLAG_ACTIVITY_NEW_TASK,系统就会搜寻是否已经存在一个由被启动的 Activity 的 taskAffinity 属性指定的任务,①若存在,则会将这个任务调度到前台。如果这个已存在的任务中不存在一个要启动的 Activity 的实例,则在这个任务栈的顶端启动一个实例。②若 taskAffinity 属性指定的任务不存在,则会启动一个新的任务,在这个新的任务中创建 Activity 的实例。

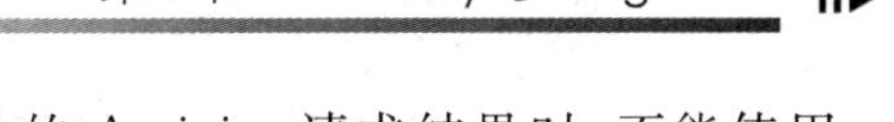

当调用者调用 startActivityForResult()从它启动的 Activity 请求结果时，不能使用 FLAG_ACTIVITY_NEW_TASK。

(2) FLAG_ACTIVITY_SINGLE_TOP 标志

如果即将被启动的 Activity 位于回退栈的顶部，则 Activity 不会被启动。

(3) FLAG_ACTIVITY_CLEAR_TOP 标志

如果正在启动的 Activity 已在当前任务中运行，则会销毁当前任务中该实例上部的所有 Activity，并通过 onNewIntent()将 Intent 传递给 Activity 已恢复的实例（现在位于栈顶），而不是启动该 Activity 的新实例。launchMode 没有与这种行为对应的属性值。如果被指定 Activity 的启动模式为 standard，则该 Activity 也会从堆栈中移除，并在其位置启动一个新实例，以便处理传入的 Intent。这是因为当启动模式为 standard 时，将始终为新 Intent 创建新实例。

FLAG_ACTIVITY_CLEAR_TOP 通常与 FLAG_ACTIVITY_NEW_TASK 结合使用。这两个标志一起使用时，可以用于定位其他任务中的已经存在的 Activity，并将其放入可以响应 Intent 的位置。

3. 清理回退栈

如果用户长时间离开任务，则系统会清除除根 Activity 外的其他 Activity，当用户再次回退到任务时，仅恢复根 Activity。系统这样做的原因是，经过很长一段时间后，用户可能已经放弃之前执行的操作，回退到该任务是为了开始执行新的操作。可以使用下列几个 Activity 属性修改此行为：

(1) alwaysRetainTaskState 属性

如果在任务的根 Activity 中将此属性设置为 true，则不会发生刚才所述的默认行为。即使在很长一段时间后，任务仍将所有 Activity 保留在其堆栈中。

(2) clearTaskOnLaunch 属性

如果在任务的根 Activity 中将此属性设置为 true，则每当用户离开任务然后回退时，系统都会将销毁除根 Activity 外的其他 Activity。换而言之，它与 alwaysRetainTaskState 正好相反，即使只离开任务片刻时间，用户也始终会回退到任务的初始状态。

(3) finishOnTaskLaunch 属性

此属性类似于 clearTaskOnLaunch，但它对单个 Activity 起作用，而非整个任务。此外，它还有可能会导致任何 Activity 停止，包括根 Activity。设置为 true 时，Activity 仍是任务的一部分，但是仅限于当前会话。如果用户离开然后回退任务，则任务将不复存在。

5.8 启动应用

通过为 Activity 提供一个以 android.intent.action.MAIN 为指定操作、以 android.intent.category.LAUNCHER 为指定类别的 Intent 过滤器，可以把 Activity 设置为任务的入口点。例如：

```
<activity … >
    <intent - filter … >
      <action android:name = "android.intent.action.MAIN" />
    <category android:name = "android.intent.category.LAUNCHER" />
   </intent - filter >
  …
</activity>
```

此类 Intent 过滤器会使 Activity 的图标和标签显示在应用启动器中，让用户能够启动 Activity 并在启动之后随时回退到创建的任务中。第二个功能非常重要：用户必须能够在离开任务后，再使用此 Activity 启动器回退到该任务。因此，只有在 Activity 具有 ACTION_MAIN 和 CATEGORY_LAUNCHER 过滤器时，才应该使用 singleTask 和 singleInstance 这两种启动模式，例如，Intent 启动一个 singleTask Activity，启动了一个新任务，并且用户花了些时间处理该任务。然后，用户按 Home 键，任务现已被发送到后台，而且不可见。如果该 Acivity 没有在过滤器中设置 CATEGORY_LAUNCHER，该任务未显示在应用启动器中，用户就无法回退到该任务。

如果不让用户能够回退到 Activity，对于这些情况，请把 Activity 元素的 finishOnTaskLaunch 设置为 true。

5.9 Fragment

android.app.Fragment 在 API level 11 被添加进来，从 API level 28 开始被废弃，应该使用支持库 V4 中的 Fragment，以便更好地兼容所有设备。android.support.v4.app.Fragment 实现了 ComponentCallbacks，View.OnCreateContextMenuListener，LifecycleOwner，ViewModelStoreOwner 接口，Fragment 主要是为了给大屏幕（如平板电脑）提供更加动态和灵活的 UI 设计支持。由于平板电脑的屏幕比手机屏幕大得多，利用 Fragment 实现此类设计时，无须管理对视图层次结构的复杂更改。通过将 Activity 布局分成 Fragment，可以在运行时修改 Activity 的外观，并在由 Activity 管理的回退栈中保留这些更改。

例如，以新闻应用为例：在平板电脑上运行时，该应用在 ActivityA 中嵌入两个 Fragment，使用 Fragment A 在左侧显示文章列表，使用 Fragment B 在右侧显示文章内容，两个 Fragment 并排显示在一个 Activity 中，在同一个 Activity 内选择文章并进行阅读。在手机上，没有足以显示两个 Fragment 的屏幕空间，因此 Activity A 只包括用于显示文章列表的 Fragment A，当用户选择文章时，它会启动 Activity B，B 包括用于阅读文章的 Fragment B。应用可通过重复使用不同组合的 Fragment 来同时支持平板电脑和手机，如图 5-9 所示。

Fragment 是可以放进 Activity 中的用户接口部分。可以把多个 Fragment 组合在一个 Activity 中来构建多窗格 UI，以及在多个 Activity 中重复使用某个 Fragment。每个 Fragment 都应该被设计为可重复使用的模块化 Activity 组件，也就是说，由于每个 Fragment 具有自己的生命周期，能接收自己的输入事件，并且可以在 Activity 运行时添加或移除 Fragment，可以将一个 Fragment 加入多个 Activity，因此，应当避免从某个

Fragment直接操纵另一个Fragment,通过更改Fragment的组合方式来适应不同的屏幕尺寸。在设计可同时支持平板电脑和手机的应用时,可以在不同的布局配置中重复使用Fragment,以根据可用的屏幕空间优化用户体验。

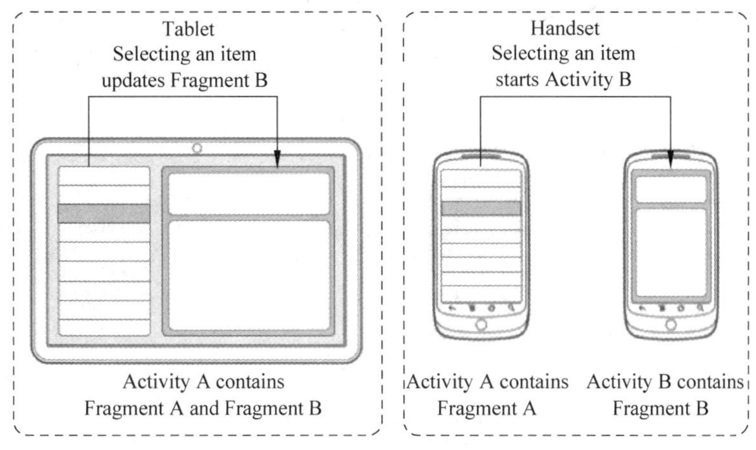

图5-9 两个Fragment适应不同大小屏幕的示例

把Fragment作为Activity布局的一部分添加时,它存在于Activity视图层次结构的某个ViewGroup内部,可以在Activity的布局文件中声明Fragment,使用< fragment >元素插入Activity布局中;或者利用代码把Fragment添加到某个现有ViewGroup中。不过,Fragment并非必须成为Activity布局的一部分;还可以将没有自己UI的Fragment用作Activity的不可见工作线程。

Fragment的所有子类必须包含一个没有参数的构造体。在必要的时候,特别是在状态恢复期间,框架会重新实例化Fragment子类,需要找到没有参数的构造体用于实例化,如果找不到,可能会抛出异常。Fragment的子类主要有:

DialogFragment:显示浮动对话框。在Activity类中,使用DialogFragment类创建对话框可有效地替代使用Dialog的helper方法,因为可以把DialogFragment纳入由Activity管理的Fragment回退栈,从而使用户能够返回清除的Fragment。

ListFragment:显示由适配器(如SimpleCursorAdapter)管理的一系列项目,类似于ListActivity。它提供了几种管理列表视图的方法,如用于处理单击事件的onListItemClick()回调。

PreferenceFragment:以列表形式显示Preference对象的层次结构,类似于PreferenceActivity。这在为应用创建"设置"Activity时很有用处。

5.9.1 Fragment生命周期

Fragment生命周期与Activity生命周期很相似,如图5-10所示。

Fragment也以三种状态存在:resumed、paused和stopped。resumed状态,Fragment在正在运行的Activity中可见。paused状态,另一个Activity位于前台并具有焦点,但此Fragment所在的Activity仍然可见(前台Activity部分透明,或未覆盖整个屏幕)。stopped状态,Fragment对用户不再可见,因为宿主Activity已停止,或者Fragment已从Activity

中移除，但已添加到回退栈，停止的 Fragment 仍然处于活动状态（系统会保留所有状态和成员信息），如果宿主 Activity 被杀死，它也会被杀死。

与 Activity 一样，假使 Activity 的进程被终止，重建 Activity 时需要恢复 Fragment 状态，可以在 Fragment 的 onSaveInstanceState(Bundle outState)方法内使用 Bundle 保存状态，并可在 Fragment 的 onCreate()、onCreateView()或 onActivityCreated()期间恢复状态。

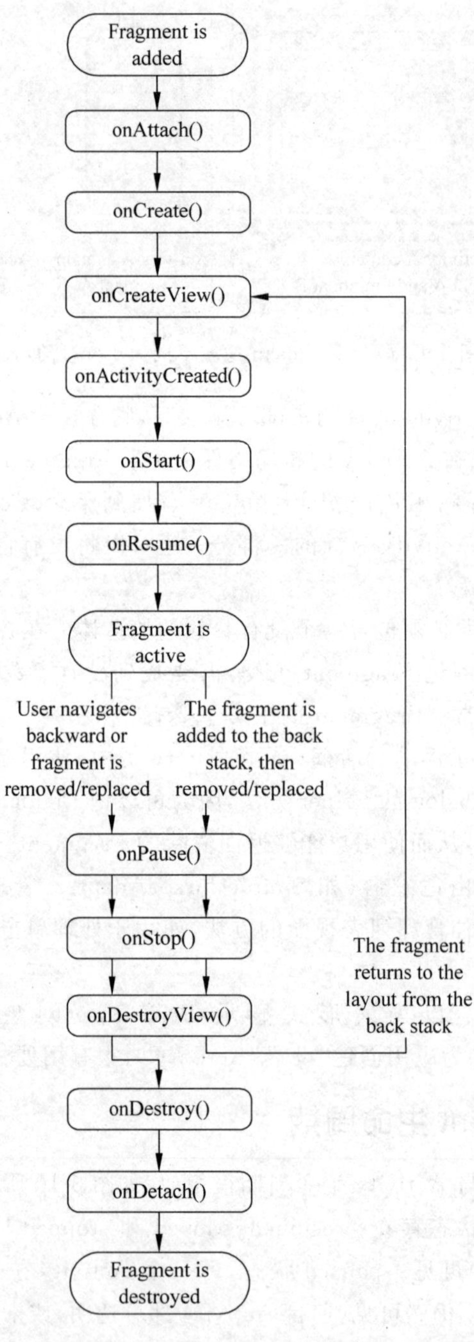

图 5-10 Fragment 生命周期

虽然 Fragment 定义了自己的生命周期，但 Fragment 所在的 Activity 的生命周期会直接影响 Fragment 的生命周期，Activity 的每次生命周期回调都会引发其中的 Fragment 的类似回调。例如，当 Activity 暂停(paused)时，其中的所有 Fragment 也会暂停；当 Activity 被销毁时，其中的 Fragment 也会被销毁。不过，当 Activity 正在运行(处于 resumed 状态)时，可以随意向 Activity 添加 Fragment、替换 Fragment 或者移除其中的 Fragment，可以独立操纵每个 Fragment。只有当 Activity 处于 resumed 状态时，Fragment 的生命周期才能独立变化。不过，Fragment 还有几个额外的生命周期回调，用于处理与 Activity 的唯一交互，以执行构建和销毁 Fragment 的 UI 等操作。

假设 Activity A 有一个 Button，单击 Button 可以启动 Activity B，Activity B 的布局资源文件中仅仅有一个 fragment。表 5-1 给出了单击 ActivityA 中的 Button 启动 Activity B、然后单击 Back 键又回退到 Activity A，Activity 和 Fragment 的方法的回调过程。

表 5-1 Activity 和其中的 Fragment 生命周期

ActivityB 的方法调用	ActivityB 中的 Fragment 的方法调用
onCreate	
	onAttach
	onCreate
	onCreateView
	onViewCreated
	onActivityCreated
	onViewStateRestored
	onStart
onStart	
onResume	
	onResume
单击手机的 Back 键从 Activity B 回退到 Activity A	
	onPause
onPause	
	onStop
onStop	
	onDestroyView
	onDestroy
	onDetach
onDestroy	

(1) onAttach(Context context)：在 Fragment 已经与 Activity 关联时调用(Activity 被传递进此方法内)。

(2) onCreate(Bundle savedInstanceState)：系统会在创建 Fragment 时调用此方法。在这个方法中，应该做一些 Fragment 的初始的创建工作、初始化必要的组件。

(3) onCreateView(LayoutInflater, ViewGroup, Bundle)：系统会在 Fragment 首次绘制其用户界面时调用此方法。要想为 Fragment 绘制 UI，必须在此方法中创建并返回与 Fragment 关联的视图层次结构，即返回一个 View。如果 Fragment 未提供 UI，也可以返回

null。这前三个方法 onAttach(Context)、onCreate(@Nullable Bundle)、onCreateView(LayoutInflater,ViewGroup,Bundle)只可能在 Activity 的 onCreate()被调用后调用。

(4) onActivityCreated(Bundle):在 Activity 的 onCreate()方法已经返回时调用,告诉 Fragment,它的宿主 Activity 已经完成了 Activity. onCreate()。

(5) onViewStateRestored(Bundle):告诉 Fragment,视图层次结构的所有的状态都已经被恢复。

(6) onStart():使 Fragment 对用户可见。

(7) onResume():使 Fragment 开始与用户交互(前提条件是它的容器 Activity 已经处于 Resumed 状态)

当 Fragment 不再被使用后,它将经历以下回调方法。

(1) onPause():Fragment 不再与用户进行交互,要么是因为它的 Activity 正处于暂停状态,要么是进行了 Fragment 操作,系统将此方法作为用户离开 Fragment 的第一个信号(但并不总是意味着此 Fragment 会被销毁)进行调用。通常应该在此方法内提交在当前用户会话结束后仍然有效的任何更改(因为用户可能不会返回)。

(2) onStop():Fragment 不再对用户可以看见,要么是因为它的 Activity 正处于停止状态,要么是进行了 Fragment 操作。

(3) onDestroyView():与 onCreateView 相对应,当与 Fragment 关联的视图层次结构被移除时调用,允许 Fragment 清除与视图相关的资源。

(4) onDestroy():做一些 Fragment 状态的最后的清除工作。

(5) onDetach():与 onAttach()相对应,当 Fragment 与它的 Activity 的关联被取消前调用。

大多数应用都应该至少为每个 Fragment 实现 onCreate()、onCreateView()和 onPause()这三个方法。

5.9.2 添加 Fragment 到 Activity

首先定义 Fragment 的子类,然后将其自己的布局融入 Activity。必须重写 onCreateView()回调方法,该方法提供了一个 LayoutInflater 对象,用于通过 XML 中定义的布局资源来扩展布局,并且返回 View。例如,以下这个 Fragment 子类从 example_fragment. xml 文件加载布局。

```
public static class ExampleFragment extends Fragment {
    @Override
    public View onCreateView(LayoutInflater inflater, ViewGroup container,
                             Bundle savedInstanceState) {
        // Inflate the layout for this fragment
        return inflater.inflate(R.layout.example_fragment, container, false);
    }
}
```

传递至 onCreateView()的 container 参数是父 ViewGroup(来自 Activity 的布局),Fragment 布局将插入到这个父 ViewGroup 中。如果 Fragment 处于 resume 状态,savedInstanceState 参数是提供 Fragment 的前一个实例相关数据的 Bundle。

ListFragment 的默认实现会从 onCreateView()返回一个 ListView,因此 ListFragment

的子类无须实现 onCreateView()方法。

在创建了 Fragment 的子类后，就可以把 Fragment 添加到 Activity 中。通常，Fragment 向宿主 Activity 贡献一部分 UI，作为 Activity 总体视图层次结构的一部分嵌入到 Activity 中。可以通过以下两种方式向 Activity 布局添加 Fragment。

1. 在 Activity 的布局文件内声明 Fragment

在本例中，把 Fragment 当作 View 来为其指定布局属性。例如，以下是一个具有两个 Fragment 的 Activity 的布局文件：

```
<?xml version = "1.0" encoding = "utf - 8"?>
< LinearLayout xmlns:android = "http://schemas.android.com/apk/res/android"
    android:orientation = "horizontal"
    android:layout_width = "match_parent"
    android:layout_height = "match_parent">
    < fragment android:name = "com.example.news.ArticleListFragment"
          android:id = "@ + id/list"
          android:layout_weight = "1"
          android:layout_width = "0dp"
          android:layout_height = "match_parent" />
    < fragment android:name = "com.example.news.ArticleReaderFragment"
          android:id = "@ + id/viewer"
          android:layout_weight = "2"
          android:layout_width = "0dp"
          android:layout_height = "match_parent" />
</LinearLayout >
```

fragment 元素中的 name 属性指定需要在布局中实例化的 Fragment 类，ID 属性是通过 Java 反射机制实例化对象所必需的，即在布局文件中一定要指定 fragment 的 ID。

当系统创建此 Activity 布局时，实例化在布局中指定了 ID 的每个 Fragment，并为每个 Fragment 调用 onCreateView()方法，以获得 Fragment 的布局，系统会直接插入 Fragment 的 onCreateView()方法返回的 View 来替代< fragment >元素。

每个 Fragment 都必须有一个唯一的标识符，重启 Activity 时，系统可以使用该标识符来恢复 Fragment(也可以使用该标识符来捕获 Fragment 以执行某些事务，如将其移除)。可以通过三种方式为 Fragment 提供 ID：①通过 ID 属性提供唯一 ID。②通过 tag 属性提供唯一字符串。③如果以上两个属性都没有定义，系统会使用容器视图的 ID。

2. 通过 Java 代码将 Fragment 添加到现有的 ViewGroup

可以在 Activity 运行期间，调用 FragmentTransaction 中的 API，随时将 Fragment 添加到 Activity 布局中。可以像下面这样从 Activity 获取一个 FragmentTransaction 实例：

```
FragmentManager fragmentManager  =  getFragmentManager();
FragmentTransaction fragmentTransaction  =  fragmentManager.beginTransaction();
```

在获取 FragmentTransaction 实例后，就可以使用 FragmentTransaction 实例的方法执行 Fragment 事务(如添加、移除或替换 Fragment)。使用 FragmentTransaction 的 add()方法可以添加一个 Fragment，指定要添加的 Fragment 以及它的容器 ViewGroup。例如：

```
ExampleFragment fragment = new ExampleFragment();
fragmentTransaction.add(R.id.fragment_container, fragment);
fragmentTransaction.commit();
```

传递到 add()的第一个参数是由资源 ID 指定的 ViewGroup，即 Fragment 视图的容器，第二个参数是要添加的 Fragment。一旦通过 FragmentTransaction 做出了更改，就必须调用 commit()以使更改生效。

3. 添加没有 UI 的 Fragment

前面的例子展示了如何向 Activity 添加 Fragment 以提供 UI。不过，也可以使用 Fragment 为 Activity 提供后台行为，而不显示额外 UI。

要想添加没有 UI 的 Fragment，请使用 add(Fragment, String) 从 Activity 添加 Fragment(String 为 Fragment 提供一个唯一的字符串“标记”，而不是父容器的视图 ID)。这会添加 Fragment，但由于它并不与 Activity 布局中的视图关联，因此不会收到对 onCreateView()的调用。因此，不需要实现 onCreateView()方法。并非只能为没有 UI 的 Fragment 提供字符串标记，也可以为具有 UI 的 Fragment 提供字符串标记，但如果 Fragment 没有 UI，则字符串标记将是标识它的唯一方式，稍后可以使用 findFragmentByTag() 找到此 Fragment。

5.9.3 管理 Fragment

可以使用 FragmentManager 管理 Activity 中的 Fragment，在 Activity 中调用 getFragmentManager()可以获取 FragmentManager。使用 FragmentManager 可以执行的操作包括：

对于在 Activity 布局中提供 UI 的 Fragment 通过 findFragmentById()，对于所有的 Fragmen，尤其是不提供 UI 的 Fragment，通过 findFragmentByTag()获取 Activity 中存在的 Fragment。

通过 popBackStack()(模拟用户发出的 Back 命令)将 Fragment 从回退栈中弹出。

通过 addOnBackStackChangedListener()注册一个侦听回退栈变化的侦听器。

可以使用 FragmentManager 打开一个 FragmentTransaction，通过它来执行某些事务，如添加和移除 Fragment。

在 Activity 中使用 Fragment 的一大优点是，可以根据用户行为执行添加、移除、替换 Fragment 以及其他操作。可以使用 FragmentTransaction 中的 API 来执行一项事务，提交给 Activity 的每组更改都称为 transaction，一个事务就是想要同时执行的一组更改，可以调用 add()、remove()和 replace()等方法为给定事务设置想要执行的所有更改，然后，必须调用 commit()把事务应用到 Activity。

不过，在调用 commit()之前，可以调用 addToBackStack()把该事务添加到由 Activity 管理的 Fragment 事务回退栈中。该回退栈由 Activity 管理，回退栈中的每个条目都是一条已发生 Fragment 事务的记录。允许用户通过按 Back 键撤销 Fragment 事务、返回到以前的 Fragment 状态。也可以将每个事务保存到由 Activity 管理的回退栈中，从而让用户能够回退 Fragment 更改(类似于回退 Activity)。

Activity 生命周期与 Fragment 生命周期之间的最显著差异是它们在其各自回退栈中的存储方式。默认情况下，Activity 停止时被放入由系统管理的 Activity 回退栈，以便用户通过 Back 键回退到 Activity。不过，仅当移除 Fragment 的事务执行期间通过调用 addToBackStack()显式请求保存实例时，系统才会将 Fragment 放入由宿主 Activity 管理的回退栈中。

例如，以下示例说明了如何将一个 Fragment 替换成另一个 Fragment，以及如何在回退栈中保存以前的状态：

```
Fragment newFragment = new ExampleFragment();
FragmentTransaction transaction = getFragmentManager().beginTransaction();
// Replace whatever is in the fragment_container view with this fragment
transaction.replace(R.id.fragment_container, newFragment);
transaction.addToBackStack(null);
transaction.commit();
```

在上面的代码中，newFragment 会替换目前在 ID 为 R.id.fragment_container 所标识的布局容器中的任何 Fragment(如果存在)。通过调用 addToBackStack()可将替换事务保存到回退栈，以便用户能够通过 Back 键撤销事务并回退到前一个 Fragment。

如果一个事务里包含多个更改（如 add、remove 或者 replace)，并且调用了 addToBackStack()，则在调用 commit()前应用的所有更改都将作为单一事务添加到回退栈，按 Back 键会将它们一并撤销。向 FragmentTransaction 添加更改的顺序无关紧要，除了以下两种情况：

(1) 必须最后调用 commit()。对于每个 Fragment 事务，都可以通过在提交前调用 setTransition()来应用过渡动画。

(2) 如果要向同一容器添加多个 Fragment，则添加 Fragment 的顺序将决定它们在视图层次结构中的出现顺序。

如果在执行移除 Fragment 的事务时没有调用 addToBackStack()，则事务提交时该 Fragment 会被销毁，用户将无法回退到该 Fragment。相反，如果在移除 Fragment 时调用了 addToBackStack()，则系统会停止该 Fragment，并在用户回退时将其恢复。

调用 commit()不会立即执行事务，而是在 Activity 的 UI 线程(主线程)可以执行该操作时再安排其在线程上运行。不过，如有必要，也可以从 UI 线程调用 executePendingTransactions()以立即执行 commit()提交的事务，通常不必这样做，除非其他线程中的作业依赖该事务。

只能在 Activity 保存它的状态之前(当用户离开 Activity 前)使用 commit()提交事务。如果试图在该时间点后提交，则会引发异常。这是因为如需恢复 Activity，则提交后的状态可能会丢失。如果 commit 丢失无关紧要，在这种情况下，请使用 commitAllowingStateLoss()。

5.9.4 与 Activity 通信

尽管 Fragment 是作为独立于 Activity 的对象实现的，并且可以在多个 Activity 内使用，但 Fragment 的给定实例会直接绑定到包含它的 Activity。

如果在 Fragment 内需要 Context 对象，可以在 Fragment 附着到 Activity 后调用 getActivity()。如果 Fragment 尚未附着到 Activity，或 Fragment 在其生命周期结束后已

经和 Activity 分离，则 getActivity() 将返回 null。具体地说，Fragment 可以通过 getActivity()访问 Activity 实例，并轻松地执行在 Activity 布局中查找视图等任务。

```
View listView = getActivity().findViewById(R.id.list);
```

同样地，Activity 也可以调用 FragmentManager 实例的 findFragmentById() 或 findFragmentByTag()获取对 Fragment 的引用。例如：

```
ExampleFragment fragment = (ExampleFragment) getFragmentManager().findFragmentById(R.id.
example_fragment);
```

5.10 Context

android.content.Context 继承关系如图 5-11 所示。Context 类本身是一个抽象类，ContextImpl 类是 Context 类的实现类，该类实现了 Context 类的功能。为了使用方便，定义了 ContextWrapper 类，ContextWrapper 类包含一个 Context 类型的对象 mBase，通过构造函数 ContextWrapper(Context base)为 mBase 初始化。调用 ContextWrapper 的方法都会被转向其所包含的真正的 Context 对象 mBase。ContextThemeWrapper 类，其内部包含了与 Theme 相关的接口，这里所说的 Theme 指在 AndroidManifest.xml 中通过 android:theme 为 Application 元素或者 Activity 元素指定的 Theme。当然，只有 Activity 才需要 Theme，Service 是不需要 Theme 的，因为 Service 是没有界面的后台场景，所以 Service 直接继承于 ContextWrapper。

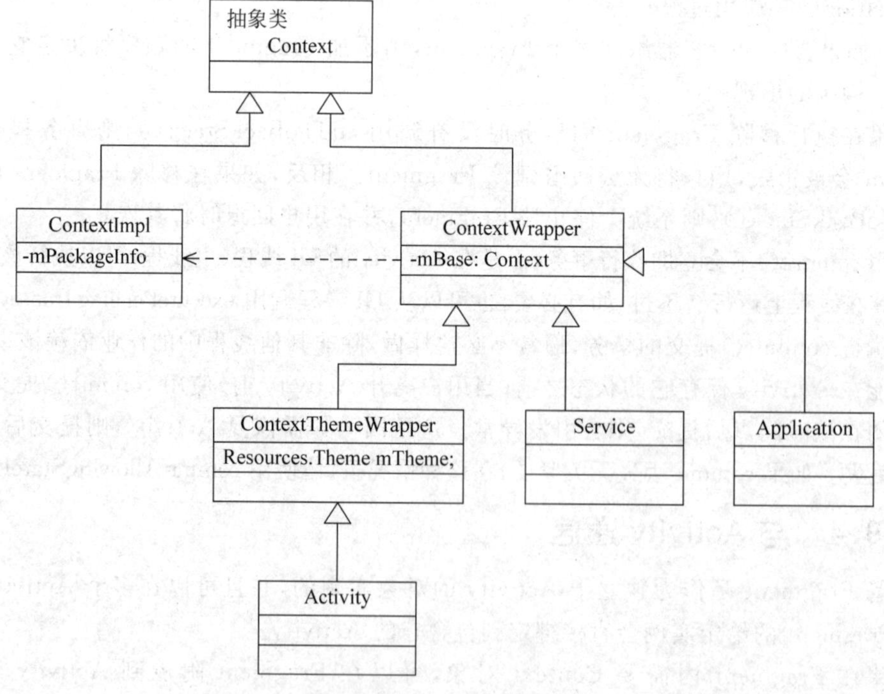

图 5-11 Context 和它的子类

凡是跟UI相关的，都应该使用Activity作为Context来处理，例如显示Dialog、启动Activity、扩展布局；Application和Service不能用来显示Dialog，也不推荐用于启动Activity、扩展布局。Activity、Application和Service都可以用来启动Service、发送广播、注册广播接收器、加载资源值。

要获取Context对象，主要有以下三种方法：①View对象的getContext方法，返回当前View对象的Context对象，通常是当前正在展显的Activity实例。②Activity. this返回当前的Activity实例，UI控件需要使用Activity作为Context对象，在Fragment中使用getActivity()返回该Fragment所依附的Activity。默认的Toast也是使用Activity实例，实际上使用ApplicationContext也可以。③Activity. getApplicationContext，获取当前Activity所在的应用进程的Context对象。只有在Activity和Service中才能调用getApplication()获取Application实例，但是如果在一些其他的场景，比如BroadcastReceiver中只能通过getApplicationContext()方法。

5.11 正则表达式

正则表达式作为一个模板，描述在搜索文本时要匹配的一个或多个字符串，将某个字符模式与所搜索的字符串进行匹配。正则表达式是由普通字符以及特殊字符(特殊字符又称为元字符)组成的文字模式，正则表达式的组件可以是单个的字符、字符集合、字符范围、字符间的选择或者所有这些组件的任意组合。普通字符包括没有显式指定为特殊字符的所有可打印和不可打印字符。可打印字符包括所有大写和小写字母、所有数字、所有标点符号和一些其他符号，非打印字符也可以是正则表达式的组成部分。表5-2列出了正则表达式的部分转义字符。

表5-2 正则表达式的部分转义字符

字　符	解　释
\cx	匹配由x指明的控制字符。例如，\cM匹配一个Control-M或回车符。x的值必须为A-Z或a-z之一。否则，将c视为一个原义的'c'字符
\f	匹配一个换页符。等价于\x0c和\cL
\n	匹配一个换行符。等价于\x0a和\cJ
\r	匹配一个回车符。等价于\x0d和\cM
\t	匹配一个制表符。等价于\x09和\cI
\v	匹配一个垂直制表符。等价于\x0b和\cK
\s	匹配任何空不可见字符，包括空格、制表符、换页符等等。等价于[\f\n\r\t\v]
\S	匹配任何可见字符。等价于[^\f\n\r\t\v]
\d	匹配一个数字字符。等价于[0-9]
\D	匹配一个非数字字符。等价于[^0-9]
\w	匹配字母，数字，下画线
\W	匹配任意不是字母，数字，下画线的字符

所谓特殊字符，就是一些有特殊含义的字符。这主要包含定位符、重复次数限定符和其他特殊字符。若要匹配这些特殊字符本身，必须首先使字符转义，即将反斜杠字符\放在它

们前面。使用定位符能够创建这样的正则表达式，这些正则表达式出现在字符串的开始或者结束、一个单词的开始或者结束或者一个单词内。正则表达式的定位符，如表 5-3 所示。

表 5-3 正则表达式的定位字符

字　符	解　释
^	匹配输入字符串开始的位置。如果设置了 RegExp 对象的 Multiline 属性，^ 还会与\n 或\r 之后的位置匹配。^ 如果在方括号中且是方括号中的第一个字符，此时它表示不接受该字符集合。要匹配^ 字符本身，请使用\^
$	匹配输入字符串结尾的位置。如果设置了 RegExp 对象的 Multiline 属性，$ 还会与\n 或\r 之前的位置匹配。要匹配 $ 字符本身，请使用\ $
\b	匹配一个单词边界，即单词与空格间的位置。例如，er\b 可以匹配 never 中的 er，但不能匹配 verb 中的 er
\B	非单词边界匹配。er\B 能匹配 verb 中的 er，但不能匹配 never 中的 er
\<	匹配单词的开始
\>	匹配单词的结束。例如正则表达式\< the\>能够匹配字符串 for the wise 中的 the，但是不能匹配字符串 otherwise 中的 the。这个元字符不是所有的软件都支持的

正则表达式的重复次数限定符，如表 5-4 所示，用来它前面组件必须要出现多少次才能满足匹配，有{n}、{n,}或{n,m}、*、+和？共 6 种。重复次数限定符不能紧跟在定位符后面，因此不允许诸如^ * 之类的表达式。重复次数总可以用花括号内的三种形式表示，而星号、加号、问号代表的重复次数也可以用花括号表示，只不过比花括号简洁一些。

表 5-4 正则表达式的重复次数限定符

字　符	解　释
{n}	n 必须是一个非负整数，匹配确定的 n 次。例如，o{2}不能匹配 Bob 中的 o，但是能匹配 food 中的两个 o
{n,}	n 必须是一个非负整数。至少匹配 n 次。例如，o{2,}不能匹配 Bob 中的 o，但能匹配 foooood 中的所有 o。o{1,}等价于 o+。o{0,}则等价于 o*
{n,m}	m 和 n 必须是非负整数，其中 n<=m，在逗号和两个数之间不能有空格。最少匹配 n 次且最多匹配 m 次。例如，o{1,3}将匹配 fooooood 中的前三个 o 为一组，后三个 o 为一组。o{0,1}等价于 o?
*	匹配前面的子表达式零次或多次。例如，zo * 能匹配 z 以及 oo。* 等价于{0,}。要匹配 * 字符，请使用\ *
+	匹配前面的子表达式一次或多次。例如，zo+能匹配 zo 以及 zoo，但不能匹配 z。+等价于{1,}。要匹配+字符，请使用\+
?	匹配前面的子表达式零次或一次，或指明一个非贪婪限定符。例如，do(es)？可以匹配 do 或 does 中的 do。？等价于{0,1}。当？紧跟在任何一个重复次数限定符({n}、{n,}、{n,m}、*、+、?)后面时，匹配模式是非贪婪的。非贪婪模式尽可能少地匹配所搜索的字符串，而默认的贪婪模式则尽可能多地匹配所搜索的字符串。例如，对于字符串 oooo，o+将尽可能多地匹配 o，得到结果["oooo"]，而 o+？将尽可能少地匹配 o，得到结果[o,o,o,o]。要匹配？字符，请使用\?

正则表达式的其他特殊字符，如表 5-5 所示。

表 5-5 正则表达式的其他特殊字符

字 符	解 释
()	标记一个子表达式的开始和结束位置。匹配子表达式的字符可以获取供以后使用。要匹配这些字符本身,请使用\(和\)
[]	[]是单个字符匹配,定义匹配的字符范围,匹配的字符定义在[]中,并且只能出现一次。 简单字符组:[123]可以匹配字符'1','2', '3'。范围字符组:[0-5]表示从 0 到 5 的任何一个字符,等价于[012345]。[a-f1-5]匹配 a～f 或者 1～5 字符。在方括号中连字符(-)表示连续字符的范围。字符范围应该这样书写:ASCII 值小的在前面,ASCII 值大的在后面。排除字符组:^如果位于方括号的第一个字符位置,则具有反向的含义:指的是匹配不在方括号内的任何字符。[^0-3],表示找到这一个位置上的字符只能是除了 0 到 3 之外的所有字符
.	匹配除换行符\n 之外的任何单字符。要匹配 . ,请使用\.
\	将下一个字符标记为特殊字符、原义字符、向后引用或八进制转义符。例如,'n'匹配字符'n'。'\n' 匹配换行符。序列'\\' 匹配 "\",而 '\(' 则匹配 "("
|	将两个匹配条件进行逻辑"或"运算,指明两项之间的一个选择。要匹配|,请使用\\|

()、[]、{}的区别

()内的内容表示的是一个子表达式,()本身不匹配任何东西,也不限制匹配任何东西,只是把括号内的内容作为一个不可分割的表达式来处理,例如(ab){1,3},就表示 ab 一起连续出现最少 1 次,最多 3 次。如果没有小括号的话,ab{1,3}就表示 a,后面紧跟的 b 出现最少 1 次,最多 3 次。另外,括号在匹配模式中也很重要。使用圆括号可以匹配多个连续的字符,而使用方括号只能匹配单个字符。(\s*)表示连续空格的字符串。

[]用于匹配单个字符,匹配的字符在[]中,并且只能出现一次,并且特殊字符写在[]内被当成普通字符来匹配。例如[(a)],会匹配(、a、)、这三个字符。[a-zA-Z0-9]表示相应位置的字符要匹配英文字符和数字。[\s*]表示空格或者*号。需要注意的是,连字符(-)只有在中括号内才是元字符,即使在中括号内、但如果在开始位置或结束位置,也只能作为一个普通字符,而不是表示范围的连字符。例如[123-]表示'1','2', '3'或者'-'。

{}一般用来表示匹配的长度,只有{n}、{n,}和{n,m}这三种形式。所以{1-9} 语法有错误。比如\s{3}表示匹配三个空格,\s{1,3}表示匹配一到三个空格。

正则表达式实例

(0-9)匹配'0-9′本身。[0-9]*匹配数字,可以为空。[0-9]+匹配数字。[0-9]{0,9}表示长度为 0 到 9 的数字字符串。[a-zA-Z0-9]表示相应位置的字符要匹配英文字符和数字。[\s*]表示空格或者*号。

5.12 Activity 的生命周期实验

实验目的:

(1) 熟悉 Activity 生命周期。

(2) 掌握 startActivity 的使用方法。

(3) 掌握 startActivityForResult()、onActivityResult()和 setResult()的用法，从而熟悉在 Activity 中启动新的 Activity，并且在新的 Activity 关闭时回传数据给原来的 Activity 的方法。

(4) 掌握通过 Intent 发送和接收数据的方法。

1. 新建工程 Add

具体步骤：运行 Android Studio，单击 Start a new Android Studio project，在 Create New Project 页中，Application name 为 Add，Company domain 为 cuc，Project location 为 C:\android\myproject\Add，单击 Next 按钮，选中 Phone and Tablet，minimum sdk 设置为 API 19：Android 4.4(minSdkVersion)，单击 Next 按钮，选择 EmptyActivity，再单击 Next 按钮，勾选 Generate Layout File 选项，勾选 Backwards Compatibility(AppCompat)选项，单击 Finish 按钮。

2. 新建三个布局文件

新建布局文件 activity_main1.xml 的具体步骤：在工程结构视图中(单击 Project 按钮可以显示或者隐藏工程结构图，使用工程结构图中的下拉列表框可以选择 Android 以显示工程逻辑结构，或者选择 Project 显示工程实际的目录结构)，右击 layout，依次选择 New|Layout resource file，File name 输入框输入 activity_main1，Root element 输入框输入 FrameLayout，Source set 为 main，单击 OK 按钮。layout/activity_main1.xml 内容如下：

```
<?xml version="1.0" encoding="utf-8"?>
<FrameLayout xmlns:android="http://schemas.android.com/apk/res/android"
    android:layout_width="match_parent" android:layout_height="match_parent"
    android:id="@+id/fragment_container"
    >
</FrameLayout>
```

布局文件 activity_main1.xml 的 FrameLayout 用作 Fragment 的容器。

同样的方法，在 layout 文件夹下，新建布局文件 add.xml，这个布局将用作 Fragment 的视图。layout/add.xml 内容如下：

```
<?xml version="1.0" encoding="utf-8"?>
<LinearLayout xmlns:android="http://schemas.android.com/apk/res/android"
    android:orientation="vertical" android:layout_width="match_parent"
    android:layout_height="match_parent">
    <LinearLayout
        android:orientation="horizontal"
        android:layout_width="match_parent"
        android:layout_height="wrap_content">
        <TextView
            android:layout_width="wrap_content"
            android:layout_height="wrap_content"
            android:text="please input operand1" />
        <EditText
            android:layout_width="wrap_content"
```

```
            android:layout_height = "wrap_content"
            android:id = "@ + id/operand1Edt"
            android:inputType = "number"
            android:layout_weight = "1" />
    </LinearLayout >
    < EditText
        android:layout_width = "match_parent"
        android:layout_height = "wrap_content"
        android:id = "@ + id/operand2Edt"
        android:hint = "please input operand2"
        android:inputType = "number" />
    < Button
        android:layout_width = "match_parent"
        android:layout_height = "wrap_content"
        android:id = "@ + id/addBtn"
        android:text = "Add"
        android:textAllCaps = "false" />
    < Button
        android:layout_width = "match_parent"
        android:layout_height = "wrap_content"
        android:id = "@ + id/divideBtn"
        android:text = "Divide" />

</LinearLayout >
```

在 layout 文件夹下，新建布局文件 result. xml，这个布局将用于显示计算的结果。layout/result. xml 内容如下：

```
<?xml version = "1.0" encoding = "utf - 8"?>
< LinearLayout xmlns:android = "http://schemas.android.com/apk/res/android"
    android:orientation = "vertical" android:layout_width = "match_parent"
    android:layout_height = "match_parent">
    < TextView
        android:layout_width = "match_parent"
        android:layout_height = "wrap_content"
        android:id = "@ + id/resultTxt"
        android:text = "result"/>
</LinearLayout >
```

3. 由 add. xml 生成 Fragment

具体步骤为：在 Android 工程结构视图中，选中布局文件 add. xml，右击，在弹出菜单中选中 generate Android Code | Fragment，Source path 选择\app\src\main\java，单击 Create File 按钮，生成了文件 AddFragment. java。然后自己在 AddFragment. java 里添上按键的响应功能以及 Fragment 生命周期有关的方法。其中单击 addBtn，调用 startActivity() 启动 ResultActivity，当从 ResultActivity 返回到 MainActivity 时，并不回传结果。单击 divideBtn 时，调用 startActivityForResult() 启动 ResultActivity，当从 ResultActivity 返回到 MainActivity 时，要求回传结果。最后 AddFragment. java 内容如下：

```
package cuc.add;

import android.content.Context;
import android.content.Intent;
import android.os.Bundle;
import android.support.annotation.Nullable;
import android.support.v4.app.Fragment;
import android.util.Log;
import android.view.LayoutInflater;
import android.view.View;
import android.view.ViewGroup;
import android.widget.Button;
import android.widget.EditText;
import static android.app.Activity.RESULT_CANCELED;
import static android.app.Activity.RESULT_OK;

public class AddFragment extends Fragment implements View.OnClickListener {
    private static final String TAG = "AddFragment";
    private static final int DIVIDE_REQUESTCODE = 1;
    private EditText operand1Edt;
    private EditText operand2Edt;
    private Button addBtn;
    private Button divideBtn;
    @Override
    public void onAttach(Context context) {
        super.onAttach(context);
        Log.i(TAG, "onAttach: ");
    }
    @Override
    public void onDetach() {
        super.onDetach();
        Log.i(TAG, "onDetach: ");
    }

    @Override
    public void onCreate(@Nullable Bundle savedInstanceState) {
        super.onCreate(savedInstanceState);
        Log.i(TAG, "onCreate: ");
    }

    @Override
    public void onDestroy() {
        super.onDestroy();
        Log.i(TAG, "onDestroy: ");
    }

    @Override
    public void onDestroyView() {
        super.onDestroyView();
```

```
            Log.i(TAG, "onDestroyView: ");
        }

        @Override
        public View onCreateView(LayoutInflater inflater, ViewGroup container,
                                 Bundle savedInstanceState) {
            Log.i(TAG, "onCreateView: ");
            return inflater.inflate(R.layout.add, null);
        }

        @Override
        public void onViewCreated(View view, Bundle savedInstanceState) {
            super.onViewCreated(view, savedInstanceState);
            Log.i(TAG, "onViewCreated: ");
            operand1Edt = (EditText) view.findViewById(R.id.operand1Edt);
            operand2Edt = (EditText) view.findViewById(R.id.operand2Edt);
            addBtn = (Button) view.findViewById(R.id.addBtn);
            divideBtn = (Button) view.findViewById(R.id.divideBtn);
            addBtn.setOnClickListener(this);
            divideBtn.setOnClickListener(this);
        }

        @Override
        public void onActivityCreated(@Nullable Bundle savedInstanceState) {
            super.onActivityCreated(savedInstanceState);
            Log.i(TAG, "onActivityCreated: ");
        }
    @Override
    public void onViewStateRestored(@Nullable Bundle savedInstanceState) {
        super.onViewStateRestored(savedInstanceState);
        Log.i(TAG, "onViewStateRestored: ");
    }

        @Override
        public void onStart() {
            super.onStart();
            Log.i(TAG, "onStart: ");
        }

        @Override
        public void onStop() {
            super.onStop();
            Log.i(TAG, "onStop: ");
        }

        @Override
        public void onResume() {
            super.onResume();
            Log.i(TAG, "onResume: ");
```

```
    }

    @Override
    public void onPause() {
        super.onPause();
        Log.i(TAG, "onPause: ");
    }

    @Override
    public void onClick(View view) {
        Intent intent = new Intent(getActivity(),ResultActivity.class);
        Bundle bundle = new Bundle();
        bundle.putString("op1",operand1Edt.getText().toString());
        bundle.putString("op2",operand2Edt.getText().toString());
        switch (view.getId()) {
            case R.id.addBtn:
                bundle.putInt("operator",0);
                intent.putExtras(bundle);
                startActivity(intent);
                break;
            case R.id.divideBtn:
                bundle.putInt("operator",1);
                intent.putExtras(bundle);
                startActivityForResult(intent,DIVIDE_REQUESTCODE);
                break;
        }
    }

    @Override
    public void onActivityResult(int requestCode, int resultCode, Intent data) {
        super.onActivityResult(requestCode, resultCode, data);
            if (resultCode == RESULT_OK){
                Log.i(TAG, "onActivityResult: " + "RESULT_OK,result is" + data.getIntExtra
("result",0) + ", requstcode:" + requestCode);
            }else if (resultCode == RESULT_CANCELED){
                Log.i(TAG, "onActivityResult: " + "RESULT_CANCELED," + "requstcode:" +
requestCode);
            }
    }
}
```

4. 新建一个 Activity

具体步骤：选中工程视图中的 java 文件夹下的 cuc.add 包，右击，在弹出菜单中选择 New|Activity|EmptyActivity，在弹出的 ConfigureActivity 界面中，输入 Activity 名字为 ResultActivity，不要勾选 Generate Layout File 选项，最后单击 Finish 按钮，框架就创建好了 Activity，并在清单文件中完成了登记。然后添加上对传送的 Intent 的数据的处理。因为调用了 startActivityForResult()启动 ResultActivity，所以需要调用 setResult()设置 Activity 返回结果。

```
package cuc.add;

import android.content.Intent;
import android.os.Bundle;
import android.app.Activity;
import android.util.Log;
import android.widget.TextView;

public class ResultActivity extends Activity {
    private static final String TAG = "ResultActivity";
    private TextView resultTxt;

    @Override
    protected void onCreate(Bundle savedInstanceState) {
        super.onCreate(savedInstanceState);
        setContentView(R.layout.result);
        Log.i(TAG, "onCreate: ");
        resultTxt = (TextView) findViewById(R.id.resultTxt);
        Intent intent = getIntent();
        Bundle bundle = intent.getExtras();
        String op1Str = bundle.getString("op1");
        String op2Str = bundle.getString("op2");
        int op1 = Integer.parseInt(op1Str);
        int op2 = Integer.parseInt(op2Str);
        Intent intent2 = new Intent();
        int result = 0;
        int operator = bundle.getInt("operator");
        if (operator == 0){
            result = op1 + op2;
            resultTxt.setText("result = " + result);
        }else if (operator == 1){
            if (op2 == 0){
                resultTxt.setText("divsor is 0,error");
                setResult(RESULT_CANCELED, intent2);
            }
            else {
                result = op1/op2;
                resultTxt.setText("result = " + result);
                intent2.putExtra("result",result);
                setResult(RESULT_OK, intent2);
            }
        }
    }

    @Override
    protected void onDestroy() {
        super.onDestroy();
        Log.i(TAG, "onDestroy: ");
    }
```

```
    @Override
    protected void onStart() {
        super.onStart();
        Log.i(TAG, "onStart: ");
    }

    @Override
    protected void onStop() {
        super.onStop();
        Log.i(TAG, "onStop: ");
    }

    @Override
    protected void onRestart() {
        super.onRestart();
        Log.i(TAG, "onRestart: ");
    }

    @Override
    protected void onResume() {
        super.onResume();
        Log.i(TAG, "onResume: ");
    }
    @Override
    protected void onPause() {
        super.onPause();
        Log.i(TAG, "onPause: ");
    }
}
```

5. 编辑 MainActivity.java

设置 MainActivity 要显示的视图为 activity_main1.xml 布局文件，并且添加上 Activity 响应的代码。最后 MainActivity.java 内容如下：

```
package cuc.add;
import android.content.Intent;
import android.os.Bundle;
import android.support.v7.app.AppCompatActivity;
import android.util.Log;

public class MainActivity extends AppCompatActivity {
    private static final String TAG = "MainActivity";
    @Override
    protected void onCreate(Bundle savedInstanceState) {
        super.onCreate(savedInstanceState);
        Log.i(TAG, "onCreate: ");
        setContentView(R.layout.activity_main1);
```

```
        getSupportFragmentManager().beginTransaction().add(R.id.fragment_container,new
AddFragment()).commit();
    }
    @Override
    protected void onDestroy() {
        super.onDestroy();
        Log.i(TAG, "onDestroy: ");
    }
    @Override
    protected void onStart() {
        super.onStart();
        Log.i(TAG, "onStart: ");
    }
    @Override
    protected void onStop() {
        super.onStop();
        Log.i(TAG, "onStop: ");
    }
    @Override
    protected void onRestart() {
        super.onRestart();
        Log.i(TAG, "onRestart: ");
    }

    @Override
    protected void onResume() {
        super.onResume();
        Log.i(TAG, "onResume: ");
    }
    @Override
    protected void onPause() {
        super.onPause();
        Log.i(TAG, "onPause: ");
    }
    @Override
    protected void onActivityResult(int requestCode, int resultCode, Intent data) {
        super.onActivityResult(requestCode, resultCode, data);
        if (resultCode == RESULT_OK){
            Log.i(TAG, "onActivityResult: " + "RESULT_OK, result is" + data.getIntExtra
("result",0) + "requstcode:" + requestCode);
        }else if (resultCode == RESULT_CANCELED){
            Log.i(TAG, "onActivityResult: " + "RESULT_CANCELED," + "requstcode:" +
requestCode);
        }
    }
}
```

6. 编译运行

编译运行,显示 MainActivity 页面时,这时 Logcat 中 log 输出如下:

```
10587 - 10587/cuc.add I/MainActivity: onCreate:
10587 - 10587/cuc.add I/AddFragment: onAttach:
10587 - 10587/cuc.add I/AddFragment: onCreate:
10587 - 10587/cuc.add I/AddFragment: onCreateView:
10587 - 10587/cuc.add I/AddFragment: onViewCreated:
10587 - 10587/cuc.add I/AddFragment: onActivityCreated:
10587 - 10587/cuc.add I/AddFragment: onStart:
10587 - 10587/cuc.add I/MainActivity: onStart:
10587 - 10587/cuc.add I/MainActivity: onResume:
10587 - 10587/cuc.add I/AddFragment: onResume:
```

在 MainActivity 页面，输入两个操作数，例如 4 和 2，然后单击 addBtn 按钮，可以看到界面跳到 ResultActivity，这时 Logcat 中 log 输出如下：

```
6156 - 6156/cuc.add I/AddFragment: onPause:
6156 - 6156/cuc.add I/MainActivity: onPause:
6156 - 6156/cuc.add I/ResultActivity: onCreate:
6156 - 6156/cuc.add I/ResultActivity: onStart:
6156 - 6156/cuc.add I/ResultActivity: onResume:
6156 - 6156/cuc.add I/AddFragment: onStop:
6156 - 6156/cuc.add I/MainActivity: onStop:
```

单击手机上的 Back 键，界面又回退到 MainActivity，这时 Logcat 中 log 输出如下：

```
6156 - 6156/cuc.add I/ResultActivity: onPause:
6156 - 6156/cuc.add I/MainActivity: onRestart:
6156 - 6156/cuc.add I/AddFragment: onStart:
6156 - 6156/cuc.add I/MainActivity: onStart:
6156 - 6156/cuc.add I/MainActivity: onResume:
6156 - 6156/cuc.add I/AddFragment: onResume:
6156 - 6156/cuc.add I/ResultActivity: onStop:
6156 - 6156/cuc.add I/ResultActivity: onDestroy:
```

把上面的日志按时间顺序制成表格，如表 5-6 所示。其中 Fragment 是在 MainActivity 的 onCreate()方法中调用 add(R.id.fragment_container,new AddFragment()).commit()方法生成的。

表 5-6 实验中 Activity 和 Fragment 的方法调用顺序

MainActivity 的方法	AddFragment 的方法	ResultActivity 的方法
onCreate		
	onAttach	
	onCreate	
	onCreateView	
	onViewCreated	
	onActivityCreated	
	onViewStateRestored	

续表

MainActivity 的方法	AddFragment 的方法	ResultActivity 的方法
	onStart	
onStart		
onResume		
	onResume	
	onPause	
onPause		
onResume		
	onResume	
	onPause	
onPause		
		onCreate
		onStart
		onResume
	onStop	
onStop		
		onPause
onRestart		
	onStart	
onStart		
onResume		
	onResume	
		onStop
		onDestroy

7. 新建布局文件

在 layout 文件夹下，新建布局文件 activity_main2. xml，在这个布局中，fragment 直接作为布局的元素。layout/activity_main2. xml 内容如下：

```
<?xml version = "1.0" encoding = "utf - 8"?>
< LinearLayout xmlns:android = "http://schemas.android.com/apk/res/android"
    android:orientation = "vertical" android:layout_width = "match_parent"
    android:layout_height = "match_parent">
    < fragment
        android:layout_width = "match_parent"
        android:layout_height = "match_parent"
        android:name = "cuc.add.AddFragment"
        android:id = "@ + id/addFragment"/>
</LinearLayout >
```

8. 编辑 MainActivity. java

修改 MainActivity 类的 OnCreate 方法、设置 MainActivity 要显示的视图为 activity_main2. xml 布局文件。最后 MainActivity 类的 OnCreate 方法内容如下：

```
@Override
protected void onCreate(Bundle savedInstanceState) {
    super.onCreate(savedInstanceState);
    Log.i(TAG, "onCreate: ");
    setContentView(R.layout.activity_main2);
}
```

9. 编译运行

编译运行，跟第六步中的操作一样：显示 MainActivity 页面时，查看 Logcat 中 log 输出；在 MainActivity 页面，输入两个操作数，例如 4 和 2，然后单击 addBtn 按钮，可以看到界面跳到 ResultActivity，再查看 Logcat 中 log 输出；再单击手机上的 Back 键，界面又回退到 MainActivity，再次查看 Logcat 中 log 输出，发现日志输出和第六步中的运行结果没有区别。

继续在 MainActivity 页面，单击 divideBtn，可以看到界面跳到 ResultActivity，这时 Logcat 中 log 输出如下：

```
6591 - 6591/cuc.add I/AddFragment: onPause:
6591 - 6591/cuc.add I/MainActivity: onPause:
6591 - 6591/cuc.add I/ResultActivity: onCreate:
6591 - 6591/cuc.add I/ResultActivity: onStart:
6591 - 6591/cuc.add I/ResultActivity: onResume:
6591 - 6591/cuc.add I/AddFragment: onStop:
6591 - 6591/cuc.add I/MainActivity: onStop:
```

再按 Back 键，返回 MainActivity 页面，这时 Logcat 中 log 输出如下：

```
6591 - 6591/cuc.add I/ResultActivity: onPause:
6591 - 6591/cuc.add I/AddFragment: onActivityResult: RESULT_OK,result is6, requstcode:1
6591 - 6591/cuc.add I/MainActivity: onActivityResult: RESULT_OK,result is6requstcode:196609
6591 - 6591/cuc.add I/MainActivity: onRestart:
6591 - 6591/cuc.add I/AddFragment: onStart:
6591 - 6591/cuc.add I/MainActivity: onStart:
6591 - 6591/cuc.add I/MainActivity: onResume:
6591 - 6591/cuc.add I/AddFragment: onResume:
6591 - 6591/cuc.add I/ResultActivity: onStop:
6591 - 6591/cuc.add I/ResultActivity: onDestroy:
```

5.13 Activity 的 launchMode 实验

实验目的：

熟悉 Activity 的 launchMode 属性与 taskAffinity 属性。

5.13.1 launchMode 为 standard 实验

1. 新建工程 ActivityLaunchMode

具体步骤：运行 Android Studio，单击 Start a new Android Studio project，在 Create

New Project 页中，Application name 为 ActivityLaunchMode，Company domain 为 cuc，Project location 为 C:\android\myproject\ActivityLaunchMode，单击 Next 按钮，选中 Phone and Tablet，minimum sdk 设置为 API 19：Android 4.4(minSdkVersion)，单击 Next 按钮，选择 EmptyActivity，再单击 Next 按钮，勾选 Generate Layout File 选项，勾选 Backwards Compatibility(AppCompat)选项，单击 Finish 按钮。

2. 新建两个布局文件

layout/activity_main1.xml 内容如下：

```
<?xml version = "1.0" encoding = "utf - 8"?>
<LinearLayout xmlns:android = "http://schemas.android.com/apk/res/android"
    android:orientation = "vertical" android:layout_width = "match_parent"
    android:layout_height = "match_parent">
    <TextView
        android:layout_width = "match_parent"
        android:layout_height = "wrap_content"
        android:textSize = "30sp"
        android:id = "@ + id/helloTxt"
        android:text = "MainActivity" />
    <Button
        android:layout_width = "match_parent"
        android:layout_height = "wrap_content"
        android:id = "@ + id/start2ActivityBtn"
        android:text = "start2Activity"/>
</LinearLayout>
```

layout/activity_main2.xml 内容如下：

```
<?xml version = "1.0" encoding = "utf - 8"?>
<LinearLayout xmlns:android = "http://schemas.android.com/apk/res/android"
    android:orientation = "vertical" android:layout_width = "match_parent"
    android:layout_height = "match_parent">
    <TextView
        android:layout_width = "match_parent"
        android:layout_height = "wrap_content"
        android:textSize = "30sp"
        android:id = "@ + id/helloTxt"
        android:text = "SecondActivity"/>
    <Button
        android:layout_width = "match_parent"
        android:layout_height = "wrap_content"
        android:id = "@ + id/start1ActivityBtn"
        android:text = "start1Activity" />
    <Button
        android:layout_width = "match_parent"
        android:layout_height = "wrap_content"
        android:id = "@ + id/start2ActivityBtn"
        android:text = "start2Activity" />
    <Button
```

```
        android:layout_width = "match_parent"
        android:layout_height = "wrap_content"
        android:id = "@ + id/start3ActivityBtn"
        android:text = "start3Activity" />
</LinearLayout>
```

3. 编辑 MainActivity.java

MainActivity.java 内容如下：

```
package cuc.activitylaunchmode;

import android.content.Intent;
import android.os.Bundle;
import android.support.v7.app.AppCompatActivity;
import android.util.Log;
import android.view.View;
import android.widget.Button;
import android.widget.TextView;

public class MainActivity extends AppCompatActivity implements View.OnClickListener {
    private static final String TAG = "MainActivity";
    private TextView helloTxt;
    private Button start2ActivityBtn;

    private void assignViews() {
        helloTxt = (TextView) findViewById(R.id.helloTxt);
        start2ActivityBtn = (Button) findViewById(R.id.start2ActivityBtn);
    }
    @Override
    public void onClick(View view) {
        Intent intent;
        switch (view.getId()){
            case R.id.start2ActivityBtn:
                intent = new Intent(this,Main2Activity.class);
                startActivity(intent);
                break;
        }
    }
    @Override
    protected void onCreate(Bundle savedInstanceState) {
        super.onCreate(savedInstanceState);
        setContentView(R.layout.activity_main1);
        assignViews();
        Log.i(TAG, "onCreate: TaskId -- >" + getTaskId());
        start2ActivityBtn.setOnClickListener(this);
    }
    @Override
    protected void onDestroy() {
```

```
        super.onDestroy();
        Log.i(TAG, "onDestroy: ");
    }
}
```

4. 新建两个 Activity

在 cuc.activitylaunchmode 包里,新建两个 Activity: Main2Activity 和 Main3Activity。Main2Activity.java 内容如下:

```
package cuc.activitylaunchmode;
import android.content.Intent;
import android.os.Bundle;
import android.support.v7.app.AppCompatActivity;
import android.util.Log;
import android.view.View;
import android.widget.Button;
import android.widget.TextView;

public class Main2Activity extends AppCompatActivity implements View.OnClickListener{
    private static final String TAG = "Main2Activity";
    private TextView helloTxt;
    private Button start1ActivityBtn;
    private Button start2ActivityBtn;
    private Button start3ActivityBtn;

    private void assignViews() {
        helloTxt = (TextView) findViewById(R.id.helloTxt);
        start1ActivityBtn = (Button) findViewById(R.id.start1ActivityBtn);
        start2ActivityBtn = (Button) findViewById(R.id.start2ActivityBtn);
        start3ActivityBtn = (Button) findViewById(R.id.start3ActivityBtn);
    }

    @Override
    public void onClick(View view) {
        Intent intent = null;
        switch (view.getId()){
            case R.id.start1ActivityBtn:
                intent = new Intent(this,MainActivity.class);
                break;
            case R.id.start2ActivityBtn:
                intent = new Intent(this,Main2Activity.class);
                break;
            case R.id.start3ActivityBtn:
                intent = new Intent(this,Main3Activity.class);
                break;
        }
        startActivity(intent);
    }
```

```
    @Override
    protected void onCreate(Bundle savedInstanceState) {
        super.onCreate(savedInstanceState);
        setContentView(R.layout.activity_main2);
        assignViews();
        helloTxt.setText("Main2Activity");
        Log.i(TAG, "onCreate: TaskId-->" + getTaskId());
        start1ActivityBtn.setOnClickListener(this);
        start2ActivityBtn.setOnClickListener(this);
        start3ActivityBtn.setOnClickListener(this);

    }
    @Override
    protected void onDestroy() {
        super.onDestroy();
        Log.i(TAG, "onDestroy: ");
    }
}
```

Main3Activity.java 内容如下：

```
package cuc.activitylaunchmode;
import android.content.Intent;
import android.os.Bundle;
import android.support.v7.app.AppCompatActivity;
import android.util.Log;
import android.view.View;
import android.widget.Button;
import android.widget.TextView;

public class Main3Activity extends AppCompatActivity implements View.OnClickListener{
    private static final String TAG = "Main3Activity";
    private TextView helloTxt;
    private Button start1ActivityBtn;
    private Button start2ActivityBtn;
    private Button start3ActivityBtn;

    private void assignViews() {
        helloTxt = (TextView) findViewById(R.id.helloTxt);
        start1ActivityBtn = (Button) findViewById(R.id.start1ActivityBtn);
        start2ActivityBtn = (Button) findViewById(R.id.start2ActivityBtn);
        start3ActivityBtn = (Button) findViewById(R.id.start3ActivityBtn);
    }

    @Override
    public void onClick(View view) {
        Intent intent = null;
        switch (view.getId()){
            case R.id.start1ActivityBtn:
```

```
                intent = new Intent(this,MainActivity.class);
                break;
            case R.id.start2ActivityBtn:
                intent = new Intent(this,Main2Activity.class);
                break;
            case R.id.start3ActivityBtn:
                intent = new Intent(this,Main3Activity.class);
                break;
        }
        startActivity(intent);
    }
    @Override
    protected void onCreate(Bundle savedInstanceState) {
        super.onCreate(savedInstanceState);
        setContentView(R.layout.activity_main2);
        assignViews();
        helloTxt.setText("Main3Activity");
        start1ActivityBtn.setOnClickListener(this);
        start2ActivityBtn.setOnClickListener(this);
        start3ActivityBtn.setOnClickListener(this);
        Log.i(TAG, "onCreate: TaskId-->" + getTaskId());
    }
    @Override
    protected void onDestroy() {
        super.onDestroy();
        Log.i(TAG, "onDestroy: ");
    }
}
```

5. 编辑清单文件

app/src/main/AndroidManifest.xml 文件内容如下：

```
<?xml version="1.0" encoding="utf-8"?>
<manifest xmlns:android="http://schemas.android.com/apk/res/android"
    package="cuc.activitylaunchmode">
    <uses-permission android:name="android.permission.GET_TASKS"/>
    <application
        android:allowBackup="true"
        android:icon="@mipmap/ic_launcher"
        android:label="@string/app_name"
        android:roundIcon="@mipmap/ic_launcher_round"
        android:supportsRtl="true"
        android:theme="@style/AppTheme">
        <activity android:name=".MainActivity">
            <intent-filter>
              <action android:name="android.intent.action.MAIN" />
              <category android:name="android.intent.category.LAUNCHER"/>
            </intent-filter>
```

```
        </activity>
        <activity android:name=".Main2Activity" android:taskAffinity="cuc.abc">
            <intent-filter>
                <action android:name="cuc.activitylaunchmode.Main2Activity"/>
                <category android:name="android.intent.category.DEFAULT"/>
            </intent-filter>
        </activity>
        <activity android:name=".Main3Activity" />
    </application>
</manifest>
```

6. 编译运行

依次单击 MainActivity 中的 start2ActivityBtn 按钮、Main2Activity 中的 start2ActivityBtn 按钮、Main2Activity 中的 start3ActivityBtn 按钮、Main3Activity 中的 start2ActivityBtn 按钮、Main2Activity 中的 start1ActivityBtn 按钮，Logcat 中 log 输出为：

```
7450-7450/cuc.activitylaunchmode I/MainActivity: onCreate: TaskId-->33
7450-7450/cuc.activitylaunchmode I/Main2Activity: onCreate: TaskId-->33
7450-7450/cuc.activitylaunchmode I/Main2Activity: onCreate: TaskId-->33
7450-7450/cuc.activitylaunchmode I/Main3Activity: onCreate: TaskId-->33
7450-7450/cuc.activitylaunchmode I/Main2Activity: onCreate: TaskId-->33
7450-7450/cuc.activitylaunchmode I/MainActivity: onCreate: TaskId-->33
```

上面实验中没有设置 Activity 的 launchMode，所以 launchMode 为默认值 standard，可以看出总是在调用 startActivity(intent)方法的任务中创建新的 Activity 实例，任务栈与 android：taskAffinity 没有关系。

5.13.2 launchMode 为 singleTop 实验

具体步骤：在清单文件中把 Main2Activity 的 launchMode 设置为 singleTop，重新编译运行。

顺序单击 MainActivity 中的 start2ActivityBtn 按钮、Main2Activity 中的 start2ActivityBtn 按钮、Main2Activity 中的 start3ActivityBtn 按钮、Main3Activity 中的 start2ActivityBtn 按钮、Main2Activity 中的 start1ActivityBtn 按钮，Logcat 中 log 输出为：

```
31343-31343/cuc.activitylaunchmode I/MainActivity: onCreate: TaskId-->36
31343-31343/cuc.activitylaunchmode I/Main2Activity: onCreate: TaskId-->36
31343-31343/cuc.activitylaunchmode I/Main3Activity: onCreate: TaskId-->36
31343-31343/cuc.activitylaunchmode I/Main2Activity: onCreate: TaskId-->36
31343-31343/cuc.activitylaunchmode I/MainActivity: onCreate: TaskId-->36
```

上面实验中设置 Main2Activity 的 launchMode 为 singleTop，即在栈顶上只能存在 Main2Activity 的唯一实例，任务栈与 android：taskAffinity 也没有关系。可以看出：如果 Main2Activity 在栈顶，就不再创建 Main2Activity 的新实例。只要 Main2Activity 不在栈顶时，就在调用 startActivity(intent)方法的任务中创建新的 Activity 实例。

5.13.3 launchMode 为 singleTask 实验

具体步骤 1：在清单文件中把 Main2Activity 的 launchMode 设置为 singleTask，这时清单文件内容如下：

```
<?xml version = "1.0" encoding = "utf - 8"?>
<manifest xmlns:android = "http://schemas.android.com/apk/res/android"
    package = "cuc.activitylaunchmode">
    <uses - permission android:name = "android.permission.GET_TASKS"/>
    <application
        android:allowBackup = "true"
        android:icon = "@mipmap/ic_launcher"
        android:label = "@string/app_name"
        android:roundIcon = "@mipmap/ic_launcher_round"
        android:supportsRtl = "true"
        android:theme = "@style/AppTheme">
        <activity android:name = ".MainActivity">
            <intent - filter>
                <action android:name = "android.intent.action.MAIN" />
                <category android:name = "android.intent.category.LAUNCHER"/>
            </intent - filter>
        </activity>
        <activity android:name = ".Main2Activity" android:taskAffinity = "cuc.abc" android:
launchMode = "singleTask">
            <intent - filter>
                <action android:name = "cuc.activitylaunchmode.Main2Activity"/>
                <category android:name = "android.intent.category.DEFAULT"/>
            </intent - filter>
        </activity>
        <activity android:name = ".Main3Activity" />
    </application>
</manifest>
```

重新编译运行，顺序单击 MainActivity 中的 start2ActivityBtn 按钮、Main2Activity 中的 start2ActivityBtn 按钮、Main2Activity 中的 start3ActivityBtn 按钮、Main3Activity 中的 start2ActivityBtn 按钮、Main2Activity 中的 start1ActivityBtn 按钮、MainActivity 中的 start2ActivityBtn 按钮，Logcat 中 log 输出为：

```
30452 - 30452/? I/MainActivity: onCreate: TaskId -- > 76
30452 - 30452/cuc.activitylaunchmode I/Main2Activity: onCreate: TaskId -- > 78
30452 - 30452/cuc.activitylaunchmode I/Main3Activity: onCreate: TaskId -- > 78
30452 - 30452/cuc.activitylaunchmode I/Main3Activity: onDestroy:
30452 - 30452/cuc.activitylaunchmode I/MainActivity: onCreate: TaskId -- > 78
30452 - 30452/cuc.activitylaunchmode I/MainActivity: onDestroy:
```

具体步骤 2：编辑清单文件，这次又设置了 MainActivity 的 android：launchMode＝"singleTask"，这时清单文件内容如下：

```
<?xml version = "1.0" encoding = "utf - 8"?>
<manifest xmlns:android = "http://schemas.android.com/apk/res/android"
    package = "cuc.activitylaunchmode">
    <uses - permission android:name = "android.permission.GET_TASKS"/>
    <application
        android:allowBackup = "true"
        android:icon = "@mipmap/ic_launcher"
        android:label = "@string/app_name"
        android:roundIcon = "@mipmap/ic_launcher_round"
        android:supportsRtl = "true"
        android:theme = "@style/AppTheme">
        <activity android:name = ".MainActivity" android:launchMode = "singleTask">
            <intent - filter>
                <action android:name = "android.intent.action.MAIN" />
                <category android:name = "android.intent.category.LAUNCHER"/>
            </intent - filter>
        </activity>
        <activity android:name = ".Main2Activity" android:taskAffinity = "cuc.abc" android:
launchMode = "singleTask">
            <intent - filter>
              <action android:name = "cuc.activitylaunchmode.Main2Activity"/>
              <category android:name = "android.intent.category.DEFAULT"/>
            </intent - filter>
        </activity>
        <activity android:name = ".Main3Activity" />
    </application>
</manifest>
```

编译运行，依次单击 MainActivity 中的 start2ActivityBtn 按钮、Main2Activity 中的 start2ActivityBtn 按钮、Main2Activity 中的 start3ActivityBtn 按钮、Main3Activity 中的 start2ActivityBtn 按钮，Main2Activity 中的 start1ActivityBtn 按钮、MainActivity 中的 start2ActivityBtn 按钮，Logcat 中 log 输出为：

```
21047 - 21047/cuc.activitylaunchmode I/MainActivity: onCreate: TaskId -- > 71
21047 - 21047/cuc.activitylaunchmode I/Main2Activity: onCreate: TaskId -- > 74
21047 - 21047/cuc.activitylaunchmode I/Main3Activity: onCreate: TaskId -- > 74
21047 - 21047/cuc.activitylaunchmode I/Main3Activity: onDestroy:
```

具体步骤 3：除了在清单文件中把 MainActivity 和 Main2Activity 的 launchMode 设置为 singleTask，再去掉 Main2Activity 的 android：taskAffinity="cuc.abc"，即 3 个 Activity 具有相同的 taskAffinity，所以即使 launchMode 设置为 singleTask，也不会创建新的任务。这时清单文件内容如下：

```
<?xml version = "1.0" encoding = "utf - 8"?>
<manifest xmlns:android = "http://schemas.android.com/apk/res/android"
    package = "cuc.activitylaunchmode">
    <uses - permission android:name = "android.permission.GET_TASKS"/>
```

```
    <application
        android:allowBackup = "true"
        android:icon = "@mipmap/ic_launcher"
        android:label = "@string/app_name"
        android:roundIcon = "@mipmap/ic_launcher_round"
        android:supportsRtl = "true"
        android:theme = "@style/AppTheme">
        <activity android:name = ".MainActivity" android:launchMode = "singleTask">
            <intent - filter>
                    <action android:name = "android.intent.action.MAIN"/>
                <category android:name = "android.intent.category.LAUNCHER"/>
            </intent - filter>
        </activity>
        <activity android:name = ".Main2Activity" android:launchMode = "singleTask">
            <intent - filter>
                <ation android:name = "cuc.activitylaunchmode.Main2Activity"/>
                <category android:name = "android.intent.category.DEFAULT"/>
            </intent - filter>
        </activity>
        <activity android:name = ".Main3Activity" />
    </application>
</manifest>
```

重新编译运行，依次单击 MainActivity 中的 start2ActivityBtn 按钮、Main2Activity 中的 start2ActivityBtn 按钮、Main2Activity 中的 start3ActivityBtn 按钮、Main3Activity 中的 start2ActivityBtn 按钮，Main2Activity 中的 start1ActivityBtn 按钮、MainActivity 中的 start2ActivityBtn 按钮，Logcat 中 log 输出为：

```
21923 - 21923/cuc.activitylaunchmode I/MainActivity: onCreate: TaskId -- > 84
21923 - 21923/cuc.activitylaunchmode I/Main2Activity: onCreate: TaskId -- > 84
21923 - 21923/cuc.activitylaunchmode I/Main3Activity: onCreate: TaskId -- > 84
21923 - 21923/cuc.activitylaunchmode I/Main3Activity: onDestroy:
21923 - 21923/cuc.activitylaunchmode I/Main2Activity: onDestroy:
21923 - 21923/cuc.activitylaunchmode I/Main2Activity: onCreate: TaskId -- > 84
```

上面实验中设置 Main2Activity 的 launchMode 为 singleTask ，在任务中实例是唯一的，任务栈与 taskAffinity 有关系。可以看出：仅仅当系统中不存在 Main2Activity 的 taskAffinity 属性指定的任务时，才创建 Main2Activity 的 taskAffinity 属性指定的任务，并创建 Main2Activity 的新实例。若系统中已经存在 Main2Activity 的 taskAffinity 属性指定的任务，这时分三种情况：①Main2Activity 在栈顶时，就不再创建新的 Main2Activity 实例(步骤 1，步骤 2)；②系统中存在 Main2Activity 的实例，但 Main2Activity 不在栈顶时，就移除 Main2Activity 上面的所有实例，不再创建新的 Main2Activity 实例(步骤 1，步骤 2)；③不存在 Main2Activity 的实例，就创建 Main2Activity 实例(步骤 3)。

具体步骤 4：除了在清单文件中把 Main2Activity 和 Main3Activity 的 launchMode 设置为 singleTask ，Main3Activity 设置 android：taskAffinity＝"cuc. xyz"，MainActivity 和

Main2Activity 不设置 android：taskAffinity 属性，这时清单文件内容如下：

```
<?xml version = "1.0" encoding = "utf - 8"?>
< manifest xmlns:android = "http://schemas.android.com/apk/res/android"
    package = "cuc.activitylaunchmode">
    < application
        android:allowBackup = "true"
        android:icon = "@mipmap/ic_launcher"
        android:label = "@string/app_name"
        android:roundIcon = "@mipmap/ic_launcher_round"
        android:supportsRtl = "true"
        android:theme = "@style/AppTheme">
       < activity android:name = ".MainActivity" android:launchMode = "singleTask">
           < intent - filter >
               < action android:name = "android.intent.action.MAIN" />
               < category android:name = "android.intent.category.LAUNCHER"/>
           </intent - filter >
        </activity >
        < activity android: name = ". Main2Activity" android: launchMode = " singleTask" >
</activity >
        < activity
              android: name = ". Main3Activity" android: launchMode = " singleTask" android:
taskAffinity = "cuc.xyz" android:exported = "true" />
    </application >
</manifest >
```

顺序单击 MainActivity 中的 start2ActivityBtn 按钮、Main2Activity 中的 start3ActivityBtn 按钮、Main3Activity 中的 start2ActivityBtn 按钮、Back 键、Back 键、Back 键，这时回退到主屏幕，Logcat 中 log 输出为：

```
11472 - 11472/cuc.activitylaunchmode I/MainActivity: onCreate: TaskId -- > 12
11472 - 11472/cuc.activitylaunchmode I/Main2Activity: onCreate: TaskId -- > 12
11472 - 11472/cuc.activitylaunchmode I/Main3Activity: onCreate: TaskId -- > 14
11472 - 11472/cuc.activitylaunchmode I/Main2Activity: onDestroy:
11472 - 11472/cuc.activitylaunchmode I/MainActivity: onDestroy:
11472 - 11472/cuc.activitylaunchmode I/Main3Activity: onDestroy:
```

从上面的日志可以看出，应用启动创建了 MainActivity，单击 MainActivity 中的 start2ActivityBtn 按钮、Main2Activity 中的 start3ActivityBtn 按钮分别创建了 Main2Activity 和 Main3Activity。再单击 Main3Activity 中的 start2ActivityBtn 按钮，含有 Main2Activity 的整个任务(TaskId 为 12)就调到前台，这时按 Back 键，并不是从 Main2Activity 回退到 Main3Activity(TaskId 为 14)，而是回退到同一任务栈的下面的 Activity，直到该任务栈的所有 Activity 都弹出来。在按 Back 键之前，是从 taskId 等于 14 的任务切换到 taskId 等于 12 的任务，所以再按 Back 键，就回退到 taskId 等于 12 的任务中。因此可以看出，任务的调度以任务栈为单位的。

5.13.4 Intent 标志为 FLAG_ACTIVITY_NEW_TASK 实验

具体步骤 1：去掉所有的 Activity 的 taskAffinity 属性设置和 launchMode，这时清单文件内容如下：

```
<?xml version = "1.0" encoding = "utf - 8"?>
<manifest xmlns:android = "http://schemas.android.com/apk/res/android"
    package = "cuc.activitylaunchmode">
    <application
        android:allowBackup = "true"
        android:icon = "@mipmap/ic_launcher"
        android:label = "@string/app_name"
        android:roundIcon = "@mipmap/ic_launcher_round"
        android:supportsRtl = "true"
        android:theme = "@style/AppTheme">
        <activity android:name = ".MainActivity" >
            <intent - filter>
                <action android:name = "android.intent.action.MAIN" />

                <category android:name = "android.intent.category.LAUNCHER" />
            </intent - filter>
        </activity>
        <activity android:name = ".Main2Activity" />
        <activity android:name = ".Main3Activity" android:exported = "true" />
    </application>

</manifest>
```

再把 MainActivity、Main2Activity 和 Main3Activity 中的启动 Main2Activity 的 Intent 加上 FLAG_ACTIVITY_NEW_TASK，start2ActivityBtn 的按钮响应方法如下：

```
@Override
public void onClick(View view) {
    Intent intent = null;
    switch (view.getId()){
        case R.id.start1ActivityBtn:
            intent = new Intent(this,MainActivity.class);
            break;
        case R.id.start2ActivityBtn:
            intent = new Intent(this,Main2Activity.class);
            intent.setFlags(Intent.FLAG_ACTIVITY_NEW_TASK);
            break;
        case R.id.start3ActivityBtn:
            intent = new Intent(this,Main3Activity.class);
            break;
    }
    startActivity(intent);
}
```

重新编译运行，顺序单击 MainActivity 中的 start2ActivityBtn 按钮、Main2Activity 中的 start2ActivityBtn 按钮、Main2Activity 中的 start3ActivityBtn 按钮、Main3Activity 中的 start2ActivityBtn 按钮，Main2Activity 中的 start1ActivityBtn 按钮、MainActivity 中的 start2ActivityBtn 按钮，Logcat 中 log 输出为：

```
9153 - 9153/cuc.activitylaunchmode I/MainActivity: onCreate: TaskId -- > 19
9153 - 9153/cuc.activitylaunchmode I/Main2Activity: onCreate: TaskId -- > 19
9153 - 9153/cuc.activitylaunchmode I/Main2Activity: onCreate: TaskId -- > 19
9153 - 9153/cuc.activitylaunchmode I/Main3Activity: onCreate: TaskId -- > 19
9153 - 9153/cuc.activitylaunchmode I/Main2Activity: onCreate: TaskId -- > 19
9153 - 9153/cuc.activitylaunchmode I/MainActivity: onCreate: TaskId -- > 19
9153 - 9153/cuc.activitylaunchmode I/Main2Activity: onCreate: TaskId -- > 19
```

从上面 log 可以看出，当所有 Activity 的 taskAffinity 相同时，设置启动 Main2Activity 的 Intent 标志为 FLAG_ACTIVITY_NEW_TASK，每次都会创建 Main2Activity 的新实例。

具体步骤 2：设置 Main2Activity 的 taskAffinity 属性为 cuc.abc，这时清单文件内容如下：

```
<?xml version = "1.0" encoding = "utf - 8"?>
< manifest xmlns:android = "http://schemas.android.com/apk/res/android"
    package = "cuc.activitylaunchmode">
    < application
        android:allowBackup = "true"
        android:icon = "@mipmap/ic_launcher"
        android:label = "@string/app_name"
        android:roundIcon = "@mipmap/ic_launcher_round"
        android:supportsRtl = "true"
        android:theme = "@style/AppTheme">
        < activity android:name = ".MainActivity" >
            < intent - filter >
                < action android:name = "android.intent.action.MAIN" />
                < category android:name = "android.intent.category.LAUNCHER" />
            </intent - filter >
        </activity>
        < activity android:name = ".Main2Activity" android:taskAffinity = "cuc.abc" />
        < activity android:name = ".Main3Activity" android:exported = "true" />
    </application >
</manifest >
```

重新编译运行，顺序单击 MainActivity 中的 start2ActivityBtn 按钮、Main2Activity 中的 start2ActivityBtn 按钮、Main2Activity 中的 start3ActivityBtn 按钮、Main3Activity 中的 start2ActivityBtn 按钮，Main2Activity 中的 start1ActivityBtn 按钮、MainActivity 中的 start2ActivityBtn 按钮，Logcat 中 log 输出为：

```
21641 - 21641/cuc.activitylaunchmode I/MainActivity: onCreate: TaskId -- > 28
21641 - 21641/cuc.activitylaunchmode I/Main2Activity: onCreate: TaskId -- > 29
21641 - 21641/cuc.activitylaunchmode I/Main3Activity: onCreate: TaskId -- > 29
```

这时从 Main3Activity 单击 start2Activity 按钮，界面并不能切换到 Main2Activity，从这点可以看出，Intent 标志为 FLAG_ACTIVITY_NEW_TASK 跟启动模式与 singleTask 的行为还是有很大区别的。

5.13.5 launchMode 为 singleInstance 实验

具体步骤 1：在清单文件中把 Main2Activity 的 launchMode 设置为 singleInstance，去掉所有 Activity 的 android：taskAffinity 设置，Java 代码不要给 Intent 添加任何标志，这时清单文件内容如下：

```
<?xml version = "1.0" encoding = "utf - 8"?>
<manifest xmlns:android = "http://schemas.android.com/apk/res/android"
    package = "cuc.activitylaunchmode">
    <uses - permission android:name = "android.permission.GET_TASKS"/>
    <application
        android:allowBackup = "true"
        android:icon = "@mipmap/ic_launcher"
        android:label = "@string/app_name"
        android:roundIcon = "@mipmap/ic_launcher_round"
        android:supportsRtl = "true"
        android:theme = "@style/AppTheme">
        <activity android:name = ".MainActivity" >
            <intent - filter >
                <action android:name = "android.intent.action.MAIN" />
                <category android:name = "android.intent.category.LAUNCHER" />
            </intent - filter >
        </activity >
        <activity android:name = ".Main2Activity" android:launchMode = "singleInstance">
            <intent - filter >
                <action android:name = "cuc.activitylaunchmode.Main2Activity"/>
                <category android:name = "android.intent.category.DEFAULT"/>
            </intent - filter >
        </activity >
        <activity android:name = ".Main3Activity" />
    </application >
</manifest >
```

重新编译运行，顺序单击 MainActivity 中的 start2ActivityBtn 按钮、Main2Activity 中的 start2ActivityBtn 按钮、Main2Activity 中的 start3ActivityBtn 按钮、Main3Activity 中的 start2ActivityBtn 按钮，Main2Activity 中的 start1ActivityBtn 按钮、MainActivity 中的 start2ActivityBtn，Logcat 中 log 输出为：

```
8102 - 8102/cuc.activitylaunchmode I/MainActivity: onCreate: TaskId -- > 90
8102 - 8102/cuc.activitylaunchmode I/Main2Activity: onCreate: TaskId -- > 92
5820 - 5820/? I/MainActivity: windowMgr ========= FLOAT_WINDOWS + _SHOW
8102 - 8102/cuc.activitylaunchmode I/Main3Activity: onCreate: TaskId -- > 90
8102 - 8102/cuc.activitylaunchmode I/MainActivity: onCreate: TaskId -- > 90
```

上面实验中设置 Main2Activity 的 launchMode 为 singleInstance，任务栈与 Main2Activity 的 taskAffinity 无关。只要没有 Main2Activity 的实例，就创建新的任务，并在新的任务中创建 Main2Activity 的实例；若已经有了 Main2Activity 的实例，就把这个实例调到前台，不再创建 Main2Activity 新的实例；在 Main2Activity 中调用 startActivity(intent)方法创建其他的 Activity 实例并不启动中 Main2Activity 所在的任务中。

具体步骤 2：在清单文件中把 Main2Activity 的 launchMode 设置为 singleInstance，并把 Main3Activity 的 taskAffinity 改为 cuc.xyz，这时清单文件内容如下：

```
<?xml version="1.0" encoding="utf-8"?>
<manifest xmlns:android="http://schemas.android.com/apk/res/android"
    package="cuc.activitylaunchmode">
    <uses-permission android:name="android.permission.GET_TASKS"/>
    <application
        android:allowBackup="true"
        android:icon="@mipmap/ic_launcher"
        android:label="@string/app_name"
        android:roundIcon="@mipmap/ic_launcher_round"
        android:supportsRtl="true"
        android:theme="@style/AppTheme">
        <activity android:name=".MainActivity" >
            <intent-filter>
                <action android:name="android.intent.action.MAIN" />
                <category android:name="android.intent.category.LAUNCHER" />
            </intent-filter>
        </activity>
        <activity android:name=".Main2Activity" android:launchMode="singleInstance" >
            <intent-filter>
                <action android:name="cuc.activitylaunchmode.Main2Activity"/>
                <category android:name="android.intent.category.DEFAULT"/>
            </intent-filter>
        </activity>
        <activity android:name=".Main3Activity" android:taskAffinity="cuc.xyz" />
    </application>
</manifest>
```

重新编译运行，顺序单击 MainActivity 中的 start2ActivityBtn 按钮、Main2Activity 中的 start2ActivityBtn 按钮、Main2Activity 中的 start3ActivityBtn 按钮、Main3Activity 中的 start2ActivityBtn 按钮、Main2Activity 中的 start1ActivityBtn 按钮、MainActivity 中的 start2ActivityBtn，Logcat 中 log 输出为：

```
6880-6880/? I/MainActivity: onCreate: TaskId-->22
7185-7185/? I/MainActivity: windowMgr==========FLOAT_WINDOWS+_SHOW
6880-6880/cuc.activitylaunchmode I/Main2Activity: onCreate: TaskId-->23
6880-6880/cuc.activitylaunchmode I/Main3Activity: onCreate: TaskId-->24
6880-6880/cuc.activitylaunchmode I/MainActivity: onCreate: TaskId-->22
```

由于 Main3Activity 是被启动模式为 singleInstance 类型的 Activity(即 Main2Activity)启动

的，framework会为它加上FLAG_ACTIVITY_NEW_TASK标志，这时framework会检索是否已经存在了一个Main3Activity的taskAffinity指定的任务cuc.xyz：如果不存在Main3Activity的taskAffinity指定的任务cuc.xyz，就会创建taskAffinity为cuc.xyz任务，并且将Main3Activity启动到这个新的任务中。如果存在这样的一个任务，就检查在这个任务中是否已经有了一个Main3Activity的实例，这包括三种情况：①如果已经存在一个Main3Activity的实例，且Mian3Activity不在栈顶，则将这个任务调到前台，销毁Main3Activity上面的其他Activity，重用这个任务和任务中的Main3Activity实例，显示Main3Activity，并调用Main3Activity的onNewIntent()；②如果已经存在一个Main3Activity的实例，且Mian3Activity在栈顶，则将这个任务调到前台，重用这个任务和任务中的Main3Activity实例，显示Main3Activity，并调用Main3Activity的onNewIntent()；③如果不存在一个Mian3Activity的实例，会在这个任务中创建Main3Activity的实例，并调用onCreate()方法。

5.13.6 不同的App中相同的taskAffinity的singleTask模式实验

两个不同的App中的两个singleTask模式的Activity具有相同的taskAffinity，这两个Activity将运行在同一个任务中。

1. 新建工程ActivityLaunchMode2

具体步骤：运行Android Studio，单击Start a new Android Studio project，在Create New Project页中，Application name为ActivityLaunchMode2，Company domain为cuc，Project location为C:\android\myproject\ActivityLaunchMode2，单击Next按钮，选中Phone and Tablet，minimum sdk设置为API 19：Android 4.4(minSdkVersion)，单击Next按钮，选择EmptyActivity；再单击Next按钮，勾选Generate Layout File选项，勾选Backwards Compatibility(AppCompat)选项，单击Finish按钮。

2. 新建布局文件activity_main1.xml

layout/activity_main1.xml内容如下：

```
<?xml version="1.0" encoding="utf-8"?>
<LinearLayout xmlns:android="http://schemas.android.com/apk/res/android"
    android:orientation="vertical" android:layout_width="match_parent"
    android:layout_height="match_parent">
    <TextView
        android:layout_width="match_parent"
        android:layout_height="wrap_content"
        android:text="MainActivity"
        android:id="@+id/helloTxt"
        android:layout_margin="10dp"/>
    <Button
        android:layout_width="match_parent"
        android:layout_height="wrap_content"
        android:id="@+id/start3ActivityInAnotherAppBtn"
        android:text="start3ActivityInAnotherApp"
        android:layout_margin="10dp" />
    <Button
```

```
        android:layout_width = "match_parent"
        android:layout_height = "wrap_content"
        android:id = "@ + id/start2ActivityBtn"
        android:text = "start2Activity"
        android:layout_margin = "10dp" />
</LinearLayout>
```

3. 新建一个 Activity

新建一个 Activity，名字为 SecondActvity，SecondActvity 内容如下：

```
package cuc.activitylaunchmode2;
import android.content.ComponentName;
import android.content.Intent;
import android.os.Bundle;
import android.support.v7.app.AppCompatActivity;
import android.util.Log;
import android.view.View;
import android.widget.Button;
import android.widget.TextView;

public class SecondActivity extends AppCompatActivity implements View.OnClickListener{
    private static final String TAG = "SecondActivity";
    private TextView helloTxt;
    private Button start3ActivityInAnotherAppBtn;
    private Button start2ActivityBtn;
    private void assignViews() {
        helloTxt = (TextView) findViewById(R.id.helloTxt);
        start3ActivityInAnotherAppBtn = (Button) findViewById(R.id.start3ActivityInAnotherAppBtn);
        start2ActivityBtn = (Button) findViewById(R.id.start2ActivityBtn);
    }
    @Override
    protected void onCreate(Bundle savedInstanceState) {
        super.onCreate(savedInstanceState);
        setContentView(R.layout.activity_main1);
        Log.i(TAG, "onCreate: " + getTaskId());
        assignViews();
        helloTxt.setText("SecondActivity");
        start3ActivityInAnotherAppBtn.setOnClickListener(this);
        start2ActivityBtn.setOnClickListener(this);
    }
    @Override
    protected void onDestroy() {
        super.onDestroy();
        Log.i(TAG, "onDestroy: ");
    }
    @Override
    public void onClick(View view) {
        Intent intent = new Intent();
```

```
        switch (view.getId()) {
            case R.id.start3ActivityInAnotherAppBtn:
                ComponentName componentName = new ComponentName("cuc.activitylaunchmode","cuc.
activitylaunchmode.Main3Activity");
                intent.setComponent(componentName);
                intent.addCategory(Intent.CATEGORY_LAUNCHER);
                break;
            case R.id.start2ActivityBtn:
                intent.setClass(this, SecondActivity.class);
                break;
        }
        startActivity(intent);
    }
}
```

4. 编辑 MainActivity.java

MainActivity.java 内容如下：

```
package cuc.activitylaunchmode2;
import android.content.ComponentName;
import android.content.Intent;
import android.os.Bundle;
import android.support.v7.app.AppCompatActivity;
import android.util.Log;
import android.view.View;
import android.widget.Button;
import android.widget.TextView;

public class MainActivity extends AppCompatActivity implements View.OnClickListener {
    private static final String TAG = "MainActivity";
    private TextView helloTxt;
    private Button start3ActivityInAnotherAppBtn;
    private Button start2ActivityBtn;
    private void assignViews() {
        helloTxt = (TextView) findViewById(R.id.helloTxt);
        start3ActivityInAnotherAppBtn = (Button) findViewById(R.id.start3ActivityInAnotherAppBtn);
        start2ActivityBtn = (Button) findViewById(R.id.start2ActivityBtn);
    }
    @Override
    protected void onCreate(Bundle savedInstanceState) {
        super.onCreate(savedInstanceState);
        setContentView(R.layout.activity_main1);
        Log.i(TAG, "onCreate: " + getTaskId());
        assignViews();
        start3ActivityInAnotherAppBtn.setOnClickListener(this);
        start2ActivityBtn.setOnClickListener(this);
    }
    @Override
```

```
    protected void onDestroy() {
        super.onDestroy();
        Log.i(TAG, "onDestroy: ");
    }
    @Override
    public void onClick(View view) {
        Intent intent = new Intent();
        switch (view.getId()) {
            case R.id.start3ActivityInAnotherAppBtn:
                ComponentName componentName = new ComponentName("cuc.activitylaunchmode","cuc.
activitylaunchmode.Main3Activity");
                intent.setComponent(componentName);
                intent.addCategory(Intent.CATEGORY_LAUNCHER);
                break;
            case R.id.start2ActivityBtn:
                intent.setClass(this, SecondActivity.class);
                break;
        }
        startActivity(intent);
    }
}
```

5. 编辑清单文件

在清单文件中，定义 SecondActivity 的启动模式为 singleTask，android：taskAffinity 属性设置为 cuc.abc，这时清单文件内容如下：

```
<?xml version="1.0" encoding="utf-8"?>
<manifest xmlns:android="http://schemas.android.com/apk/res/android"
    package="cuc.activitylaunchmode2">
    <application
        android:allowBackup="true"
        android:icon="@android:drawable/ic_input_add"
        android:label="@string/app_name"
        android:roundIcon="@mipmap/ic_launcher_round"
        android:supportsRtl="true"
        android:theme="@style/AppTheme">
        <activity android:name=".MainActivity">
            <intent-filter>
                <action android:name="android.intent.action.MAIN" />
                <category android:name="android.intent.category.LAUNCHER"/>
            </intent-filter>
        </activity>
        <activity
            android:name=".SecondActivity"
            android:launchMode="singleTask"
            android:taskAffinity="cuc.abc" />
    </application>
</manifest>
```

6. 编译运行

单击 MainAcivity 的 start2ActivityBtn，Logcat 中 log 输出如下：

```
27391 - 27391/cuc.activitylaunchmode2 I/MainActivity: onCreate: 63
27391 - 27391/cuc.activitylaunchmode2 I/SecondActivity: onCreate: 65
```

再把前一个工程 ActivityLaunchMode 的清单文件做如下修改：

```
<?xml version = "1.0" encoding = "utf - 8"?>
< manifest xmlns:android = "http://schemas.android.com/apk/res/android"
    package = "cuc.activitylaunchmode">
    < uses - permission android:name = "android.permission.GET_TASKS"/>
    < application
        android:allowBackup = "true"
        android:icon = "@mipmap/ic_launcher"
        android:label = "@string/app_name"
        android:roundIcon = "@mipmap/ic_launcher_round"
        android:supportsRtl = "true"
        android:theme = "@style/AppTheme">
        < activity android:name = ".MainActivity" >
            < intent - filter >
                < action android:name = "android.intent.action.MAIN" />
                < category android:name = "android.intent.category.LAUNCHER" />
            </intent - filter >
        </activity >
        < activity android: name = ". Main2Activity" android: launchMode = " singleTask"
android:taskAffinity = "cuc.abc">
            < intent - filter >
                < action android:name = "cuc.activitylaunchmode.Main2Activity"/>
                < category android:name = "android.intent.category.DEFAULT"/>
            </intent - filter >
        </activity >
        < activity android:name = ".Main3Activity" />
    </application >
</manifest >
```

再编译运行，单击 MainAcivity 的 start2ActivityBtn 按钮，Logcat 中 log 输出如下：

```
28630 - 28630/cuc.activitylaunchmode I/MainActivity: onCreate: TaskId -- > 67
28630 - 28630/cuc.activitylaunchmode I/Main2Activity: onCreate: TaskId -- > 65
```

5.13.7 allowTaskReparenting = "true"实验

先把前一个应用 ActivityLaunchMode 的 Main3Activity 设置为 android: exported="true" android: allowTaskReparenting ="true"，ActivityLaunchMode 的清单文件内容如下：

```
<?xml version = "1.0" encoding = "utf - 8"?>
< manifest xmlns:android = "http://schemas.android.com/apk/res/android"
```

```
    package = "cuc.activitylaunchmode">
    < uses - permission android:name = "android.permission.GET_TASKS"/>
    < application
        android:allowBackup = "true"
        android:icon = "@mipmap/ic_launcher"
        android:label = "@string/app_name"
        android:roundIcon = "@mipmap/ic_launcher_round"
        android:supportsRtl = "true"
        android:theme = "@style/AppTheme">
        < activity android:name = ".MainActivity" >
            < intent - filter >
                < action android:name = "android.intent.action.MAIN" />

                < category android:name = "android.intent.category.LAUNCHER" />
            </intent - filter >
        </activity >
        < activity android:name = ".Main2Activity" >
            < intent - filter >
                < action android:name = "cuc.activitylaunchmode.Main2Activity"/>
                < category android:name = "android.intent.category.DEFAULT"/>
            </intent - filter >
        </activity >
         < activity android: name = ".Main3Activity" android: exported = "true" android:
allowTaskReparenting = "true" />
    </application >
</manifest >
```

单击 ActivityLaunchMode2 工程的 MainActivity 中的 start3ActivityInAnotherAppBtn，可以启动 ActivityLaunchMode(另一个应用)的 Main3Activity，这时按下手机的 Home 键切换到主屏幕，若这时启动另一个应用 ActivityLaunchMode，那么这个时候并不会启动 ActivityLaunchMode 的 MainActivity，而是直接显示已被应用 ActivityLaunchMode2 启动的 Main3Activity，也可以认为 Main3Activity 从 ActivityLaunchMode 的任务栈转移到了 ActivityLaunchMode2 的任务栈中，这是由于 Main3Activity 设置了属性 android:allowTaskReparenting＝"true"。

5.14 Fragment 实验

实验目的：

(1) 熟悉在布局资源文件中使用< fragment >元素的方法。

(2) 熟悉执行 Fragment 事务、并添加到 Fragment 回退栈的方法。

1. 新建工程 Fragments1

具体步骤：运行 Android Studio，单击 Start a new Android Studio project，在 Create New Project 页中，Application name 为 Fragments1，Company domain 为 cuc，Project location 为 C:\android\myproject\Fragments1，单击 Next 按钮，选中 Phone and Tablet，minimum sdk 设置为 API 19：Android 4.4(minSdkVersion)，单击 Next 按钮，选择

EmptyActivity，再单击 Next 按钮，勾选 Generate Layout File 选项，勾选 Backwards Compatibility(AppCompat)选项，单击 Finish 按钮。

2. 在 res\layout 新建 5 个 layout 文件

在工程结构视图中(单击 Project 按钮可以显示或者隐藏工程结构图，使用工程结构图中的下拉列表框可以选择 Android 以显示工程逻辑结构，或者选择 Project 显示工程实际的目录结构)，右击 layout，选择 New|Layout resource file，File name 输入框输入 fragment1，Root element 输入框输入 LinearLayout，Source set 为 main，单击 OK 按钮。layout/fragment1.xml 内容如下：

```
<?xml version = "1.0" encoding = "utf - 8"?>
<LinearLayout xmlns:android = "http://schemas.android.com/apk/res/android"
    android:background = "#800000"
    android:orientation = "horizontal" android:layout_width = "match_parent"
    android:layout_height = "match_parent">
    <TextView
        android:layout_gravity = "top"
        android:id = "@ + id/fragment1_Txt"
        android:layout_width = "match_parent"
        android:layout_height = "wrap_content"
        android:text = "fragemt1"
        android:textColor = "#ffffff"
        android:textSize = "20sp" />
</LinearLayout>
```

同样的方法创建另外 4 个布局文件。layout/fragment2.xml 内容如下：

```
<?xml version = "1.0" encoding = "utf - 8"?>
<LinearLayout xmlns:android = "http://schemas.android.com/apk/res/android"
    android:background = "#008000"
    android:orientation = "horizontal" android:layout_width = "match_parent"
    android:layout_height = "match_parent">
    <TextView
        android:layout_gravity = "center_vertical"
        android:id = "@ + id/fragment2_Txt"
        android:layout_width = "match_parent"
        android:layout_height = "wrap_content"
        android:text = "fragemt2"
        android:textColor = "#ffffff"
        android:textSize = "20sp"/>
</LinearLayout>
```

layout/fragment3.xml 内容如下：

```
<?xml version = "1.0" encoding = "utf - 8"?>
<LinearLayout xmlns:android = "http://schemas.android.com/apk/res/android"
    android:background = "#000080"
    android:orientation = "horizontal" android:layout_width = "match_parent"
```

```
    android:layout_height = "match_parent">
    <TextView
        android:layout_gravity = "bottom"
        android:id = "@ + id/fragmentb_Txt"
        android:layout_width = "match_parent"
        android:layout_height = "wrap_content"
        android:text = "fragemt3"
        android:textColor = "#ffffff"
        android:textSize = "20sp"/>
</LinearLayout>
```

在 fragment1. xml、fragment2. xml 和 fragment3. xml 中，根元素都采用了 LinearLayout，都设置了方向为水平，因而只能设置 LinearLayout 的子视图的垂直方向上的 android: layout_gravity 对齐方式。另外这 3 个 LinearLayout，背景颜色分别为红色、绿色和蓝色。这三个布局文件显示在手机屏幕的右半部分。

layout/fragmenta. xml 内容如下：

```
<?xml version = "1.0" encoding = "utf - 8"?>
<LinearLayout xmlns:android = "http://schemas.android.com/apk/res/android"
    android:orientation = "vertical" android:layout_width = "match_parent"
    android:layout_height = "match_parent">
    <Button
        android:id = "@ + id/addFrag1Btn"
        android:text = "add Fragment1"
        android:layout_width = "match_parent"
        android:layout_height = "wrap_content" />
    <Button
        android:id = "@ + id/addFrag2Btn"
        android:text = "add Fragment2"
        android:layout_width = "match_parent"
        android:layout_height = "wrap_content" />
    <Button
        android:id = "@ + id/addFrag3Btn"
        android:text = "add Fragment3"
        android:layout_width = "match_parent"
        android:layout_height = "wrap_content" />
    <Button
        android:id = "@ + id/removeFrag1Btn"
        android:text = "remove Fragment1"
        android:layout_width = "match_parent"
        android:layout_height = "wrap_content" />
    <Button
        android:id = "@ + id/removeFrag2Btn"
        android:text = "remove Fragment2"
        android:layout_width = "match_parent"
        android:layout_height = "wrap_content" />
    <Button
        android:id = "@ + id/removeFrag3Btn"
```

```
            android:text = "remove Fragment3"
            android:layout_width = "match_parent"
            android:layout_height = "wrap_content" />
        <Button
            android:id = "@ + id/replaceFrag1Btn"
            android:text = "replace Fragment1"
            android:layout_width = "match_parent"
            android:layout_height = "wrap_content" />
        <Button
            android:id = "@ + id/replaceFrag2Btn"
            android:text = "replace Fragment2"
            android:layout_width = "match_parent"
            android:layout_height = "wrap_content" />
        <Button
            android:id = "@ + id/replaceFrag3Btn"
            android:text = "replace Fragment3"
            android:layout_width = "match_parent"
            android:layout_height = "wrap_content" />
</LinearLayout>
```

fragmenta.xml 显示在手机屏幕的左半部分，单击这个布局中的按钮，能够改变手机屏幕的右半部分显示的内容。

layout/activity_main1.xml 内容如下：

```
<?xml version = "1.0" encoding = "utf - 8"?>
<LinearLayout xmlns:android = "http://schemas.android.com/apk/res/android"
        android:orientation = "horizontal" android:layout_width = "match_parent"
        android:layout_height = "match_parent">
        <fragment
            android:layout_width = "0dip"
            android:layout_height = "match_parent"
            android:layout_weight = "1"
            android:id = "@ + id/fragmenta"
            android:name = "cuc.fragments1.FragmentaFragment"/>
        <FrameLayout
            android:id = "@ + id/fragment_container"
            android:layout_weight = "1"
            android:layout_width = "0dp"
            android:layout_height = "match_parent">
        </FrameLayout>
</LinearLayout>
```

activity_main1.xml 用作 Activity 的布局，根元素采用了 LinearLayout，并且设置了方向为水平。LinearLayout 左边放置了 fragmenta.xml，还放置了一个 FrameLayout，用作 Fragment1Fragment、Fragment2Fragment 和 Fragment3Fragment 的容器。<fragment>元素必须添加 ID 属性，否则 Android 系统不能根据 Java 反射机制创建 Fragment 对象。

3. 由布局文件生成 Fragment 类

从布局文件生成 Fragment 类使用了 Android Code Generator 插件，可以通过 Android Studio 菜单中的 File|Settings|Plugins 查看该插件是否已经安装，若没有安装，请参看 1.5 Android Studio 常用插件的安装。

具体步骤：在 Android 工程结构视图中，选中布局文件 fragment1.xml，右击，在弹出菜单中选中 generate Android Code|Fragment，Source path 选择\app\src\main\java，包名采用默认的 cuc.fragments1，单击 Create File 按钮，就生成了文件 Fragment1Fragment.java，文件内容如下：

```
package cuc.fragments1;
import android.support.v4.app.Fragment;
import android.view.View;
import android.view.LayoutInflater;
import android.view.ViewGroup;
import android.os.Bundle;
import android.widget.TextView;
public class Fragment1Fragment extends Fragment {
    private TextView fragment1Txt;
    @Override
    public View onCreateView(LayoutInflater inflater, ViewGroup container,
                             Bundle savedInstanceState) {
        return inflater.inflate(R.layout.fragment1, null);
    }
    @Override
    public void onViewCreated(View view, Bundle savedInstanceState) {
        super.onViewCreated(view, savedInstanceState);
        fragment1Txt = (TextView) view.findViewById(R.id.fragment1_Txt);
    }
}
```

同样的方法产生跟 fragment2.xml 对应的 fragment 类 Fragment2Fragment.java，生成的文件 Fragment2Fragment.java 内容如下：

```
package cuc.fragments1;
import android.support.v4.app.Fragment;
import android.view.View;
import android.view.LayoutInflater;
import android.view.ViewGroup;
import android.os.Bundle;
import android.widget.TextView;
public class Fragment2Fragment extends Fragment {
    private TextView fragment2Txt;
    @Override
    public View onCreateView(LayoutInflater inflater, ViewGroup container,
                             Bundle savedInstanceState) {
        return inflater.inflate(R.layout.fragment2, null);
    }
```

```
    @Override
    public void onViewCreated(View view, Bundle savedInstanceState) {
        super.onViewCreated(view, savedInstanceState);
        fragment2Txt = (TextView) view.findViewById(R.id.fragment2_Txt);
    }
}
```

同样的方法产生跟 fragment3.xml 对应的 fragment 类 Fragment3Fragment.java，生成的文件 Fragment3Fragment.java 内容如下：

```
package cuc.fragments1;
import android.support.v4.app.Fragment;
import android.view.View;
import android.view.LayoutInflater;
import android.view.ViewGroup;
import android.os.Bundle;
import android.widget.TextView;
public class Fragment3Fragment extends Fragment {
    private TextView fragmentbTxt;
    @Override
    public View onCreateView(LayoutInflater inflater, ViewGroup container,
                             Bundle savedInstanceState) {
        return inflater.inflate(R.layout.fragment3, null);
    }
    @Override
    public void onViewCreated(View view, Bundle savedInstanceState) {
        super.onViewCreated(view, savedInstanceState);
        fragmentbTxt = (TextView) view.findViewById(R.id.fragmentb_Txt);
    }
}
```

同样的方法产生跟 fragmenta.xml 对应的 fragment 类 FragmentaFragment.java，并添加按键的监听方法，最后文件 FragmentaFragment.java 内容如下：

```
package cuc.fragments1;
import android.os.Bundle;
import android.support.v4.app.Fragment;
import android.view.LayoutInflater;
import android.view.View;
import android.view.ViewGroup;

public class FragmentaFragment extends Fragment implements View.OnClickListener {
        private Fragment1Fragment fragment1;
        private Fragment2Fragment fragment2;
        private Fragment3Fragment fragment3;

    @Override
    public View onCreateView(LayoutInflater inflater, ViewGroup container,
```

```
                    Bundle savedInstanceState) {
        return inflater.inflate(R.layout.fragmenta, null);
    }

    @Override
    public void onViewCreated(View view, Bundle savedInstanceState) {
        super.onViewCreated(view, savedInstanceState);
        view.findViewById(R.id.addFrag1Btn).setOnClickListener(this);
        view.findViewById(R.id.addFrag2Btn).setOnClickListener(this);
        view.findViewById(R.id.addFrag3Btn).setOnClickListener(this);
        view.findViewById(R.id.removeFrag1Btn).setOnClickListener(this);
        view.findViewById(R.id.removeFrag2Btn).setOnClickListener(this);
        view.findViewById(R.id.removeFrag3Btn).setOnClickListener(this);
        view.findViewById(R.id.replaceFrag1Btn).setOnClickListener(this);
        view.findViewById(R.id.replaceFrag2Btn).setOnClickListener(this);
        view.findViewById(R.id.replaceFrag3Btn).setOnClickListener(this);
        fragment1 = new Fragment1Fragment();
        fragment2  = new Fragment2Fragment();
        fragment3 = new Fragment3Fragment();
    }
    @Override
    public void onClick(View view) {
         android.support.v4.app.FragmentTransaction transaction = getFragmentManager().
beginTransaction();
        switch (view.getId()) {
            case R.id.addFrag1Btn:
                if(!fragment1.isAdded()){
                    transaction.add(R.id.fragment_container,fragment1,"frag1");
                }
                break;
            case R.id.addFrag2Btn:
                if (!fragment2Fragment.isAdded()){
                  transaction.add(R.id.fragment_container,fragment2,"frag2");
                }
                break;              case R.id.addFrag3Btn:
                if (!fragment3Fragment.isAdded()){
                    transaction.add(R.id.fragment_container,fragment3,"frag3");
                }
                break;
            case R.id.removeFrag1Btn:
                transaction.remove(fragment1);
                break;
            case R.id.removeFrag2Btn:
                transaction.remove(fragment2);
                break;
            case R.id.removeFrag3Btn:
                transaction.remove(fragment3);
                break;
            case R.id.replaceFrag1Btn:
```

```
            transaction.replace(R.id.fragment_container,fragment1,"frag1");
                    break;
                case R.id.replaceFrag2Btn:
        transaction.replace(R.id.fragment_container,fragment2,"frag2");
                    break;
                case R.id.replaceFrag3Btn:
        transaction.replace(R.id.fragment_container,fragment3,"frag3");
                    break;
            }
            transaction.addToBackStack(null);
            transaction.commit();
        }
    }
```

4. 编辑 MainActivity.java

MainActivity.java 内容如下：

```
package cuc.fragments1;
import android.support.v7.app.AppCompatActivity;
import android.os.Bundle;
public class MainActivity extends AppCompatActivity {
    @Override
    protected void onCreate(Bundle savedInstanceState) {
        super.onCreate(savedInstanceState);
        setContentView(R.layout.activity_main1);
    }
}
```

5. 编译运行

连接手机，单击工具栏上的绿色箭头，或者按 Shift+F10 键编译运行。运行后可以发现：FragmentTransaction 事务不能多次添加同一个 Fragment 对象，在采用 Remove 方法移除后可以再次添加同一个 Fragment 对象。

5.15 本章主要参考文献

1. https://developer.android.google.cn/guide/components/activities.html
2. https://developer.android.google.cn/reference/android/content/Intent
3. https://developer.android.google.cn/guide/topics/manifest/activity-element.html
4. https://developer.android.google.cn/guide/components/tasks-and-back-stack.htm
5. https://developer.android.google.cn/guide/components/fragments.html
6. https://developer.android.google.cn/reference/android/support/v4/app/Fragment
7. https://developer.android.google.cn/reference/android/app/Fragment.html
8. https://developer.android.google.cn /training/basics/fragments/creating.html
9. https://developer.android.google.cn/topic/libraries/architecture/livedata

第 6 章 Intent 和 IntentFilter

CHAPTER 6

Intent 提供了一种通用的消息系统，它允许数据在不同的应用程序间传递。一个 Intent 类型的对象是对一次操作的抽象描述，使用 Intent 对象向其他的应用组件请求操作。使用 Intent 可以激活 Android 应用的三个核心组件：Activity、Service 和 BroadcastReceiver。通过将 Intent 对象传递给 startActivity()，可以启动新的 Activity 实例。如果希望在 Activity 完成后收到结果，可以调用 startActivityForResult()，在 Activity 的 onActivityResult() 回调方法中，Activity 将接收用于传递结果的 Intent 对象。通过把 Intent 对象传递给 startService()，可以启动 Service 执行一次操作(例如，下载文件)。如果 Service 是使用客户端/服务器接口设计的，则通过把 Intent 对象传递给 bindService()，从而绑定到此 Service。通过将 Intent 传递给 sendBroadcast()、sendOrderedBroadcast()，从而把广播传递给其他应用。

6.1 Intent 对象的主要信息

一个 Intent 对象主要包含 6 类信息：ComponentName、Action、Category、Data、Extras 和 Flags。下面是 Intent 类的源代码，摘自于 sdk\sources\android-24\android\content\intent.java。

```
public class Intent implements Parcelable, Cloneable {
...
private String mAction;                              //动作
    private Uri mData;                               //Uri
    private String mType;                            //MIME 类型
    private String mPackage;
    private ComponentName mComponent;                //被启动的组件名称
    private int mFlags;
    private ArraySet < String > mCategories;         //种类
    private Bundle mExtras;                          //普通数据
    private Rect mSourceBounds;
    private Intent mSelector;
private ClipData mClipData;
...
}
```

1. 被启动的组件名称

Intent 的成员变量 mComponent 是一个 ComponentName 类型的对象，可以通过完全限定类名指定目标组件，例如 com. example. ExampleActivity。可以调用 Intent 对象的 setComponent()、setClass()、setClassName()方法或 Intent 构造函数设置组件名称。

ComponentName 是可选项，但它是构建显式 Intent 的一项必要信息，这意味着 Intent 将传递到由组件名称定义的应用组件。如果没有定义组件名称，则 Intent 是隐式的，且系统将根据其他 Intent 信息(例如 action、category、data)决定哪个组件应当接收 Intent。因此，如需在应用中启动指定的组件，则应指定待启动组件的 ComponentName。

应该使用显式 Intent 启动 Service，即启动 Service 时指定组件名称。否则，无法确定哪个 Service 会响应 Intent，且用户无法看到哪个 Service 已经被启动。

2. Action

Action 字符串用于指定要执行的操作，Action 在很大程度上决定了 Intent 的构成，可以调用 Intent 对象的 setAction()方法或 Intent 构造函数为 Intent 指定 action。开发人员可以定义自己的 action，自己定义 action 时，请确保将应用的软件包名称作为前缀。例如：

```
static final String ACTION_TIMETRAVEL = "com.example.action.TIMETRAVEL";
```

但是，通常应该使用在 Intent 类或其他框架类定义的 action 常量。例如，Settings 中定义了系统设置应用中打开特定屏幕的 action。如果有一些信息供某个 Activity 向用户显示(比如使用地图应用查看地址)时，请把具有 ACTION_VIEW 的 Intent 传递给 startActivity()作为参数。如果有一些可通过其他应用(例如，电子邮件应用或社交共享应用)共享的数据，请把具有 ACTION_SEND 的 Intent 传递给 startActivity()作为参数。

3. Category

Category 是一个包含组件类型的附加信息的字符串。可以调用 Intent 对象的 addCategory()方法把任意数量的 category 描述添加到一个 Intent 对象中，但大多数 Intent 均不需要 category。Intent 类中定义了大量的 category。例如：

CATEGORY_BROWSABL：目标 Activity 允许本身通过网络浏览器启动，以显示链接指向的数据，如图像或电子邮件。

CATEGORY_LAUNCHER：该 Activity 在系统的应用启动器中列出，是任务的初始 Activity。

4. URI 及 MIME 类型

URI 类型的 mData 是指向待操作数据的 URI。String 类型的 mType 是待操作数据的 MIME 类型，通常由 Intent 对象的 Action 决定。例如，如果操作是 ACTION_EDIT，则数据应包含待编辑文档的 URI。

创建 Intent 时，除了指定 URI 以外，指定 MIME 数据类型往往也很重要。调用 setData()仅仅设置 URI，调用 setType()仅仅设置 MIME 类型，调用 setDataAndType()可以同时设置 URI 和 MIME 类型。若要同时设置 URI 和 MIME 类型，请不要分别调用 setData()和 setType()，因为它们都会使对方的取值无效，请始终调用 setDataAndType()同时设置 URI 和 MIME 类型。

组件名称、action、category、URI 和 MIME 类型这些属性用于 Android 系统解析出哪个应用组件应当被启动。但是，Intent 也有可能会携带一些信息，不是用于限定哪些组件被启动。例如 Intent 还可以提供：Bundle 类型的 mExtras 和 int 类型的 mFlags。

5. 标志

int 类型的 mFlags 指示 Android 系统如何启动 Activity(例如,Activity 应属于哪个任务),以及启动之后如何处理(例如,它是否属于最近的 Activity 列表)。调用 Intent 对象的 setFlags()方法设置标志。

6. Bundle 类型的 mExtras

Bundle 类型的 mExtras 是携带附加信息的键值对。正如某些 action 使用特定类型的数据 URI 一样,某些 action 也使用特定的 mExtras。可以调用 Intent 对象的各种 putExtra()方法向 mExtras 添加数据,每种方法均接受两个参数:键名和值。也可以先创建一个 Bundle 类型对象,再往该 Bundle 对象添加数据,最后使用 putExtras()将 Bundle 插入 Intent 中。

Intent 类将为标准化的数据类型指定多个 EXTRA_ * 常量。例如,使用 ACTION_SEND 创建用于发送电子邮件的 Intent 时,可以使用 EXTRA_EMAIL 键指定目标收件人,并使用 EXTRA_SUBJECT 键指定主题。如需声明自己的 extra 键,请确保使用应用的软件包名称作为前缀。例如:

```
static final String EXTRA_GIGAWATTS = "com.example.EXTRA_GIGAWATTS";
```

6.2 Intent 传递对象的两种方法

数据序列化主要有实现 Serializable 接口和 Parcelable 接口两种实现方式。类实现 Serializable 接口,只需要在声明类的时候,在类名后添加 implements Serializable 就可以了。如需创建实现 Parcelable 接口的类,必须执行以下操作:①让类实现 Parcelable 接口;②实现 writeToParcel,它会获取对象的当前状态并将其写入 Parcel;③给类添加一个名为 CREATOR 的静态字段,这个字段是一个实现 Parcelable.Creator 接口的对象。可以使用 Android Studio 的 Live Templates 实现 Parcelable 接口。具体方法是:单击 Android Studio 的菜单 File | Settings | Editor | Live Templates,勾选 AndroidParcelable,启用 AndroidParcelable 模板。在类的实现里面完整地输入 Parcelable 后,就可以看见 Create a parcelable block for your current class 的提示,按回车键,为当前的类创建一个 Parcelable 模块,生成的 writeToParcel(@NonNull Parcel dest, int flags)方法是空的,另外生成的以 Parcel 作为参数的构造函数也是空的。还可以使用插件 android parcelable code generator 自动实现 Parcelable 接口,该插件的使用方法如下:新建一个实体类,在声明类的时候,在类名后添加 implements Parcelable,添加实例变量后,光标放在类的实现里的任何位置按下 Alt+Insert 键,选择 Palcelable 选项,选择所有的实例变量,单击 OK 按钮,这个类就实现了 Parcelable 接口。

Java 提供了一种对象序列化的机制:实现 Serializable 接口。一个对象可以被表示为一个字节序列,该字节序列包括该对象的数据、有关对象的类型的信息和存储在对象中数据的类型。将序列化对象写入文件之后,可以从文件中读取出来,并且对它进行反序列化,也就是说,对象的类型信息、对象的数据,还有对象中的数据类型可以用来在内存中新建对象。

在 Android 系统中,所有需要跨进程传递的复杂数据对象都必须实现 Parceable 接口。

实现 Parcelable 接口是为了让对象序列化，以进行进程间的数据传递。Parcelable 是为程序内不同组件之间以及不同 Android 程序间(AIDL)高效的传输数据而设计，这些数据仅在内存中存在。Parcelable 使用时要用到一个 Parcel，可以简单地把 Parcel 看成一个容器，序列化时将数据写入 Parcel，反序列化时从中读出。与 Serializable 相比，Parcelable 更高效、内存开销更小，在内存间数据传输时推荐使用 Parcelable，Parcelable 支持 Intent 的数据传递，也支持进程间通信(IPC)。

Serializable 可用于保存对象的属性到本地文件、数据库或者网络传输，这种场合不推荐使用 Parcelable 进行数据持久化，因为 Android 不同版本 Parcelable 可能不同，所以在需要保存对象的属性到本地文件和数据库或者网络传输时应该选择 Serializable。

Intent 通过 Bundle 传递数据，Bundle 类实现了 Parcelable 接口，Bundle 类的继承关系如下：

```
public final class Bundle extends BaseBundle implements Cloneable, Parcelable
```

Intent 中 Bundle 类型的 mExtras 可以携带附加信息，形式为键值对，值可以是简单类型，也可以是对象。Intent 通过 Bundle 传递复杂数据对象主要有两种方法，一种是传送实现了 Parcelable 接口的对象，调用 Bundle. putParcelable(String key，Parcelable value)或者 Intent. putExtra(String name，Parcelable value)写入对象；另一种是传送实现了 Serializable 接口的对象，调用 Bundle. putSerializable(String key，Serializable value)或者 Intent. putExtra (String name，Serializable value)写入对象。Intent 使用 Parcelable 传递复杂数据对象更加高效。

6.3　显式 Intent 和隐式 Intent

Intent 对象可以分为两种类型：显式 Intent 和隐式 Intent。如果一个 Intent 明确指定了要启动的组件的完全限定类名，那么这个 Intent 就是显式 Intent，否则就是隐式 Intent。当使用显式 Intent 去启动组件时，系统会根据 Intent 对象所提供的 ComponentName 直接找到要启动的组件。如果知道要启动的 Activity 或 Service 的类名，应该使用显式 Intent 来启动组件。

显式 Intent 用于启动某个指定应用组件(例如，应用中的某个特定 Activity 或 Service)。要创建显式 Intent，请为 Intent 对象定义全限定的组件名称，显式 Intent 始终会传递给其类名指定的目标组件，无论该组件是否声明了 IntentFilter。例如，如果在应用中构建了一个名为 DownloadService、旨在从网页下载文件的 Service，则可使用以下代码启动该服务：

```
String fileUrl = "http://www.example.com/image.png";
Intent intent = new Intent(this, DownloadService.class);
intent.setData(Uri.parse(fileUrl));
startService(intent);
```

创建显式 Intent 启动 Activity 或 Service 时，系统将立即启动 Intent 对象中指定的应用组件。创建隐式 Intent 时，Android 系统通过把 Intent 对象的内容与系统中已经注册的组件的 IntentFilter 进行比较，从而找到要启动的相应组件。如果只有一个 IntentFilter 匹配，则系统将启动该组件，并向其传递 Intent 对象。如果有多个 IntentFilter 匹配，则系统会显示一个对话框，支持用户选取要使用的应用。图 6-1 说明了 Activity A 通过隐式 Intent 启动 Activity B 的过程。

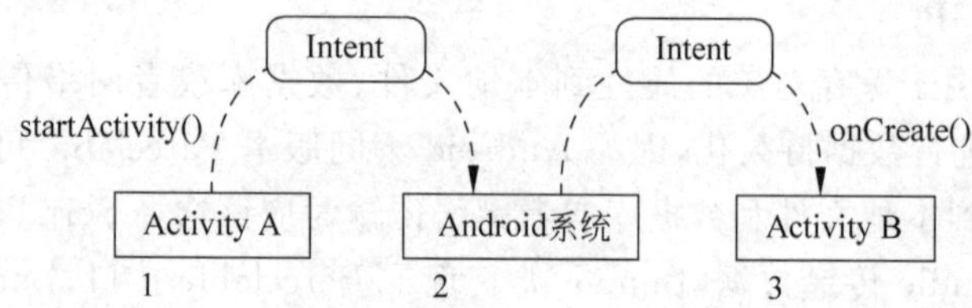

图 6-1　隐式 Intent 如何通过系统传递以启动其他 Activity 的图解

Activity A 通过隐式 Intent 启动 Activity B 的过程如下：①在 Activity A 创建包含 action 的 Intent 对象，并将其传递给 startActivity()；②Android 系统搜索所有应用中与 Intent 对象匹配的 IntentFilter；③找到匹配项之后，系统调用匹配的 Activity(即 Activity B)的 onCreate()方法并将 Intent 对象传递给它。

当 Android 系统接收到一个隐式 Intent 要启动一个 Activity(或其他组件)时，Android 会根据 Intent 对象中的 action、category、Uri 以及 MIME 类型这四个信息和已经注册的组件的 IntentFilter 比较，即要分别通过 action 测试、category 测试以及 data 测试。如果隐式 Intent 对象同时通过了某个 IntentFilter 的 action 测试、category 测试以及 data 测试，这个 Intent 对象就可以匹配该 IntentFilter。由于一个组件可有任意数量的 IntentFilter，其中每个 IntentFilter 描述一种不同的功能，一个隐式 Intent 对象只要通过了某个组件的任何一个 IntentFilter 的测试，这个 Intent 对象就能启动该组件。如果隐式 Intent 对象没有通过系统中任何组件的 IntentFilter 测试，Android 系统就无法找到该 Intent 对象要启动的组件。

使用 Intent 过滤器时，无法安全地防止其他应用启动组件。如果必须确保某一组件只能被同一个应用内的组件启动，可以把该组件的 exported 属性设置为 false。

在 AndroidManifest. xml 文件中，组件内的每个< intent-filter >元素代表一个 IntentFilter。关于 intent-filter 元素的语法，请参看 2. 8intent-filter。

6.4　接收隐式 Intent

应用组件应当为自身可执行的每个功能声明单独的 IntentFilter，也就是说，一个应用组件可以有一个或者多个 IntentFilter，每个 IntentFilter 描述一个功能。只要隐式 Intent 可以匹配某个应用组件的任何一个 IntentFilter 系统就会将该 Intent 传递给该应用组件。例如，图像应用中的一个 Activity 可能会有两个 IntentFilter，分别用于查看图像和编辑图像。当 Activity 启动时，它将根据 Intent 对象中的信息决定具体的行为(例如是否显示编辑器控件)。

为了避免无意中运行其他应用的 Service，请始终使用显式 Intent 启动应用内的 Service，且不必为该 Service 声明 IntentFilter。可以通过调用 registerReceiver(BroadcastReceiver receiver, IntentFilter filter)方法注册 BroadcastReceiver 和它的 IntentFilter。调用 unregisterReceiver

(BroadcastReceiver receiver)注销该接收器。这样,应用仅在动态注册到注销的这段时间内侦听特定的广播。对于所有 Activity,为了接收隐式 Intent,必须在清单文件中声明 IntentFilter。系统认为传递给 startActivity()和 startActivityForResult()的所有 Intent 对象都已经申明 CATEGORY_DEFAULT 类别,因此如果没有在 IntentFilter 中声明此类别,则隐式 Intent 不会传递给 Activity。为了接收隐式 Intent,必须将 CATEGORY_DEFAULT 类别包括在 < activity >元素内的< intent-filter >元素中。

隐式 Intent 若要匹配某个 IntentFilter,必须通过 action、data 和 category 这三项测试。如果 Intent 不能通过任何一项匹配测试,则 Android 系统不会将 Intent 传递给组件。但是,由于一个组件可能有多个 IntentFilter,因此未能通过某一 IntentFilter 的 Intent 可能会通过另一 IntentFilter。下面介绍隐式 Intent 对象和 IntentFilter 的匹配规则。

6.4.1 Action 测试

一个 IntentFilter 应该声明最少支持 1 个 action。每个隐式 Intent 对象必须设置唯一的一个 action 值,否则该 Intent 对象不能通过任何 IntentFilter 的 action 测试。只有 Intent 对象设置了唯一的 action,并且组件的 IntentFilter 中包含了 Intent 对象指定的 action,action 测试才能通过。如果 IntentFilter 没有声明任何 action 或者或者 Intent 对象没有设置 Action,Intent 对象就无法通过 IntentFilter 的 action 测试。可以通过 Intent 构造函数或者调用 Intent 对象的 setAction()方法为 Intent 对象设置唯一的 action。

实例：Intent 对象内容如下：

```
Intent intent = new Intent();
intent.setAction("cuc.intent1.action.Test1");
intent.addCategory("cuc.intent1.category.Test1");
Uri uri = Uri.parse("http://www.sina.com.cn/");
intent.setData(uri);
```

上例中的 Intent 对象可以通过下面的 intent-filter 里面的 action 测试。该 Intent 能通过 action 测试是因为 IntentFilter 中包含该 Intent 的 action 值 cuc.intent1.action.Test1。

```
<intent-filter>
    <action android:name = "cuc.intent1.action.Test1"/>
    <category android:name = "android.intent.category.DEFAULT"/>
    <category android:name = "cuc.intent1.category.Test1"/>
    <data android:scheme = "http" android:host = "www.sina.com.cn" android:path = "/" />
</intent-filter>
```

下面的 Intent 对象无法通过上面 IntentFilter 里面的 action 测试：

```
Intent intent = new Intent();
intent.setAction("cuc.intent1.action.Test2");
intent.addCategory("cuc.intent1.category.Test1");
Uri uri = Uri.parse("http://www.sina.com.cn/");
intent.setData(uri);
```

该 Intent 对象无法通过 action 测试，是因为 intent-filter 中不包含该 Intent 对象的 action 值 cuc.intent1.action.Test2。

6.4.2 category 测试

一个 IntentFilter 可以声明其支持 0 个或多个 category。Intent 对象可以 0 次或者多次调用 addCategory()方法添加 category，也就是说一个 Intent 对象可以关联 0 个或多个 category。Intent 对象的所有 category 都要在 IntentFilter 中找到对应项，也就是说，IntentFilter 中必须要包含 Intent 对象中的所有 category(即 IntentFilter 中声明的 category 可以超出 Intent 中指定的数量)，Intent 对象才能通过 category 测试，否则无法通过测试。特别地，当一个隐式的 Intent 对象传递给 startActivity()或 startActivityForResult()方法用于启动 Activity 时，Android 系统会自动给该隐式 Intent 添加值为 android.intent.category.DEFAULT 的 category，所以为了能让 IntentFilter 包含 Intent 对象中全部的 category，Activity 的 IntentFilter 还必须额外包含值为 android.intent.category.DEFAULT 的 category。如果 Intent 对象不包含任何 category，并且该 Intent 不是用来启动 Activity 的，那么该 Intent 对象总是能通过任意的 IntentFilter 的 category 测试。

如果 Intent 对象没有调用过 addCategory()方法，那么 Intent 对象就不包含任何的 category。这种情形下，如果该 Intent 对象不是用来启动 Activity 的话，那么无论 IntentFilter 中有没有 category，Intent 都能通过 category 测试。

一个实例：

```
    Intent intent = new Intent();
    intent.setAction("cuc.intent1.action.Test1");
    intent.addCategory("cuc.intent1.category.Test1");
intent.addCategory("cuc.intent1.category.Test2");
     Uri uri = Uri.parse("http://www.sina.com.cn/");
    intent.setData(uri);
```

上例中的 Intent 对象可以通过下面的 intent-filter 里面的 category 测试。

```
<intent-filter>
    <action android:name = "cuc.intent1.action.Test1"/>
    <action android:name = "cuc.intent1.action.Test1A"/>
    <category android:name = "android.intent.category.DEFAULT"/>
    <category android:name = "cuc.intent1.category.Test1"/>
<category android:name = "cuc.intent1.category.Test2"/>
    <data android:scheme = "http" android:host = "www.sina.com.cn" android:path = "/" />
</intent-filter>
```

这里需要特别说明的是，在上面的实例中，都给 Activity 的 intent-filter 元素添加了值为 android.intent.category.DEFAULT 的 category。上面的 Intent 对象之所以可以通过 category 测试是因为 intent-filter 包含了该 Intent 对象中所有的 category 值：cuc.intent1.category.Test1 和 cuc.intent1.category.Test2，还包括 startActivity()时系统给 Intent 添加的 android.intent.category.DEFAULT。

6.4.3 data 测试

URI 匹配规则如下：

如果 IntentFilter 仅指定 scheme，则具有该 scheme 的所有 URI 均与该过滤器匹配。

如果 IntentFilter 指定 scheme 和 host，但未指定 path，则具有相同 scheme 和 host 的所有 URI 都会通过过滤器，无论其路径如何均是如此。

如果 IntentFilter 指定了 scheme、host 和 path，则仅具有相同 scheme、host 和 path 的 URI 才会通过过滤器。IntentFilter 的 path 可以包含星号通配符（*），一个星号匹配它前面的字符的 0 次或者多次出现。

data 测试会将 Intent 中的 URI 和 MIME 类型与 IntentFilter 的 URI 和 mimeType 进行比较。包含以下 4 种情况：

1. Intent 对象既不包含 URI 又不包含 MIME 类型

对于不含 URI 和 MIME 类型的 Intent 对象，仅当 IntentFilter 既没有指定任何 URI，也没有指定任何 MIME 类型时，才能通过测试。

2. Intent 对象包含 URI 但不包含 MIME 类型

对于包含 URI 但不含 MIME 类型(既没有显式声明，也无法通过 URI 推断得出 MIME 类型)的 Intent 对象，只有当 intent-filter 的所有 data 元素都没有指定 MIME 类型(即整个 intent-filter 中完全没有 mimeType 字样)，并且 intent-filter 中的其中一个 data 标签的 URI 格式能够匹配 Intent 对象的 URI，才能通过 data 测试。大家从下面的几个实例中可以体会到这点。

实例：Intent 对象内容如下：

```
Intent intent = new Intent();
intent.setAction("cuc.intent1.action.Test1");
intent.addCategory("cuc.intent1.category.Test1");
intent.addCategory("cuc.intent1.category.Test2");
Uri uri = Uri.parse("http://www.sina.com.cn/");
intent.setData(uri);
```

上面的 Intent 对象也可以通过下面的 intent-filter 的 data 测试：

```
<intent-filter>
    <action android:name = "cuc.intent1.action.Test1"/>
    <action android:name = "cuc.intent1.action.Test1A"/>
    <category android:name = "android.intent.category.DEFAULT"/>
    <category android:name = "cuc.intent1.category.Test1"/>
    <category android:name = "cuc.intent1.category.Test2"/>
    <data android:scheme = "http" android:host = "www.sina.com.cn" android:path = "/" />
</intent-filter>
```

上面的 Intent 对象也可以通过下的 intcnt-filtcr 的 data 测试：

```
<intent-filter>
    <action android:name = "cuc.intent1.action.Test1"/>
```

```
        <action android:name = "cuc.intent1.action.Test1A"/>
        <category android:name = "android.intent.category.DEFAULT"/>
        <category android:name = "cuc.intent1.category.Test1"/>
        <category android:name = "cuc.intent1.category.Test2"/>
        <data android:scheme = "http" android:host = "www.sina.com.cn" android:path = "/" />
        <data android:scheme = "https" android:host = "www.baidu.com" />
    </intent-filter>
```

上面的Intent对象虽然不能通过host为www.baidu.com的URI测试，但是可以通过<data android: scheme="http"android:host="www.sina.com.cn"android: path="/"/>的URI测试。

上面的Intent对象无法通过下面的intent-filter中的data测试，是因为Intent对象没有设置MIME类型，但是intent-filter的第二个data标签设置了MIME类型，即不是所有的data标签都没有指定MIME类型，因而不能通过data测试。

```
    <intent-filter>
        <action android:name = "cuc.intent1.action.Test1"/>
        <action android:name = "cuc.intent1.action.Test1A"/>
        <category android:name = "android.intent.category.DEFAULT"/>
        <category android:name = "cuc.intent1.category.Test1"/>
        <category android:name = "cuc.intent1.category.Test2"/>
        <data android:scheme = "http" android:host = "www.sina.com.cn" android:path = "/" />
        <data android:scheme = "https" android:host = "www.baidu.com" android:mimeType = "text/
plain"/>
    </intent-filter>
```

3. Intent对象包含MIME类型但不包含URI

这种情况下，只有当intent-filter中的所有data标签都没有指定任何格式的URI(即整个intent-filter中完全没有android: scheme、android: host、android: port以及android: path字样)、并且intent-filter中的其中一个data标签的MIME类型与Intent对象的MIME类型匹配，才能通过data测试。大家在下面的几个实例中可以体会到这点。

实例：Intent对象内容如下：

```
Intent intent = new Intent();
intent.setAction("cuc.intent1.action.Test1");
intent.addCategory("cuc.intent1.category.Test1");
intent.addCategory("cuc.intent1.category.Test2");
intent.setType("text/plain");
```

上面的Intent对象可以通过下面的intent-filter的data测试。intent-filter中的其中一个data标签的MIME类型与Intent对象的MIME类型匹配，并且Intent对象和intent-filter都没有指定任何URI格式，因此该Intent能通过下面的intent-filter的data测试。

```
    <intent-filter>
        <action android:name = "cuc.intent1.action.Test1"/>
```

```
        <action android:name="cuc.intent1.action.Test1A"/>
        <category android:name="android.intent.category.DEFAULT"/>
        <category android:name="cuc.intent1.category.Test1"/>
        <category android:name="cuc.intent1.category.Test2"/>
        <data android:mimeType="text/plain" />
        <data android:mimeType="image/*" />
    </intent-filter>
```

上面的 Intent 对象不能通过以下 IntentFilter 中的 data 测试：

```
    <intent-filter>
        <action android:name="cuc.intent1.action.Test1"/>
        <action android:name="cuc.intent1.action.Test1A"/>
        <category android:name="android.intent.category.DEFAULT"/>
        <category android:name="cuc.intent1.category.Test1"/>
        <category android:name="cuc.intent1.category.Test2"/>
        <data android:mimeType="text/plain"/>
        <data android:scheme="http"/>
    </intent-filter>
```

上面的 Intent 对象没有指定 URI 信息，但是上面的 intent-filter 中第二个 data 标签设置了 URI 中的 scheme 信息，即在 intent-filter 中的其中一个 data 标签指定了 URI，所以 Intent 对象无法通过该 IntentFilter 的 data 测试。

4. Intent 对象同时包含 URI 和 mimeType

这种情况下，要分别测试 URI 以及 mimeType 测试是否通过，只有 URI 以及 MIME 测试都通过了，data 测试才能通过。

MIME 测试：如果 Intent 的 MIME 类型能够匹配 intent-filter 的某一个 data 标签中的 MIME 类型，那么 MIME 类型测试就通过了。也就是说，无论 intent-filter 中有多少个 data 标签，只要有一个 data 标签的 mimeType 能够匹配 Intent 对象中的 mimeType，跟这个 data 标签是否有 URI 或者 URI 是否匹配没有关系，就通过了 mimeType 测试。

URI 测试：URI 测试又细分两种情况，满足下面的任何一种情况都可以通过 URI 测试：①如果 intent-filter 的某一个 data 标签中的 URI 能够匹配 Intent 对象的 URI 格式，URI 测试就通过了，跟这个 data 标签是否指定了 mimeType 或者指定的 mimeType 是否匹配 Intent 的 mimeType 都没关系。②如果整个 intent-filter 的所有 data 标签都没有指定 URI，这时如果 Intent 对象的 URI 是 content：协议或 file：协议，就能通过 URI 测试。换句话说，如果一个 intent-filter 只列出了 MIME 类型，没有列出任何 URI 相关的格式的话，那么这个 intent-filter 就默认是支持 content：协议或 file：协议的。

实例：Intent 对象内容如下：

```
Intent intent = new Intent();
intent.setAction("cuc.intent1.action.Test1");
intent.addCategory("cuc.intent1.category.Test1");
intent.addCategory("cuc.intent1.category.Test2");
```

```
Uri uri = Uri.parse("http://www.sina.com.cn/");
intent.setDataAndType(uri, "text/plain");
```

上面的 Intent 同时设置了 URI 和 MIMETYPE，能通过下面的 intent-filter 的测试。intent-filter 中的第一个 data 标签匹配 Intent 的 mimeType，但不能匹配 URI；第二个 data 标签匹配 URI，但不能匹配 mimeType。也就是说，虽然没有任何一个 data 标签能同时匹配 mimeType 和 URI，但有一个 data 标签匹配了 Intent 的 URI，也有一个 data 标签匹配了 Intent 的 mimeType，因而 Intent 对象能够通过 intent-filter 的 data 测试。

```
<intent-filter>
    <action android:name = "cuc.intent1.action.Test1"/>
    <action android:name = "cuc.intent1.action.Test1A"/>
    <category android:name = "android.intent.category.DEFAULT"/>
    <category android:name = "cuc.intent1.category.Test1"/>
    <category android:name = "cuc.intent1.category.Test2"/>
    <data android:mimeType = "text/plain" android:scheme = "https" android:host = "www.baidu.com" />
    <data android:mimeType = "image/*" android:scheme = "http" android:host = "www.sina.com.cn"/>
</intent-filter>
```

上面的 Intent 能通过下面的 intent-filter 的 data 测试，第二个 data 标签能同时匹配 Intent 的 URI 和 mimeType。

```
<intent-filter>
    <action android:name = "cuc.intent1.action.Test1"/>
    <action android:name = "cuc.intent1.action.Test1A"/>
    <category android:name = "android.intent.category.DEFAULT"/>
    <category android:name = "cuc.intent1.category.Test1"/>
    <category android:name = "cuc.intent1.category.Test2"/>
    <data android:mimeType = "image/*" android:scheme = "https" android:host = "www.baidu.com" />
    <data android:mimeType = "text/plain" android:scheme = "http" android:host = "www.sina.com.cn"/>
</intent-filter>
```

上面的 Intent 不能通过下面的 intent-filter 的 data 测试，没有一个 data 标签能够匹配 Intent 对象的 mimeType，因而 mimeType 不能通过测试。

```
<intent-filter>
    <action android:name = "cuc.intent1.action.Test1"/>
    <action android:name = "cuc.intent1.action.Test1A"/>
    <category android:name = "android.intent.category.DEFAULT"/>
    <category android:name = "cuc.intent1.category.Test1"/>
    <category android:name = "cuc.intent1.category.Test2"/>
    <data android:mimeType = "image/*" android:scheme = "https" android:host = "www.baidu.com" />
```

```
    <data android:scheme = "http" android:host = "www.sina.com.cn"/>
</intent - filter>
```

上面的 Intent 不能通过下面的 intent-filter 的 data 测试，没有一个 data 标签指定了 Intent 对象的 mimeType(因为整个 IntentFilter 中没有出现 mimeType)，因而 mimeType 不能通过测试。

```
<intent - filter>
    <action android:name = "cuc.intent1.action.Test1"/>
    <action android:name = "cuc.intent1.action.Test1A"/>
    <category android:name = "android.intent.category.DEFAULT"/>
    <category android:name = "cuc.intent1.category.Test1"/>
    <category android:name = "cuc.intent1.category.Test2"/>
    <data android:scheme = "http" android:host = "www.sina.com.cn"/>
</intent - filter>
```

上面的 Intent 不能通过下面的 intent-filter 的 data 测试，所有的 data 标签都没有指定 URI，这时 intent-filter 默认的 URI 是 content：协议或 file：协议，不能匹配 Intent 的 http 协议，因此不能通过 URI 测试。

```
<intent - filter>
    <action android:name = "cuc.intent1.action.Test1"/>
    <action android:name = "cuc.intent1.action.Test1A"/>
    <category android:name = "android.intent.category.DEFAULT"/>
    <category android:name = "cuc.intent1.category.Test1"/>
    <category android:name = "cuc.intent1.category.Test2"/>
<data android:mimeType = "text/plain"/>
</intent - filter>
```

6.5 隐式 Intent 示例

隐式 Intent 指定了 action，能够启动能执行相应 action 的组件。注意，可能没有任何应用能够处理发送到 startActivity()的隐式 Intent。如果出现这种情况，则调用将会失败，且应用会崩溃。要验证是否会有 Activity 会接收 Intent，请调用 Intent 对象的 resolveActivity (@NonNull PackageManager pm)方法。如果结果为空，则不应使用该 Intent，如有可能，应该停用该 Intent 的功能。如果结果为非空，则至少有一个应用能够处理该 Intent，就可以安全调用 startActivity()，如果只有一个应用能够处理，则该应用将立即打开并为其提供 Intent。如果多个 Activity 接受 Intent，则系统将显示一个对话框，使用户能够选取要使用的应用。

例如，如果希望用户与他人分享内容，请使用 ACTION_SEND 创建 Intent，并添加指定分享内容的 extra。使用该 Intent 调用 startActivity()时，用户可以选取分享内容所使用的应用。

```
Intent sendIntent = new Intent();
sendIntent.setAction(Intent.ACTION_SEND);
sendIntent.putExtra(Intent.EXTRA_TEXT, textMessage);
intent.setType("text/plain");
// Verify that the intent will resolve to an activity
if (intent.resolveActivity(getPackageManager()) != null) {
    startActivity(intent);
}
```

上面的 Intent 对象并没有使用 URI，但指定了数据 MIME 类型，用于指定 extra 携带的内容。

6.6 强制使用应用选择器

如果有多个应用能够响应隐式 Intent，则用户可以选择要使用的应用，并将其设置为该操作的默认选项，以便今后一直使用相同的应用执行某项操作(例如，打开网页时，用户往往倾向于仅使用一种网络浏览器)。但是，如果多个应用可以响应 Intent，且用户可能希望每次使用不同的应用，则应显示选择器对话框。选择器对话框每次都要求用户选择用于操作的应用，用户无法为该操作选择默认应用。例如，当应用使用 ACTION_SEND 进行分享时，用户可能需要每次选择不同的应用，因此应当始终使用选择器对话框。要显示选择器，请使用 createChooser()创建 Intent，并将其传递给 startActivity()。例如：

```
Intent sendIntent = new Intent(Intent.ACTION_SEND);
String title = getResources().getString(R.string.chooser_title);
// Create intent to show the chooser dialog
Intent chooser = Intent.createChooser(sendIntent, title);
// Verify the original intent will resolve to at least one activity
if (sendIntent.resolveActivity(getPackageManager()) != null) {
    startActivity(chooser);
}
```

这将显示一个对话框，对话框中列出了能响应传递给 createChooser 方法作为参数的 Intent 的应用，并且将提供的文本用作对话框标题。

6.7 本章主要参考文献

1. https://developer.android.google.cn/guide/components/intents-filters.html

第 7 章 线性、表格、栅格、相对布局与帧布局

CHAPTER 7

7.1 LinearLayout

android.widget.LinearLayout 是 ViewGroup 的一个子类，在 XML 布局文件中使用 <LinearLayout>标签表示。必须设置 orientation 属性，该属性的值是 vertical 或者 horizontal，即 LinearLayout 的布局方向可以设置为垂直或者水平。LinearLayout 按照垂直或者水平的顺序依次排列子元素，每一个子元素都位于前一个元素之后。当线性布局的方向是 vertical 时，LinearLayout 的所有直接子视图都沿着垂直方向线性排列、形成 N 行单列的结构，因此无论直接子视图有多宽，每一行只能有一个子视图。当线性布局的方向是 horizontal 时，LinearLayout 的所有直接子视图都沿着水平方向线性排列、形成单行 N 列的结构，高度为最高子视图的高度。如果搭建两行两列的结构，通常的方式是先垂直排列两个元素，每一个元素里再包含一个 LinearLayout 进行水平排列。LinearLayout 遵守子视图之间的 margins 以及每个子视图的 gravity(右对齐、居中对齐、左对齐)。LinearLayout 类的继承关系，如图 7-1 所示。从图中可以看出，LinearLayout 是 ViewGroup 的直接子类，它继承了 ViewGroup 的属性和 View 的属性。下面先介绍 View 的 xml 属性，再介绍 ViewGroup 的 xml 属性，最后介绍 LinearLayout 的 xml 属性。View 的属性，如表 7-1 所示。ViewGroup 的 xml 属性，如表 7-2 所示。LinearLayout 的 xml 属性，如表 7-3 所示。

```
java.lang.Object
  android.view.View
    android.view.ViewGroup
      android.widget.LinearLayout
```

图 7-1 LinearLayout 类的继承关系

表 7-1 View 的属性

View 的 xml 属性	解释
accessibilityLiveRegion	AccessibilityService 是一个可以响应 AccessibilityEvent 事件的服务类，主要用于一些辅助功能的实现，让丧失视力的人能够通过辅助功能使用应用程序。AccessibilityService 运行在后台，能够收到 AccessibilityEvent(这些事件表示用户界面一系列的状态变化，比如焦点改变、输入内容变化、按钮被单击)。该选项设置 AccessibilityService 是否应自动通知，例如视图的内容描述或文本的改变，或者子视图的文字描述或文本的改变。取值为：none(不通知)、polite(通知)、assertive(组件马上中断当前工作，并通知用户)

续表

View 的 xml 属性	解　释
accessibilityTraversalAfter	取值是某个 View 组件(组件 B)的 ID,当屏幕阅读器阅读了指定的组件(组件 B)后,才阅读本组件(组件 A)
accessibilityTraversalBefore	取值是某个 View 组件(组件 B)的 ID,屏幕阅读器必须在阅读指定的组件(组件 B)前,阅读本组件(组件 A)
alpha	设置该组件的透明度,值在 0 和 1 之间。0 代表完全透明,1 代表完全不透明
background	设置该组件的背景,可以是一幅完整绘图资源的引用(如.png 图片),也可以是纯色,如#ff000000(黑色)
backgroundTint	指定渲染背景使用的颜色,必须是一个颜色值,形式如#rgb、#argb、、#rrggbb 或者#aarrggbb
backgroundTintMode	指定背景渲染所使用的混色模式。默认值是 SRC_IN
clickable	指定这个 View 是否响应单击事件
contentDescription	设置简要描述 View 内容的文本
contextClickable	布尔量,指定这个 View 是否响应上下文相关的单击事件,例如单击鼠标右键,出现上下文相关的菜单
drawingCacheQuality	设置该组件所使用的半透明的绘制缓存的质量
duplicateParentState	布尔量,如果设置为 true,将从直接父容器中而不是从它自己获取绘图状态(光标,按下等)。注意,仅仅是获取绘图状态,而不是事件,例如点一下 LinearLayout 时 Button 有被单击的效果,但是并不执行单击事件
elevation	尺寸值,即一个浮点数,加上长度单位,如 14.5sp。设置该组件“浮”起来的高度,通过设置该属性可以让该组件呈现 3D 效果
fadeScrollbars	定义当 scrollbars 不再使用时,是否使用淡出效果(逐渐消失)
fadingEdgeLength	设置淡出的边沿的长度
filterTouchesWhenObscured	设置当 View 窗口被其他可见窗口遮住时,是否过滤触摸事件。如果设置为 true,当有 Toast、对话或其他视窗显示在 View 上面时,该 View 就不能接受触摸事件了
fitsSystemWindows	布尔量,让 View 可以根据系统窗口(例如 statusbar)来调整自己的布局,如果值为 true,就会调整 View 的 padding 属性来给系统窗口留出空间。当 status bar 为透明或半透明时(Android 4.4 以上),系统会设置 View 的 paddingTop 值为一个适合的值(status bar 的高度)让 View 的内容不被上拉到状态栏,在不占据 status bar 的情况下(4.4 以下)设置 paddingTop 值为 0(因为没有占据 status bar 所以不用留出空间)
focusable	布尔量,控制 View 是否可以获得焦点,这个值不影响直接调用 requestFocus()的行为
focusableInTouchMode	布尔量,控制 View 在触摸模式时是否可以获得焦点。true 代表这个 View 能在单击时获取焦点,如果另外一个设置此属性为 false 的 View 被单击,此 View 仍能保持焦点
forceHasOverlappingRendering	布尔量,设置画图时是否有覆盖的元素
foreground	设置绘制在内容上的前景图片,可以用作覆盖,可以是绘图资源的引用(如.png 图片),也可以是纯色,如#ff000000(黑色)

续表

View 的 xml 属性	解　释
foregroundGravity	定义前景图片相对于父容器的位置，如 top、bottom、left、right 等，默认值是 fill
foregroundTint	指定渲染前景使用的颜色，必须是一个颜色值，形式如 #rgb、#argb、#rrggbb、或 #aarrggbb。Tint 是 Android5.0 引入的一个属性
foregroundTintMode	指定前景渲染所使用的混色模式
hapticFeedbackEnabled	单击视图时，为了引起用户的注意，系统提供一个触力反馈(使手机震动一下)，就要设置该属性为 true
id	为 View 提供唯一标识符 id，以便后来通过 View. findViewById()或者 Activity. findViewById()获取 View
importantForAccessibility	控制视图对于 AccessibilityService 的重要性，例如发生访问事件报告给屏幕阅读器
isScrollContainer	定义布局是否作为一个滚动容器、可以缩小整个窗体，以便给输入方法腾出空间
keepScreenOn	控制 View 所在的窗口可见时，是否让屏幕一直亮着。手机通常都带有省电模式，比如设置无操作 15s 后屏幕自动关闭
layerType	取值为：none(默认值，按一般方式绘制，不使用离屏缓冲)，software(能更好地兼容硬件)、hardware(可以提供更快的动画)。可以使用 View 的绘制 cache，或使用 Canvas. saveLayer()离屏缓冲，为复杂 View 动画或组合效果提高性能. 例如，可以使用 Canvas. saveLayer()实现淡出效果，这个方法临时地把一个 View 画到一个 layer 中然后使用一个透明系数把它组合回屏幕上
layoutDirection	定义 layout 画图的方向，跟使用的语言的书写方向有关：如 ltr(left to right)、rtl(right to left)、locale(语言环境)、inherit(继承父控件)
longClickable	设置 View 是否能响应长单击事件
minHeight、minWidth	定义该 View 的最小高度/宽度
nextFocusDown/ nextFocusForward/nextFocusLeft/ nextFocusRight/nextFocusUp	指定当焦点在该视图上，下一个获取焦点的 View 的 ID。如果引用的 View 不存在或者是不可见的分级视图的一部分，当引用被访问时，将会发生运行时例外。系统默认会按照布局从上到下、从左到右传递 focus。绝对方向包括 Up、Down、Left 和 Right，是键盘上的 4 个方向键；或者相对方向：Forward 和 Backward，通常 Tab 代表 Forward，Shift+Tab 代表 Backward
onClick	为该视图的单击事件绑定监听器，当视图被单击时，从上下文中调用的方法的名字
padding	取值为尺寸值。padding 设置 View 内部所有四个方向的 padding 的默认值。padding 指的是视图的边沿和视图的内容之间的距离。视图的大小包含 padding
paddingBottom/paddingEnd/ paddingLeft/paddingRight/ paddingStart/paddingTop	取值为尺寸值，分别设置 View 各个方向上的 padding。对于从左往右的布局，开始方向是左、结束方向是右；对于从右往左的布局，开始方向是右、结束方向是左
requiresFadingEdge	定义滚动时哪些边沿逐渐淡出：取值为 horizontal，vertical，none
rotation	视图的旋转角度(单位：度)，取值是浮点数
rotationX、rotationY	取值是浮点数，分别设置视图绕 X/Y 轴的旋转角度(单位：度)

续表

View 的 xml 属性	解　释
saveEnabled	在配置改变等情况出现时是否保存 View 的状态数据。如果设置为 true、并且该 View 指定了 ID,系统就会保存该 View。如果值为 false,当被冻结时,该视图的状态将不会被保存,但不包括可能被保存的子视图
scaleX/scaleY	取值必须是浮点数,设置视图在水平/垂直方向的缩放比例
scrollIndicators	设置当视图滚动时,显示哪个方向的滚动标识,可能取值：none, top,bottom,left,right,start,end,也可以把多个值用 | 组合起来
scrollX/scrollY	必须是尺寸值,即一个浮点数,再附加单位。设置初始时 X/Y 方向的滚动偏移
scrollbarAlwaysDrawHorizontalTrack/ scrollbarAlwaysDrawVerticalTrack	定义是否总是显示水平/垂直滚动条的轨道
scrollbarDefaultDelayBeforeFade	整数,单位毫秒。定义滚动条淡出前需要等待的时间
scrollbarFadeDuration	整数,例如 100,单位毫秒。定义滚动条淡出需要消耗的时间
scrollbarSize	必须是个尺寸值。设置垂直滚动条的宽度和水平滚动条的高度
scrollbarStyle	控制滚动条的风格和位置,可选值有 insideOverlay、insideInset、outsideOverlay、outsideInset 四种。其中 inside 和 outside 分别表示滚动条在 View 的 padding 区域内或者 padding 区域外,overlay 和 Insert 分别表示覆盖在 View 上和放在 View 后面
scrollbarThumbHorizontal	必须是对另一资源的引用。设置该组件的水平滚动条的滑块对应的 Drawable 对象
scrollbarThumbVertical	必须是对另一资源的引用。设置该组件的垂直滚动条的滑块对应的 Drawable 对象
scrollbarTrackHorizontal	必须是对另一资源的引用。设置该组件的水平滚动条的轨道对应的 Drawable 对象
scrollbarTrackVertical	设置该组件的垂直滚动条的轨道对应的 Drawable 对象
scrollbars	定义在滚动时显示哪个滚动条,取值 horizontal,vertical,none 或者二者的或
soundEffectsEnabled	布尔量,设置该组件被单击、触摸等事件时是否使用音效
stateListAnimator	必须是对另一资源的引用。给视图设以状态为基础的动画
tag	为该视图设置一个字符串类型的 tag 值。接下来可通过 View. getTag()获取该字符串,或通过 View. findViewWithTag()查找该视图
textAlignment	定义文字的对齐方式,比如对齐到段落的开始(textStart),段落的结束(textEnd),视图的开始 viewStart,视图的结束 viewEnd,等等
textDirection	定义段落中文字的方向。例如：从左到右(LTR),从右到左(RTL),文字方向取决于系统 Locale(locale)
theme	theme 就是显示界面的样式,是一些资源属性值的集合。默认情况下,使用 View(Context context)构造体中 Context 对象的 theme 创建 View,也可以使用 theme 指定不同的主题
transformPivotX	必须是个尺寸值。设置该组件旋转和伸缩时旋转中心的 X 坐标
transformPivotY	必须是个尺寸值,设置该组件旋转和伸缩时旋转中心的 Y 坐标

续表

View 的 xml 属性	解　释
transitionName	命名一个视图，以便以后用于过渡动画(Transition)时能唯一标识。自 KitKat 起，Android 加入了 Transition 框架，用于动画前后需要改变布局的场景，实现 Activity 级别或 View 级别的动画效果，有别于传统的 Animation，Transition 需要对动画前后设置不同布局，通过相应的 API 实现两个布局的切换动画。而传统 Animation 只对应一个布局文件
translationX/translationY/translationZ	必须是个尺寸值，设置该视图在 X/Y/Z 方向上的平移
visibility	控制视图初始时的可见性

表 7-2　ViewGroup 的 XML 属性

属　性	解　释
addStatesFromChildren	设置这个 ViewGroup 的 drawable 状态是否包括子 View 的状态。若设为 true，当子 View 如 EditText 或 Button 获得焦点时，整个 ViewGroup 也会获得焦点
alwaysDrawnWithCache	设置 ViewGroup 在绘制子 View 时是否一直使用绘图缓存。默认为 true
animateLayoutChanges	布尔量。如果设置为 true，当在 ViewGroup 中添加或者移除子 View 引起布局变化时，子 View 就会呈现出默认的过渡的动画效果
animationCache	设置布局在绘制动画效果时是否为其子 View 创建绘图缓存。若设为 true，将会消耗更多的内存，初始化时间更长，但性能更好。默认为 true
clipChildren	设置子视图是否受限于在自己的边界内绘制。若设为 false，当设置动画缩放子视图时、子视图可以绘制在边界外。默认为 true
clipToPadding	如果 padding 不等于 0，定义 ViewGroup 是否裁剪它的子视图、重新调整子视图的边沿效果到 padding 位置。当 clipToPadding 为 true 时，表示布局不能绘制到 padding 区域。当 clipToPadding 为 false 时，表示布局能够被绘制到 padding 区域。默认为 true
descendantFocusability	定义当寻找一个即将获得焦点的视图的时候，ViewGroup 与其子视图之间的关系。可选项为：①beforeDescendants：ViewGroup 会比其子视图更先获得焦点；②afterDescendants：只有当没有子视图想要获取焦点时，ViewGroup 才会获取焦点；③blockDescendants：ViewGroup 会阻止子视图获取焦点
layoutAnimation	必须是对另一资源的引用。定义当 ViewGroup 第一次布局时要使用的动画，也可在第一次布局后人为地调用
layoutMode	定义这个 ViewGroup 的布局模式
persistentDrawingCache	定义绘图缓存的持久性，有如下可选项：①none：当使用过后不保留绘图缓存；②animation：在 layoutAnimation 之后保留绘图缓存；③scrolling：在 Scroll 操作后保留绘图缓存；④all：总是保留绘图缓存
splitMotionEvents	布尔量，定义布局是否分割 MotionEvents 到多个子视图。若设置为 false，触摸事件仅被派发到第一个指针按下去的视图，直到最后指针抬起。若设置为 true，触摸事件仅被派发到 ACTION_DOWN 事件发生的多个视图

当把 addStatesFromChildren 设为 true 时，若 ViewGroup 组中的某个 EditText 或是 Button 获取焦点时，可以把 ViewGroup 的 Background 设置成相应 EditText 或的 Button 的 Drawable，这样看上去该 ViewGroup 中的子 view 是一个整体。在 Android 中 Button 可

以根据选中、单击等状态切换图片，但是默认情况下继承于 ViewGroup 的类（例如 LinearLayout）是接收不到 pressed 状态的，即 LinearLayout 的按下是没有效果的。ViewGroup 可以通过设置 addStatesFromChildren＝"true"或者调用方法 linearLayout.setAddStatesFromChildren(true)获得内部 View 的状态，就可以取得 pressed 状态。还有一个相反的方法 setDuplicateParentStateEnabled，内部类获得外部容器类的状态，这个方法和 setAddStatesFromChildren 不能同时设置，同时设置会产生异常。

表 7-3 LinearLayout 的 XML 属性

属　性	解　释
baselineAligned	布尔量。当设定为 flase，阻止布局对齐它的子视图的基准线。baselineAligned 只对可以显示文字的 View（例如 TextView、Button、EditText）起作用
baselineAlignedChildIndex	必须是一个整数。如果当前的 LinearLayout 是另一个 baseline（基准线）对齐的布局的一部分时，baselineAlignedChildIndex 属性指定当前的 LinearLayout 中的第几个子视图（这个子视图必须是可以显示文字的 View，计数从 0 开始）用于基线对齐
divider	指定一个 Drawable 作为按钮之间的垂直分界线
gravity	指定该对象在它自己的边界内，怎样放置它的内容
measureWithLargestChild	当设置为 true，有同一个 weight 的所有子视图将和最大的子视图的最小尺寸一样大。当设置为 false，所有子视图正常测量
orientation	定义布局的方向是水平还是垂直：horizontal 或者 vertical
weightSum	定义了 weight 总和。如果没指定，总和是所有子视图的 layout_weight 的总和

英文字符的基准线，主要是针对字母 g，p，y 等字母的部分笔画低于基线，如图 7-2 所示。当 baselineAligned 属性设置为 false 时，布局中的子视图的基准线不对齐。baselineAligned 的默认值为 true。

图 7-2 英文字符的基准线

下边是 baselineAligned 属性的一个例子：

```
<?xmlversion = "1.0" encoding = "utf - 8"?>
<LinearLayout xmlns:android = "http://schemas.android.com/apk/res/android"
    android:orientation = "horizontal"
    android:baselineAligned = "true"
    android:layout_width = "match_parent"
    android:layout_height = "match_parent">
    <TextView
        android:layout_width = "wrap_content"
        android:layout_height = "wrap_content"
        android:text = "TextView:p"
```

```
            android:layout_marginEnd = "5dp"/>
        < TextView
            android:layout_width = "wrap_content"
            android:layout_height = "wrap_content"
            android:text = "LargeTextView:p"
            android:textSize = "30sp" />
    </LinearLayout >
```

上面的布局文件的布局效果如图 7-3 所示。

在上面的布局文件中，如果设置 android：baselineAligned＝"false"，布局效果如图 7-4 所示。

TextView:p LargeTextView:p

图 7-3 android：baselineAligned＝"true"的效果

TextView:p LargeTextView:p

图 7-4 android：baselineAligned＝"false"的效果

7.1.1 LinearLayout.LayoutParams

上面描述的 LinearLayout 的属性，是 LinearLayout 元素自身的属性；LinearLayout 布局中的子视图有些属性跟 LinearLayout 有关联，这些属性定义在 LinearLayout.LayoutParams 中，它描述的是 LinearLayout 的子视图的跟 LinearLayout 有关联的属性。LinearLayout.LayoutParams 类的继承关系，如图 7-5 所示。

java.lang.Object
　　android.view.ViewGroup.LayoutParams
　　　　android.view.ViewGroup.MarginLayoutParams
　　　　　　android.widget.LinearLayout.LayoutParams

图 7-5 LinearLayout.LayoutParams 类的继承关系

LinearLayout.LayoutParams 包括两个属性：layout_gravity 和 layout_weight。

1. layout_gravity 属性

LinearLayout 的子视图可以使用 layout_gravity 属性为自己指定相对父容器(LinearLayout)的位置，layout_gravity 的取值包括：top、bottom、left、right、start、end、center_vertical、center_horizontal、center(这些取值不改变视图大小)，还包括 fill_vertical、fill_horizontal、fill(必要的时候增加对象的垂直、水平或者两个方向，以便填充父容器)，还包括 clip_vertical、clip_horizontal。clip_vertical 基于其垂直方向上的 gravity，当 gravity 为 top 时，剪切底部；当 gravity 为 bottom 时剪切顶部；除此之外剪切顶部和底部。clip_horizontal 基于其水平方向上的 gravity：当 gravity 为 left 时，剪切右侧；当 gravity 为 right 时剪切左侧；除此之外剪切左侧和右侧。①当 android：orientation＝"vertical"时，android：layout_gravity 只有水平方向的设置(left，right，center_horizontal)才起作用，垂直方向的设置不起作用；②当 android：orientation＝"horizontal"时，android：layout_gravity 只有垂直方向的设置(top，bottom，center_vertical)才起作用，水平方向的设置不起作用。layout_gravity 的取值可以包含多个值，需用“|”分开。

因为 layout_gravity 属性的含义是相对父容器的位置，所以该属性没有直接放在控件类里，而是放在了容器类的内部类 LinearLayout. LayoutParams。比如一个 Button 在 Linearlayout 里，可以通过 layout_gravity 属性设置放在 Linearlayout 里靠左、靠右等位置。

2. gravity 与 layout_gravity 的区别

gravity 属性是设置视图自身的属性，是设置自己边界内的内容或者子视图的位置。例如 LinearLayout 的 gravity 属性用于设置该 LinearLayout 的内容或者子视图相对于该 LinearLayout 自身的显示位置。TextView 的 gravity 属性是设置它的内容即文本在这个 TextView 中的位置。

LinearLayout. LayoutParams 中的 android: layout_gravity 用于设置 LinearLayout 的子元素的属性，设置该控件相对于它的父容器的 LinearLayout 的位置。比如只有一个 Button 在 LinearLayout 里，可以通过该 Button 的 layout_gravity 属性设置把该 Button 放在 LinearLayout 的靠左靠右等位置，也可以通过 Button 的父容器 LinearLayout 的 gravity 属性来设置。

3. layout_weight 属性

LinearLayout 的子视图可以使用 layout_weight 属性为自己分配在父容器的权重，以控制各个控件在布局中的相对大小。线性布局会根据该控件 layout_weight 值与其所处布局中所有控件 layout_weight 值之和的比值为该控件分配剩余空间。剩余空间不包括子视图的 layout_width、layout_height 指定的部分，即把所有子视图的 layout_width、layout_height 指定的空间除开后 LinearLayout 中还剩下的空间。子视图可以指定权重值，然后系统会按照子视图声明的权重值的比例，将 LinearLayout 中的剩余空间分配给子视图，权重值更大的视图可以扩展父视图中更多剩余的空间。子视图的默认权重为零。

下面的例子中，LinearLayout 的布局方向是 vertical，有三个 EditText 和一个 Button，除第三个 EditText 声明权重为 1 外，其他的视图都没有声明权重。没有声明权重的视图默认的权重为 0，因而这些视图的高度就是 layout_height 指定的高度。第三个 EditText 声明权重为 1，所以高度除了 layout_height 指定的高度外，还要扩展，占据 LinearLayout 分配所有四个子视图的高度后的所有的剩余高度。

```
<?xml version = "1.0" encoding = "utf - 8"?>
<LinearLayout xmlns:android = "http://schemas.android.com/apk/res/android"
    android:layout_width = "match_parent"
    android:layout_height = "match_parent"
    android:paddingLeft = "16dp"
    android:paddingRight = "16dp"
    android:orientation = "vertical" >
    <EditText android:layout_width = "match_parent"
        android:layout_height = "wrap_content" android:hint = "@string/to" />
    <EditText android:layout_width = "match_parent"
        android:layout_height = "wrap_content" android:hint = "@string/subject" />
    <EditText android:layout_width = "match_parent"
        android:layout_height = "0dp" android:layout_weight = "1"
        android:gravity = "top" android:hint = "@string/message" />
    <Button android:layout_width = "100dp"
```

```
        android:layout_height = "wrap_content" android:layout_gravity = "right"
        android:text = "@string/send" />
</LinearLayout>
```

上面的 xml 文件的布局效果如图 7-6 所示。

图 7-6 四个控件的 layout_weight 分别为 0、0、1 和 0 的布局效果

假设某个垂直 LinearLayout 的所有子视图得权重加起来、总权重为 Wt，第 n 个子视图的权重为 Wn，则在所有的子视图按照 layout_height 指定的高度分配后的剩余高度里，第 n 个子视图扩展的高度为剩下高度的 Wn/Wt。

下面的线性布局中，四个控件的 layout_height 都是 0dp，layout_weight 分别是 1、2、3、4。因此这四个控件的高度除了指定的 0dp 外，额外扩展的高度分别是剩余高度的 1/(1+2+3+4)、2/(1+2+3+4)、3/(1+2+3+4)、4/(1+2+3+4)。

```
<?xml version = "1.0" encoding = "utf - 8"?>
<LinearLayout xmlns:android = "http://schemas.android.com/apk/res/android"
    android:layout_width = "match_parent"
    android:layout_height = "match_parent"
    android:paddingLeft = "16dp"
    android:paddingRight = "16dp"
    android:orientation = "vertical" >
    <EditText android:layout_width = "match_parent"
android:layout_height = "0dp"
        android:layout_weight = "1" android:hint = "to" />
    <EditText android:layout_width = "match_parent"
android:layout_height = "0dp"
        android:layout_weight = "2" android:hint = "subject" />
    <EditText android:layout_width = "match_parent"
android:layout_height = "0dp" android:layout_weight = "3"
```

```
android:gravity = "top" android:hint = "message" />
    < Button android:layout_width = "100dp" android:layout_height = "0dp"
        android:layout_weight = "4" android:layout_gravity = "right"
android:text = "send" />
</LinearLayout >
```

上面的 xml 文件的布局效果如图 7-7 所示。

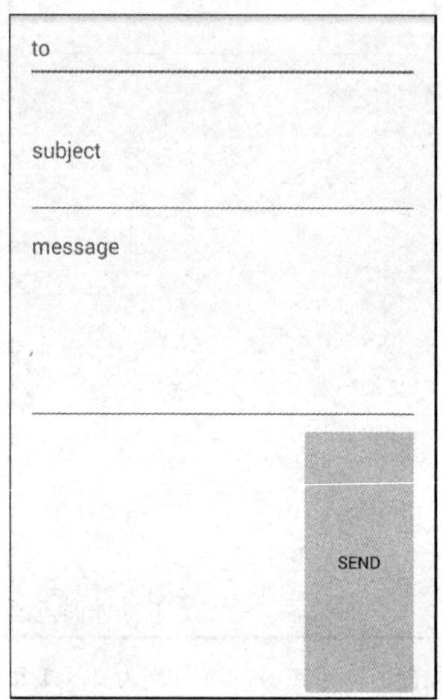

图 7-7　四个控件的 layout_weight 分别为 1、2、3 和 4 的布局效果

7.1.2　ViewGroup.LayoutParams

ViewGroup.LayoutParams 用于设置 ViewGroup 的子视图的属性，用于告诉它的容器它想怎么布局。ViewGroup.LayoutParams 的属性有 layout_height 和 layout_width，用于设置 ViewGroup 的子元素的属性，告诉它的父容器自己期望的高度和宽度，取值可以是一个尺寸值(即一个浮点数附加一个单位)、match_parent 或者 wrap_content。match_parent，指定视图的大小是父容器的尺寸减去父容器的 padding 后的大小，这个值在 API 级别 8 被引入。wrap_content 指定视图的大小应该足够大，大到可以完全包围它的内容和它的 padding。

7.1.3　ViewGroup.MarginLayoutParams

ViewGroup.MarginLayoutParams 包含的属性有 layout_margin、layout_marginTop、layout_marginBottom、layout_marginLeft、layout_marginRight、layout_marginStart、layout_marginEnd。除 layout_margin 用于指定了上下左右四个方向的默认值外，其他六个属性分别用于指定各个具体方向的 margin，它将覆盖 layout_margin 属性设置的默认值。属性 layout_margin 指定了在该视图的边界以外(不是在视图的边界内)保留的额外空间。Margin 是一个尺寸值，取值应该是一个正的浮点数，后面附加尺寸单位，单位可以是 px (pixels)、dp(density-independent pixels)、sp(scaled pixels based on preferred font size)、in (inches)和 mm(millimeters)。

7.1.4　layout_margin 和 padding 的区别

View 的属性 padding，设置 View 的内容相对 view 的边框的距离。padding 指的是在控件的边界内、把控件内的子控件或者文字内容除开，为控件保留的额外空间。而 layout_margin 指的是在控件的外部为该控件额外保留的空间，控件所占的空间除了 layout_width、layout_height 外，还包括 margin 所占的空间。关于 layout_margin 和 padding 的区别，请参看图 7-8。

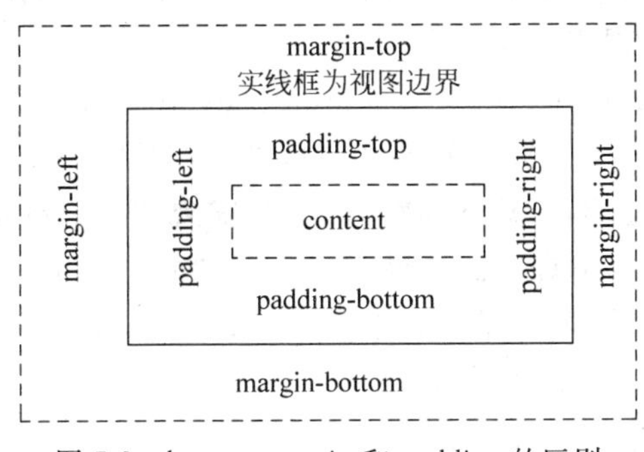

图 7-8　layout_margin 和 padding 的区别

7.1.5　视图的大小

一个视图的大小用宽度和高度来表达。一个视图实际拥有两对宽和高的值。第一对被称为测量的宽和高(MeasuredWidth 和 MeasuredHeight)，这些尺寸定义了一个视图在它的父视图里想要是多大，可以通过 getMeasuredWidth()和 getMeasuredHeight()获得测量的宽和高。第二对宽和高被简单地称为宽和高(width and height)，有时也称为画图宽度和画图高度(drawing width 和 drawing height)，这对宽和高定义了在布局后画图时屏幕上视图的实际大小。视图的实际大小可能和期望的大小(MeasuredWidth 和 MeasuredHeight)不相等。实际的宽和高可以通过 getWidth()and getHeight()来获得。

MeasuredWidth 和 MeasuredHeight 必须考虑 padding。Padding 用来表示视图里的内容相对视图边沿的偏移。例如左边的 padding 为 2 像素，即视图的左边额外保留 2 像素的空间，视图的内容相对于视图的左边沿往视图中心偏移 2 像素。Padding 可以通过 setPadding(int，int，int，int)或者 setPaddingRelative(int，int，int，int)来设置，可以通过 getPaddingLeft()，getPaddingTop ()，getPaddingRight()，getPaddingBottom()，getPaddingStart()，getPaddingEnd()来查询视图的各个方向的 padding。视图能够定义 padding，但不能定义 margin。只有 ViewGroup 能定义 margin。

7.1.6　从右到左的布局

中国人的阅读习惯是从左往右阅读，阿拉伯国家的阅读习惯是从右往左阅读。所以同一个布局在不同的语言环境下，显示的样式是不一样的。在 Android 4.2 系统之后，Google 在 Android 中引入了 RTL(right to left)布局，更好了支持了由右到左文字布局的显示。例如，在线性布局中第 1 个子视图默认都是在左上角的，如果采用 RTL 布局，默认就在右上角了。为了解决 RTL 和 LTR 两种布局模式中 UI 排列的混乱，在 Android 4.2 中新加了 android：layout_marginStart，android：layout_marginEnd 两个布局属性，用来替代 MarginLeft 和 MarginRight。在 LTR 布局中，layout_marginStart 和 layout_marginEnd 分别等同于 layout_marginLeft 和 ayout_marginRight。在 RTL 布局中，layout_marginStart 和 layout_marginEnd 分别等同于 ayout_marginRight 和 layout_marginLeft。

为了更精确地控制应用程序在 UI 上的文字书写顺序（从左到右，或者从右到左），Android 4.2 引入了：layoutDirection 该属性控制组件的布局排列方向、textDirection 属性控制组件的文字排列方向、textAlignment 属性控制文字的对齐方式。getLayoutDirectionFromLocale() 方法用于获取指定地区的惯用布局方式。

若采用 android：layout_marginLeft，android：layout_marginRight，对于从左到右的布局和从右到左的布局，设置不一样。如果当前是默认布局方式是 LTR（从左到右，Left-to-Right），TextView1 在左边，TextView2 在右边，只需要将 TextView2 的 android：layout_marginLeft 属性值设为 100dp 即可。如果当前布局方式是 RTL（从右到左，Right-to-Left），TextView1 在右边，TextView2 在左边，只需要将 TextView2 的 android：layout_marginRight 属性值设为 100dp 即可。

7.1.7 尺寸单位

首先了解下面三个基本概念：手机屏幕大小、分辨率、像素密度。

屏幕大小：屏幕大小是手机屏对角线的尺寸，以英寸（inch）为单位。比如某手机屏为 5 英寸屏，就是指对角线的尺寸是 5 英寸。

分辨率：分辨率就是手机屏幕的像素点数，一般描述成屏幕的"宽×高"，安卓手机屏幕常见的分辨率有 480×800、720×1280、1080×1920 等。720×1280 表示此屏幕在宽度方向有 720 个像素，在高度方向有 1280 个像素。

像素密度（dpi，dots per inch；或 PPI，pixels per inch）：就是每英寸的像素点数，数值越高显示越细腻。假如一部手机的分辨率是 1080×1920，屏幕大小是 5 英寸，能否算出此屏幕的像素密度呢？先利用勾股定理计算出屏幕对角线上的像素：$(1080^2+1920^2)^{1/2}=2203$，再用屏幕对角线上的像素除以对角线长度就得到此屏幕的密度了，用 2203/5 是 440。同一个分辨率在不同的屏幕尺寸上像素密度并不相同，为了解决这个问题，Android 系统内置了几个默认的 dpi，在特定的分辨率下自动调用，也可以手动在配置文件中修改。在 Android 系统中使用的全部都是系统 dpi，对界面元素进行缩放的比例依据是系统密度，而不是计算出来的屏幕实际密度。1920×1080 分辨率的手机默认就使用 480 的 dpi，不管实际尺寸是多大。物理 dpi 主要用于厂家对于手机的参数描述，系统 dpi 跟物理 dpi 并不一定相同。Android 定义的系统密度，如表 7-4 所示。

表 7-4 Android 系统定义的系统密度

密度名称	ldpi	mdpi	hdpi	xhdpi	xxhdpi
密度值	120	160	240	320	480
密度基比例	0.75	1	1.5	2	3
代表分表率	240×320	320×480	480×800	720×1280	1080×1920

从图 7-9 可以看出，以像素为单位设计用户界面，如果在一个每英寸像素点数（dots per inch）更高的新显示器上运行该程序，则用户界面会显得很小。在有些情况下，用户界面可能会小到难以看清内容。

dp 也可写为 dip，即 density-independent pixel。dp 更类似一个物理尺寸，比如一张宽和高均为 100dp 的图片在分辨率为 320×480 和 480×800 的手机上看起来一样大小。不管这个屏幕的密度是多少，相同 dp 大小的尺寸在不同的屏幕上具有同样的物理长度。

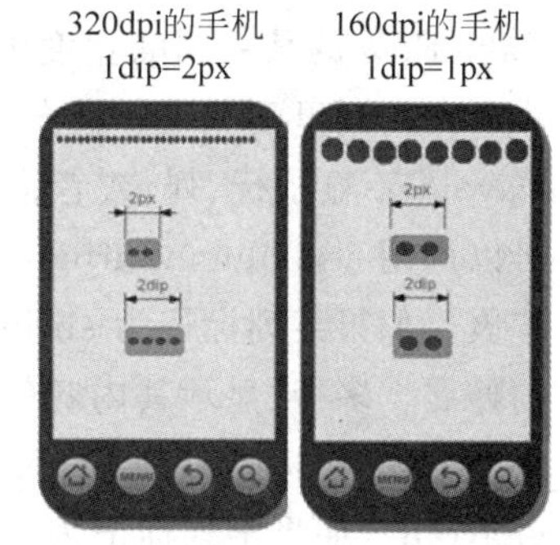

图 7-9　不同像素密度的手机的 px 和 dip 的实际尺寸

不管开发者使用哪个度量单位，最后经过 Android 系统的处理，都是要转换成像素单位（px）的。Google 把 160dpi 定为像素密度标准，1 个 dip 的物理尺寸为 $\frac{1}{160}$ 英寸。对于 160dpi 的设备，1dp＝1 像素。定义

密度比例因子＝系统像素密度/标准像素密度＝系统像素密度/160dpi

对于不同密度比例因子的设备，使用 dp 乘以密度比例因子计算实际的像素数，比如对于系统像素密度是 240dpi 的设备，密度基准因子＝240/160＝1.5，100dp 就有 100×1.5＝150 个像素。对于 320dpi 的设备，像素密度比 160dpi 高了一倍，相同的物理尺寸上的像素数是 160dpi 的 2 倍，这时 1dp＝2 像素。

sp 主要用作文字大小的单位，除了与密度无关外，还与 scale 有关。如果屏幕密度为 160dpi，这时 dp 和 sp 和 px 是一样的。1dp＝1sp＝1px，也就是说，如果使用 dp 和 sp，系统会根据屏幕密度的变化自动进行转换。如果想使得不管用户如何设置偏好字体大小，应用的字体大小都不变，文本大小的单位就应该使用 dp，而不是 sp。期望改变系统字体大小时，应用中的文字的大小跟着改变，这时应该使用 sp 作为文本大小的单位。

7.2　TableLayout

android.widget.TableLayout 类以行和列的形式对子控件进行管理，每一行为一个 TableRow 对象，或一个 View 控件，TableLayout 的直接子视图不能指定 layout_width 属性，layout_width 的值总是 MATCH_PARENT，但可以指定 layout_height 属性。但如果直接子视图是 TableRow，layout_height 的值总是为 WRAP_CONTENT。也就是说，Tablelayout 的直接子元素是 TableRow，或者是任何 View 的子类。当 TableLayout 的直接子元素为 View 时，该 View 将独占一行，横跨所有的列；当 TableLayout 的直接子元素为 TableRow 时，可在 TableRow 下添加子控件，默认情况下，每个子控件占据一列。

android.widget.TableLayout 和 android.widget.TableRow 都是 LinearLayout 的直接子类，TableLayout 相当于一个垂直方向的 LinearLayout，TableRow 相当于水平方向的线性布局。TableRow 通常用作 TableLayout 的子视图，如果 TableRow 的父不是 TableLayout，TableRow 就相当于水平方向的线性布局。TableRow 的子视图并不需要在 xml 布局文件

中指定 android：layout_width 和 android：layout_height 属性。

TableRow 的某一列的宽度由该列中最宽的格子决定，每一个 TableRow 的每一个控件都和所在的列里面 width 最大的那个控件对齐，并且任何一列中的所有控件都相互对齐。Tablelayout 不显示行、列、或者格子(cell)的边界线。

可以调用 setColumnShrinkable()或者 setColumnStretchable()来设置某些列可收缩或者可扩展。如果某列标记为 shrinkable，当水平方向内容太多时，该列的宽度可收缩，向垂直方向伸展、“多行”显示其内容(这里不是真正的多行，而是系统根据需要自动调节该行的 layout_height)，以便整个 TableLayout 能适应父对象，在父对象里完整地显示整个 TableLayout。如果某列标记为 stretchable，该列的宽度可扩展，以适应额外的空间。某列可同时标记为可收缩和可扩展，则该列的宽度可以改变到最大可得到的空间，但永远不超过最大可得到的空间。如果需要隐藏某列，可以调用 setColumnCollapsed()。

1. TableLayout 行列数的确定

TableLayout 的行数为 TableRow 对象的数量加上 TableLayout 的直接子视图的数量。每一个 TableRow 可以有一个或者多个格子(cell)，每个格子可以容纳一个 View 对象。每一个 TableRow 中列序号的起始编号是 0，格子按增加的列序号添加到行中，如果没有为一个格子指定列序号，列序号自动增加到下一个可得到的列。如果跳过了某个列序号，它被认为是那一行里空的格子。TableRow 的某个格子可以是空的，也可以像 HTML 那样横跨多列。可以使用 TableRow.LayoutParams 的 span 属性指定横跨的列。

每个 TableLayout 的列数量和包含最多格子的行的格子数量一样，即 TableLayout 的列数等于含有最多子控件的 TableRow 的列数。例如某个 TableLayout 的 3 个 TableRow 分别含 2 个、3 个和 4 个子控件，那么该 TableLayout 的列数就是 4。TableLayout 元素可设置的属性，如表 7-5 所示。

表 7-5 TableLayout 的 XML 属性

属　性	解　释
stretchColumns	指定可以沿水平方向伸展的列索引(起始编号是 0)，最多可占据一整行，以填满水平方向剩下的空白空间，若有多列需要设置为可伸展，请用逗号将需要伸展的列序号隔开，重复的或者不合法的索引将被忽略。可以使用“＊”表示伸展所有的列
shrinkColumns	指定可以沿水平方向收缩的列索引(起始编号是 0)。当该列子控件的内容太多，已经挤满所在行，那么该子控件的内容将向垂直方向伸展。当 TableRow 里面的控件在水平方向上还没有充满整个行时，shrinkColumns 不起作用。当需要设置多列为可收缩时，将列序号用逗号隔开，重复的或者不合法的索引将被忽略。可以使用“＊”表示收缩所有的列
collapseColumns	将 TableLayout 里面指定的列隐藏，若有多列需要隐藏，请用逗号将需要隐藏的列序号用逗号隔开

例如，android：stretchColumns＝"0"设置第 0 列可伸展，android：shrinkColumns＝"1,2"设置第 1 列和第 2 列皆可收缩，android：collapseColumns＝"＊"隐藏所有行。

2. TableLayout.LayoutParams

android.widget.TableLayout.LayoutParams 是 android.widget.LinearLayout.LayoutParams

的直接子类，相对于 LinearLayout. LayoutParams，并没有增加新的属性，这套参数主要强制每个子视图的宽度为 MATCH_PARENT，如果子视图的高度没有指定，强制子视图的高度为 WRAP_CONTENT。

3. TableRow. LayoutParams

相对于 LinearLayout，TableRow 并没有增加属性，但 TableRow 的子元素的属性定义在定义在 TableRow.LayoutParams 中。android.widget.TableRow.LayoutParams 是 android.widget. LinearLayout. LayoutParams 的直接子类，相对于 LinearLayout. LayoutParams，TableRow. LayoutParams 增加了 android：layout_column 和 android：layout_span 两个属性。layout_column 属性设置视图的列位置的索引，即指定该单元格在第几列显示。layout_span 属性设置视图跨越多少个列，即指定该单元格占据的列数(未指定时，为 1)，一个格子不可能占据多行。一个控件也可以同时具备 layout_column 和 layout_span 这两个属性。例如：android：layout_column＝"1"android：layout_span＝"2"，该控件从第 1 列开始、占据 2 列。

4. TableLayout 实例

```
<?xml version = "1.0" encoding = "utf - 8"?>
< TableLayout xmlns:android = "http://schemas.android.com/apk/res/android"
      android:layout_width = "match_parent"
     android:layout_height = "match_parent"
     android:stretchColumns = "1"
     android:shrinkColumns = "0">
     < Button android:text = "direct child is button " />
     < TableRow >
         < Button android:layout_column = "1"
             android:text = "layout_\ncolumn = 1" />
     </TableRow >
     < TableRow >
         < Button android:text = "77777777777" />
         < Button android:text = "1" />
         < Button android:text = "2" />
         < Button android:text = "3" />
     </TableRow >
     < TableRow >
         < Button
             android:text = "layout_column = 1\nlayout_span = 3 "
             android:layout_column = "1"
             android:layout_span = "3" />
     </TableRow >
</TableLayout >
```

该布局文件的布局效果如图 7-10 所示。

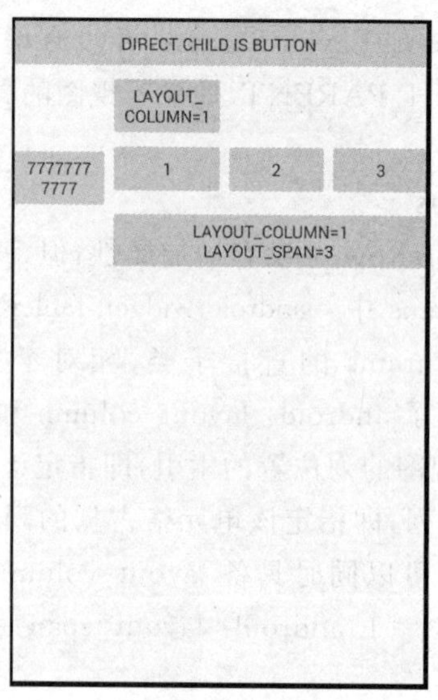

图 7-10　上面 TableLayout 实例的布局效果

7.3　GridLayout

android. widget. GridLayout 是 android. view. ViewGroup 的直接子类，除了继承的 ViewGroup 的属性外，GridLayout 特有的属性如表 7-6 所示。GridLayout 使用无穷细的线把屏幕的视图区域分开成网格，网格线可以通过网格索引引用。一个有 N 列的网格有 N+1 个网格索引，从 0 到 N(包括 0 和 N)，网格索引 0 总是容器的开始边沿，网格索引 N 总是容器的结束边沿。GridLayout 把它所有的子视图放在矩形的网格中，也支持一个子视图件在行、列上都有交错排列。

表 7-6　GridLayout 的 XML 属性

属　　性	解　　释
alignmentMode	取值为 alignBounds 或者 alignMargins。当设置成 alignMargins，使子视图的外边界(由 margin 定义)对齐。当设置成 alignBounds 时，对齐子视图的原始边沿。对齐是指同一行的子视图的上边对齐，同一列的子视图的左边对齐
columnCount	设置 GridLayout 总共有多少列。当自动定位子视图时，创建的列的最大数量
columnOrderPreserved	当设置为 true，使列边界显示的顺序和列索引的顺序相同。默认是 true
orientation	取值是 horizontal 或者 vertical。布局的时候并不使用 orientation 属性。仅仅当 GridLayout 的子视图没有使用布局参数指定行和列的索引时，Orientation 属性才被用来分配行和列的索引，在这种情况下，GridLayout 工作非常类似于 LinearLayout。如果 orientation 取值是 horizontal，水平方向被首先处理，当某一行已经充满 columnCount 列时创建新的一行。如果 orientation 取值是 vertical，当某一列已经充满 rowCount 行时创建新的一列
rowCount	当自动定位子视图时，设置创建的行的数量
rowOrderPreserved	当设置为 true 时，强制行边界显示的顺序和行索引顺序相同
useDefaultMargins	当设置为 true 时，如果视图的布局参数里没有指定 margins，GridLayout 使用默认的 margins

1. 行和列的规格

子视图占用一个或者多个连续单元格，rowSpec 定义了单元格所在组的垂直特征，包括沿着垂直轴方向的索引和沿着垂直方向的对齐方式(每个单元格的高度值)。columnSpec 定义了单元格所在组的水平特征，包括沿着水平轴方向的索引和和沿着水平方向的对齐方式(每个单元格的宽度值)。在一个 GridLayout 中，尽管各个单元格正常情况下不重叠，GridLayout 并没有阻止几个子视图被定义为占据相同的单元格或者单元格组，在这种情况下，不能保证子视图在布局操作完成后自己不会重叠。

若要指定视图显示在固定的行或列，只需设置该子控件的 layout_row 和 layout_column 属性即可，android：layout_row="0"表示从第一行开始，android：layout_column="0"表示从第一列开始，这与编程语言中一维数组的赋值情况类似。

2. 默认单元格分配

如果一个子视图没有指定占据的行和列索引，GridLayout 会使用 orientation、rowCount 和 columnCount 自动指定单元格位置，这时它与 LinearLayout 布局一样，也分为水平和垂直两种方式。当设置 orientation 为水平时，控件从左到右依次排列，但是达到由 columnCount 设置的列数后，控件会自动换行进行排列。同时，对于 GridLayout 布局中的子控件，layout_width 和 layout_height 的默认值为 wrap_content。

如果需要设置某控件跨越多行或多列，只需将该子控件的 layout_rowSpan 或者 layout_columnSpan 属性设置为数值，再设置其 layout_gravity 属性为 fill 即可，前一个设置表明该控件跨越的行数量或列数量，后一个设置表明该控件将填满所跨越的行或列。

3. Space

android.widget.Space 是 android.view.View 的直接子类，用于在通用的布局中的组件之间创建缝隙，是一个轻量级的视图。相邻子视图之间的空间可以使用专用的 Space 视图的实例，或者设置 leftMargin，topMargin，rightMargin 和 bottomMargin 布局参数来指定。当设置为 android：useDefaultMargins="true"时，子视图周围的默认边距将根据当前平台的用户界面风格自动分配。默认边距可通过相应的布局参数来独立覆盖。默认值通常在不同组件之间会产生一个合理的间距，但在不同平台版本之间可能会改变。

4. 多余的空间分配

从 API 21 开始，GridLayout 的多余的空间是基于 weight 分配的。当没有定义 weight 时，按照以前的惯例，如果视图使用 gravity 属性定义了某种形式的对齐就认为是灵活的。也就是说，如果一个组件沿给定的轴定义了 weight 或者 gravity，则该组件在这个方向就是灵活的。如果 weight 和 gravity 都没有定义，该组件就是不灵活的。在同一个行组或列组内的多个组件被认为并行操作的，如果组中所有的组件是灵活的，那么这个小组就是灵活的。为了使一列可以伸展，必须确保它里面所有的组件都定义 gravity 或者 weight 属性。为了阻止某列伸展，确保列中的至少有一个组件既没有定义 gravity，也没有定义 weight 属性。位置在一个共同的边界两侧的行组或者列组，称为复合组，是串行操作的。如果复合组的任何一个元素是灵活的，则这个复合组是灵活的。

5. android：visibility="gone"的解释

对于布局用途，GridLayout 把可见性状态为 gone 的视图当成具有宽度和高度都为零的视图处理，这和直接忽略可见性状态为 gone 的视图的策略有细微不同。例如，如果有一个可见性状态为 gone 的视图单独在一列中，当且仅当这个视图没有指定 gravity 参数，列的宽度将折叠为 0；如果指定了 gravity，可见性状态为 gone 的视图将对布局没有影响，父容器将好像这个视图从来没有添加进去一样进行布局。可见性状态为 gone 的视图在剩余空间分配时被认为具有零权重。这些规则同等地适用于行和列，也适用于行组和列组。

6. GridLayout. LayoutParams

上面描述的 GridLayout 的属性，是 GridLayout 布局自身的属性。android. widget. GridLayout. LayoutParams 是 android. view. ViewGroup. MarginLayoutParams 的直接子类，描述了 GridLayout 布局的子视图的布局信息。GridLayout 布局的子视图跟 GridLayout 有关的属性，定义在 GridLayout. LayoutParams 中。表 7-7 列出了 GridLayout. LayoutParams 的 xml 属性。

表 7-7 GridLayout. LayoutParams 的 XML 属性

属　性	解　释
layout_column	指定了这个视图占据的栅格组的左边的列边界
layout_columnSpan	列跨度，指定了视图占据的栅格组的右边的列边界和左边的列边界的差值
layout_columnWeight	指定了在剩余空间分配过程中分配给视图的水平空间的相对比例
layout_gravity	gravity 指定组件怎样放置在它的格子组中。默认值是 LEFT|BASELINE
layout_row	指定了这个视图占据的栅格组的上边的行边界
layout_rowSpan	行跨度，指定了视图占据的栅格组的下边的行边界和上边的行边界的差值
layout_rowWeight	指定了在剩余空间分配过程中分配给视图的垂直空间的相对比例

GridLayout 既支持行跨度，也支持列跨度，还支持任意形式的对齐。与栅格组关联的基本参数聚集成垂直和水平组件，存储在 GridLayout. Spec 类型的对象 rowSpec 和 columnSpec 中。rowSpec 描述栅格组的垂直特征，columnSpec 描述栅格组的水平特征。

7. GridLayout 的默认值

GridLayout 的子元素的 layout_width 属性和 layout_height 属性的默认值都是 WRAP_CONTENT，因此在 GridLayout 的子视图的布局参数中没有必要显式地声明为 WRAP_CONTENT。另外，当 useDefaultMargins 为 false 时，topMargin、bottomMargin、leftMargin 和 rightMargin 的默认值都为 0；否则，默认值为 UNDEFINED，默认值根据需要计算。其他默认值如下，rowSpec. row＝UNDEFINED，columnSpec. column＝UNDEFINED，rowSpec. rowSpan＝1，columnSpec. columnSpan＝1，rowSpec. alignment＝BASELINE，columnSpec. alignment＝START，rowSpec. weight＝0，columnSpec. weight＝0。

GridLayout 实例 1：alignmentMode

这个例子主要是演示 alignmentMode 取值为 alignMargins 和 alignBounds 布局效果的差别。

```
<?xml version = "1.0" encoding = "utf - 8"?>
<GridLayout xmlns:android = "http://schemas.android.com/apk/res/android"
    android:columnCount = "2"
     android:alignmentMode = "alignMargins"
     android:layout_width = "match_parent"
```

```
android:layout_height = "match_parent">
    <TextView android:text = "1" android:textSize = "50sp" />
    <TextView android:text = "222" android:textSize = "50sp"/>
    <TextView android: text = "3456" android: textSize = "50sp" android: layout_margin = "
30dp"/>
    <TextView android:text = "4" android:textSize = "50sp"/>
</GridLayout>
```

在上面的代码中，android: alignmentMode＝"alignMargins"，第一列所有视图的左边的外边界对齐了，第二行所有视图的上边的外边界对齐了，上面的布局文件的布局效果如图 7-11 所示。

把上面的布局文件中的 alignmentMode 修改为 alignBounds，第二行第一列的视图设置了 margin，为了让第一列的所有视图的左边的原始边界对齐，第一行第一列的视图位置向右移动了。为了让第二行所有视图的上边的原始边界对齐，第二行第二列的视图位置向下移动了。alignmentMode 为 alignBounds 的对齐效果如图 7-12 所示。

图 7-11　GridLayout 实例 1 中 alignMargins 的布局效果

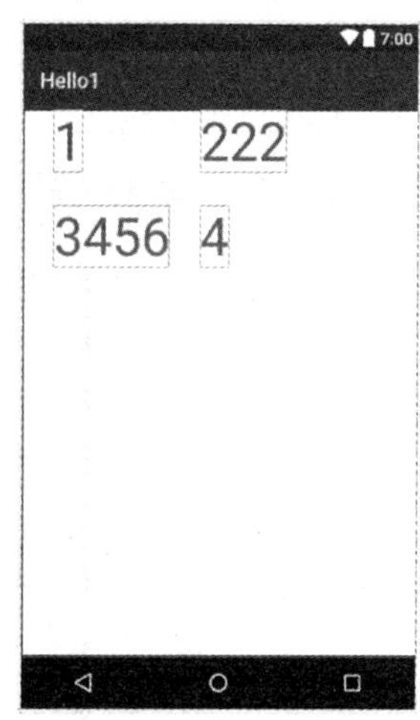

图 7-12　GridLayout 实例 1 中 alignBounds 的布局效果

GridLayout 实例 2：orientation 为 horizontal

在下面的布局文件中，GridLayout 中有 9 个 Button，每个 Button 都没有指定 layout_row 和 layout_column，因此需要自动给这些 Button 分配行索引和列索引，orientation 为 horizontal，所以所有的子视图先水平方向排列，columnCount＝3，控件超过 columnCount 指定的数量后，就创建新的一行。

```
<?xml version = "1.0" encoding = "utf - 8"?>
<GridLayout xmlns:android = "http://schemas.android.com/apk/res/android"
android:orientation = "horizontal"
android:layout_width = "match_parent"
    android:layout_height = "match_parent"
```

```
    android:columnCount = "3" >
    <Button android:layout_columnWeight = "1" android:layout_rowWeight = "1" android:text =
"1" />
    <Button android:layout_columnWeight = "1" android:layout_rowWeight = "1" android:text =
"2"/>
    <Button android:layout_columnWeight = "1" android:layout_rowWeight = "1" android:text =
"3"/>
    <Button android:layout_columnWeight = "1" android:layout_rowWeight = "1" android:text =
"4"/>
    <Button android:layout_columnWeight = "1" android:layout_rowWeight = "1"android:text =
"5"/>
    <Button android:layout_columnWeight = "1" android:layout_rowWeight = "1" android:text =
"6"/>
    <Button android:layout_columnWeight = "1" android:layout_rowWeight = "1" android:text =
"7"/>
    <Button android:layout_columnWeight = "1" android:layout_rowWeight = "1" android:text =
"8"/>
    <Button android:layout_columnWeight = "1" android:layout_rowWeight = "1" android:text =
"9"/>
</GridLayout>
```

上面的布局文件的效果如图 7-13 所示。

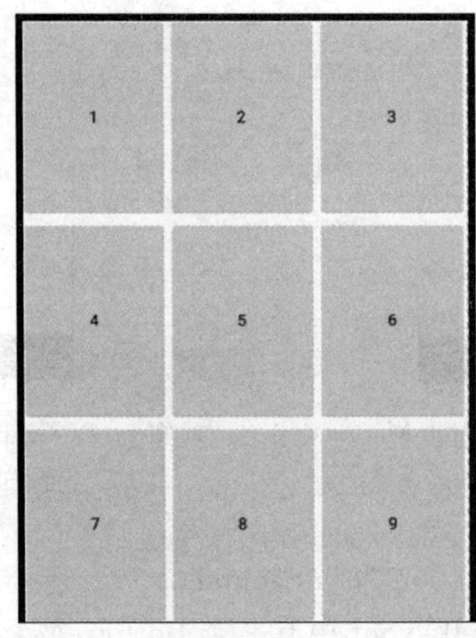

图 7-13 GridLayout 实例 2 中 orientation="horizontal"的布局效果

GridLayout 实例 3：orientation 为 vertical

在下面的布局文件中，GridLayout 中有 9 个 Button，每个 Button 仍然没有指定 layout_row 和 layout_column，因此需要自动给这些 Button 分配行索引和列索引，orientation 为"vertical"，所以所有的子视图先垂直方向排列，rowCount 为 2，控件超过 rowCount 指定的数量后，就创建新的一列。这个例子与实例 1 唯一不同的是 GridLayout 的 android：orientation 属性不一样。

```
<?xml version = "1.0" encoding = "utf - 8"?>
< GridLayout xmlns:android = "http://schemas.android.com/apk/res/android"
    android:orientation = "vertical" android:layout_width = "match_parent"
    android:layout_height = "match_parent"
    android:rowCount = "3">
    < Button android:layout_columnWeight = "1" android:layout_rowWeight = "1" android:text =
"1" />
    < Button android:layout_columnWeight = "1" android:layout_rowWeight = "1" android:text =
"2"/>
    < Button android:layout_columnWeight = "1" android:layout_rowWeight = "1" android:text =
"3"/>
    < Button android:layout_columnWeight = "1" android:layout_rowWeight = "1" android:text =
"4"/>
    < Button android:layout_columnWeight = "1" android:layout_rowWeight = "1"android:text =
"5"/>
    < Button android:layout_columnWeight = "1" android:layout_rowWeight = "1" android:text =
"6"/>
    < Button android:layout_columnWeight = "1" android:layout_rowWeight = "1" android:text =
"7"/>
    < Button android:layout_columnWeight = "1" android:layout_rowWeight = "1" android:text =
"8"/>
    < Button android:layout_columnWeight = "1" android:layout_rowWeight = "1" android:text =
"9"/>
</GridLayout >
```

上面的布局文件的效果如图 7-14 所示。

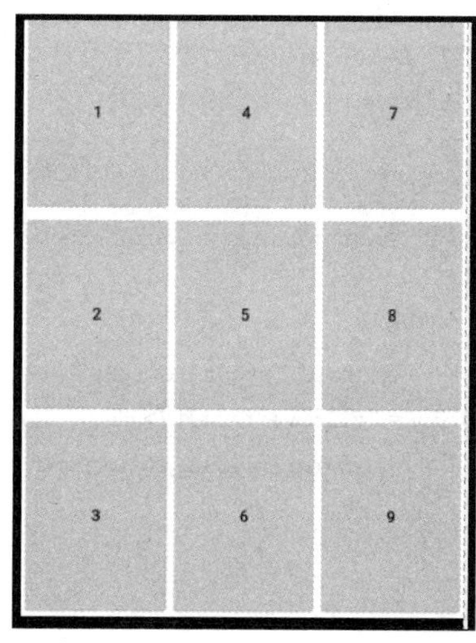

图 7-14　GridLayout 实例 3 中 orientation="horizontal"的布局效果

布局的方向顺序(即先 x 轴方向还是先 y 轴方向)是非常重要的。例如一个子视图的高度可能依赖于它的宽度。例如,如果 GridlLayout 中包含一个 TextView(或者它的派生类,例如 Button,EditText,CheckBox 等等),并且处于多行模式,这时最好把 GridLayout 的方向设置为水平,因为 TextView 如果给定了宽度值,就能得到高度值,但是反过来却不行。除了上面提到的问题，方向并不影响 GridlLayout 实际的布局效果，最好还是设置 GridLayout 的方向为水平。

GridLayout 实例 4

在下面的布局中，android: text="0"的按钮，layout_columnSpan="2"表示这个按钮占据两列的位置，layout_gravity="fill"的含义是必要的时候增加对象的横纵向大小，以完全充满其父容器，表示把它占据的两列位置都填充满，因而实际宽度是 2 列，它占据了(3,0)和(3,1)两个网格。若去掉 layout_gravity="fill"，layout_gravity 的默认值是 LEFT | BASELINE，是不改变大小的，因而宽度是 1 列，占据(3,0)位置，(3,1)位置是空着的。

```
<?xml version = "1.0" encoding = "utf - 8"?>
<GridLayout xmlns:android = "http://schemas.android.com/apk/res/android"
    android:layout_width = "match_parent"
    android:layout_height = "match_parent"
    android:columnCount = "4"
    android:alignmentMode = "alignMargins"
android:orientation = "horizontal">
    <Button android:text = "1" />
    <Button android:text = "2" />
    <Button android:text = "3" />
    <Button android:text = "/" />
    <Button android:text = "4" />
    <Button android:text = "5" />
    <Button android:text = "6" />
    <Button android:text = " * " />
    <Button android:text = "7" />
    <Button android:text = "8" />
    <Button android:text = "9" />
    <Button android:text = " - " />
    <Button android: text = "0" android: layout_columnSpan = "2" android: layout_gravity =
"fill" />
    <Button android:text = "." />
    <Button android:text = " + " android:layout_rowSpan = "2"
        android:layout_gravity = "fill_vertical"/>
    <Button android: text = " = " android: layout_columnSpan = "3" android: layout_gravity =
"fill_horizontal"/>
    <Space />
</GridLayout>
```

上面的布局文件的效果如图 7-15 所示。

GridLayout 实例 5

在下面布局文件中，大部分 Button 指定了 layout_row 和 layout_column，还有几个没有指定，没有指定的，将根据 orientation 为 horizontal 和 columnCount 为 4，沿水平方向增加索引，当超过 columnCount 指定的列数后转到下一行。

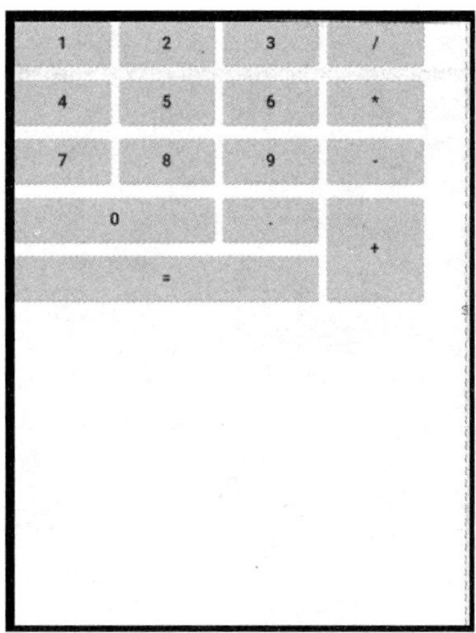

图 7-15 GridLayout 实例 4 中加号占 2 行的布局效果

```
<?xml version = "1.0" encoding = "utf - 8"?>
< GridLayout xmlns:android = "http://schemas.android.com/apk/res/android"
     android:layout_width = "match_parent"
     android:layout_height = "match_parent"
     android:columnCount = "4"
     android:alignmentMode = "alignMargins"
     android:orientation = "horizontal" >
< Button android:text = "1" android:layout_row = "0" android:layout_column = "0" />
< Button android:text = "2" android:layout_row = "0" android:layout_column = "1"/>
< Button android:text = "3" android:layout_row = "0" android:layout_column = "2"/>
    < Button android:text = "/" />
    < Button android:text = "4" />
< Button android:text = "5" android:layout_row = "1" android:layout_column = "1"/>
< Button android:text = "6" android:layout_row = "1" android:layout_column = "2"/>
< Button android:text = " * " android:layout_row = "1" android:layout_column = "3"/>
< Button android:text = "7" android:layout_row = "2" android:layout_column = "0"/>
< Button android:text = "8" android:layout_row = "2" android:layout_column = "1"/>
< Button android:text = "9" android:layout_row = "2" android:layout_column = "2"/>
< Button android:text = " - " android:layout_row = "2" android:layout_column = "3"/>
    < Button android:text = "0" android:layout_row = "3" android:layout_column = "0"
        android:layout_columnSpan = "2" android:layout_gravity = "fill" />
< Button android:text = "." android:layout_row = "3" android:layout_column = "2" />
    < Button android:text = " + " android:layout_row = "3" android:layout_column = "3"
        android:layout_rowSpan = "2"
        android:layout_gravity = "fill_vertical" />
    < Button android:text = " = " android:layout_row = "4" android:layout_column = "0"
        android:layout_columnSpan = "3" android:layout_gravity = "fill_horizontal"/>
    < Space android:layout_row = "5" android:layout_column = "0"
        android:layout_columnSpan = "4" />
</GridLayout >
```

上面的布局文件的效果如图 7-16 所示。

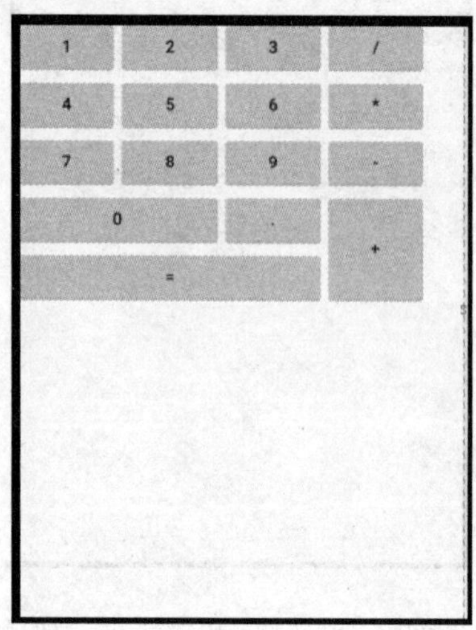

图 7-16　GridLayout 实例 4 中等号占 3 列的布局效果

7.4　相对布局(RelativeLayout)

android. support. percent. PercentFrameLayout 在 API 26 中已经被废弃,可以使用 ConstraintLayout 来替代 PercentFrameLayout 的功能。

android. widget. RelativeLayout 是 android. view. ViewGroup 的直接子类,子视图的位置可以用相对位置(相对于彼此或者它们的父视图)来描述,每个子视图能够指定它相对于兄弟视图的位置(比如在其他视图的左边、右边、上边或是下边)或者与父视图的某条边对齐(比如与 RelativeLayout 底部对齐、或者顶部对齐等)。RelativeLayout 的大小与子视图的位置不能有循环依赖关系。表 7-8 列出了 RelativeLayout 的属性。

表 7-8　RelativeLayout 的属性

属　性	解　释
gravity	指定该对象在它的边界内,怎样放置它的内容
ignoreGravity	指定哪个视图应该不受 gravity 的影响

RelativeLayou 层级结构比较扁平,可以消除嵌套视图组,这可以提高运行时性能。一个 RelativeLayout 可以替代多个嵌套的 LinearLayout。

RelativeLayout. LayoutParams

上面描述的 RelativeLayout 的属性,是 RelativeLayout 元素自身的属性。android. widget. RelativeLayout. LayoutParams 是 android. view. ViewGroup. MarginLayoutParams 的直接子类,描述 RelativeLayout 的子元素所支持的属性。在 RelativeLayout 中,子视图默认在布局的左上角显示,所以必须使用 RelavieLayout. LayoutParams 提供的布局属性来定义每个子视图的位置。属性可以分成两类,第一类指定该视图与兄弟视图的位置关系,属性的取值为兄弟视图的 id,第二类指定该视图与父视图的位置关系,属性的取值为布尔量。第一类属性包括：①layout_above、layout_below、layout_toLeftOf、layout_toRightOf 把该视图放置在由 ID 指定的视图的上边、下边、左边、右边；② layout_alignLeft、layout_alignRight、layout_alignStart、layout_alignEnd、layout_alignTop、layout_alignBottom 把该视图和由 ID 指定的视图对齐一条边：左边、右边、开始、结束、上边、下边；③ layout_alignBaseline 让两个视图(该视图和由 ID 指定的视图)的 baseline 对齐。书写英语单词时为了规范书写会设有四条线，从上至下第三条就是基线。基线对齐主要是为了两个控件中显示的英文单词的基线对齐。第二类属性包括：① layout_alignParentLeft、layout_alignParentRight、layout_alignParentStart、layout_alignParentEnd、layout_alignParentTop、layout_alignBottom 设置该视图与父视图是否对齐某一边：左边、右边、开始、结束、上边、下边。②layout_centerHorizontal、layout_centerVertical、layout_centerInParent 设置该视图是否放在父视图的水平居中、垂直居中、同时居中。

RelativeLayout 实例

```
<?xml version = "1.0" encoding = "utf - 8"?>
< RelativeLayout xmlns:android = "http://schemas.android.com/apk/res/android"
    android:layout_width = "match_parent"
    android:layout_height = "match_parent"
    android:paddingLeft = "16dp"
    android:paddingRight = "16dp" >
    < Button
        android:id = "@ + id/button1"
        android:layout_width = "wrap_content"
        android:layout_height = "wrap_content"
        android:layout_alignParentLeft = "true"
        android:layout_alignParentTop = "true"
        android:layout_marginLeft = "20dp"
        android:layout_marginTop = "20dp"
        android:text = "Button1" />
    < Button
        android:id = "@ + id/button2"
        android:layout_width = "wrap_content"
        android:layout_height = "wrap_content"
        android:layout_below = "@ + id/button1"
        android:layout_toRightOf = "@ + id/button1"
        android:layout_marginTop = "15dp"
        android:text = "Button2" />
    < Button
        android:id = "@ + id/button3"
        android:layout_width = "wrap_content"
        android:layout_height = "wrap_content"
        android:layout_below = "@ + id/button2"
        android:layout_toLeftOf = "@ + id/button2"
        android:layout_marginTop = "15dp"
        android:text = "Button3" />
</RelativeLayout >
```

上面布局文件的效果如图 7-17 所示。

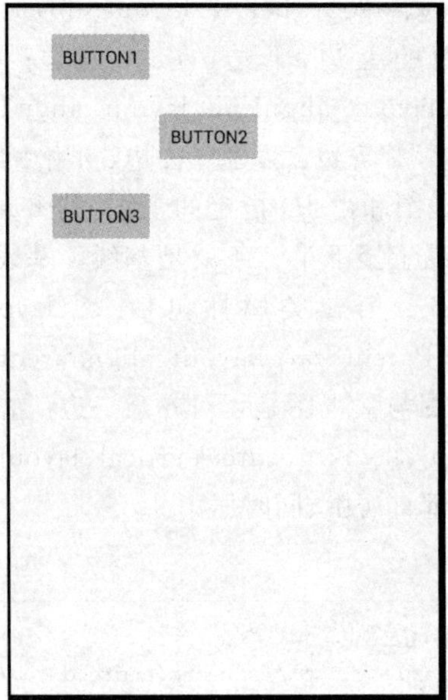

图 7-17 RelativeLayout 实例的布局效果

PercentRelativeLayout

这个类在 API 26 中已经被废弃，下面的示例演示怎样使用 ConstraintLayout 来替代 PercentRelativeLayout 的功能。

```
<android.support.constraint.ConstraintLayout
     xmlns:android="http://schemas.android.com/apk/res/android"
     xmlns:app="http://schemas.android.com/apk/res-auto"
     android:layout_width="match_parent"
     android:layout_height="match_parent">

  <android.support.constraint.Guideline
     android:layout_width="wrap_content"
android:layout_height="wrap_content"
     android:id="@+id/left_guideline"
app:layout_constraintGuide_percent=".15"
     android:orientation="vertical"/>
  <android.support.constraint.Guideline
     android:layout_width="wrap_content"
android:layout_height="wrap_content"
     android:id="@+id/right_guideline"
app:layout_constraintGuide_percent=".85"
     android:orientation="vertical"/>
  <android.support.constraint.Guideline
     android:layout_width="wrap_content"
android:layout_height="wrap_content"
     android:id="@+id/top_guideline"
app:layout_constraintGuide_percent=".15"
     android:orientation="horizontal"/>
```

```
    <android.support.constraint.Guideline
        android:layout_width = "wrap_content"
android:layout_height = "wrap_content"
        android:id = "@ + id/bottom_guideline"
app:layout_constraintGuide_percent = ".85"
        android:orientation = "horizontal"/>
    <Button android:text = "Button"
android:layout_width = "0dp" android:layout_height = "0dp"
        android:id = "@ + id/button"
        app:layout_constraintLeft_toLeftOf = "@ + id/left_guideline"
        app:layout_constraintRight_toRightOf = "@ + id/right_guideline"
        app:layout_constraintTop_toTopOf = "@ + id/top_guideline"
        app:layout_constraintBottom_toBottomOf = "@ + id/bottom_guideline" />
</android.support.constraint.ConstraintLayout>
```

7.5 FrameLayout

android.widget.FrameLayout 是 android.view.ViewGroup 的直接子类，用于在屏幕上画出一块区域，以便显示单一的视图。通常，FrameLayout 应该用来容纳单个子视图，因为很难有一种方式组织子视图、能够伸缩到不同的屏幕大小，同时子视图始终相互不覆盖。可以添加多个子视图到 FrameLayout，所有的子视图默认在屏幕的左上角；每个子视图都可以使用 layout_gravity 属性控制它在 FrameLayout 中的位置。子视图的绘制基于栈，最近添加的子视图在栈的最上面。FrameLayout 的大小是最大的子视图的大小(加上 padding)，不管它是否可见。处于 GONE 状态的视图，只有 setConsiderGoneChildrenWhenMeasuring()设置为 true，才考虑它的大小。

FrameLayout 的属性包括 foregroundGravity 和 measureAllChildren，foregroundGravity 设置前景图像的显示位置，measureAllChildren 设置是仅测量所有的子视图，还是仅仅测量那些处于 VISIBLE 或者 INVISIBLE 状态的视图。

android.widget.FrameLayout.LayoutParams 是 android.view.ViewGroup.MarginLayoutParams 的直接子类，layout_gravity 属性用于设置子视图相对父容器的位置。因为该属性是相对父容器的位置，所以该属性没有直接放在控件类里，而是放在了容器类的内部类 LayoutParams 里。一个 FrameLayout 实例，布局文件的内容如下：

```
<?xml version = "1.0" encoding = "utf - 8"?>
<FrameLayout xmlns:android = "http://schemas.android.com/apk/res/android"
    android:orientation = "vertical" android:layout_width = "match_parent"
    android:layout_height = "match_parent">
   <Button android:layout_width = "200dp" android:layout_height = "40dp"
      android:text = "button1" android:gravity = "bottom|right"/>
    <Button android:layout_width = "100dp" android:layout_height = "100dp"
       android:text = "button2"android:gravity = "bottom|right"/>
    <Button android:layout_width = "120dp" android:layout_height = "120dp"
       android:text = "button3" android:gravity = "bottom|right"
```

```
        android:layout_gravity = "bottom|left"/>
    <Button android:layout_width = "120dp" android:layout_height = "120dp"
        android:text = "button4" android:gravity = "bottom|right"
        android:layout_gravity = "bottom|right"/>
</FrameLayout>
```

上面的布局文件的效果如图 7-18 所示。

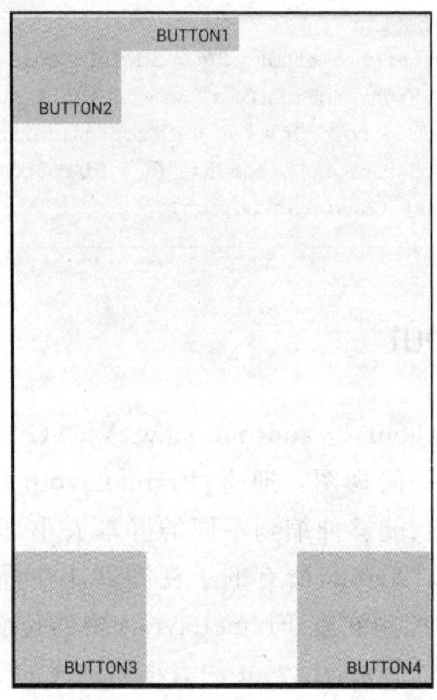

图 7-18 FrameLayout 实例的布局效果

7.6 CardView

Android 5.0 引入了 android.support.v7.widget.CardView，它是 android.widget.FrameLayout 的直接子类，可以把它作为一个带有圆角背景和阴影的 FrameLayout。CardView 一般被用在 ListView 或者 RecylerView 的 item 中，用来显示不同的内容，因为卡片的边界会帮助用户快速扫描列表。Cardview 的属性，如表 7-9 所示，在 XML 文件中使用这些属性时，并没有必要使用硬编码的包名，只需要在属性前添加"app："，app 用来代替相应的包名 android.support.v7.cardview。

表 7-9 CardView 的 XML 属性

属　性	解　释
cardBackgroundColor	设置 CardView 的背景颜色
cardCornerRadius	设置 CardView 的圆角半径
cardElevation	设置 CardView 的阴影高度
cardMaxElevation	设置 CardView 的最大阴影高度
cardPreventCornerOverlap	往 API 20 及以前的版本添加 padding 阻止内容和圆角相交

续表

属　性	解　释
cardUseCompatPadding	往 API 21 及以后的版本添加 padding 以便和以前的版本有相同的测量
contentPadding	设置在卡片边沿和 CardView 的子视图之间的内部 padding
contentPaddingBottom	设置卡片的下部边沿和 CardView 的子视图之间的 padding
contentPaddingLeft	设置卡片的左边沿和 CardView 的子视图之间的 padding
contentPaddingRight	设置卡片的右边沿和 CardView 的子视图之间的 padding
contentPaddingTop	设置卡片的上边沿和 CardView 的子视图之间的 padding

CardView 使用 elevation 属性设置阴影大小，由于圆角裁剪代价极高，在 Android 5.0 以前的平台上，CardView 并不裁剪跟圆角相交的子视图。替代的方法是在它的内容上添加 padding，并在 padding 上加上阴影。既然 padding 用于设置阴影，因而在 XML 文件中使用 contentPadding 属性(而不是 padding 属性)或者在代码中使用 setContentPadding(int,int,int,int)设置 CardView 的子视图与 CardView 的边沿间的 padding。

如果指定了 CardView 的精确尺寸，它的内容面积在 Android 5.0 以前和以后的版本是不同的(由于阴影)。如果希望在 Android 5.0 及以后的版本中添加内部 paddings，可以调用 setUseCompatPadding(boolean)，并传递 true 作参数。

为了以反向兼容的方式改变 CardView 的阴影大小 elevation，使用 setCardElevation(float)方法。为了避免当阴影大小改变时移动视图，应该使用 getMaxCardElevation()限制住阴影大小。如果希望动态改变阴影大小，应该在初始化时调用 setMaxCardElevation(float)。

7.7　SeekBar

android.widget.SeekBar 类的继承关系，如图 7-19 所示。SeekBar 是 ProgressBar 的扩展，它添加了可拖动的滑块。用户可以触摸滑块、向左向右拖动来设置进度。现在已经不鼓励在 SeekBar 的左边或者右边放置可获得焦点的控件。SeekBar 的客户端可以附着一个 SeekBar.OnSeekBarChangeListener，用于通知用户的动作。SeekBar 的 thumb 属性用于设置 SeekBar 上的滑块使用的图片，必须是对另一资源的引用，形式为"@[+][package:]type:name"；或者是主题属性的引用，形式为"?[package:][type:]name"。ProgressBar 用于显示进度，在一个进度条上也可以显示一个次要的进度，例如流播放器的缓冲进度。ProgressBar 支持两种表示进度的模式：确定模式和不确定模式。应该在应用的用户接口里或者通知里采用不间断的方式显示 ProgressBar，而不是在 Dialog 中显示进度条。

如果不知道操作需要花费多长时间时，就选用 ProgressBar 的不确定模式，图像可以是一个旋转的圆环或一个水平条。ProgressBar 的默认模式是不确定模式，默认的进度条为一个环形的动画，并且没有特定的进度指示。

```
java.lang.Object
  android.view.View
    android.widget.ProgressBar
      android.widget.AbsSeekBar
        android.widget.SeekBar
```

图 7-19　SeekBar 类的继承关系

当希望显示特定数量的进度已经发生时，采用进度条的确定模式，例如正在播放的音频文件的剩余百分数。进度条默认的式样是一个旋转的圆环(不确定的指示器)，为了显示实际的进度，需要设置 ProgressBar 的式样为 Widget_ProgressBar_Horizontal(水平进度条)。可以通过调用 setProgress(int)方法更新显示的进度的百分数，也可以调用 incrementProgressBy(int)给现在的进度增加一定数量。可以通过设置 max 属性调整最大值。max 的默认值为 100。

不确定进度条的示例如下：

```
<ProgressBar android:id = "@ + id/indeterminateBar"
      android:layout_width = "wrap_content" android:layout_height = "wrap_content"/>
```

确定进度条的示例如下：

```
<ProgressBar
     android:id = "@ + id/determinateBar"
     style = "@android:style/Widget.ProgressBar.Horizontal"
     android:layout_width = "wrap_content"
android:layout_height = "wrap_content"
     android:progress = "25"/>
```

系统中提供的进度条式样还包括：Widget. ProgressBar. Horizontal、Widget. ProgressBar. Small、Widget. ProgressBar. Large、Widget. ProgressBar. Inverse、Widget. ProgressBar. Small. Inverse、Widget. ProgressBar. Large. Inverse。

7.8 AddStatesFromChildren 实验

1. 新建工程 AddStatesFromChildren

具体步骤：运行 Android Studio，单击 Start a new Android Studio Project，在 Create New Project 页中，Application Name 为 AddStatesFromChildren，Company Domain 为 cuc，Project Location 为 C:\android\myproject\AddStatesFromChildren，单击 Next 按钮，选中 Phone and Tablet，minimum sdk 设置为 API 19：Android 4.4(minSdkVersion)，单击 Next 按钮，选择 Empty Activity，再单击 Next 按钮，勾选 Generate Layout File 选项，勾选 Backwards Compatibility(AppCompat)选项，单击 Finish 按钮。

2. 编辑 colors. xml 文件中，添加一些颜色的定义

具体步骤：在工程结构视图中(若没有看到工程视图，单击 Android Studio 界面左侧垂直方向的 project，在 project 视图的列表框中选择 Android)，展开 app(App 展开后前面的箭头方向变成向下，若没有展开，箭头方向则是水平向右，单击水平向右箭头就可以展开)，再展开 res|values，values 下有三个文件，单击 colors. xml，打开该文件进行编辑。编辑后，colors. xml 的文件内容如下：

```
<?xml version = "1.0" encoding = "utf - 8"?>
<resources>
```

```
    <color name = "colorPrimary"># 3F51B5 </color>
    <color name = "colorPrimaryDark"># 303F9F </color>
    <color name = "colorAccent"># FF4081 </color>
    <color name = "white"># ffffff </color>
    <color name = "black"># 000000 </color>
    <color name = "green"># 00ff00 </color>
    <color name = "gray"># cccccc </color>
</resources>
```

3. 在 drawable 文件夹下，新建两个资源文件

具体步骤：在左边的工程视图中，依次展开 app | res，右击 drawable，选择 New | Drawable resource file，File name 输入框中输入 button_drawable_selector，Source set 为默认值 main，单击 OK 按钮，这时 Android Studio 创建并打开了 button_drawable_selector. xml 文件，编辑后，button_drawable_selector. xml 内容如下：

```
<?xml version = "1.0" encoding = "utf - 8"?>
<selector xmlns:android = "http://schemas.android.com/apk/res/android">
    <item android:drawable = "@color/colorAccent" android:state_focused = "true"/>
    <item android:drawable = "@color/green" android:state_pressed = "true"/>
    <item android:drawable = "@color/gray"/>
</selector>
```

同样的方式，新建 viewgroup_drawable_selector. xml，内容如下.

```
<?xml version = "1.0" encoding = "utf - 8"?>
<selector xmlns:android = "http://schemas.android.com/apk/res/android">
    <item android:drawable = "@color/colorAccent" android:state_focused = "true"/>
    <item android:drawable = "@color/green" android:state_pressed = "true"/>
</selector>
```

4. 新建布局文件 viewgroup. xml

如果 res 里没有 Layout 文件夹，就手动创建 Layout 文件夹，否则跳过这一步；创建 Layout 文件夹的步骤如下：右击 res，选择 New | Android Resource Directory，resource type 选择 Layout，再单击 OK 按钮，Layout 文件夹就创建好了。

在工程结构视图中，右击 Layout，选择 New | Layout resource file，File name 输入框输入 viewgroup，Root element 输入框输入 LinearLayout，Source set 为 main，单击 OK 按钮。layout/viewgroup. xml 内容如下：

```
<?xml version = "1.0" encoding = "utf - 8"?>
<LinearLayout xmlns:android = "http://schemas.android.com/apk/res/android"
    android:orientation = "vertical" android:layout_width = "match_parent"
    android:layout_height = "match_parent">
    <LinearLayout
        android:orientation = "vertical"
        android:layout_width = "match_parent"
        android:layout_height = "wrap_content"
```

```
        android:addStatesFromChildren = "true"
        android:background = "@drawable/viewgroup_drawable_selector"
        android:layout_weight = "1">
        <TextView
            android:layout_width = "match_parent"
            android:layout_height = "wrap_content"
            android:text = "this TextView belong to group 1"
            android:layout_weight = "1" />
        <Button
            android:focusable = "true"
            android:layout_width = "match_parent"
            android:layout_height = "wrap_content"
            android:text = "this Button belong to group 1"
            android:layout_weight = "1"
            android:background = "@drawable/button_drawable_selector" />
    </LinearLayout>
    <LinearLayout
        android:orientation = "vertical"
        android:layout_width = "match_parent"
        android:layout_height = "wrap_content"
        android:addStatesFromChildren = "true"
        android:background = "@drawable/viewgroup_drawable_selector"
        android:layout_weight = "1">
        <TextView
            android:layout_width = "match_parent"
            android:layout_height = "wrap_content"
            android:text = "this TextView belong to group 2"
            android:layout_weight = "1" />
        <Button
            android:focusable = "true"
            android:layout_width = "match_parent"
            android:layout_height = "wrap_content"
            android:text = "this Button belong to group 2"
            android:layout_weight = "1"
            android:background = "@drawable/button_drawable_selector" />
    </LinearLayout>
</LinearLayout>
```

5. 编辑 MainActivity. Java

编辑后，MainActivity. Java 内容如下：

```
package edu.cuc.addstatesfromchildren;
import android.support.v7.app.AppCompatActivity;
import android.os.Bundle;
public class MainActivity extends AppCompatActivity {
    @Override
    protected void onCreate(Bundle savedInstanceState) {
        super.onCreate(savedInstanceState);
        setContentView(R.layout.viewgroup);
    }
}
```

6. 编译运行

具体步骤：把Android手机通过USB数据线连接到计算机，在手机的弹出窗口上勾选允许USB调试。单击工具栏上的Run app图标，选中已经连接的手机，单击OK按钮，在手机的App运行界面上，当用手指按住界面中的某个按钮不松开时，可以看到它的父LinearLayout的背景改变为press状态指定的绿色。

7.9 实验：CardView及SeekBar的使用

1. 新建工程SeekBar1

具体步骤：运行Android Studio，单击Start a New Android Studio Project，在Create New Project页中，Application name为SeekBar1，Company domain为cuc，Project location为C:\android\myproject\SeekBar1，单击Next按钮，选中Phone and Tablet，minimum sdk设置为API 19：Android 4.4(minSdkVersion)，单击Next按钮，选择Empty Activity，再单击Next按钮，勾选Generate Layout File选项，勾选Backwards Compatibility(AppCompat)选项，单击Finish按钮。

2. 添加cardview-v7库，并同步

因为要使用android.support.v7.widget.CardView，所以需要添加com.android.support：cardview-v7的库依赖。具体步骤为：

单击Android Studio的菜单File|Project Structure|app|Dependencies，单击右边的"+"，选择1 Library dependency选项，在输入栏中输入cardview-v7，并且按下回车键进行搜索，搜索到cardview库后，如果没有版本信息，就添加版本信息，例如28.+，单击OK按钮完成库的添加，再单击OK按钮，关闭Project Structure界面。

也可以在build.gradle(module：app)文件的dependencies中直接添加依赖：

```
implementation 'com.android.support:cardview-v7:28.+'
```

由于build.gradle文件内容已被修改，这个文件窗口的上面就出现提示：

"Gradle files have changed since last project sync. A project sync may be necessary for the IDE to work properly. sync now"，单击sync now按钮进行同步。

3. 添加DataBinding支持，并同步

在文件build.gradle(module：app)的android配置的花括号内添加：

```
dataBinding.enabled = true
```

由于修改了build.gradle文件内容，这个文件窗口的上面就出现提示：

"Gradle files have changed since last project sync. A project sync may be necessary for the IDE to work properly. sync now"，单击sync now按钮进行同步。

4. 新建布局文件activity_main1.xml

```
<?xml version="1.0" encoding="utf-8"?>
<layout>
    <data/>
```

```
<LinearLayout xmlns:android = "http://schemas.android.com/apk/res/android"
    xmlns:app = "http://schemas.android.com/apk/res - auto"
    android:orientation = "vertical" android:layout_width = "match_parent"
    android:layout_height = "match_parent">
    <android.support.v7.widget.CardView
        android:id = "@ + id/cardView"
        android:layout_width = "wrap_content"
        android:layout_height = "wrap_content"
        app:cardBackgroundColor = "@android:color/white"
        app:cardElevation = "24dp"
        android:layout_gravity = "center_horizontal"
        android:layout_margin = "40dp">
        <LinearLayout
            android:layout_width = "match_parent"
            android:layout_height = "match_parent"
            android:orientation = "vertical">
            <ImageView
                android:layout_width = "wrap_content"
                android:layout_height = "wrap_content"
                android:src = "@mipmap/ic_launcher" />
            <TextView
                android:layout_width = "wrap_content"
                android:layout_height = "wrap_content"
                android:text = "I am a TextView in CardView" />
            <Button
                android:layout_width = "wrap_content"
                android:layout_height = "wrap_content"
                android:text = "I am a TextView in CardView" />
        </LinearLayout>
    </android.support.v7.widget.CardView>
    <TextView
        android:id = "@ + id/elevationTxt"
        android:layout_marginTop = "50sp"
        android:layout_width = "wrap_content"
        android:layout_height = "wrap_content"
        android:text = "change elevation shallow"/>
    <SeekBar
        style = "@style/Widget.AppCompat.ProgressBar.Horizontal"
        android:id = "@ + id/elevationSeekBar"
        android:layout_width = "match_parent"
        android:layout_height = "40dp" />
    <TextView
        android:id = "@ + id/radiusTxt"
        android:layout_width = "wrap_content"
        android:layout_height = "wrap_content"
        android:text = "change Radius"/>
    <SeekBar
        android:id = "@ + id/radiusSeekBar"
        android:layout_width = "match_parent"
```

```
            android:layout_height = "40sp" />
    </LinearLayout>
</layout>
```

5. 编辑 MainActiviy. java

编辑后，MainActiviy. java 内容如下：

```
package cuc.seekbar1;
import android.databinding.DataBindingUtil;
import android.support.v7.app.AppCompatActivity;
import android.os.Bundle;
import android.widget.SeekBar;
import cuc.seekbar1.databinding.ActivityMain1Binding;

public class MainActivity extends AppCompatActivity {
    @Override
    protected void onCreate(Bundle savedInstanceState) {
        super.onCreate(savedInstanceState);
        final ActivityMain1Binding binding = DataBindingUtil.setContentView(this, R.
layout.activity_main1 );
        binding.elevationSeekBar.setOnSeekBarChangeListener(new SeekBar.OnSeekBarChangeListener() {
            @Override
            public void onProgressChanged(SeekBar seekBar, int progress, boolean fromUser) {
                if (fromUser) {
                    binding.cardView.setCardElevation((float) progress);
                    binding.elevationTxt.setText("change elevation shallow,progress ->" +
progress);
                }
            }
            @Override
            public void onStartTrackingTouch(SeekBar seekBar) {
            }
            @Override
            public void onStopTrackingTouch(SeekBar seekBar) {
            }
        });
        binding.radiusSeekBar.setOnSeekBarChangeListener(new SeekBar.OnSeekBarChangeListener() {
            @Override
            public void onProgressChanged(SeekBar seekBar, int progress, boolean fromUser) {
                if(fromUser){
                    binding.cardView.setRadius((float) progress);
                    binding.radiusTxt.setText("change Radius,progress ->" + progress);
                }
            }
            @Override
            public void onStartTrackingTouch(SeekBar seekBar) {
            }
```

```
            @Override
            public void onStopTrackingTouch(SeekBar seekBar) {
            }
        });
    }
}
```

6. 完成上述编辑后，编译运行

7.10 本章主要参考文献

1. https://developer.android.google.cn/reference/android/view/View.html
2. https://developer.android.google.cn/reference/android/widget/LinearLayout.html
3. https://developer.android.google.cn/reference/android/widget/LinearLayout.LayoutParams.html
4. https://developer.android.google.cn/reference/android/view/ViewGroup.LayoutParams.html
5. https://developer.android.google.cn/reference/android/view/ViewGroup.MarginLayoutParams.html
6. https://developer.android.google.cn/reference/android/view/ViewGroup.html
7. https://developer.android.google.cn/reference/android/widget/TableLayout.html
8. https://developer.android.google.cn/reference/android/widget/TableRow.html
9. https://developer.android.google.cn/reference/android/widget/GridLayout
10. https://developer.android.google.cn/reference/android/widget/RelativeLayout.html
11. https://developer.android.google.cn/reference/android/widget/FrameLayout
12. https://developer.android.google.cn/reference/android/widget/FrameLayout.LayoutParams.html
13. https://developer.android.google.cn/reference/android/support/v7/widget/CardView.html
14. https://developer.android.google.cn/reference/android/widget/SeekBar.html
15. https://material.io/guidelines/components/progress-activity.html#progress-activity-behavior
16. https://developer.android.google.cn/reference/android/widget/ProgressBar.html

第8章 ConstraintLayout

CHAPTER 8

android. support. constraint. ConstraintLayout 是 android. view. ViewGroup 的直接子类，用来替代 RelativeLayout，在新版 Android Studio 中，ConstraintLayout 已替代 RelativeLayout，成为 HelloWorld 项目的默认布局。ConstraintLayout，在 2016 年由 GoogleI/O 推出，将成为主流布局样式，完全代替其他布局，减少布局的层级，优化渲染性能。从 API 9 开始，Android 系统上可以使用 ConstraintLayout 支持库。ConstraintLayout 作为非绑定的支持库，命名空间是 App:，即来源于本地的包命名空间。使用 ConstraintLayout 布局，需要在项目的 build. gradle 中添加依赖

```
compile 'com.android.support.constraint:constraint - layout:1.0.2'
```

android. support. constraint. ConstraintLayout. LayoutParams 是 android. view. ViewGroup. MarginLayoutParams 的直接子类，包含设置视图怎样放进 ConstraintLayout 的很多属性。如果需要在运行时建立约束，推荐使用 ConstraintSet。ConstraintLayout 的含义就是根据布局中的其他视图，确定视图在屏幕中的位置，主要有三类约束：其他视图、父容器(parent)和基准线(Guideline)。

8.1 单条边约束(相对定位)

单条边约束(相对定位)是指一个控件相对于另一个控件进行定位，可以在水平方向和垂直方向添加约束关系，水平方向的约束有：Left、Right、Start、End，垂直方向的约束有：Top、Bottom、Baseline(文本底部的基准线)，如图 8-1 所示。

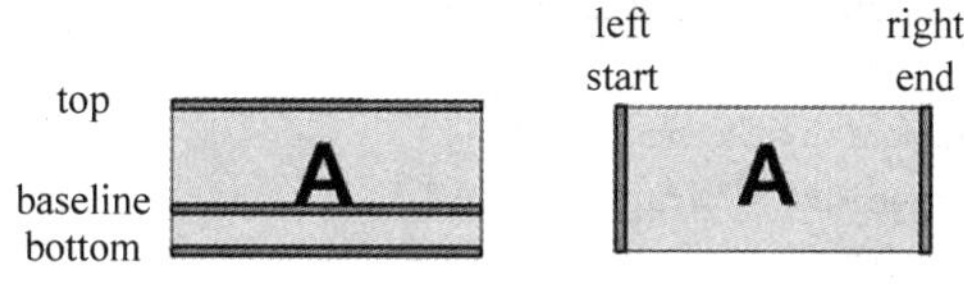

图 8-1 ConstraintLayout 的相对位置约束

水平方向的约束关系有：layout_constraintLeft_toLeftOf、layout_constraintLeft_toRightOf、layout_constraintRight_toLeftOf、layout_constraintRight_toRightOf、layout_constraintStart_toStartOf、layout_constraintStart_toEndOf、layout_constraintEnd_toStartOf、layout_constraintEnd_toEndOf。垂直向的约束关系有 layout_constraintTop_ toTopOf、layout_constraintTop_toBottomOf、layout_constraintBottom_toTopOf、layout_constraintBottom_toBottomOf、layout_constraintBaseline_toBaselineOf。这些约束的形式为：

```
layout_constraint<当前视图的某条边>_<目标视图的某条边>="[目标视图的 ID]"
```

这些属性的取值是目标视图的 ID,目标视图用于约束当前视图。目标视图可以是其他视图或父容器,若目标视图是父容器,也可以用字符串 parent 代替 ID。这些属性设置当前视图的某条边与目标视图的某条边对齐。

实例 1：水平方向的单条边约束如下：

```
<Button android:id="@+id/buttonA" … />
<Button android:id="@+id/buttonB" …
        app:layout_constraintLeft_toRightOf="@+id/buttonA" />
```

上面的代码中,当前视图 buttonB 的左边沿被约束到目标视图 buttonA 的右边沿,即 buttonB 的左边沿和 buttonA 的右边沿在相同的位置,布局效果如图 8-2 所示。

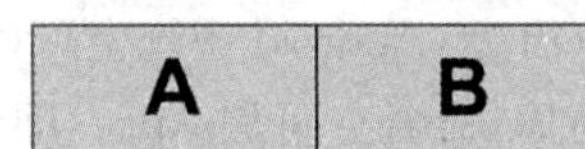

图 8-2　ConstraintLayout 实例 1 的水平方向单条边约束的布局效果

1. margin

水平方向的 margin 有：layout_marginStart、layout_marginEnd、layout_marginLeft 和 layout_marginRight。垂直方向的 margin 有：layout_marginTop 和 layout_marginBottom。margin 的取值只能是正数或者 0。若当前视图指定了某个方向的 margin,并在该方向有约束条件,将在当前视图的指定位置与目标视图的指定位置保留 margin 指定的空间,即 margin 指定了该视图的某一边与对应的约束位置应保留的空间；若当前视图指定了某个方向的 margin,但在对应的方向没有约束条件,这时将忽略指定的 margin。

实例 2：布局文件内容如下：

```
<?xml version="1.0" encoding="utf-8"?>
<android.support.constraint.ConstraintLayout
xmlns:android="http://schemas.android.com/apk/res/android"
    android:layout_width="match_parent"
    android:layout_height="match_parent"
    xmlns:app="http://schemas.android.com/apk/res-auto">
    <Button
        android:layout_width="wrap_content"
        android:layout_height="wrap_content"
        android:text="A1"
        android:id="@+id/buttonA1"
        android:layout_marginLeft="50dp"/>
    <Button
        android:layout_width="wrap_content"
        android:layout_height="wrap_content"
        android:text="B1"
```

```
        android:id = "@ + id/buttonB1"
        app:layout_constraintLeft_toRightOf = "@id/buttonA1"
        android:layout_marginRight = "50dp" />
    <Button
        android:layout_width = "wrap_content"
        android:layout_height = "wrap_content"
        android:text = "C1"
        android:id = "@ + id/buttonC1"
        app:layout_constraintLeft_toRightOf = "@id/buttonB1" />
</android.support.constraint.ConstraintLayout>
```

上面的例子中，buttonA1 指定了 layout_marginLeft，但没有左边的约束条件，所以指定的 layout_marginLeft 将被忽略。buttonB1 指定了 layout_marginRight，但没有右边的约束条件，所以指定的 layout_marginRight 将被忽略。上面的布局文件的布局效果如图 8-3 所示。

图 8-3 ConstraintLayout 实例 2 的布局效果

实例 3：把上面的布局文件中的 buttonA1 的左边同时指定约束和 margin，buttonC1 的左边也同时指定约束和 margin，修改后的布局文件如下：

```
<?xml version = "1.0" encoding = "utf - 8"?>
<android.support.constraint.ConstraintLayout
xmlns:android = "http://schemas.android.com/apk/res/android"
    android:layout_width = "match_parent"
    android:layout_height = "match_parent"
    xmlns:app = "http://schemas.android.com/apk/res - auto">
    <Button
        android:layout_width = "wrap_content"
        android:layout_height = "wrap_content"
        android:text = "A1"
        android:id = "@ + id/buttonA1"
        android:layout_marginLeft = "50dp"
```

```
        app:layout_constraintLeft_toLeftOf = "parent"/>
    <Button
        android:layout_width = "wrap_content"
        android:layout_height = "wrap_content"
        android:text = "B1"
        android:id = "@ + id/buttonB1"
        app:layout_constraintLeft_toRightOf = "@id/buttonA1"
        android:layout_marginRight = "50dp" />
    <Button
        android:layout_width = "wrap_content"
        android:layout_height = "wrap_content"
        android:text = "C1"
        android:id = "@ + id/buttonC1"
        android:layout_marginLeft = "50dp"
        app:layout_constraintLeft_toRightOf = "@id/buttonB1"/>
</android.support.constraint.ConstraintLayout>
```

这时,buttonA1 同时设置了左边的 margin 和左边的约束,因此 buttonA1 的左边与 parent 的左边保留 layout_marginLeft 指定的空间。同样,buttonC1 同时设置了左边的 margin 和左边的约束,因此 buttonC1 的左边与 buttonB1 的右边保留了 buttonC1 的 layout_marginLeft 指定的空间。布局效果如图 8-4 所示。

2. 视图的可见性

对标记为 GONE 的控件,像通常一样,它们的实际尺寸是不会改变的,将不会被显示。但在布局计算时,标记为 GONE 的控件仍然是布局的一部分,它的尺寸好像是 0,就像一个点,同时它的任何 margin 也好像等于 0 一样。如果标记为 GONE 的控件有对其他控件的约束,仍然遵守这些约束,只不过标记 GONE 的控件的尺寸和 margin 都相当于 0。标记为 GONE 的控件的布局效果,如图 8-5 所示。

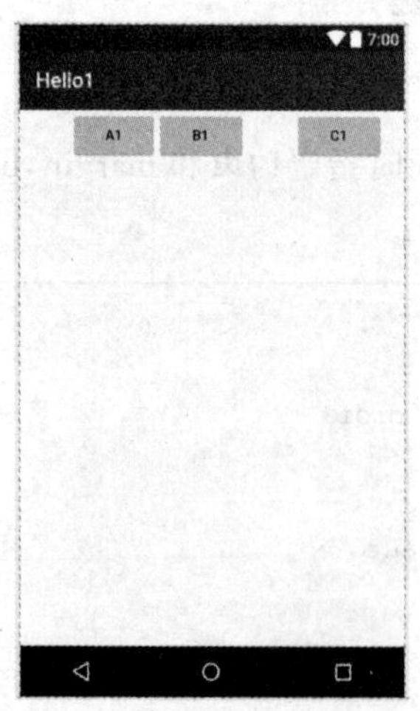

图8-4 ConstraintLayout 实例 3 的布局效果

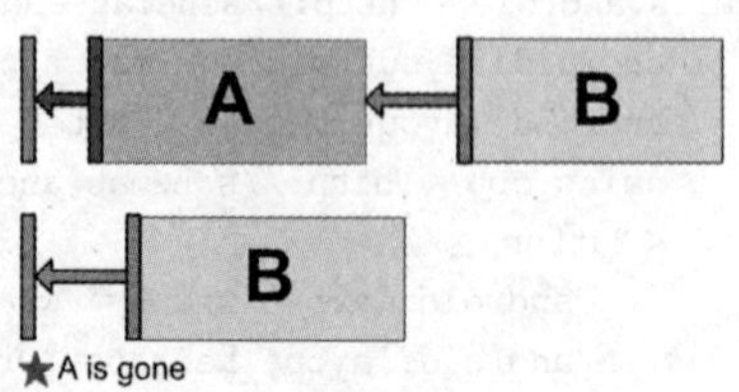

图 8-5 GONE 控件的行为

实例 4：布局文件内容如下。

```
<?xml version = "1.0" encoding = "utf - 8"?>
< android.support.constraint.ConstraintLayout xmlns:android = "http://schemas.android.com/
apk/res/android"
    xmlns:app = "http://schemas.android.com/apk/res - auto"
    xmlns:tools = "http://schemas.android.com/tools"
    android:layout_width = "match_parent"
    android:layout_height = "match_parent">
    < Button
        android:layout_width = "100dp"
        android:layout_height = "wrap_content"
        android:id = "@ + id/buttonA"
        android:text = "A"
        app:layout_constraintLeft_toLeftOf = "parent"
        android:layout_marginLeft = "50dp" />
    < Button
        android:layout_width = "wrap_content"
        android:layout_height = "wrap_content"
        android:id = "@ + id/buttonB"
        android:text = "B"
        android:layout_marginLeft = "50dp"
        app:layout_constraintLeft_toRightOf = "@id/buttonA"/>
</android.support.constraint.ConstraintLayout >
```

上面代码的布局效果如图 8-6 的(a)所示，buttonA 的 layout_marginLeft 是 50dp，layout_width 是 100dp，buttonB 的 layout_marginLeft 是 50dp，这时 buttonB 相对于父容器的左边沿是 50+100+50=200dp。

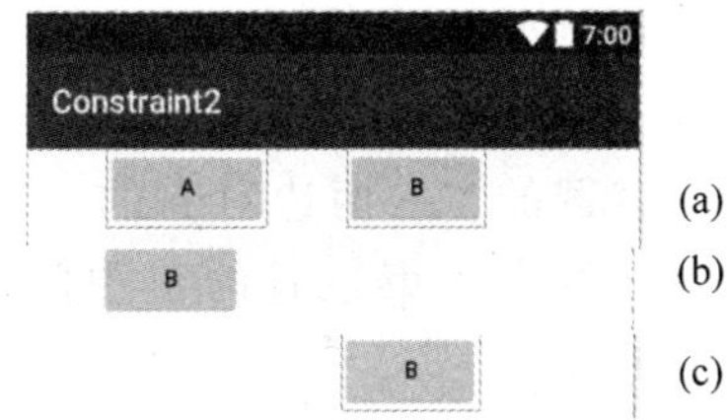

图 8-6 ConstraintLayout 实例 3 的布局效果

在上面的 XML 布局文件中，buttonA 的属性里添加：

```
android:visibility = "gone"
```

这时 buttonA 的 layout_marginLeft 和 layout_width 都相当于 0，因而 buttonB 相对于父容器左边沿的距离是 50dp，布局后的效果如图 8-6 中的(b)所示。

为了让 A 变为 GONE 以后，B 位置不变，可以在 B 的属性中再加上：

```
app:layout_goneMarginLeft = "200dp"
```

这时布局效果如图 8-6 中的(c)所示。这样无论 buttonA 是否可见，buttonB 的位置固定不变，相对于相对于父容器的左边沿都是 200dp。

8.2 不可能约束

如果视图在水平方向的左右两边同时有约束，或者垂直方向的上下两边同时有约束，除非视图的大小为某个特定的值，否则同一个方向(水平方向或者垂直方向)的两条边同时满足约束是不可能的，这种约束可简称为不可能约束。对于这种不可能约束，视图的默认位置是在两个约束位置的正中间，如果期望视图不是在两个约束位置的正中间，可以添加偏移。属性 layout_constraintHorizontal_bias 设置视图到左边的约束边沿的偏移，属性 layout_constraintVertical_bias 设置视图到上边的约束边沿的偏移。

实例 5：不可能约束，布局文件内容如下：

```
<?xml version = "1.0" encoding = "utf - 8"?>
<android.support.constraint.ConstraintLayout
xmlns:android = "http://schemas.android.com/apk/res/android"
    android:layout_width = "match_parent"
    android:layout_height = "match_parent"
    xmlns:app = "http://schemas.android.com/apk/res - auto">
    <Button
        android:layout_width = "wrap_content"
        android:layout_height = "wrap_content"
        android:text = "A"
        android:id = "@ + id/buttonA"
        app:layout_constraintLeft_toLeftOf = "parent"
        app:layout_constraintRight_toRightOf = "parent" />
</android.support.constraint.ConstraintLayout >
```

上面的布局文件中的 Button 的水平方向上的左右两边都有约束，而 Button 的宽度是 wrap_content，除非 Button 的大小刚好等于父容器的宽度，Button 的左右两边是不可能和父容器的左右两边对齐的。因此，这种左右两边都有约束的条件下，视图的默认位置是在两个约束位置的正中间，即左右两边的偏移都是 0.5。视图的实际位置如图 8-7 所示。

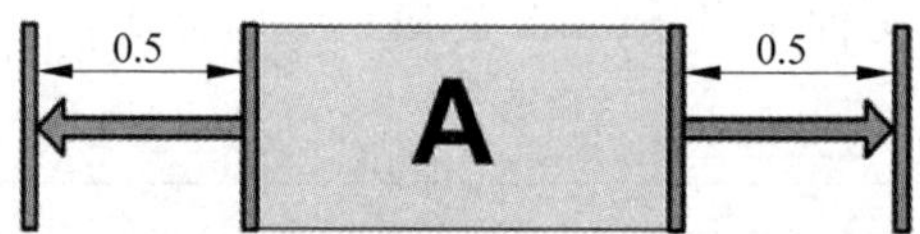

图 8-7 ConstraintLayout 实例 5 的仅仅不可能约束的布局效果

如果期望视图不是在两个约束位置的正中间，可以添加偏移。在前面的 XML 布局文件中，加上水平偏移属性：

```
app:layout_constraintHorizontal_bias = "0.3"
```

Button 左边的偏移设置成了 0.3，布局文件的效果如图 8-8 所示。

图 8-8 ConstraintLayout 实例 5 的不可能约束并且设置了偏移后的布局效果

实例 6：不可能约束，布局文件内容如下：

```
<?xml version = "1.0" encoding = "utf - 8"?>
< android.support.constraint.ConstraintLayout
xmlns:android = "http://schemas.android.com/apk/res/android"
    android:layout_width = "match_parent"
    android:layout_height = "match_parent"
    xmlns:app = "http://schemas.android.com/apk/res - auto">
    < Button
        android:layout_width = "wrap_content"
        android:layout_height = "wrap_content"
        android:text = "A6"
        android:id = "@ + id/buttonA6"
        app:layout_constraintLeft_toLeftOf = "parent" />
    < Button
        android:layout_width = "wrap_content"
        android:layout_height = "wrap_content"
        android:text = "B6"
        android:id = "@ + id/buttonB6"
        app:layout_constraintLeft_toRightOf = "@id/buttonA6"
        app:layout_constraintRight_toRightOf = "parent" />
    < Button
        android:layout_width = "wrap_content"
        android:layout_height = "wrap_content"
        android:text = "A8"
        android:id = "@ + id/buttonA8"
        app:layout_constraintLeft_toLeftOf = "parent"
        app:layout_constraintTop_toBottomOf = "@id/buttonA6" />
    < Button
        android:layout_width = "wrap_content"
        android:layout_height = "wrap_content"
        android:text = "B8"
        android:id = "@ + id/buttonB8"
        app:layout_constraintLeft_toRightOf = "@id/buttonA6"
        app:layout_constraintRight_toRightOf = "parent"
        app:layout_constraintHorizontal_bias = "0.2"
        app:layout_constraintTop_toBottomOf = "@id/buttonB6"/>
</android.support.constraint.ConstraintLayout >
```

上面的布局文件中，buttonB6 的左右两边所有约束，因而是“不可能约束”，但没有指定偏移，因而 buttonB6 在它的左右两个约束位置的正中间。buttonB8 的左右两边所有约束，因而是“不可能约束”，它指定了水平方向偏移为 0.2，因而 buttonB8 距离左边的约束位置是 0.2，距离右边的约束位置是 1－0.2＝0.8。这时布局效果如图 8-9 所示。

给 B6，B8 都添加 layout_marginLeft 属性：

```
android:layout_marginLeft = "100dp"
```

添加 margin 后，B6 到左边的约束边沿的距离，应该留出 layout_marginLeft，剩余的空间再按偏移的默认值 0.5 来分配。同样，B8 到左边的约束边沿的距离，应该留出 layout_marginLeft，剩余的空间再按偏移值 0.2 来分配。这时布局效果如图 8-10 所示。

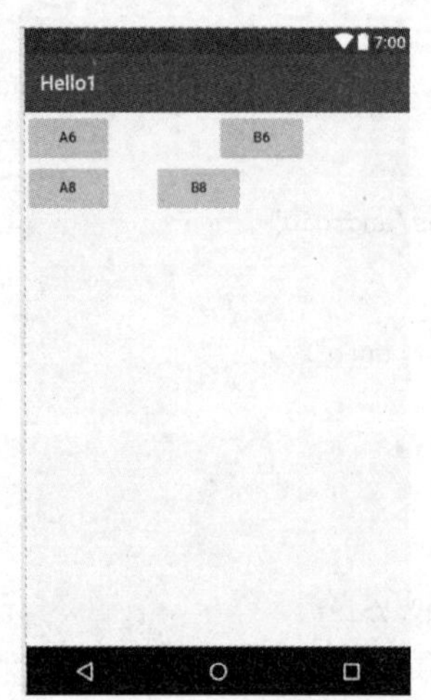

图 8-9 ConstraintLayout 实例 6 的不可能约束的布局效果

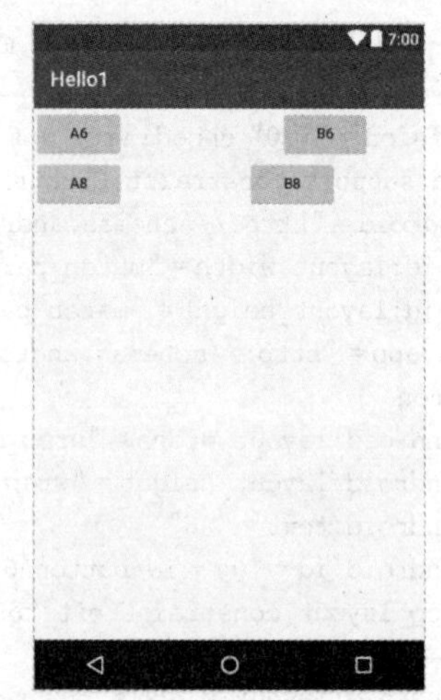

图 8-10 ConstraintLayout 实例 6 的有 margin 的布局效果

总结：在 ConstraintLayout 布局中，一个视图通常需要约束它在水平方向的相对位置和垂直方向的相对位置。若没有水平方向的约束，则水平位置在父视图的左边；若没有垂直方向的约束，则垂直位置在父视图的上边。如果一个视图在水平方向上有一个约束（或者约束了该视图的左边与某个位置对齐，或者约束了该视图的右边与某个位置对齐），这通常是没问题的。但是若水平方向上同时对该视图的左边和右边进行约束，这通常是"不可能的"，这个视图默认将放在左边约束位置和右边约束位置的正中间，即该视图与左右两个约束位置的间距是一样的，若期望左右间距不一样，就采用 app：layout_constraintHorizontal_bias 来指定左边距离约束位置的间距（把该视图与左右两边的约束位置的间距的和为 1）。

8.3 视图的尺寸

视图的尺寸可以通过 layout_width 和 layout_height 来设置，layout_width 和 layout_height 的取值有三种方式：使用固定的尺寸值（例如 100dp 或者对某个尺寸的引用）、使用 WRAP_CONTENT（让控件自己计算它的大小）、使用 0dp（仅当控件的左右两边或者上下两边都有约束的情况下，0dp 才相当于 MATCH_CONSTRAINT）。先扣除尺寸是固定的尺寸值的控件、尺寸是和 WRAP_CONTENT 值的控件、并且扣除了 margin 后，最后剩下的空间是尺寸为 MATCH_CONSTRAINT 控件的尺寸。

还可以使用 minWidth 属性设置视图的最小宽度，仅当 layout_width 设置为 wrap_content 时才起作用。也可以使用 minHeight 设置视图的最小高度，仅当 layout_height 设置为 wrap_content 时才起作用。

通过 app：layout_constraintDimensionRatio 属性设置宽高比，让宽度和高度中的中一个尺寸跟随另一个尺寸变化。为了实现宽高比，需要把控件的 layout_width 和 layout_height 中的一个设置为 0dp，当然也可以都设置为 0dp。比例的设置有两种格式：比例形式和浮点数形式。比例形式为宽度：高度，浮点数形式是宽度/高度得到的浮点数值。

当控件的宽度和高度中的一个设置为 0dp 时，这个尺寸就跟随另一个的尺寸变化。例如当 layout_width 设置为 0dp，高度设置为 WRAP_CONTENT 或者固定尺寸时，layout_width 的尺寸就由 layout_height 的尺寸和宽高比来确定。同样，当 layout_height 设置为 0dp，宽度设置为 WRAP_CONTENT 或者固定尺寸时，layout_height 的尺寸就由 layout_width 的尺寸和宽高比来确定。当 layout_width 和 layout_height 都设置为 0dp 时，系统将该视图设置为满足所有约束条件和宽高比的最大尺寸。也可以在宽高比的前面加上 W 或者 H，并且用逗号隔开，W(Width)表示宽度由宽高比确定，H 表示高度由宽高比。

实例 7：水平方向只有一个控件，布局文件内容如下：

```
<?xml version = "1.0" encoding = "utf - 8"?>
<android.support.constraint.ConstraintLayout
xmlns:android = "http://schemas.android.com/apk/res/android"
    xmlns:app = "http://schemas.android.com/apk/res - auto"
    xmlns:tools = "http://schemas.android.com/tools"
    android:layout_width = "match_parent"
    android:layout_height = "match_parent"
    tools:context = "cuc.constraint2.MainActivity">
    <Button
        android:layout_width = "wrap_content"
        android:layout_height = "wrap_content"
        android:id = "@ + id/buttonA"
        android:text = "A"
        app:layout_constraintLeft_toLeftOf = "parent"
        app:layout_constraintRight_toRightOf = "parent" />
</android.support.constraint.ConstraintLayout >
```

上面代码的布局效果如图 8-11 中的(a)。

(b) 把上面代码中 buttonA 的 layout_width 设置为 0dp，即设置为 MATCH_CONSTRAINT，相应的布局效果见图 8-11 中的(b)。

(c) 把上面代码中 buttonA 的 layout_width 设置为 0dp，另外设置 android:layout_marginLeft="50dp"后，相应的布局效果见图 8-11 中的(c)。

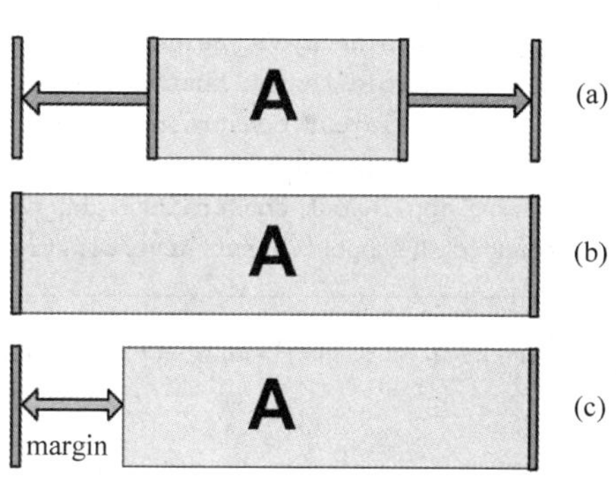

图 8-11　ConstraintLayout 实例 7 的布局效果

实例 8：水平方向超过一个控件，布局文件内容如下：

```
<?xml version = "1.0" encoding = "utf - 8"?>
<android.support.constraint.ConstraintLayout
xmlns:android = "http://schemas.android.com/apk/res/android"
    xmlns:app = "http://schemas.android.com/apk/res - auto"
    xmlns:tools = "http://schemas.android.com/tools"
    android:layout_width = "match_parent"
    android:layout_height = "match_parent"
    tools:context = "cuc.constraint2.MainActivity">
    <Button
```

```
        android:layout_width = "wrap_content"
        android:layout_height = "wrap_content"
        android:id = "@ + id/buttonA"
        android:text = "A"
        app:layout_constraintLeft_toLeftOf = "parent"
        app:layout_constraintRight_toRightOf = "parent" />
</android.support.constraint.ConstraintLayout>
```

buttonA 由于左边没有约束，所以 android: layout_marginLeft="50dp"不起作用。buttonB 左右两边都有约束，buttonB 的 android: layout_width="0dp"就相当于 android: layout_width 为 MATCH_CONSTRAINT，布局效果如 8-12 所示。

图 8-12　ConstraintLayout 实例 8 的 0dp 左右都有约束的布局效果

去掉 buttonB 的右边约束语句 app: layout_constraintRight_toRightOf="parent"后，布局效果如图 8-13 所示。

实例 9：高度由宽高比确定，布局文件如下：

```
<?xml version = "1.0" encoding = "utf - 8"?>
<android.support.constraint.ConstraintLayout
xmlns:android = "http://schemas.android.com/apk/res/android"
    android:layout_width = "match_parent"
    android:layout_height = "match_parent"
    xmlns:app = "http://schemas.android.com/apk/res - auto">
    <Button android:layout_width = "0dp"
        android:layout_height = "0dp"
        android:text = "buttonA"
        app:layout_constraintDimensionRatio = "H,10:1"
         app:layout_constraintLeft_toLeftOf = "parent"
        app:layout_constraintRight_toRightOf = "parent"/>
</android.support.constraint.ConstraintLayout>
```

上面的布局文件中，高度 H 由宽高比 10∶1 确定。布局效果如 8-14 所示。

图 8-13　ConstraintLayout 实例 8 的 0dp 右边没有约束的布局效果

图 8-14　ConstraintLayout 实例 9 的布局效果

8.4 Guideline

android.support.constraint.Guideline 是 android.view.View 的直接子类，仅仅用在 ConstraintLayout 中，是一个帮助对象，仅仅在布局时使用，并不显示在设备上。Guideline 可以是水平方向的，也可以是垂直方向的。水平方向的 Guideline 宽度等于父 ConstraintLayout 的宽度、高度为 0。垂直方向的 Guideline 高度等于父 ConstraintLayout 的高度、宽度为 0。Guideline 的属性包括：layout_constraintGuide_begin、layout_constraintGuide_end、layout_constraintGuide_percent，属性的取值为尺寸。layout_constraintGuide_begin 设置 Guideline 离容器左边或者上边的距离、layout_constraintGuide_end 设置 Guideline 离容器右边或者下边的距离。layout_constraintGuide_percent 设置 Guideline 离容器左边或者上边的百分比。控件可以使用 Guideline 来做约束，以便多个控件使用同一条 Guideline 来定位。

实例 10：使用垂直方向的 Guideline。

在下面的代码中，Guideline 设置了 orientation 为 vertical，即垂直方向的线，因而 app：layout_constraintGuide_begin="50dp"就是离父窗口左边沿的距离是 50dp，buttonA 的垂直方向的两条边都有约束，都是父窗口，因而垂直位置由 app：layout_constraintVertical_bias="0.2"确定。buttonA 的水平方向上只有一条边有约束，左边约束到 Guideline，因而 buttonA 的左边与 Guideline 对齐。buttonB 的约束与 buttonA 类似，区别是 buttonB 的垂直偏移是 0.5。布局文件的内容如下：

```
<?xml version = "1.0" encoding = "utf - 8"?>
<android.support.constraint.ConstraintLayout xmlns:android = "http://schemas.android.com/
apk/res/android"
    xmlns:app = "http://schemas.android.com/apk/res - auto"
    xmlns:tools = "http://schemas.android.com/tools"
    android:layout_width = "match_parent"
    android:layout_height = "match_parent"
    tools:context = "cuc.constraint2.MainActivity">
    <android.support.constraint.Guideline
        android:id = "@ + id/guideLine"
        android:layout_width = "wrap_content"
        android:layout_height = "wrap_content"
        android:orientation = "vertical"
      app:layout_constraintGuide_begin = "50dp"/>
      <Button
        android:layout_width = "wrap_content"
        android:layout_height = "wrap_content"
        android:id = "@ + id/buttonA"
        android:text = "A"
        app:layout_constraintStart_toStartOf = "@id/guideLine"
        app:layout_constraintTop_toTopOf = "parent"
        app:layout_constraintBottom_toBottomOf = "parent"
        app:layout_constraintVertical_bias = "0.2"/>
```

```
    <Button
        android:layout_width = "wrap_content"
        android:layout_height = "wrap_content"
        android:id = "@ + id/buttonB"
        android:text = "B"
        app:layout_constraintStart_toStartOf = "@id/guideLine"
        app:layout_constraintTop_toTopOf = "parent"
        app:layout_constraintBottom_toBottomOf = "parent"
        app:layout_constraintVertical_bias = "0.5"/>
</android.support.constraint.ConstraintLayout>
```

上边布局文件的内容的布局效果如图 8-15 所示。

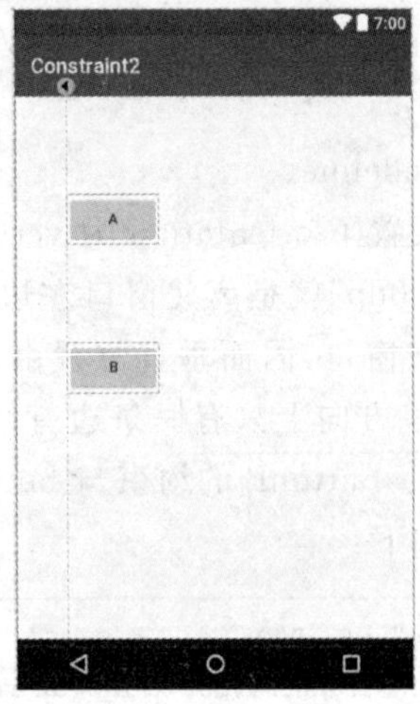

图 8-15　ConstraintLayout 实例 10 的布局效果

8.5　链条

链条(Chains)在同一个轴上(水平或者垂直)提供一个类似群组的统一表现,另一个轴可以独立控制。如果一组控件通过双向连接链接起来,则将其视为链条。对于水平方向的链条,最左边的视图称为链条头(head),水平链条里的每个控件的左边沿和右边沿都要求有约束,除链条头的左边沿约束是 parent,其他控件的左边沿约束是它左边紧邻控件的右边沿;除链条尾的右边沿约束是 parent,其他控件的右边沿约束是它右边紧邻控件的左边沿。与水平方向类似,对于垂直方向的链条,最上边的视图称为链条头(head),垂直链条里的每个控件的上边沿和下边沿都要求有约束,除链条头的上边沿约束是 parent,其他控件的上边沿约束是它上边紧邻控件的下边沿;除链条尾的下边沿约束是 parent,其他控件的下边沿约束是它下边紧邻控件的上边沿。链条中相邻两个视图之间的 margin 将被忽略,只有链条两端外侧的 margin(链条头外侧的 margin 和链条尾外侧的 margin)被考虑。

图 8-16 是只有两个控件的水平链条,图 8-17 是有 3 个控件的水平链条。

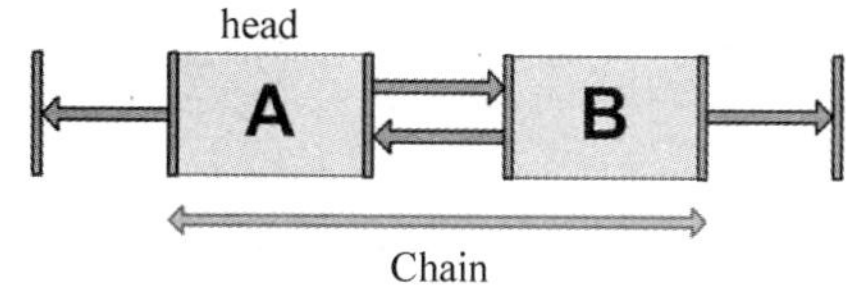

图 8-16 只有两个控件的水平链条

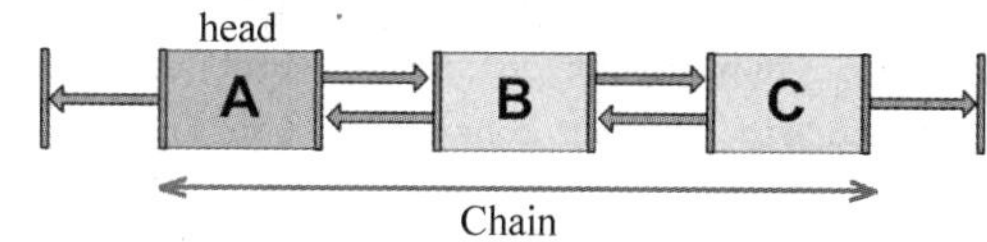

图 8-17 3 个控件的水平链条

链条的式样是通过设置链条头的 layout_constraintHorizontal_chainStyle 或者 layout_constraintVertical_chainStyle 属性来控制的，取值有三种：CHAIN_SPREAD、CHAIN_SPREAD_INSIDE 和 CHAIN_PACKED，在 XML 中分别对应于 spread、spread_inside 和 packed。默认的样式为 CHAIN_SPREAD。

8.5.1 CHAIN_SPREAD 链模式

CHAIN_SPREAD 是链条的默认样式，它会让链条上每个控件的间隙是相等的。以水平链条为例，链条头到父控件左边的间距、链条尾部到父控件右边的间距，以及链条内的每个控件的间距都是一样的，平分剩余空间。

实例 11：CHAIN_SPREAD。

```
<?xml version = "1.0" encoding = "utf - 8"?>
< android.support.constraint.ConstraintLayout
xmlns:android = "http://schemas.android.com/apk/res/android"
    xmlns:app = "http://schemas.android.com/apk/res - auto"
    xmlns:tools = "http://schemas.android.com/tools"
    android:layout_width = "match_parent"
    android:layout_height = "match_parent"
    tools:context = "cuc.constraint2.MainActivity">
    < Button
        android:layout_width = "wrap_content"
        android:layout_height = "wrap_content"
        android:id = "@ + id/buttonA"
        android:text = "A"
        app:layout_constraintLeft_toLeftOf = "parent"
        app:layout_constraintRight_toLeftOf = "@ + id/buttonB"
        app:layout_constraintHorizontal_chainStyle = "spread" />
    < Button
        android:layout_width = "wrap_content"
        android:layout_height = "wrap_content"
        android:id = "@id/buttonB"
        android:text = "B"
        app:layout_constraintLeft_toRightOf = "@id/buttonA"
        app:layout_constraintRight_toLeftOf = "@ + id/buttonC"/>
    < Button
        android:layout_width = "wrap_content"
        android:layout_height = "wrap_content"
        android:id = "@id/buttonC"
        android:text = "C"
```

```
        app:layout_constraintLeft_toRightOf = "@id/buttonB"
        app:layout_constraintRight_toRightOf = "parent" />
</android.support.constraint.ConstraintLayout>
```

上面布局文件中的 layout_constraintHorizontal_chainStyle 为"spread",可以去掉。上面布局文件的效果如图 8-18 所示。

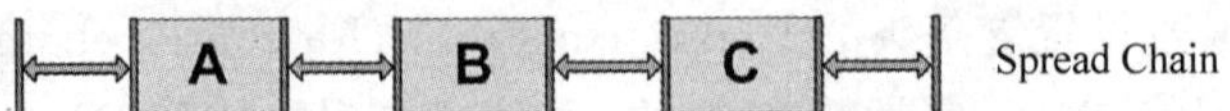

图 8-18 ConstraintLayout 实例 11 的布局效果

1. CHAIN_SPREAD 链条两边的 margin

在 CHAIN_SPREAD 模式中,还可以设置整个链条两边的 margin(链条内的 margin 设置将被忽略),则布局时,先分配 margin,剩余的空间再被所有的间隙(包括链条两端向外的间隙)均分。

在上面的布局文件中,给 buttonC 添加属性如下:

```
android:layout_marginRight = "50dp"
```

添加 layout_marginRight 属性后,布局效果如 8-19 所示。

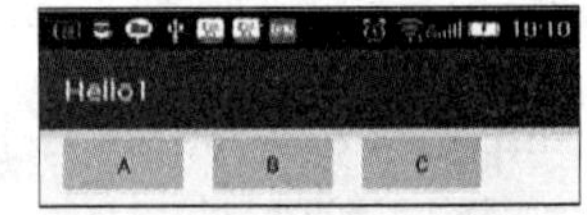

图 8-19 ConstraintLayout 实例 11 链条尾有 margin 的布局效果

2. CHAIN_SPREAD 中的加权链

在 CHAIN_SPREAD 链模式中,如果一些控件沿链条方向的尺寸设置为 MATCH_CONSTRAINT(即 0dp),则实际尺寸将根据对应方向的权重来分配。权重的默认值为 0,若设置了尺寸为 MATCH_CONSTRAINT 的控件的权重都为 0,则这些控件的权重一样,控件的尺寸均分剩余的空间,这时链条内各个控件之间、链条的两边都没有间隙。

实例 12:CHAIN_SPREAD 中的加权链。

buttonA、buttonB 和 buttonC 左右相接,形成水平链条,buttonA 的宽度是 wrap_content,buttonB 和 buttonC 的宽度都是 0dp,所以,在给 buttonA 的宽度按照 wrap_content 分配完空间后,剩余的空间将根据 buttonB 和 buttonC 的 layout_constraintHorizontal_weight 来给 buttonB 和 buttonC 分配宽度。buttonB 和 buttonC 的 app:layout_constraintHorizontal_weight 都是 1,因而它们的宽度比是 1∶1。

```
<?xml version = "1.0" encoding = "utf - 8"?>
<android.support.constraint.ConstraintLayout
xmlns:android = "http://schemas.android.com/apk/res/android"
    xmlns:app = "http://schemas.android.com/apk/res - auto"
    xmlns:tools = "http://schemas.android.com/tools"
    android:layout_width = "match_parent"
    android:layout_height = "match_parent"
    tools:context = "cuc.constraint2.MainActivity">
```

```
    <Button
        android:layout_width = "wrap_content"
        android:layout_height = "wrap_content"
        android:id = "@ + id/buttonA"
        android:text = "A"
        app:layout_constraintLeft_toLeftOf = "parent"
        app:layout_constraintRight_toLeftOf = "@ + id/buttonB"
        app:layout_constraintHorizontal_chainStyle = "spread" />
    <Button
        android:layout_width = "0dp"
        android:layout_height = "wrap_content"
        android:id = "@id/buttonB"
        android:text = "B"
        app:layout_constraintLeft_toRightOf = "@id/buttonA"
        app:layout_constraintRight_toLeftOf = "@ + id/buttonC"
        app:layout_constraintHorizontal_weight = "1" />
    <Button
        android:layout_width = "0dp"
        android:layout_height = "wrap_content"
        android:id = "@id/buttonC"
        android:text = "C"
        app:layout_constraintLeft_toRightOf = "@id/buttonB"
        app:layout_constraintRight_toRightOf = "parent"
        app:layout_constraintHorizontal_weight = "1"/>
</android.support.constraint.ConstraintLayout>
```

上面布局的效果如图 8-20 所示。

图 8-20 ConstraintLayout 实例 12 加权链的布局效果

8.5.2 CHAIN_SPREAD_INSIDE 链模式

CHAIN_SPREAD_INSIDE 链模式把链条两端的视图向外到父组件边沿的距离设为0，链条上所有控件之间的间距一样，平分除控件外的剩余空间。

实例 13：CHAIN_SPREAD_INSIDE。

把实例 12 的代码中的 CHAIN_SPREAD 模式修改为 spread_inside，修改后布局文件如下：

```
<?xml version = "1.0" encoding = "utf - 8"?>
<android.support.constraint.ConstraintLayout
xmlns:android = "http://schemas.android.com/apk/res/android"
    xmlns:app = "http://schemas.android.com/apk/res - auto"
    xmlns:tools = "http://schemas.android.com/tools"
    android:layout_width = "match_parent"
    android:layout_height = "match_parent" >
```

```
    <Button
        android:layout_width = "wrap_content"
        android:layout_height = "wrap_content"
        android:id = "@ + id/buttonA"
        android:text = "A"
        app:layout_constraintLeft_toLeftOf = "parent"
        app:layout_constraintRight_toLeftOf = "@ + id/buttonB"
        app:layout_constraintHorizontal_chainStyle = "spread_inside" />
    <Button
        android:layout_width = "wrap_content"
        android:layout_height = "wrap_content"
        android:id = "@id/buttonB"
        android:text = "B"
        app:layout_constraintLeft_toRightOf = "@id/buttonA"
        app:layout_constraintRight_toLeftOf = "@ + id/buttonC" />
    <Button
        android:layout_width = "wrap_content"
        android:layout_height = "wrap_content"
        android:id = "@id/buttonC"
        android:text = "C"
        app:layout_constraintLeft_toRightOf = "@id/buttonB"
        app:layout_constraintRight_toRightOf = "parent" />
</android.support.constraint.ConstraintLayout>
```

上面的布局文件的布局效果如图 8-21 所示。

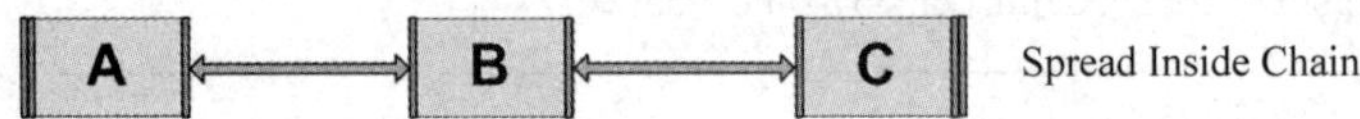

图 8-21 ConstraintLayout 实例 13 的布局效果

在 CHAIN_SPREAD_INSIDE 模式中，还可以设置整个链条两端的 margin(链条内的 margin 设置将被忽略)，则布局时，先分配 margin，剩余的空间再被链条内部所有的间隙(不包括链条两端向外的间隙)平均分配。

实例 14：再给实例 13 的布局文件中的链条末端添加 margin 属性，即给 buttonC 添加 layout_marginRight 属性：

```
android:layout_marginRight = "50dp"
```

添加 layout_marginRight 属性后布局效果如图 8-22 所示。

8.5.3 CHAIN_PACKED 链模式

它将所有 Views 打包到一起，链条上的各个控件之间没有间隙，整个控件组在父容器的位置由 layout_constraintHorizontal_bias 或者 layout_constraintVertical_bias 确定。bias 的默认值是0.5，即整个控件组在父容器的居中位置，只有链条头和链条尾向外到父容器边沿的距离相等。

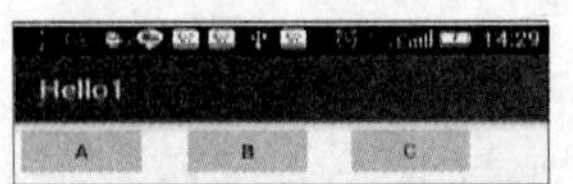

图 8-22 ConstraintLayout 实例 14 的布局效果

实例 15：CHAIN_PACKED。

把实例 11（CHAIN_SPREAD 模式）代码中的 app：layout_constraintHorizontal_chainStyle 设置为 packed，同时添加 app：layout_constraintHorizontal_bias="0.5"，修改后的代码如下：

```
<?xml version = "1.0" encoding = "utf - 8"?>
<android.support.constraint.ConstraintLayout
xmlns:android = "http://schemas.android.com/apk/res/android"
    xmlns:app = "http://schemas.android.com/apk/res - auto"
    xmlns:tools = "http://schemas.android.com/tools"
    android:layout_width = "match_parent"
    android:layout_height = "match_parent"
    tools:context = "cuc.constraint2.MainActivity">
    <Button
        android:layout_width = "wrap_content"
        android:layout_height = "wrap_content"
        android:id = "@ + id/buttonA"
        android:text = "A"
        app:layout_constraintLeft_toLeftOf = "parent"
      app:layout_constraintRight_toLeftOf = "@ + id/buttonB"
        app:layout_constraintHorizontal_chainStyle = "packed"
        app:layout_constraintHorizontal_bias = "0.5"/>
    <Button
        android:layout_width = "wrap_content"
        android:layout_height = "wrap_content"
        android:id = "@id/buttonB"
        android:text = "B"
        app:layout_constraintLeft_toRightOf = "@id/buttonA"
        app:layout_constraintRight_toLeftOf = "@ + id/buttonC" />
    <Button
        android:layout_width = "wrap_content"
        android:layout_height = "wrap_content"
        android:id = "@id/buttonC"
        android:text = "C"
        app:layout_constraintLeft_toRightOf = "@id/buttonB"
        app:layout_constraintRight_toRightOf = "parent" />
</android.support.constraint.ConstraintLayout >
```

上面文件的布局效果如图 8-23 所示。

图 8-23 ConstraintLayout 实例 15 的布局效果

把上面代码中的 app：layout_constraintHorizontal_bias 从 0.5 改为 0.3，即

```
app:layout_constraintHorizontal_bias = "0.3"
```

修改后，布局效果如图 8-24 所示。

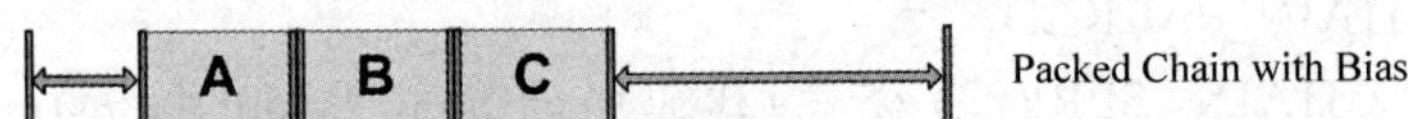

图 8-24 ConstraintLayout 实例 15 偏移为 0.3 的布局效果

在 CHAIN_PACKED 模式中，还可以设置整个链条两边的 margin（链条内的 margin 设置将被忽略），则布局时，先分配 margin，剩余的空间根据 app：layout_constraintHorizontal_bias 或者 app：layout_constraintVertical_bias 来分配给链条左右两边的间隙。

实例 16：在实例 15 的布局文件中，给 buttonC 添加属性如下：

```
android:layout_marginRight = "50dp"
```

这时布局文件内容如下：

```
<?xml version = "1.0" encoding = "utf - 8"?>
<android.support.constraint.ConstraintLayout
xmlns:android = "http://schemas.android.com/apk/res/android"
   mlns:app = "http://schemas.android.com/apk/res - auto"
   xmlns:tools = "http://schemas.android.com/tools"
   android:layout_width = "match_parent"
   android:layout_height = "match_parent">
   <Button
       android:layout_width = "wrap_content"
       android:layout_height = "wrap_content"
       android:id = "@ + id/buttonA"
       android:text = "A"
       app:layout_constraintLeft_toLeftOf = "parent"
       app:layout_constraintRight_toLeftOf = "@ + id/buttonB"
       app:layout_constraintHorizontal_chainStyle = "packed"
       app:layout_constraintHorizontal_bias = "0.5" />
   <Button
       android:layout_width = "wrap_content"
       android:layout_height = "wrap_content"
       android:id = "@id/buttonB"
       android:text = "B"
       app:layout_constraintLeft_toRightOf = "@id/buttonA"
       app:layout_constraintRight_toLeftOf = "@ + id/buttonC" />
   <Button
       android:layout_width = "wrap_content"
       android:layout_height = "wrap_content"
       android:id = "@id/buttonC"
       android:text = "C"
       app:layout_constraintLeft_toRightOf = "@id/buttonB"
       app:layout_constraintRight_toRightOf = "parent"
       android:layout_marginRight = "50dp" />
</android.support.constraint.ConstraintLayout>
```

布局效果如图 8-25 所示。

各种链条样式的对比，如图 8-26 所示。

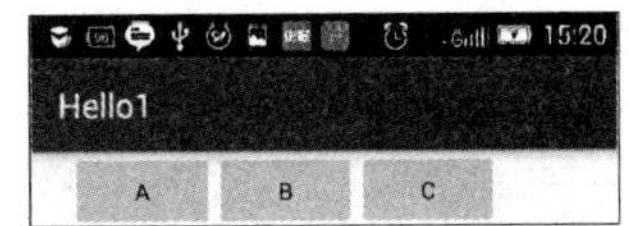

图 8-25 ConstraintLayout 实例 16 的布局效果

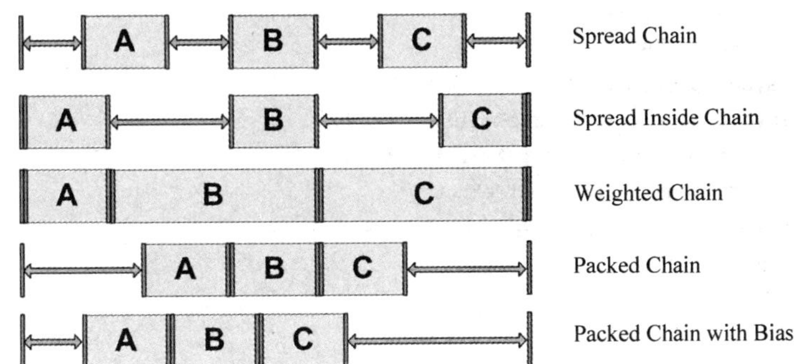

图 8-26 各种链条样式的对比

8.6 圆形定位

圆形定位允许把一个视图的中心放在以另一个视图的中心为圆心的圆上面，可以指定圆心、半径和角度，如图 8-27 所示。角度的单位为度，取值范围为 0 到 360，从圆心垂直向上的位置定为 0 度，角度沿顺时针方向增大。

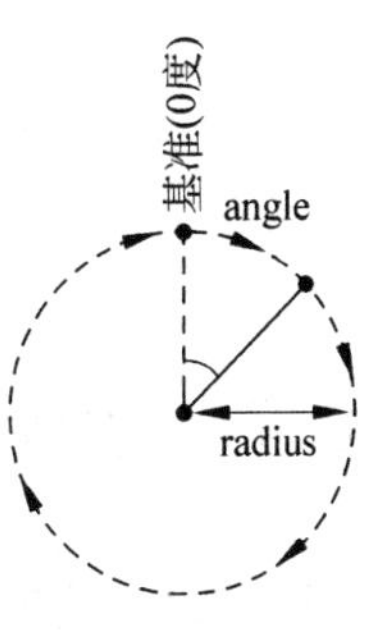

图 8-27 圆形定位

layout_constraintCircle 的取值是 id，设置圆心的位置为由 id 指定的视图的中心。

layout_constraintCircleRadius 设置半径，即该视图的中心到另一个视图的中心的距离。

layout_constraintCircleAngle 设置该视图的中心所在的位置。

实例 17：圆形定位。

```
<?xml version = "1.0" encoding = "utf - 8"?>
< android.support.constraint.ConstraintLayout
xmlns:android = "http://schemas.android.com/apk/res/android"
    xmlns:app = "http://schemas.android.com/apk/res - auto"
    xmlns:tools = "http://schemas.android.com/tools"
    android:layout_width = "match_parent"
    android:layout_height = "match_parent" >
    < Button
        android:layout_width = "wrap_content"
        android:layout_height = "wrap_content"
        android:id = "@ + id/buttonA"
        android:text = "A"
        app:layout_constraintLeft_toLeftOf = "parent"
```

```
        app:layout_constraintRight_toRightOf = "parent"
        app:layout_constraintTop_toTopOf = "parent"
        app:layout_constraintBottom_toBottomOf = "parent" />
    <Button
        android:layout_width = "wrap_content"
        android:layout_height = "wrap_content"
        android:id = "@ + id/buttonB"
        android:text = "B"
        app:layout_constraintCircle = "@ + id/buttonA"
        app:layout_constraintCircleRadius = "100dp"
        app:layout_constraintCircleAngle = "45" />
</android.support.constraint.ConstraintLayout>
```

布局效果如图 8-28 所示。

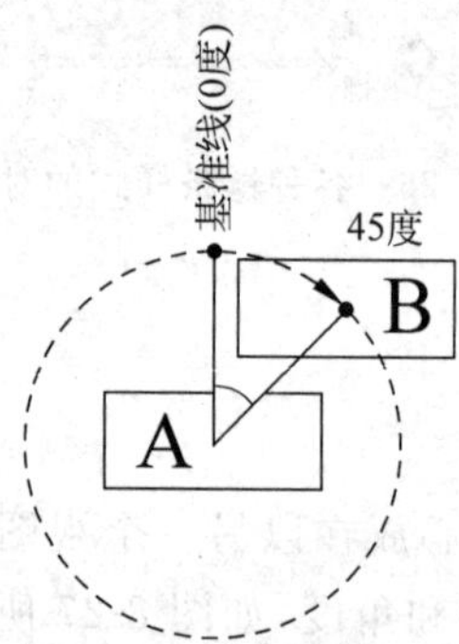

图 8-28　ConstraintLayout 实例 17 的布局效果

8.7　本章主要参考文献

1. https://developer.android.google.cn/reference/android/support/constraint/ConstraintLayout.html
2. https://developer.android.google.cn/reference/android/support/constraint/Guideline.html

第9章 TextView

CHAPTER 9

android. widget. TextView 是 android. view. View 的直接子类，用来显示文字给用户。TextView 被配置成不允许编辑。EditText 是一个把文字视图配置成可以编辑的子类。TextView 的 xml 属性如表 9-1 所示。

表 9-1　TextView 的 xml 属性

属　　性	解　　释
autoLink	控制是否自动在 text 中找到链接（例如电话号码、Web 地址等），并转化为可以单击的链接。这个属性的取值可以为：none，web，email，phone，map，all 中的一个或者多个，多个值用"|"隔开。其中 map 目前只适用于美国的地址，默认值是 none，禁止这个特性
autoText	已经过时，不推荐使用，请使用 android：inputType
breakStrategy	设置段落布局时行折断的策略，取值为 simple，high_quality 或者 balanced。若设置为 high_quality，行折断包含连字符。若设置为 balanced，行折断时平衡行的长度
bufferType	定义了 getText() 方法返回的最小类型。取值为 normal，spannable 或者 editable。Normal 可以返回任何 CharSequence。Spannable 则可在给定的字符区域使用样式。Editable 继承自 Spannable，同样也可以在给定的字符区域内使用各种样式；还可以增加、删除、修改字符
capitalize	已经过时，不推荐使用，请使用 android：inputType
cursorVisible	布尔量，默认值 true，设置指定光标是否可见
digits	指定可以接受的输入字符。例如 digits＝"123abc"，则 EditText 可以接受的字符只能是这 6 个字符：1、2、3、a、b 和 c
drawableBottom、drawableEnd、drawableLeft、drawableRight、drawableStart、drawableTop	设置在 TextView 的文本的四周的图像，可以是格式为"@[＋][package：]type：name"的对另一资源的引用或者格式为"？[package：][type：]name"的主题属性，也可以是一个颜色值
drawablePadding	指定图片和文字之间的 padding，必须是一个尺寸值
drawableTint	为文本的图像设置颜色，必须是颜色值，形式为＃rgb、＃argb、＃rrggb 或者＃aarrggbb
drawableTintMode	混合模式用于调节文本图像的颜色，如颜色渐变之类的

续表

属　性	解　释
editable	设置文本框是否允许编辑
editorExtras	必须是对另一资源或者 theme 属性的引用。引用一个 input-extras 资源，它包含了提供给输入方法的附加数据，这对输入方法的实现是私有的，它简单地填充 EditorInfo. extras 域
elegantTextHeight	取值为布尔量，主要针对不太复杂的文字
ellipsize	文本超出 TextView 的长度时，如何处理该内容。取值为：none 不做任何处理；start 在开头显示省略号；end 在结尾显示省略号；middle 在中间显示省略号；marquee 以跑马灯的方式显示(动画横向移动)，还需要配合其他属性使用
ems	整数值，用于指定 TextView 的精确宽度是多少个 ems，即设置 TextView 能容纳多少个字母“M”，而不管实际的文本大小。Em 是印刷排版领域的字宽的单位，例如在 16 点阵的字体中，1 个 Em 就是 16 点。EM 也通常指字母 M 的宽度，因为字母 M 通常占据一个方框的全部宽度，在现代印刷体中，字母 M 的宽度比 1 个 EM 略小
fontFamily	必须是一个字符串值，设置字体系列
fontFeatureSettings	必须是一个字符串值，设置字体特征
freezesText	布尔量，默认值为 false。如果设置 true，冻结文本和元数据(如当前光标位置)，这可能在 TextView 的内容还没有存储在持久位置(如 ContentProvider)前是有用的。对于 EditText 总是被使能的，不管该属性如何设置
gravity	当文本比视图小时，该属性设置文本沿着 x 轴和(或者)y 轴的对齐方式，指定该视图的文本在该视图中的位置
height	必须是一个尺寸值。设置 TextView 的精确高度
hint	设置当文本框内的内容为空时，文本框内默认显示的提示文本
hyphenationFrequency	设置自动产生连字符的频率，取值为：none 不产生连字符，normal 产生较少的连字符，full 产生标准数量的连字符
imeActionId	必须是一个整数值，例如 100，当 TextView 和一个输入方法相连时，用于为 EditorInfo. actionId 提供一个值
imeActionLabel	必须是一个字符串值，当 TextView 和一个输入方法相连时，用于为 EditorInfo. actionLabel 提供一个值
imeOptions	可以在与编辑器相连的输入方法使能附加功能：默认情况下 android 软键盘右下角的按钮为“下一个”，单击会到下一个输入框，保持软键盘设置。通过 imeOptions 可以设置软键盘的右下角的按钮与输入方法和编辑框相关的动作，设置 imeOptions="actionDone"，软键盘下方变成“完成”，单击后光标保持在原来的输入框上，并且软键盘关闭
includeFontPadding	布尔量，默认为 true。设置文本是否包含顶部和底部额外空白
inputMethod	必须是完全限定的类名，设置文本输入方法

续表

属　性	解　释
inputType	设置文本的类型，在 EditText 输入文本时，inputType 属性用于帮助输入法显示合适的键盘类型。none：文本不可编辑。text：输入普通旧式文本。textCapCharacters：请求大写所有的字符。textCapWords：请求大写每个单词的第一个字母。textCapSentences 请求大写每个句子的第一个字母。textAutoCorrect：请求自动矫正输入的文本。textAutoComplete：进行自动完成功能。textMultiLine：允许输入多行文本，如果没有设置这个标志，该文本域将限制成单行文本。textImeMultiLine：虽然常规的 TextView 不是多行，如果输入法能够支持多行，就提供多行输入的支持。textNoSuggestions：输入方法不应该显示任何以字典为基础的单词的建议。textUri：文本为 URI。textEmailAddress：文本为电子邮件地址。textEmailSubject：文本为电子邮件主题。textShortMessage：文本将用作短消息的内容。textLongMessage：文本将用作长消息的内容。textPersonName：文本是人的名字。textPostalAddress：文本是邮政邮寄地址。textPassword：文本是密码。textVisiblePassword：文本是可以看得见的密码。textWebEditText：文本将作为网页表单的文本。textFilter：文本过滤一些数据。textPhonetic：文本为拼音输入。Number：文本是纯数字。numberSigned：文本是有符号数字。numberDecimal：文本是可以带小数点的十进制数。Phone：用于输入电话号码。datetime：用于输入日期时间。date：用于输入日期。time：用于输入时间
letterSpacing	必须是以浮点值，例如"1.2"。设置文本的字母间距
lineSpacingExtra	必须是尺寸值，即浮点数附加一个单位，例如"14.5sp"，用于控制两行文本的额外间距。单位是 px、dp、sp、in(inches)，mm
lineSpacingMultiplier	必须是浮点值，例如 1."。两行文本的实际间距为：默认间距 * lineSpacingMultiplier ＋lineSpacingExtra
lines	必须是整数值，用于指定 TextView 的精确高度是多少行
linksClickable	取值为布尔量，如果设置为 false，即使指定的 autoLink 属性正确识别出了链接，单击也不会发生任何动作
marqueeRepeatLimit	必须是一个整数值，或者 marquee_forever。用于设置动画重复滚动的次数，仅当 ellipsize 取值为"marquee"的情况下生效
maxEms	整数值，用于指定 TextView 的最大宽度是多少个 ems，即设置最多能容纳多少个字母“M”，而不管实际的文本大小
maxHeight	尺寸值，即浮点数附加一个单位，用于指定 TextView 的最大高度
maxLength	整数值，例如 5，用于设置输入过滤器，把输入文本的字符数限制在指定的数量内
maxLines	整数值，用于指定 TextView 的高度最高多少行文字，当用于 EditText 时，inputType 必须和 textMultiLine 组合使用
maxWidth	尺寸值，即浮点数附加一个单位，用于设置 TextView 的最大宽度。
minEms	整数值，用于指定 TextView 的最小宽度是多少个 ems，即设置最少能容纳多少个字母“M”，而不管实际的文本大小
minHeight	尺寸值，即浮点数附加一个单位，用于设置 TextView 的最小高度
minLines	整数值，用于指定 TextView 的最小高度是多少行文字，当用于 EditText 时，inputType 必须和 textMultiLine 组合使用
minWidth	尺寸值，即浮点数附加一个单位，用于设置 TextView 的最小宽度
numeric	已经过时，使用 inputType 代替

续表

属 性	解 释
password	已经过时,使用 inputType="textPassword"代替
phoneNumber	已经过时,使用 inputType="phone"代替
privateImeOptions	必须是一个字符串值,当连接文本输入方法时,用该参数填充 EditorInfo.privateImeOptions 域,给文本输入方法提供附加内容
scrollHorizontally	取值为布尔量,设置是否可以水平滚动
selectAllOnFocus	布尔量,当设置为 true 时,如果文本是可以选择的,当前组件在得到焦点的时候,自动选取该组件内的所有的文本内容
shadowColor	必须是颜色值,形式为 #rgb、#argb、#rrggbb 或者 #aarrggbb。设置文本下边文本阴影的颜色。文本阴影与产生实时阴影的属性(elevation 和 translationZ)并不相互作用
shadowDx	文本阴影的水平偏移,必须是以浮点值,例如 1.2
shadowDy	文本阴影的垂直偏移,必须是以浮点值,例如 1.2
shadowRadius	文本阴影的模糊半径,必须是以浮点值,例如 1.2
singleLine	用于约束文本为单一的水平滚动行,而不是包裹多行,遇到回车键,强制不让换行。当文本超出 1 行时,以省略号或者滚动时的形式显示文本。配合 ellipsize 使用。如果要截取字符串并追加省略号显示,尽量用 singleLine 属性。如果要控制显示行数,尽量用 maxLines 属性
text	设置要显示的文字,必须是一个字符串值
textAllCaps	布尔量,若设置为 true,文本中的英文字母(包含小写字母)全部以大写字母显示
textAppearance	必须是对另一资源或者 theme 属性的引用,用于设置基本的文本颜色、字体、大小、风格。格式分别为:@[+][package:]type:name 或者?[package:][type:]name。可设置的值如下:textAppearanceButton、textAppearanceInverse,textAppearanceLarge,textAppearanceLargeInverse, textAppearanceMedium, textAppearanceMediumInverse,textAppearanceSmall,textAppearanceSmallInverse
android:textColor	设置文字的颜色
textColorHighlight	设置文本框文本被选中时的底色,默认为蓝色
textColorHint	设置提示信息文字的颜色,默认为灰色。与 hint 一起使用
textColorLink	设置链接的文本颜色
textIsSelectable	布尔量,设置不可编辑的文本的内容是否可以被选择
textScaleX	设置文字的水平伸缩因子,取值是一个浮点数值
textSize	设置文字大小,推荐文本类型的尺寸单位为"sp",如"15sp"
textStyle	设置文本的风格,取值为 normal,bold(粗体),italic(斜体)。可以设置一个或多个,用"|"隔开
typeface	设置文本字体类型,必须是以下常量值之一:normal、sans、serif 和 monospace(等宽字体)
width	设置 TextView 的精确宽度,也可以在布局参数里指定这个数值达到同样的效果。android:layout_width 可以是精确的宽度,也可以是 wrap_content、match_parent

TextView 的直接子类有:AppCompatTextView、Button、CheckedTextView、Chronometer、DigitalClock、EditText、RowHeaderView 和 TextClock。

9.1 EditText

android. widget. EditText 是 android. widget. TextView 的直接子类。EditText 是输入和修改文字的用户接口，定义一个 EditText 控件时，必须指定 inputType 属性。可以添加 TextWatcher 到 EditText 以便接收用户改变文字时的回调。当希望添加文字变化时自动保存功能时，或者验证用户输入的格式有效时，可以使用 addTextChangedListener(TextWatcher)添加 TextWatcher。TextWatcher 定义如下：

```
public interface TextWatcher extends NoCopySpan {
    public void beforeTextChanged(CharSequence s, int start, int countBefore, int countAfter);
    public void onTextChanged(CharSequence s, int start, int countBefore, int countAfter);
    public void afterTextChanged(Editable s);
}
```

EditText 里的内容变化包括：新增加字符、删除字符、替换输入框中的若干个字符。方法 beforeTextChanged 在 EditText 里的内容即将发生变化之前触发，第一个参数 s 是变化之前的字符串，无论何种内容变化方式，都可以理解为：在变化前的字符串 s 里，从索引位置 start 开始的 countBefore 个字符即将被 countAfter 字符代替。

方法 onTextChanged 在 EditText 里的内容变化时触发，第一个参数 s 是变化之后的字符串，无论何种内容变化方式，都可以理解为：在变化后的字符串 s 里，从索引位置 start 开始的 countAfter 字符已经代替了原来的 countBefore 个字符。

方法 afterTextChanged 在 EditText 里的内容变化后触发，参数 s 是变化之后的字符串。

实例 1：在布局文件 activity_main. xml 中，有一个 id 为 nameEdt 的 EditText，相应的 Activity 代码如下：

```
public class MainActivity extends AppCompatActivity {
    private static final String TAG = "MainActivity";
    @Override
    protected void onCreate(Bundle savedInstanceState) {
        super.onCreate(savedInstanceState);
        setContentView(R.layout.activity_main);
        ((EditText) findViewById(R.id.nameEdt)).addTextChangedListener(new TextWatcher() {
            @Override
            public void beforeTextChanged(CharSequence s, int start, int countbefore, int
countAfter) {
                Log.i(TAG, "beforeTextChanged: s:" + s + ", " + countbefore + " chars from " +
start + " are about to be replaced by "
                        + countAfter + " chars");
            }
            @Override
            public void onTextChanged(CharSequence s, int start, int countbefore, int
countAfter) {
                Log.i(TAG, "onTextChanged: s:" + s + ", " + countbefore + " chars from " + start +
" have been replaced by "
```

```
                + countAfter + " chars");
        }

        @Override
        public void afterTextChanged(Editable s) {
            Log.i(TAG, "afterTextChanged: s:" + s);
        }
    });
  }
}
```

编译执行后，在输入框中，输入字符、删除字符、复制字符，观察 log 输出。

9.2 Button

android.widget.Button 是 android.widget.TextView 的直接子类，Button 是可以单击执行动作的用户接口。为了在 Activity 中显示 Button，把 Button 添加到 Activity 的布局 XML 文件中。Button 可以包含文字或者图标，或者同时包含文字和图标，当用户触摸时，执行特定的动作。如果单击的视图仅仅包含文字，使用 Button 类。如果单击的视图同时包含文字和图片，使用 Button 类，然后使用 drawableLeft、drawableRight、drawableStart、drawableEnd、drawableTop、drawableBottom 属性中说明图片放在文本的哪个位置。如果单击的视图仅仅包含图片，使用 ImageButton 类。

```
<Button android:layout_width = "wrap_content"
      android:layout_height = "wrap_content" android:text = "click me"/>
<Button
android:layout_width = "wrap_content" android:layout_height = "wrap_content"
      android:text = "click me" android:drawableLeft = "@mipmap/ic_launcher"/>
<ImageButton android:layout_width = "wrap_content"
     android:layout_height = "wrap_content" android:src = "@mipmap/ic_launcher"/>
```

图 9-1 是 ImageButton 类的继承关系。

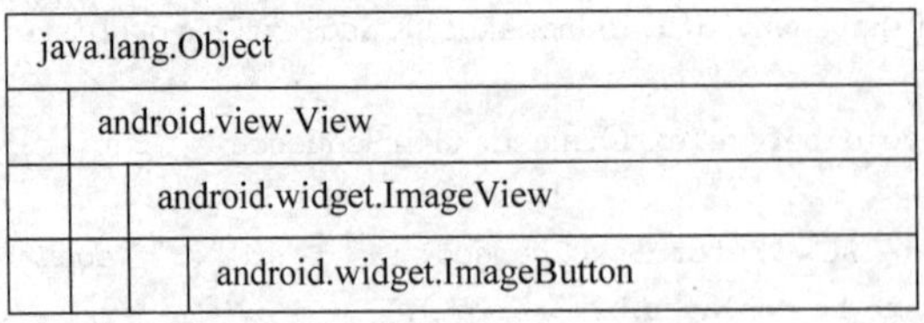

```
java.lang.Object
  android.view.View
    android.widget.ImageView
      android.widget.ImageButton
```

图 9-1 ImageButton 类的继承关系

为了设置 Button 被按下后的动作，主要有以下两种方法：

方法 1：在 XML 布局文件的 Button 元素添加 onClick 属性，这个属性的值就是响应 click 事件调用的方法的名字，持有该布局文件的 Activity 必须实现该方法。特别地，这个方法必须：①声明为 public，②返回 void，③定义一个 View 作为方法的唯一参数，这个

View 就是被单击的 View。例如：

```
<Button
android:layout_width = "wrap_content" android:layout_height = "wrap_content"
android:id = "@ + id/sendBtn" android:text = "send Message" android:onClick = "onClickMe"/>
```

在持有布局文件的 Activity 中实现 onClick 属性定义的方法 onClickMe。

```
public void onClickMe(View view) {
  // Do something in response to button click
}
```

方法 2：在持有 Button 所在的布局文件的 Activity 代码中调用 Button 对象的 setOnClickListener(View. OnClickListener)方法。

```
Button button = (Button) findViewById(R.id. sendBtn);
button.setOnClickListener(new View.OnClickListener() {
    public void onClick(View v) {
      // Do something in response to button click
    }
})
```

系统在主线程里执行 Button 响应，这意味着 onClick 必须执行得很快，以避免延时应用对下一步用户行为的响应。

每个 Button 使用系统默认的 Button 背景，系统默认的 Button 背景随着平台版本的不同而不同。如果不满意默认的 Button 风格，可以使用 drawable 资源或者颜色资源来设置 Button 的 background 属性。在 Android 4.0 和更高的版本上，可以在清单文件的 application 元素中，设置属性 android：theme="@android：style/Theme. Holo"。

无边框的按钮除了没有边框或者背景外，其他方面很像基本的按钮，仍然可以在不同的状态改变外观。

```
<Button android:id = "@ + id/button_send"
     android:layout_width = "wrap_content" android:layout_height = "wrap_content"
                                        android: text = " @ string/button _ send" style =
"?android:attr/borderlessButtonStyle" />
```

9.3 width 与 layout_width 的关系

TextView 及其子类具有 android：width 和 android：height 这两种属性，但不是所有的视图都具有这两种属性，这两个属性一般用来控制视图的精确大小，如 64dp，1px 等。width 和 height 属性设置了视图的初始大小，但是如果这个视图是 android. view. ViewGroup 的一部分，ViewGroup 将根据它自己的布局规则重新调整它的子视图的大小。对于每个 ViewGroup，ViewGroup. LayoutParams 用作 ViewGroup 的子视图的属性，用于告诉它的容器它想怎么布局。ViewGroup. LayoutParams 中所有的属性都以 layout_开始，ViewGroup 将按照自己的布局规则来布局。对于一个 ViewGroup 的子视图，layout_width 和 layout_height 是 ViewGroup 的子视图必需定义的属性，并且它更加灵活，除了可以跟 width 和 height 一样可以指定精确的尺寸值外，还可以取值 match_parent 或者 wrap_content 随着父控件或者随着内容改变大小。当视图放在 ViewGroup 中，可以不使用 width 和 height 属性，而使用 android：minHeight 和 android：maxHeight 来指定最小尺寸和最大尺寸。

在一个 ViewGroup 中，对子视图的大小起决定作用的是 layout_开始的属性，也就是说子视图的大小是由它的 ViewGroup 决定的。如果子视图的 layout_width 和 layout_height 的取值为尺寸值，则视图的实际尺寸就是由 layout_width 和 layout_height 指定的尺寸值。如果子视图的 layout_width 和 layout_height 的取值为 match_parent，子视图的大小是父容器的尺寸减去父容器的 padding 后的大小。如果子视图的 layout_width 和 layout_height 的取值为 wrap_content，并且没有指定 width 和 height 属性，则子视图的大小应该足够大，大到可以完全包围它的内容和它的 padding；如果子视图的 layout_width 和 layout_height 的取值为 wrap_content，并且指定了 width 和 height 属性，则子视图的尺寸是它的 width 或者 height 加上它对应的 padding。

实例 2：layout_width＝"wrap_content"受 width 的影响。

```
<?xml version = "1.0" encoding = "utf - 8"?>
<LinearLayout xmlns:android = "http://schemas.android.com/apk/res/android"
    android:orientation = "vertical" android:layout_width = "match_parent"
    android:layout_height = "match_parent">
    <Button android:id = "@ + id/buttonA" android:layout_width = "wrap_content"
        android:layout_height = "wrap_content" android:width = "200dp"
        android:text = "11111111111111111111111111111111111111111111111111" />
</LinearLayout>
```

在上面的布局文件中，buttonA 的 layout_width＝"wrap_content"，buttonA 没有指定任何 padding，所以 buttonA 的实际宽度就是 width 属性的取值。布局效果如图 9-2 所示。

给上面的布局文件中，buttonA 加上 paddingRight 属性：

```
android:paddingRight = "50dp"
```

由于 buttonA 的 layout_width 为 wrap_content，且 buttonA 指定了 padding，所以 buttonA 的实际宽度就是 width 属性的取值加上 buttonA 的 padding。布局效果如图 9-3 所示。

实例 3：layout_width＝"match_parent"不受 width 的影响，布局文件如下：

```
<?xml version = "1.0" encoding = "utf - 8"?>
<LinearLayout xmlns:android = "http://schemas.android.com/apk/res/android"
    android:orientation = "vertical" android:layout_width = "match_parent"
    android:layout_height = "match_parent">
    <Button android:id = "@ + id/buttonA" android:layout_width = "match_parent"
        android:layout_height = "wrap_content" android:width = "200dp"
        android:text = "11111111111111111111111111111111111111111111111111" />
</LinearLayout>
```

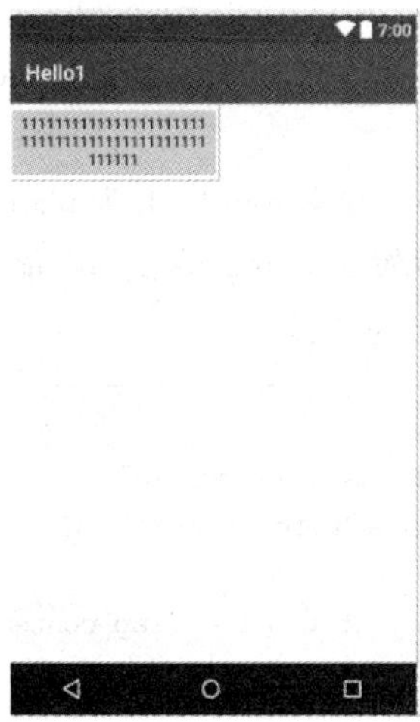

图 9-2 实例 2 的布局效果

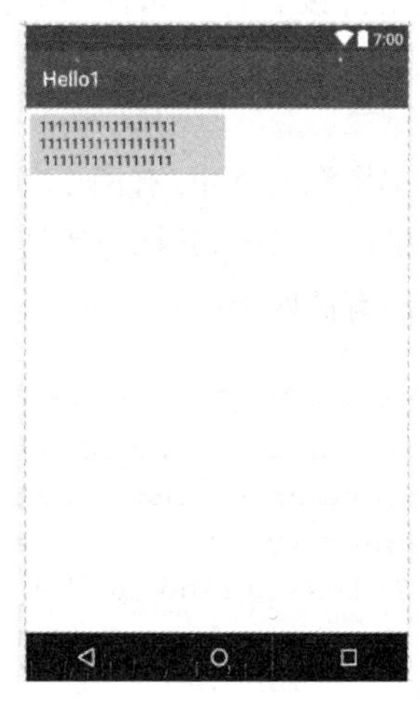

图 9-3 实例 2 buttonA 具有 padding 的布局效果

在上面的布局文件中，buttonA 的 layout_width 为 match_parent，buttonA 的实际宽度就是父容器的大小减去父容器的 padding，与 buttonA 指定的 width 没有关系。布局效果如图 9-4 所示。

给上面文件中的 Linearlayout 加上 padding 属性：

```
android:padding = "100dp"
```

加上 padding 属性后的布局效果如图 9-5 所示。

图 9-4 实例 3 layout_width 为 match_parent 的布局效果

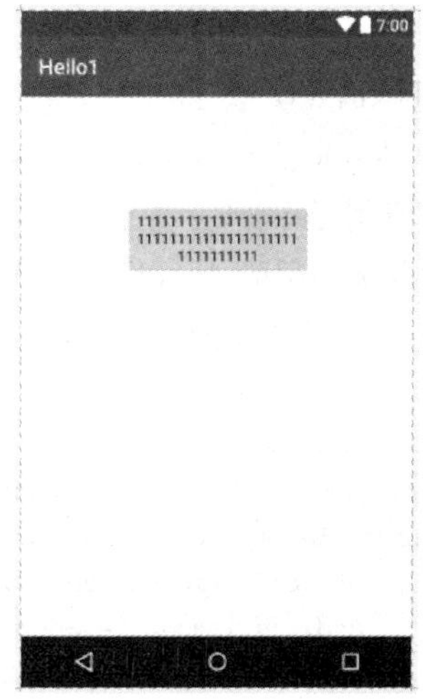

图 9-5 实例 3 buttonA 具有 padding 的布局效果

实例 4：autoLink 属性。

autoLink 属性控制是否自动在 text 中找到链接（例如电话号码、web 地址等），并转化

为可以单击的链接。这个属性的取值可以为：none、web、email、phone、map 和 all 中的一个或者多个，多个值用"|"隔开。其中 map 目前只适用于美国的地址，默认值是 none，禁止这个特性。

下面布局资源文件中，有两个 TextView，第一个 TextView 设置 autoLink 为 phone，仅能识别出文本中的电话号码，并标记为链接；第二个 TextView 设置了 autoLink 为 all，能够自动找出文本中的所有的匹配(包括 web，email，phone，map)，并标记为链接。

```
?xml version = "1.0" encoding = "utf - 8"?>
<LinearLayout xmlns:android = "http://schemas.android.com/apk/res/android"
    android:orientation = "vertical" android:layout_width = "match_parent"
    android:layout_height = "match_parent">
    <TextView android:id = "@ + id/textView1" android:layout_width = "wrap_content"
        android:layout_height = "wrap_content"
        android:text = "Tel:65779093cuc12345678"
        android:autoLink = "phone" android:textSize = "30sp"/>
    <TextView android:id = "@ + id/textView2" android:layout_width = "wrap_content"
        android:layout_height = "wrap_content" android:text = "www.sina.com"
        android:autoLink = "all" android:textSize = "30sp" />
</LinearLayout>
```

执行后，单击第一个链接 65779093，出现的界面如图 9-6 所示，可以选择手机上的一个应用来打电话。

图 9-6　实例 4 的布局效果

实例 5：走马灯效果。

要在 TextView 实现走马灯效果，文字一定是单行显示，如果多行显示，那走马灯效果也就失去了存在的意义。另外，在 EditText 中使用走马灯没有必要，也不合理。走马灯效果需要 TextView 获得当前的焦点，TextView 的 Clickable、LongClickable、Focusable、FocusableInTouchMode 这四个属性的默认值都是 false。要让 TextView 得到焦点，这里主要涉及 focusable 和 focusableInTouchMode 这两个属性，简单来说把这两个属性都设置成 true，那么在运行程序以后，走马灯效果就显示出来了。

```
<?xml version = "1.0" encoding = "utf - 8"?>
<LinearLayout xmlns:android = "http://schemas.android.com/apk/res/android"
    android:orientation = "vertical" android:layout_width = "match_parent"
    android:layout_height = "match_parent">
    <TextView android:textSize = "40sp" android:layout_width = "wrap_content"
        android:layout_height = "wrap_content" android:singleLine = "true"
        android:ellipsize = "marquee" android:marqueeRepeatLimit = "marquee_forever"
        android:clickable = "true" android:focusable = "true"
        android:focusableInTouchMode = "true"
        android:text = "123456789\nabcdefghijklmnopqrst" />
</LinearLayout>
```

实例 6：ellipsize。

下面布局文件，第一个 TextView，没有指定 android：maxLines＝"1"，所以默认是多行。第二个 TextView，指定了 android：maxLines＝"1"，但 android：ellipsize＝"none"，超出一行的文字没有显示。后面三个都指定了单行，第三个 TextView 指定 android：ellipsize＝"start"，省略号显示在开始；第四个 TextView 指定 android：ellipsize＝"middle"，省略号显示在中间，第五个 TextView 指定 android：ellipsize＝"end"，省略号显示在结束。不过当省略号显示在开始或者中间，使用 android：maxLines＝"1"指定 TextView 高度为 1 行会出现错误"java.lang.ArrayIndexOutOfBoundsException：length＝61；index＝－1"，所以第三个 TextView（android：ellipsize＝"start"）、第四个 TextView（android：ellipsize＝"middle"）都采用 android：singleLine＝"true"指定 TextView 高度为 1 行。

```
<?xml version = "1.0" encoding = "utf - 8"?>
< LinearLayout xmlns:android = "http://schemas.android.com/apk/res/android"
android:orientation = "vertical" android:layout_width = "match_parent"
android:layout_height = "match_parent">
    < TextView android:id = "@ + id/textView1"
        android:layout_width = "match_parent"
  android:layout_height = "wrap_content"
        android:ellipsize = "none" android:text = "0123456789abcdefghijklmn"
        android:textSize = "30sp" android:layout_marginBottom = "40dp" />
    < TextView android:id = "@ + id/textView2"
        android:layout_width = "match_parent"
android:layout_height = "wrap_content"
        android:singleLine = "true" android:ellipsize = "none"
        android:text = "0123456789abcdefghijklmn"
        android:textSize = "30sp" android:layout_marginBottom = "40dp" />
    < TextView android:id = "@ + id/textView3"
        android:layout_width = "match_parent"
android:layout_height = "wrap_content"
        android:singleLine = "true" android:ellipsize = "start"
        android:text = "0123456789abcdefghijklmn"
        android:textSize = "40sp" android:layout_marginBottom = "40dp"/>
    < TextView android:id = "@ + id/textView4"
        android:layout_width = "match_parent"
android:layout_height = "wrap_content"
        android:singleLine = "true" android:ellipsize = "middle"
        android:text = "0123456789abcdefghijklmn"
        android:textSize = "30sp" android:layout_marginBottom = "40dp" />
    < TextView android:id = "@ + id/textView5"
        android:layout_width = "match_parent"
android:layout_height = "wrap_content"
        android:singleLine = "true" android:ellipsize = "end"
        android:text = "0123456789abcdefghijklmn"
        android:textSize = "30sp" />
</LinearLayout >
```

在手机上的运行效果如图 9-7 所示。

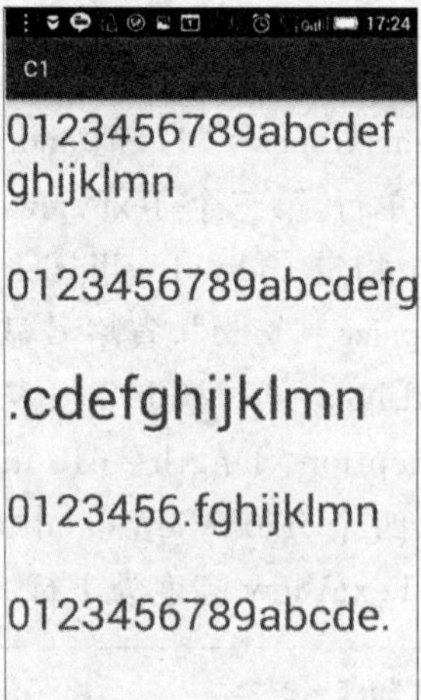

图 9-7　实例 6 的布局效果

实例 7：imeOptions，布局文件内容如下：

```
<?xml version = "1.0" encoding = "utf - 8"?>
<LinearLayout xmlns:android = "http://schemas.android.com/apk/res/android"
    android:orientation = "vertical" android:layout_width = "match_parent"
    android:layout_height = "match_parent">
    <EditText
        android:layout_width = "match_parent"
        android:layout_height = "wrap_content"
        android:inputType = "phone"
        android:imeOptions = "actionNone" />
    <EditText
        android:layout_width = "match_parent"
        android:layout_height = "wrap_content"
        android:imeOptions = "actionNext"
        android:inputType = "phone" />
    <EditText
        android:layout_width = "match_parent"
        android:layout_height = "wrap_content"
        android:imeOptions = "actionDone"
        android:inputType = "numberSigned" />
</LinearLayout>
```

把布局放入 activity 显示，并在手机上执行，当输入焦点在第一个 EditText(android: imeOptions＝"actionNone")时，输入软键盘有一个回车键，效果如图 9-8 所示。

单击软键盘上的回车键，光标将移动到下一个 EditText 的上次的光标位置。当输入焦点在布局中的第二个 EditText(android: imeOptions＝"actionNext")时，输入软键盘有一个键"下个"，效果如图 9-9 所示。

单击软键盘上的键"下个"，光标将移动到下一个 EditText 的上次的光标位置。当输入焦点在第三个 EditText(android: imeOptions＝"actionDone")时，输入软键盘有一个键"完成"，效果如图 9-10 所示。

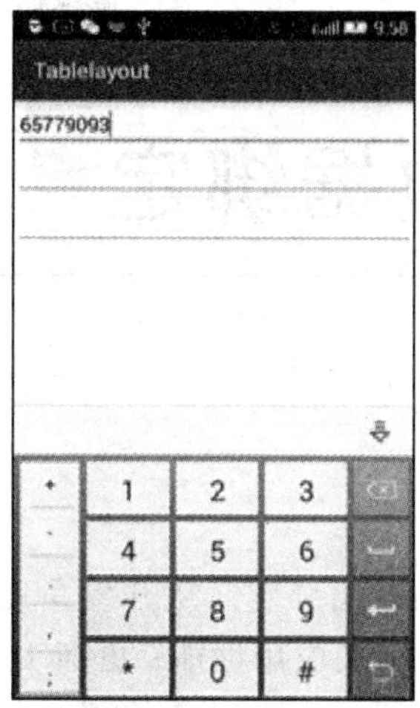

图 9-8 实例 7 imeOptions="actionNone"的布局效果

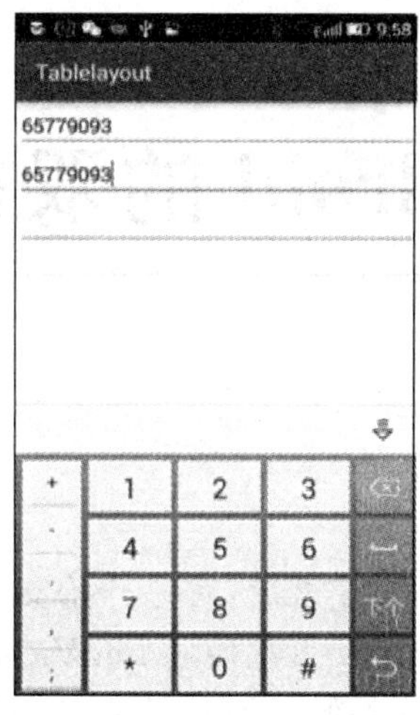

图 9-9 实例 7 imeOptions="actionNext"的布局效果

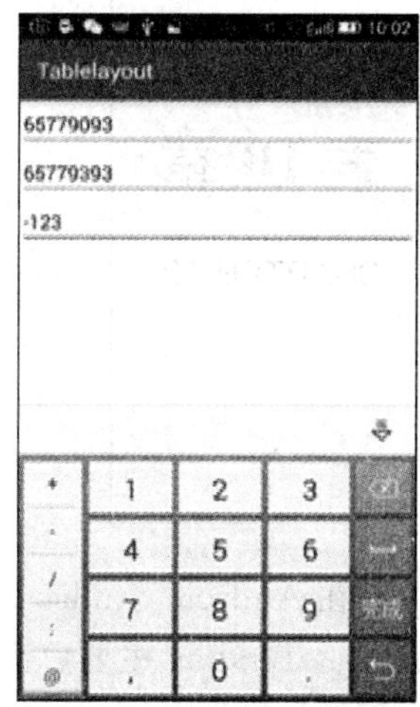

图 9-10 实例 7 imeOptions="actionDone"的布局效果

单击软键盘上的键“完成”，单击后光标保持在原来的输入框上，同时软键盘关闭。

9.4 本章主要参考文献

1. https://developer.android.google.cn/reference/android/widget/TextView
2. https://developer.android.google.cn/reference/android/widget/EditText.html
3. https://developer.android.google.cn/reference/android/text/TextWatcher
4. https://developer.android.google.cn/reference/android/widget/Button.html
5. https://developer.android.google.cn/guide/topics/ui/controls/button.html
6. https://developer.android.google.cn/reference/android/widget/ImageButton

第 10 章 Android 的双向数据绑定

CHAPTER 10

在 Android Studio 2.1 Preview 3 之后，官方开始支持数据的双向绑定。从开发角度看，DataBinding 主要解决了两个问题：多次使用 findViewById，损害了应用性能；更新 UI 数据需切换至 UI 线程，将数据分解映射到各个 view 比较麻烦。目前 Android 系统支持数据双向绑定的控件，如表 10-1 所示。

表 10-1　Android 系统支持数据双向绑定的控件

控　　件	可以绑定的属性
AbsListView	android：selectedItemPosition
CalendarView	android：date
CompoundButton	android：checked
DatePicker	android：year，android：month，android：day
NumberPicker	android：value
RadioGroup	android：checkedButton
RatingBar	android：rating
SeekBar	android：progress
TabHost	android：currentTab
TextView	android：text
TimePicker	android：hour，android：minute

DataBinding 处理过程为：针对每个 Activity 的布局，在编译阶段，生成一个 ViewDataBinding 类的对象，该对象持有 Activity 要绑定的变量和布局中的带有 ID 的视图的引用。该对象主要的功能有：将数据分解到各个视图、在 UI 线程上更新数据、监控数据的变化，实时更新。为了使用数据绑定，首先必须在模块的 build.gradle 文件的 Android 语句块中设置：

```
android {…
    dataBinding{ enabled = true}
 }
```

或者

```
android {...
    dataBinding.enabled = true
}
```

另外，要使用数据绑定，必须确保在工程的 build. gradle 中设置的 android gradle 插件的版本不低于 2.1-alpha3。

10.1 可观察的数据对象

普通的数据对象绑定以后，当数据改变时，与该数据绑定的图形 UI 中的显示不能自动更新，因而应该采用可以观测的数据。可以观测的数据的含义是：数据改变时，绑定的 UI 随着自动更新。观察者模式包含观察者和被观察者，Observable 指被观察者，Observer 指观察者。android. databinding. Observable 接口定义如下：

```
public interface Observable {
    void addOnPropertyChangedCallback(OnPropertyChangedCallback callback);
    void removeOnPropertyChangedCallback(OnPropertyChangedCallback callback);
    abstract class OnPropertyChangedCallback {
        public abstract void onPropertyChanged(Observable sender, int propertyId);
}
```

Observable 接口有添加和移除监听器的机制，但通知由开发人员自己决定。当类的被观察的属性变化时，类对象应该通知 Observable. OnPropertyChangedCallback。观察者模式定义了对象间的一对多的依赖关系，当一个对象的状态发生改变的时候，所有依赖于它的对象都将得到通知，并自动更新。实现了 android. databinding. Observable 接口的类允许把一个监听器附着在绑定对象上，监听对象的所有属性的变化。android. databinding. BaseObservable 实现了 Observable 接口，提供的其他方法，如表 10-2 所示。

表 10-2 android. databinding. BaseObservable 的方法

方　　法	解　　释
void notifyChange()	通知监听者这个实例的所有属性都发生了变化
void notifyPropertyChanged(int fieldId)	通知监听者一个特定的属性已经发生变化。变化属性的 getter 方法前必须加上@Bindable 注解，以便在 BR 类中，生成一个域，用作 fieldId

数据类可以直接继承 BaseObservable，并实现在成员属性变化的时候发出通知。定义可以观察的数据类的具体步骤如下：①首先实体类继承 BaseObservable 类，或者自己实现 Observable。②给需要自动更新 UI 的成员变量的 getter 方法添加@Bindable 注解，然后编译模块以便生成 BR 类。具体方法如下：单击 Android Studio 的菜单 Build，Make Module'App'，编译模块，编译生成的 BR 类源文件位于工程文件夹的 app\build\generated\source\apt\debug 文件夹里的数据观察对象指定的包里，功能与 R. java 类似，每个用@Bindable 注解过的 getter 方法在编译生成的 BR 类中对应一个入口。Android annotation process tool(apt)是一种处理注释的工具，在编译时对源代码文件进行检测找出其中的 Annotation，把注解生成新的 Java 源文件，减少手动的代码输入，而不是运行时利用反射，这样大大优化了性能。③在需要自动更新 UI 的成员变量的 set 方法中调用方法 notifyPropertyChanged(BR. xxx)，即当数据发生变化时还是需要调用 notifyPropertyChanged 手动发出通知需要更新 UI，该方法的参数是在 BR 类中

记录的变量。

实例 1：一个可观察的类，文件内容如下：

```
package cuc.mvvm1;
/* 生成的 BR.java 位于 app\build\generated\source\apt\debug\ cuc\mvvm1 文件夹里
 */
import android.databinding.BaseObservable;
import android.databinding.Bindable;
public class User extends BaseObservable {
    private String name;
    private int age;
    public User () {
    }
    public User (String name, int age) {
        this.name = name;
        this.age = age;
    }
@Bindable
    public String getName() {
        return name;
    }
    public void setName(String name) {
        this.name = name;
        notifyPropertyChanged(BR.name);
    }
@Bindable
    public int getAge() {
        return age;
    }
    public void setAge(int age) {
        this.age = age;
        notifyPropertyChanged(BR.age);
    }
}
```

1. ObservableField

除了让数据类继承 BaseObservable，还有一种更细粒度的绑定方式，可以具体到成员变量，成员变量使用 android.databinding.ObservableField< T >，泛型可以填入自己需要的类型。ObservableField 继承自 BaseObservable，且实现了 Serializable 接口。这种类型的域必须被声明为 public final 类型，且必须要初始化，通过 set 和 get 方法为 ObservableField 设置和读取域的值。因为绑定仅仅检测域的值的变化，而不是域自身。对于一个 final 变量，如果是基本数据类型的变量，则其数值一旦在初始化之后便不能更改；如果是引用类型的变量，则在对其初始化之后便不能再让其指向另一个对象。系统提供了所有的基本数据类型所对应的 Observable 类，就是基本数据类型前面加上 Observable 组成，例如 ObservableBoolean，ObservableByte，ObservableChar，ObservableShort，ObservableInt，ObservableLong，ObservableFloat，ObservableDouble 和 ObservableParcelable，可以直接使用，这些类都继承自 BaseObservable，且实现了 Parcelable 和 Serializable 接口。而 String

类型对应的是 ObservableField< String >,ObservableField 是一个泛型。

实例 2：ObservableField。

```
package cuc.mvvm1;
import android.databinding.ObservableField;
import android.databinding.ObservableInt;
public class User {
    public final ObservableField< String > name = new ObservableField<>();
    public final ObservableInt age = new ObservableInt();
    public User (String name, int age) {
        this.name.set(name);
        this.age.set(age);
    }
}
```

上面类 User 中,name 和 age 这两个域都是可以被观察的,这种类型的域必须被声明为 public final 类型,因为绑定仅仅检测域的值的变化,而不是域自身。

2. 可观察的集合

一些应用使用动态的结构保存数据,例如键值对。当键是引用类型(例如字符串)时,可以使用 ObservableArrayMap。当键是整数时,可以使用 ObservableArrayList。

实例 3：ObservableArrayMap。

```
ObservableArrayMap< String,Object > userMap = new ObservableArrayMap<>();
userMap.put("name","zhangsan");
userMap.put("age",30);
activityMain1Binding.setUser(userMap);
```

布局文件 activity_main1.xml 内容如下：

```
<?xml version = "1.0" encoding = "utf - 8"?>
< layout xmlns:android = "http://schemas.android.com/apk/res/android">
    < data class = ".ActivityMain1Binding">
        < import type = "android.view.View"/>
        < import type = "android.databinding.ObservableMap"/>
        < variable name = "user" type = "ObservableMap&lt;String,Object&gt;"/>
    </data >
        < TextView
            android:text = '@{user["name"] + ", " + user["age"]}'
            android:visibility = "@{((Integer)user[`age`]).intValue()> 18 ? View.VISIBLE:
View.INVISIBLE}"
            android:layout_width = "match_parent"
            android:layout_height = "wrap_content" />
</layout >
```

实例 4：ObservableArrayList。

```
ObservableArrayList< Object > userList = new ObservableArrayList<>();
```

```
userList.add("adult");
userList.add(30);
activityMain1Binding.setUser(userList);
```

布局文件 activity_main1.xml 内容如下：

```
<?xml version = "1.0" encoding = "utf - 8"?>
<layout xmlns:android = "http://schemas.android.com/apk/res/android">
    <data class = ".ActivityMain1Binding">
        <import type = "android.view.View"/>
        <import type = "android.databinding.ObservableList"/>
        <variable name = "user" type = "ObservableList&lt;Object&gt;"/>
    </data>
        <TextView
            android:text = '@{"name:" + user[0] + ", age:" + user[1]}'
            android:visibility = "@{((Integer)user[1]).intValue()> 18 ? View.VISIBLE:
View.INVISIBLE}"
            android:layout_width = "match_parent"
            android:layout_height = "wrap_content" />
</layout>
```

10.2 XML 布局文件

相对于普通布局资源文件(不支持数据绑定)，支持数据绑定的布局资源文件新增了 layout 和 data 标签，并且新增加的 layout 为根元素，data 元素作为 layout 的第一个子元素，并把普通的布局资源文件的根元素(View)作为 layout 的第二个子元素。也就说，layout 包含 data 元素和视图两个元素，data 包含要绑定的数据模型。

使用 DataBinding 编写布局，系统会自动生成一个继承 ViewDataBinding 的类，而 data 标签的 class 属性可以指定这个类的名字，class 属性的值可以是完全限定的类名(字符串中有点，但点不在开头，形式为 class="packagename.className")，也可以是点开头的相对包名(字符串中有点，并且点在开头，形式为 class=".className"，生成的类直接放在模块的包中)，也可以仅仅指定类名(即字符串中没有点，开头也没有点，形式为 class="className"，则生成的类放在模块包下的 databinding 子包下)。如果 data 元素没有指定 class 属性，绑定类名则会根据 xml 文件的名字自动生成，形式为{Layout}Binding，自动生成规则是：将布局文件的首字母大写，并且去掉下画线，将下画线后面的字母大写，最后加上 Binding，生成的类放在模块包下的 databinding 子包下。例如，模块包名为 cuc.mvvm1，若<data>元素指定 class=".ActivityMainBinding"，则生成的类的全限定名字就是 cuc.mvvm1.ActivityMainBinding；若指定 class="ActivityMainBinding"，则生成的类的全限定名字就是 cuc.mvvm1.databinding.ActivityMainBinding。如果布局文件名为 activity_main.xml，<data>元素没有指定 class，则生成的类的全限定名字就是 cuc.mvvm1.databinding.ActivityMainBinding。

当 Android Studio 的版本在 3.2 以下，仅仅生成文件名为{Layout}Binding 形式的类，

并且不是抽象类，源代码位于 app\build\generated\source\apt\debug 文件夹下。当 Android Studio 的版本在 3.2 时，生成文件名为{Layout}Binding 和{Layout}BindingImpl 两种形式的类，其中{Layout}Binding 是抽象类，源代码位于 app\build\generated\data_binding_base_class_source_out\debug\dataBindingGenBaseClassesDebug\out\文件夹下；它的实现类：类名为{Layout}BindingImpl，源码位于 app\build\generated\source\apt\debug 文件夹下。例如，模块包名为 cuc.mvvm1，布局文件名为 activity_main.xml，<data>元素没有指定 class，Android Studio 的版本在 3.2 以下，则自动生成的全限定类名为 cuc.mvvm1.databinding.ActivityMainBinding，文件 ActivityMainBinding.java 位于 app\build\generated\source\apt\debug\cuc\mvvm1\databinding 目录下，类 ActivityMainBinding 定义如下：

```
public class ActivityMainBinding extends android.databinding.ViewDataBinding implements
android.databinding.generated.callback.OnClickListener.Listener {…}
```

当 Android Studio 升级到 3.2 以后，自动生成的 ActivityMainBinding 只是一个抽象类，源文件 ActivityMainBinding.java 位于 app\build\generated\data_binding_base_class_source_out\debug\dataBindingGenBaseClassesDebug\out\cuc\mvvm1\databinding 目录下，ActivityMainBinding 继承自 ViewDataBinding：

```
public abstract class ActivityMainBinding extends ViewDataBinding {…}
```

它的实现类(类名后加 Impl)的源文件 ActivityMainBindingImpl.java 位于 app\build\generated\source\apt\debug\cuc\mvvm1\databinding 目录下。

```
public class ActivityMainBindingImpl extends ActivityMainBinding {…}
```

在 Android gradle 插件 3.1.0 Cannary 6 中附带了可选的编译器 V2。如果想使用 DataBinding 编译器 V2，可以在工程的根目录里 gradle.properties 文件中添加：

```
android.databinding.enableV2 = true
```

在 V2 编译器中，ViewBinding 系列的类将在 Java 编译器编译之前由 Android Gradle Plugin 生成，这样可以避免因为不相关的原因导致 Java 编译失败，从而导致得到太多误报错误。而在 V1 编译器中，binding 系列的类将会在 app 编译完成后再次生成(去分享生成的代码并关联常量 BR 和 R'文件)。在 V2 中，绑定库为多模块项目将保留生成的 binding 系列类和映射信息，从而显著提高数据绑定性能。

注意：新的 V2 编译器向后不兼容，所以使用 v1 编译的库不能被 V2 使用，反之亦然。在 V1 中，一个应用程序能够提供绑定适配器，可以覆盖依赖项中的适配器。在 V2 中，它只会在自己的模块/应用程序及其依赖项中生效。在 V2 中，不同的模块不能在清单文件中使用相同的包名，因为数据绑定将使用该包名来生成绑定映射类。

10.2.1 variable 元素

data 元素的内部可以拥有任意多个 variable 元素，variable 元素用作 data 元素的子元素，来声明在布局文件中使用的变量。variable 的 name 属性设置变量名。variable 的 type 属性设置变量的类型，可以是基本类型如 int，也可以为集合或者对象；type 属性可以采用全限定类名，也可以仅仅使用类名（但需要使用 import 元素像在 java 代码中那样导入包）。

根据布局文件自动生成的 ViewDataBinding 类的规则如下，布局中的每个 variable 元素都会生成相应的 getter 和 setter 方法，布局的视图中每个带有 id 的视图都会生成一个 public final 域。视图的 id 对于数据绑定不是必需的，但是如果需要在 Java 代码中访问这些视图，id 属性仍然是必需的。抽取这些带有 id 的视图，在绑定过程中仅仅遍历一遍。例如 variable 的名字是 xyz，则自动生成的 ViewDataBinding 类中包含 setXyz 和 getXyz 方法。这些变量在 setter 方法调用以前将取默认的 Java 值，引用类型的默认值为 null，int 的默认值为 0，布尔量的默认值为 false，等等。如果希望数据改变时，图形界面中的显示也自动更新，相应的数据就应该使用可观察的对象。

实例 5：以下是支持数据绑定的一个布局文件实例：

```
<?xml version = "1.0" encoding = "utf - 8"?>
< layout xmlns:android = "http://schemas.android.com/apk/res/android">
    < data >
        < variable name = "user" type = "cuc. mvvm1.User"/>
    </data >
    < TextView
        android:text = "@{user.name}"
        android:layout_width = "match_parent"
        android:layout_height = "wrap_content" />
</layout >
```

实例 6：以下是一个布局文件及根据布局文件生成的数据绑定类的成员：

```
<?xml version = "1.0" encoding = "utf - 8"?>
< layout xmlns:android = "http://schemas.android.com/apk/res/android">
   < data >
< import type = "android.graphics.drawable.Drawable"/>
   < variable name = "user" ?type = "cuc.mvvm1.User"/>
   < variable name = "image" type = "Drawable"/>
   < variable name = "note" ?type = "String"/>
  </data >
  < LinearLayout
     android:orientation = "vertical"
     android:layout_width = "match_parent"
     android:layout_height = "match_parent">
     < TextView android:layout_width = "wrap_content"
        android:layout_height = "wrap_content"
        android:text = "@{user.name}"
        android:id = "@ + id/nameTxt"/>
     < TextView android:layout_width = "wrap_content"
        android:layout_height = "wrap_content"
        android:text = "@{user.age}"
        android:id = "@ + id/ageTxt"/>
```

```
    </LinearLayout>
</layout>
```

根据上面布局文件生成的数据绑定类的成员如下：

```
public final TextView nameTxt;
public final TextView ageTxt;
public abstract cuc.mvvm1.User getUser();
public abstract void setUser(cuc.mvvm1.User user);
public abstract Drawable getImage();
public abstract void setImage(Drawable image);
public abstract String getNote();
public abstract void setNote(String note);
```

自动生成的类中还包含一个名字叫 context 的特殊变量，它的值就是调用根视图的 getContext()返回的 Context，如果有显式的变量声明名字为 context，则覆盖上面的值。

10.2.2 import 元素

data 元素可以拥有 0 个或者多个 import 元素，import 元素用作 data 元素的子元素。import 元素的 type 属性的取值必须是全限定类名，import 并不能像 Java 一样可以 import xx.xxx.*，必须具体到写清楚每个要导入的类名。导入的类可以在 variable 元素和表达式中直接使用类名用作类型引用，不需要使用全限定类名。当有类名冲突时，就可以使用 import 元素的 alias 属性给导入的类起别名。当 import 元素设置了 alias 属性后，在 variable 元素和表达式中就应该直接使用别名，不能再用原来的类名。在表达式中可以使用导入的类的静态变量和方法。

实例 7：以下是包括 import 元素的一个布局文件实例：

```
<?xml version="1.0" encoding="utf-8"?>
<layout xmlns:android="http://schemas.android.com/apk/res/android">
    <data class=".ActivityMain1Binding">
        <import type="cuc.mvvm1.User" alias="UserA"/>
        <import type="android.view.View"/>
        <import type="java.util.List"/>
        <variable name="user" type="UserA"/>
        <variable name="userAList" type="List&lt;UserA&gt;"/>
    </data>
    <LinearLayout
        android:orientation="vertical"
        android:layout_width="match_parent"
        android:layout_height="match_parent">
        <TextView
            android:text='@{user.name + ", " + user.age}'
            android:visibility="@{user.age > 18?View.VISIBLE:View.INVISIBLE}"
            android:layout_width="match_parent"
            android:layout_height="wrap_content" />
```

```
        <TextView
            android:text = "@{userAList[0].name +`, `+ userAList[0].age}"
            android:visibility = " @ { userAList [ 0 ]. age &lt; = 18 ? View. VISIBLE : View.
INVISIBLE}"
            android:layout_width = "match_parent"
            android:layout_height = "wrap_content" />
    </LinearLayout>
</layout>
```

上面 xml 文件使用了 XML 语法中预定义的实体引用："<"代表小于号，">"代表大于号，"""代表双引号，所以上面变量 userList 的类型为 List < user >。上面的代码中使用 import 导入了 android. view. View，在 variable 元素和@{}表达式中就可以使用 View 代替 android. view. View 类型（像 Java 代码中一样）。上面代码中使用 import 导入了类 cuc. mvvm1. User，并给该类起了别名 UserA，以后 variable 元素和@{}表达式就应该使用 UserA 代替 cuc. mvvm1. User 类。

10.2.3 include 元素

在布局文件中，可以使用 include 元素包含另一布局文件。如果 include 元素和 include 的布局文件的根视图只有一个指定了 ID，包含进来的视图就使用这个 ID，如果两者都定义了 ID，使用 include 元素的 ID。被包含进来的布局文件中的变量必须从容器布局中传递进来，使用应用的名称空间和变量名字。

实例 8：布局文件 usergroup. xml 通过 include 元素包含了布局文件 user. xml，user. xml 文件内容如下：

```
<?xml version = "1.0" encoding = "utf - 8"?>
<layout xmlns:android = "http://schemas.android.com/apk/res/android">
    <data>
        <variable
            name = "user"
            type = "cuc.mvvm1.User"/>
    </data>
    <TextView
        android:text = '@{"name:" + user.name + ",age:" + user.age}'
        android:layout_width = "match_parent"
        android:layout_height = "wrap_content" />
</layout>
```

usergroup. xml 文件内容如下：

```
<?xml version = "1.0" encoding = "utf - 8"?>
<layout xmlns:android = "http://schemas.android.com/apk/res/android"
    xmlns:app = "http://schemas.android.com/apk/res - auto">
    <data class = ".UserGroupBinding">
        <import type = "cuc.mvvm1.User"/>
```

```
        <import type = "java.util.List"/>
        <variable name = "userList" type = "List&lt;User&gt;"/>
    </data>
    <LinearLayout
        android:layout_width = "match_parent"
        android:layout_height = "match_parent"
        android:orientation = "vertical">
        <include layout = "@layout/user" app:user = "@{userList[0]}"/>
        <include layout = "@layout/user" app:user = "@{userList[1]}"/>
    </LinearLayout>
</layout>
```

include 元素中必须包含 user.xml 文件中的变量的取值，它的取值必须来自于容器布局 usergroup.xml 的 data 结点内定义的变量。

10.2.4　属性的取值

当使用单引号包围属性值时，属性取值中的字符串只能使用双引号。例如：

```
android:text = '@{map["firstName"]}'
```

也可以使用双引号包围属性值，这时属性取值中的字符串只能使用反引号，反引号就是计算机键盘上数字 1 左边的符号。例如：

```
android:text = "@{map[`firstName`}"
```

10.2.5　表达式语言

正向绑定使图形界面可以自动更新显示 Java 代码中定义的对象，视图属性的取值形式为"@{表达式}"。双向绑定不仅使图形界面可以自动更新显示 Java 代码中定义的对象，还可以把图形界面中的状态及文本的改变通知给 Java 代码中的对象视图属性的取值形式为"@={表达式}"。正向绑定与双向绑定微小的差别就是"@{表达式}"改变成了"@={表达式}"。

表达式语法支持大部分的 Java 操作，当然最好不要写太复杂的语句，如果有这个需求，可以在 Java 类中写一个方法，然后在此调用。表达式支持的操作符有：数学运算符＋ － / * %、字符串级联符＋、逻辑运算符 &&、||、二进制运算符 &、|、^，一元操作符＋、－、!、~、移位符≫、≪，比较运算符＝＝、＞、＜、＞、＝、＜、＝，判断类的实例运算符 instanceof、分组符号()、类型转换、res 资源访问、方法调用、成员访问、数组访问[]，三元运算符?:，NULL 合并操作符??。在 Java 语句中可以使用但不能用在属性表达式中的运算符有：this，super，new。Null 合并运算符 ?? 的形式为：

```
表达式 1 ?? 表达式 2
```

含义为，如果 ?? 左边的表达式不为 NULL，取 ?? 左边表达式的值；否则，取 ?? 右边表达式的值。例如：

```
android:text = "@{user.displayName ?? user.lastName}"
```

这个式子在功能上等价于：

```
android:text = "@{user.displayName != null ? user.displayName :user.lastName}"
```

实例 9：双向绑定，activity_main.xml 内容如下：

```
<?xml version = "1.0" encoding = "utf - 8"?>
<layout xmlns:android = "http://schemas.android.com/apk/res/android">
    <data>
        <variable name = "user" type = "cuc.mvvm1.UserPojo"/>
    </data>
    <LinearLayout
        android:layout_width = "match_parent"
        android:layout_height = "match_parent"
        android:orientation = "vertical">
        <EditText
            android:layout_width = "match_parent"
            android:layout_height = "wrap_content"
            android:hint = "name"
            android:inputType = "text"
            android:text = "@{user.name}" />
        <EditText
            android:layout_width = "match_parent"
            android:layout_height = "wrap_content"
            android:hint = "name"
            android:inputType = "text"
            android:text = "@ = {user.name}" />
     </LinearLayout>
</layout>
```

上面的例子通过"@={user.name}"双向绑定了 cuc.mvvm1.UserPojo 类型的对象 user 的 name 到第二个 EditText 的文本域。当 user.name 改变时，两个 EditText 的显示将得到更新(正向绑定)。当用户在第一个 EditText 中输入时，由于它没有使用双向绑定，user.name 将不会更新，因而第二个 EditText 也不会更新。当用户在第二个 EditText 中输入时，由于它使用了双向绑定，user.name 将更新(反向绑定)，user.name 的更新将导致第一个 EditText 显示的更新(正向绑定)。

10.2.6 属性的绑定

在 XML 中使用 databinding 方式赋值的属性，在对应的类中都有相应的 setter，如果在 XML 中属性值的类型与对应的 setter 的参数类型不符，这时 dataBinding 就会去寻找可以让属性值转换为正确类型的方法，而寻找的根据就是所有被@BindingConversion 注解标记的方法。例如，对于名为 example 的属性，DataBinding 库会自动尝试查找接受兼容类型作为参数的 setExample(arg)方法，匹配时忽略属性的自定义命名空间，只使用属性的名称和类型。例如，给定 android：text="@{user.name}"表达式，DataBinding 库将查找接受 user.getName()返回的类型的 setText(arg)方法；如果 user.getName()的返回类型是 String，则库将查找接受 String 参数的 setText()方法；如果该表达式返回一个 int，则库将搜索接受 int 参数的 setText()方法。

@BindingConversion 注解用于自动地把表达式类型转换为 setter 方法的参数类型。@BindingConversion 作用于方法，方法必须为公共静态(public static)方法，且有且只能有 1 个参数。被该注解标记的方法，被视为 dataBinding 的转换方法。例如 setBackground (Drawable background)的参数类型是 Drawable，即 android：background 接收的值是 drawable，若通过 Data Binding 把 R.color.white 赋值给 android：background 属性，而 R.color.white 是 int，所以需要把 int 转换为 Drawable。

```
@BindingConversion
public static Drawable convertColorToDrawable(int drawable) {
  return new ColorDrawable(drawable);
}
```

@BindingAdapter 用于修饰方法，定义了 XML 属性赋值的 Java 实现。@BindingAdapter 修饰的方法必须为公共静态(public static)方法，可以有一到多个参数。@BindingAdapter 主要有以下几种使用场景：①属性在类中没有对应的 setter，如 ImageView 的 android：src，ImageView 中没有 setSrc()方法；②属性在类中有 setter，但是接收的参数不是自己想要的，如 android：background 属性，对应的 setter 是 setBackgound(drawable)，但是绑定的属性值是 int 类型 id，这时候 android：background="@{id}"就不行；③没有对应的属性，但是却要实现相应的功能。

用@BindingAdapter 修饰的静态方法可以解决上面的问题。

```
public class CardViewBindingAdapter {
    @BindingAdapter("contentPadding")
    public static void setContentPadding(CardView view, int padding) {
        view.setContentPadding(padding, padding, padding, padding);
    }
}
```

@BindingAdapter 注解的括号内是属性名称，attribute_name 可以自由定义，不过最好与 method 名称对应。属性默认是自定义的命名空间，在匹配时会被忽略。也可以重写 android 的命名空间，只有 android 是特殊的命名空间。实例如下：

```
@BindingAdapter("android:contentPadding")
```

app：contentPadding 与 android：contentPadding 处理行为可以不一样。app：contentPadding 与 custom：contentPadding 处理行为是一致的。当自定义的适配器属性与系统默认的产生冲突时，自定义适配器将会覆盖掉系统原先定义的注解。

绑定适配器方法的参数类型很重要，第一个参数是控件本身的 View 类型，第二个参数是属性的绑定表达式可以接受的数据类型。

适配器接收多个属性，如下所示：

```
@BindingAdapter({"imageUrl", "error"})
public static void loadImage(ImageView view, String url, Drawable error) {
  Picasso.with(view.getContext()).load(url).error(error).into(view);
}
```

可以在布局中使用适配器，如下例所示：

```
<ImageView
        app:imageUrl = "@{venue.imageUrl}"
        app:error = "@{@drawable/venueError}" />
```

注意，@drawable/venueError 是指应用中的资源，用@{}括住资源使其成为有效的绑定表达式。

10.2.7 Java 类型签名和方法签名

为了介绍视图事件的处理，需要用到 Java 签名的一些知识，下面简单介绍一下 Java 签名。Java 基本数据类型的签名采用单个字母：byte：B，char：C，short：S，int：I，float：F，double：D，void：V，boolean：Z，long：J。除 boolean 和 long 外，其他基本类型，都用大写的首字母表示。引用类型的签名规则为"L＋全限定类名＋；"，即类的签名由三部分组成，第一部分是字母 L，第二部分是全限定类名，类名以"/"分隔，而不是用"."或"_"分隔，第三部分是分号。数组的签名为"[＋类型描述符"，即数组的签名由两部分组成，第一部分是左中括号，左中括号的数量表示数组的维数，n 维数组就包含 n 个左中括号；第二部分是类型描述符。例如：数组 int[]的签名为[I，int[][]的签名为[[I，String[]的签名为[Ljava/lang/String；，Object[]的签名为[Ljava/lang/Object；。方法的签名就是其参数类型签名和返回值类型签名的字符串，其形式如下：

```
(类型签名 1 类型签名 2...)返回值类型签名
```

即将参数类型的描述符按照申明顺序放入一对小括号中、后跟返回值类型的描述符，各个签名之间没有空格。对于没有返回值的，用 V(表示 void 型)表示。表 10-3 给出了几个例子。

表 10-3 Java 方法签名的例子

Java 方 法	签 名
String test ( )	Ljava/lang/String；
int f (int i, Object object)	(ILjava/lang/Object；)I
void set (byte[ ] bytes)	([B)V

javap 是 JDK 自带的反汇编器，可以查看 java 编译器生成的字节码。通过它，可以对照源代码和字节码，从而了解很多编译器内部的工作。cmd 进入控制台，改变到要生成签名的那个类的目录下，以 User 类为例，输入：javap -s -private User，所有方法签名都会被输出，关于 javap 的一些参数可以输入 javap -help 查看。

10.2.8 处理事件

数据绑定除了可以绑定视图的属性值，还可以绑定视图中发出的事件，视图事件绑定的是一个方法,类似于视图的 onClick 属性可以指定 Activity 中的方法。事件的属性名通常由监听器的方法名决定,但也有一些例外。View 的 onClick 和 onLongClick 事件,对应的监听器定义如下:

```
public interface OnClickListener { void onClick(View v); }
public interface OnLongClickListener {boolean onLongClick(View v); }
```

如上所示,View. OnClickListener 有一个方法叫 onClick(View v),View. OnLongClickListener 有一个方法叫 onLongClick(View v),事件的属性名分别为 onClick 和 onLongClick。绑定事件有两种方式: Method References(方法引用)和 Listener Bindings(监听器绑定)。方法引用的优点是方法引用的表达式在编译时处理,如果方法不存在或者方法签名不正确,将会产生编译时错误。方法引用和监听器绑定的主要区别在于: 在方法引用中,实际的监听器实现在数据绑定的时候创建,而不是在事件被触发时; 而监听器绑定在事件发生时才评估表达式。在方法引用中,方法的参数类型和返回值类型必须与监听器对象的参数类型和返回值类型相匹配。而在监听器绑定中,只要方法的返回值类型与监听器对象的预期返回值类型相匹配即可。

1. 方法引用

如果在布局文件中定义 android: onClick 属性,然后在 Activity 中定义相关的方法,就可能在运行时报错,直接崩溃。采用方法引用时,表达式在编译时处理,如果方法不存在或者方法签名不正确,将会产生编译错误。在方法引用中,表达式中方法的签名(signature)必须精确匹配处理事件用的监听器对象方法签名,即方法的参数顺序和类型以及返回值的类型完全一致。

实例 10: 方法引用。

MyHandlers 类中,myOnClick 方法的签名和 OnClickListener 接口中 onClick 方法的签名完全一样,myOnLongClick 方法的签名和 OnLongClickListener 接口中 onLongClick 方法的签名完全一样。

```
public class MyHandlers {
      public void myOnClick(View view){
        Toast.makeText(view.getContext(),"hello",Toast.LENGTH_SHORT).show();
    }
    public boolean myOnLongClick(View view){
        Toast.makeText(view.getContext(),"hello world",Toast.LENGTH_SHORT).show();
        return true;
    }
}
```

以下布局文件 activity_main. xml 绑定了视图事件,文件内容如下:

```
<?xml version = "1.0" encoding = "utf - 8"?>
< layout xmlns:android = "http://schemas.android.com/apk/res/android">
   < data class = ".ActivityMainBinding">
      < variable name = "handlers" type = " cuc.mvvm1.MyHandlers"/>
      < variable name = "user" type = "cuc.mvvm1.User"/>
```

```
    </data>
    <LinearLayout
        android:orientation = "vertical"
        android:layout_width = "match_parent"
        android:layout_height = "match_parent">
        <TextView android:layout_width = "wrap_content"
            android:layout_height = "wrap_content"
            android:text = " Method References "
                android:onClick = "@{handlers::myOnClick}"
                android:onLongClick = "@{handlers::myOnLongClick}"/>
    </LinearLayout>
</layout>
```

在 Activity 的 onCreate 方法可以这样获取绑定对象：

```
@Override
    protected void onCreate(Bundle savedInstanceState) {
        super.onCreate(savedInstanceState);
        ActivityMainBinding activityMainBinding  =  DataBindingUtil.setContentView(this, R.layout.
activity_main);
        activityMainBinding.setHandlers(new MyHandlers());
    }
```

2. 监听器绑定

监听器绑定功能仅在 Android Gradle 插件 2.0 或更新版本上可用，形式为 lambda 表达式。这种绑定是在运行时，事件发生以后才绑定的。监听器绑定不再要求方法签名一致，而只是要求返回值类型一样，方法的参数可以定制。

实例 11：监听器绑定。

```
public class MyHandlers {
        public void myOnClick(View view,String name){
Toast.makeText(view.getContext(),"hello" + name,Toast.LENGTH_SHORT).show();
        }
       public boolean myOnLongClick(View view){
           Toast.makeText(view.getContext(),"hello world",Toast.LENGTH_SHORT).show();
           return true;
       }
}
```

布局文件 activity_main.xml 文件内容如下：

```
<?xml version = "1.0" encoding = "utf - 8"?>
<layout xmlns:android = "http://schemas.android.com/apk/res/android">
    <data class = ".ActivityMainBinding">
        <variable name = "handlers" type = " cuc.mvvm1.MyHandlers"/>
        <variable name = "user" type = "cuc.mvvm1.User"/>
    </data>
```

```
    <LinearLayout
        android:orientation = "vertical"
        android:layout_width = "match_parent"
        android:layout_height = "match_parent">
        <TextView android:layout_width = "wrap_content"
            android:layout_height = "wrap_content"
            android:text = " Method References "
            android:onClick = "@{(view) -> handlers.myOnClick(view,&quot;CUC&quot;)}"
        </LinearLayout>
</layout>
```

上面的代码中，绑定 click 事件到 MyHandlers 的方法 myOnClick(View view,String name)上，myOnClick 方法的返回值和 OnClickListener 接口中的 onClick 方法一样，但方法的参数不一样。

在 Activity 的 onCreate 方法中可以这样绑定对象：

```
@Override
    protected void onCreate(Bundle savedInstanceState) {
        super.onCreate(savedInstanceState);
        ActivityMainBinding activityMainBinding = DataBindingUtil.setContentView(this, R.
layout.activity_main);
        activityMainBinding.setHandlers(new MyHandlers());
    }
```

整个响应表达式是一个 lambda 表达式，在编译的时候数据绑定会创建一个默认响应方法，当事件触发时，默认响应方法执行这个 lambda 表达式。lambda 表达式分成两部分：箭头左边的圆括号部分和箭头右边的方法调用部分，箭头左边的圆括号内部列出的是原来监听器的方法中的参数，必须全部列出或者全部不列出，取决于方法的参数中是否包含原来监听器的方法中的参数。

两种方法的比较如下：

①Method References 要求响应方法的签名和事件 listener 中的签名完全一致，Listener Bindings 只要求返回值一致；②Method References 不可以自定义参数，Listener Bindings 可以自定义参数，可以编写简单的代码逻辑；③Method References 在编译时绑定，Listener Bindings 在运行时才会执行绑定 lambda 表达式。

10.3 在 Java 代码中使用数据绑定

在 Activity 类的 onCreate 方法中，可以使用 DatabindingUtil. setContentView(Activity activity,int layoutId) 来替换掉 setContentView(int layoutResID)，它的返回值类型是框架根据布局文件生成的绑定类。Fragment 类并不具有 setContentView(int layoutResID)方法，因此绑定方法与 Activity 略有不同。下面先介绍一些相关的类。

1. DataBindingUtil 类

android. databinding. DataBindingUtil 是一个从布局文件创建 ViewDataBinding 对象

的工具类,它的主要方法,如表 10-4 所示。

表 10-4 DataBindingUtil 的主要方法

方 法	解 释
static < T extends ViewDataBinding > T bind(View root)	返回给定的根视图的绑定,如果绑定不存在,就创建一个绑定
static < T extendsViewDataBinding > T bind (View root, DataBindingComponent bindingComponent)	返回给定的根视图的绑定,如果绑定不存在,就创建一个绑定
staticString convertBrIdToString(int id)	把给定的 BR. id 转换为字符串,这可能在日志用途时有用。
static < T extends ViewDataBinding > T findBinding(View view)	返回给定的视图的绑定
static < T extends ViewDataBinding > T getBinding(View view)	获得给定根视图的绑定
staticDataBindingComponent getDefaultComponent()	返回在数据绑定中被使用的默认的 DataBindingComponent
static < T extends ViewDataBinding > T inflate (LayoutInflater inflater, int layoutId, ViewGroup parent, boolean attachToParent, DataBindingComponent bindingComponent)	扩展一个绑定的布局,返回为那个布局新创建的绑定
static < T extendsViewDataBinding > T inflate (LayoutInflater inflater, int layoutId, ViewGroup parent, boolean attachToParent)	扩展一个绑定的布局,返回为那个布局新创建的绑定
static < T extendsViewDataBinding > T	setContentView (Activity activity, int layoutId) 把 Activity 的内容设置为给定的布局,并返回关联的绑定
static < T extends ViewDataBinding > T setContentView (Activity activity, int layoutId, DataBindingComponent bindingComponent)	把 Activity 的内容设置为给定的布局,并返回关联的绑定
static void setDefaultComponent (DataBindingComponent bindingComponent)	设置用于数据绑定的默认的 DataBindingComponent

2. ViewDataBinding 类

android. databinding. ViewDataBinding 是 android. databinding. BaseObservable 的子类,它是生成的所有数据绑定类的基类。如果可能,生成的绑定应该使用生成的静态的方法 bind(View)或者 inflate(LayoutInflater,int,ViewGroup,boolean)进行实例化。ViewDataBinding 的 public 方法,如表 10-5 所示。

表 10-5 ViewDataBinding 的 Public 方法

方 法	解 释
void addOnRebindCallback(OnRebindCallback listener)	当重新评估脏的域时,添加一个被调用的监听器
void executePendingBindings()	评估还在等待中的绑定时,更新那些有表达式绑定到被修改的变量的视图
View getRoot()	返回跟绑定关联的布局文件中的最外层的视图

续表

方　　法	解　　释
abstract boolean hasPendingBindings()	返回是否有 UI 需要被刷新，以便显示现在的数据
abstract void invalidateAll()	使得所有的现在的绑定表达式无效，请求一次重新绑定以便刷新 UI
void removeOnRebindCallback(OnRebindCallback listener)	移除用 addOnRebindCallback(OnRebindCallback)添加的监听器
abstract boolean setVariable(int variableId, Object value)	在绑定的类中，设置一个变量的值
void unbind()	移除到表达式变量的绑定的监听器

3. 数据绑定布局文件用作 Fragment 的视图

绑定布局最常用方式是采用生成的绑定类的静态方法，这种扩展是类型安全的，也可以调用 DataBindingUtil 类的静态方法绑定布局。在 Fragment 中绑定布局主要有以下两种方法：

第一种方法：在 Fragment 类的 onCreateView 方法中调用生成的绑定类的静态方法 inflate(android. view. LayoutInflater inflater)绑定布局；或者 inflate(android. view. LayoutInflater inflater, android. view. ViewGroup root, boolean attachToRoot)方法把当前视图添加到 root 视图中；或者调用 DataBindingUtil. inflate(LayoutInflater inflater, int layoutId, ViewGroup parent, boolean attachToParent)方法绑定布局。最后通过 getRoot()返回 View。

第二种方法：在 Fragment 类的 onViewCreated 方法中调用 DataBindingUtil 类的静态方法 bind(View root)绑定布局。

实例 12：Fragment 的绑定。

假设 CucFragment 采用 fragment. xml 作为视图，没有采用数据绑定时，Fragment 定义如下：

```
package cuc.mvvm1;
import android.support.v4.app.Fragment;
import android.view.View;
import android.view.LayoutInflater;
import android.view.ViewGroup;
import android.os.Bundle;
public class CucFragment extends Fragment {
   @Override
   public View onCreateView(LayoutInflater inflater, ViewGroup container,
                        Bundle savedInstanceState) {
      return inflater.inflate(R.layout.fragment,container,false);
   }
   @Override
   public void onViewCreated(View view, Bundle savedInstanceState) {
       super.onViewCreated(view, savedInstanceState);
   }
}
```

布局文件 fragment. xml 内容如下：

```
<?xml version = "1.0" encoding = "utf - 8"?>
< layout xmlns:android = "http://schemas.android.com/apk/res/android">
    < data class = ".FragmentBinding">
        < variable
            name = "user"
            type = "cuc.mvvm1.User"/>
    </data >
    < TextView
        android:layout_width = "match_parent"
        android:layout_height = "wrap_content"
        android:text = "@{user.name +`->`+ user.age}" />
</layout >
```

调用生成的绑定类的 inflate 方法绑定数据：

```
public class CucFragment extends Fragment {
    private cuc.mvvm1.FragmentBinding fragmentBinding;
    @Override
    public View onCreateView(LayoutInflater inflater, ViewGroup container,
                             Bundle savedInstanceState) {
        fragmentBinding = FragmentBinding.inflate(inflater,container,false);
        User user = new User("zhangsan",30);
        fragmentBinding.setUser(user);
        return fragmentBinding.getRoot();
    }
    @Override
    public void onViewCreated(View view, Bundle savedInstanceState) {
        super.onViewCreated(view, savedInstanceState);
    }
}
```

也可以调用 DataBindingUtil 类的 inflate 方法绑定数据：

```
public class CucFragment extends Fragment {
    private cuc.mvvm1.FragmentBinding fragmentBinding;
    @Override
    public View onCreateView(LayoutInflater inflater, ViewGroup container,
                             Bundle savedInstanceState) {
        fragmentBinding =  DataBindingUtil.inflate(inflater,R.layout.fragment,container,false);
        User user = new User("zhangsan",30);
        fragmentBinding.setUser(user);
        return fragmentBinding.getRoot();
    }
    @Override
    public void onViewCreated(View view, Bundle savedInstanceState) {
        super.onViewCreated(view, savedInstanceState);
    }
}
```

也可以在 Fragment 类的 onViewCreated 方法中调用 DataBindingUtil 类的静态方法绑定布局：

```
public class CucFragment extends Fragment {
    private cuc.mvvm1.FragmentBinding fragmentBinding;
    @Override
    public View onCreateView(LayoutInflater inflater, ViewGroup container,
                             Bundle savedInstanceState) {
        return inflater.inflate(R.layout.fragment,container,false);
    }
    @Override
    public void onViewCreated(View view, Bundle savedInstanceState) {
        super.onViewCreated(view, savedInstanceState);
        fragmentBinding = DataBindingUtil.bind(view);
        User user = new User("zhangsan",30);
        fragmentBinding.setUser(user);
    }
}
```

10.4　数据双向绑定实验

实验目的

(1) 了解 EditText 的 text 属性的绑定。

(2) 了解 Button 的 OnClick 事件处理的两种方法：方法引用和监听器绑定。

1. 新建工程 Mvvm1

具体步骤：运行 Android Studio，单击 Start a new Android Studio project，在 Create New Project 页中，Application name 为 Mvvm1，Company domain 为 cuc，Project location 为 C:\android\myproject\Mvvm1，单击 Next 按钮，选中 Phone and Tablet，minimum sdk 设置为 API 19：Android 4.4(minSdkVersion)，单击 Next 按钮，选择 Empty Activity，再单击 Next 按钮，勾选 Generate Layout File 选项，勾选 Backwards Compatibility(AppCompat) 选项，单击 Finish 按钮。

2. 在 build.gradle(module：app)中，添加 DataBinding 支持，并同步

在文件 build.gradle(module：app)的 android 配置的花括号内添加：

```
dataBinding.enabled = true
```

由于修改了 build.gradle 文件内容，这个文件窗口的上面就出现提示：

Gradle files have changed since last project sync. A project sync may be necessary for the IDE to work properly. sync now。单击 sync now 按钮进行同步。

编辑后，build.gradle(module：app)的内容如下：

```
apply plugin: 'com.android.application'
android {
```

```
    dataBinding.enabled = true
    compileSdkVersion 28
    defaultConfig {
        applicationId "cuc.mvvm1"
        minSdkVersion 19
        targetSdkVersion 28
        versionCode 1
        versionName "1.0"
        testInstrumentationRunner "android.support.test.runner.AndroidJUnitRunner"
    }
    buildTypes {
        release {
            minifyEnabled false
            proguardFiles getDefaultProguardFile('proguard-android.txt'), 'proguard-rules.pro'
        }
    }
}
dependencies {
    implementation fileTree(dir: 'libs', include: ['*.jar'])
    implementation 'com.android.support:appcompat-v7:28.0.0-beta01'
    implementation 'com.android.support.constraint:constraint-layout:1.1.2'
    testImplementation 'junit:junit:4.12'
    androidTestImplementation 'com.android.support.test:runner:1.0.2'
    androidTestImplementation 'com.android.support.test.espresso:espresso-core:3.0.2'
}
```

3. 新建一个名为 UserPojo 的 Java 类

UserPojo.Java 内容如下：

```
package cuc.mvvm1;
public class UserPojo {
    private String name;
    private int age;
    public UserPojo() {
    }
    public UserPojo(String name, int age) {
        this.name = name;
        this.age = age;
    }
    public String getName() {
        return name;
    }
    public void setName(String name) {
        this.name = name;
    }
    public int getAge() {
        return age;
    }
```

```
    public void setAge(int age) {
        this.age = age;
    }
}
```

4. 新建布局文件 activity_main1. xml

新建布局文件 activity_main1. xml 的具体步骤：在工程结构视图中(单击 Project 按钮可以显示或者隐藏工程结构图，使用工程结构图中的下拉列表框可以选择 Android 以显示工程逻辑结构，或者选择 Project 显示工程实际的目录结构)，右击 layout，依次选择 New | Layout resource file，File name 输入框输入 activity_main1，Root element 输入框输入 layout，Source set 为 main，单击 OK 按钮。layout/activity_main1. xml 内容如下：

```
<?xml version = "1.0" encoding = "utf - 8"?>
< layout xmlns:android = "http://schemas.android.com/apk/res/android">
    < data >
        < variable name = "user" type = "cuc.mvvm1.UserPojo"/>
    </data >
    < LinearLayout
        android:layout_width = "match_parent"
        android:layout_height = "match_parent"
        android:orientation = "vertical">
        < EditText
            android:id = "@ + id/name1Edt"
            android:layout_width = "match_parent"
            android:layout_height = "wrap_content"
            android:hint = "name"
            android:inputType = "text"
            android:text = "@{user.name}" />
        < EditText
            android:id = "@ + id/name2Edt"
            android:layout_width = "match_parent"
            android:layout_height = "wrap_content"
            android:hint = "name"
            android:inputType = "text"
            android:text = "@ = {user.name}" />
        < TextView
            android:layout_width = "match_parent"
            android:layout_height = "wrap_content"
            android:text = "@{String.valueOf(user.age) }" />
        < Button
            android:onClick = "onClick"
            android:id = "@ + id/changeAgeBtn"
            android:layout_width = "match_parent"
            android:layout_height = "wrap_content"
            android:text = "change age by increase 1" />
    </LinearLayout >
</layout >
```

使用 DataBinding 时，xml 的布局文件中除了 UI 元素外，还需要定义 UI 元素用到的变量。所以，它的根元素不再是一个 ViewGroup，而是变成了< layout >，并且新增了一个 data 结点。data 元素提供了数据与 UI 进行绑定的桥梁，在 data 元素内添加 variable 元素，这个变量为 UI 元素(例如 TextView 的 android: text)提供数据，然后在 Java 代码中把某个数据与 variable 元素声明的变量进行绑定。

在上面的代码中，data 结点中声明一个变量 user，type 属性定义了变量类型：UserEntity 类，databinding 插件会根据 XML 文件的名称生成一个继承自 ViewDataBinding 的类，自动生成类名的规则是：把布局文件的首字母大写，并且去掉下画线，将下画线后面的字母大写，最后加上 Binding 组成。根据 activity_main. xml 生成的类名为 ActivityMain1Binding，自动生成的 Java 文件位置为 app\build\generated\source\apt\debug\包名称\databinding 文件夹内。

5. 编辑 MainActivity. java，实现数据绑定

编辑 MainActivity 类的 onCreate 方法，用 DatabindingUtil. setContentView(Activity activity,int layoutId)来替换掉 setContentView()，返回值的类型是框架根据布局文件生成的类。框架根据 activity_main1. xml 生成的类名为 ActivityMain1Binding，里面包括了布局中的 View 以及声明的变量，可以通过这个对象获取到布局中的 UI 元素，它的所有的 set 方法都是根据 variable 名称生成的，布局文件的 data 结点中有几个 variable 就会自动生成几个 set 方法。

然后创建一个 UserPojo 类型的对象 userPojo，通过 activityMainBinding. setUser (userPojo)与布局文件中名为 user 的变量进行绑定。MainActivity. java 代码如下：

```
package cuc.mvvm1;
import android.databinding.DataBindingUtil;
import android.os.Bundle;
import android.support.v7.app.AppCompatActivity;
import android.view.View;
import cuc.mvvm1.databinding.ActivityMain1Binding;

public class MainActivity extends AppCompatActivity {
  private UserPojo userPojo;
  private ActivityMain1Binding binding;
  @Override
  protected void onCreate(Bundle savedInstanceState) {
    super.onCreate(savedInstanceState);
    binding = DataBindingUtil.setContentView(this,R.layout.activity_main1 );
    userPojo = new UserPojo("zhangsan",25);
    binding.setUser(userPojo);
  }
  public void onClick(View view) {
     userPojo.setAge(userPojo.getAge() + 1);
  }
}
```

6. 编译运行

单击 changeAgeBtn 按钮，可以改变 userPojo 对象的数据，图形界面上的数据并不能自动刷新，只有通过再次调用 setUser 进行设置才可以刷新了。把上面 MainActivity.java 代码中的 onclick 方法修改为

```
public void onClick(View view) {
  userPojo.setAge(userPojo.getAge() + 1);
  binding.setUser(userPojo);
}
```

再次编译运行。这次发现单击 changeAgeBtn 按钮，图形界面中的 age 就可以改变了。

7. 编辑 UserPojo，继承 BaseObservable 类

UserPojo.java 内容如下：

```
package cuc.mvvm1;
import android.databinding.BaseObservable;
import android.databinding.Bindable;
public class UserPojo extends BaseObservable {
   private String name;
   private int age;
   public UserPojo() {
   }
   public UserPojo(String name, int age) {
      this.name = name;
      this.age = age;
   }
@Bindable
    public String getName() {
       return name;
    }
    public void setName(String name) {
       this.name = name;
       notifyPropertyChanged(cuc.mvvm1.BR.name);
    }
@Bindable
    public int getAge() {
       return age;
    }
    public void setAge(int age) {
       this.age = age;
       notifyPropertyChanged(cuc.mvvm1.BR.age);
    }
}
```

单击 Android Studio 菜单中的 Build|Make Module 'app'对模块进行编译，以便生成 BR 类。生成的源代码 BR.java 位于工程文件夹\app\build\generated\source\apt\debug\cuc\mvvm1 文件夹下，用@Bindable 标记过 getter 方法会在 BR 中生成一个 entry。

8. 编辑 MainActivity,编译运行

编辑 MainActivity.java,MainActivity 类的 onClick 方法代码如下:

```
public void onClick(View view) {
  userPojo.setAge(userPojo.getAge() + 1);
}
```

编译运行。单击 changeAgeBtn 按钮,可以改变 userPojo 对象的数据,图形界面上的数据可以自动刷新。

9. 编辑 UserPojo,使用更加简单的写法

UserPojo.java 内容如下:

```
package cuc.mvvm1;
import android.databinding.ObservableField;
import android.databinding.ObservableInt;
public class UserPojo {
    public final ObservableField<String> name = new ObservableField<>();
    public final ObservableInt age = new ObservableInt();
    public UserPojo(String name, int age) {
        this.name.set(name);
        this.age.set(age);
    }
}
```

ObservableInt,ObservableField 都不是基本数据类型,而是引用类型变量,在声明后必须通过 new 实例化开辟数据空间,才能对变量所指向的对象进行访问。由于在 XML 文件中使用了 user.name 和 user.age,以及在下面的 MainActivity.java 中使用了 user.age,所以 UserPojo 中的 name 和 age 都要使用 public 修饰。

10. 编辑 MainActivity.java,并编译运行

把上面 MainActivity.java 代码中的 onclick 方法修改为

```
public void onClick(View view) {
 userPojo.age.set(userPojo.age.get() + 1);
}
```

编译运行后,单击 changeAgeBtn 按钮,可以发现:图形界面上的年龄可以自动刷新。

关于双向绑定的效果:当在第一个 EditText 中输入时,第二个 EditText 的文本不会更新,这是由于第一个 EditText 没有使用双向绑定,EditText 输入变化时,数据 user.name 不会更新,因而第二个 EditText 显示不会更新。当用户在第二个 EditText 中输入时,第一个 EditText 显示也会跟着改变,这是由于第二个 EditText 使用了双向绑定,EditText 输入变化时,user.name 将会更新(反向绑定),因而第一个 EditText 显示将会随着 user.name 更新而更新。

11. 方法引用

给 MainActivity 类添加两个方法 onClickMy1 和 onClickMy2,onClickMy1 用于方法引

用，onClickMy2 用于监听器绑定；并在 onCreate 方法的最后添加布局文件中 listener 变量的初始化。最后 MainActivity.java 文件内容如下：

```
package cuc.mvvm1;
import android.databinding.DataBindingUtil;
import android.os.Bundle;
import android.support.v7.app.AppCompatActivity;
import android.view.View;
import cuc.mvvm1.databinding.ActivityMain1Binding;
public class MainActivity extends AppCompatActivity {
 private UserPojo userPojo;
 private ActivityMain1Binding binding;
 @Override
 protected void onCreate(Bundle savedInstanceState) {
  super.onCreate(savedInstanceState);
  binding = DataBindingUtil.setContentView(this,R.layout.activity_main1 );
  userPojo = new UserPojo("zhangsan",25);
  binding.setUser(userPojo);
   binding.setListener(this);
 }
 public void onClick(View view) {
  userPojo.age.set(userPojo.age.get() + 1);
 }
 public void onClickMy1(View view) {
  userPojo.age.set(userPojo.age.get() + 1);
 }
 public void onClickMy2(View view,int i) {
  userPojo.age.set(userPojo.age.get() + i);
 }
}
```

在布局文件 activity_main1.xml 中，给 data 元素添加一个 variable 子元素，内容如下：

```
<variable name = "listener" type = "cuc.mvvm1. MainActivity "/>
```

在布局文件 activity_main1.xml 中，把 changeAgeBtn 的 onClick 属性修改为：

```
android:onClick = "@{listener::onClickMy1}"
```

编译运行，可以看到，单击 changeAgeBtn 按钮，可以改变 userPojo 对象的数据，图形界面上的 age 每次加 1。

12. 监听器绑定

在布局文件 activity_main1.xml 中，把 changeAgeBtn 的 onClick 属性修改为：

```
android:onClick = "@{(view) -> listener.onClickMy2(view,5)}"
```

onClick 属性采用 lamba 表达式，onClickMy2 的参数与 View.OnClickListener 接口中的参数不同。编译运行，可以看到，单击 changeAgeBtn 按钮，可以改变 userPojo 对象的数据，图形界面上的 age 每次加 5。

10.5 本章主要参考文献

1. https://developer.android.google.cn/topic/libraries/data-binding/
2. https://developer.android.google.cn/reference/android/databinding/package-summary.html
3. https://developer.android.google.cn/reference/android/databinding/Observable
4. https://developer.android.google.cn/reference/android/databinding/BaseObservable.html
5. https://developer.android.google.cn/reference/android/databinding/ObservableField
6. https://developer.android.google.cn/reference/android/databinding/DataBindingUtil.html
7. https://developer.android.google.cn/reference/android/databinding/ObservableInt.html

第11章 滚动与翻页

CHAPTER 11

android. widget. ScrollView 是 android. widget. FrameLayout 的直接子类，仅仅支持垂直滚动，在 ScrollView 中只能放置一个直接子视图，子视图里包含要垂直滚动的全部内容，这个子视图可以是具有复杂的对象层次结构的布局管理器，最经常使用的子视图是垂直方向的 LinearLayout。允许放在它里面的视图层次结构由用户进行滚动，允许它的内容比物理显示屏更大。使用 ScrollView，所有的视图都在存储器中，没有视图的再利用，也不支持适配器。ScrollView 仅仅支持垂直滚动，若需要水平滚动，请使用 HorizontalScrollView。

ListView 是显示可以滚动的列表项的 ViewGroup。android. support. v7. widget. RecyclerView 是 android. view. ViewGroup 的直接子类，实现了 ScrollingView，NestedScrollingChild2 接口，是 ListView 的更加先进和灵活的版本，可以代替 ListView 和 GridView。RecyclerView 和 ListView 中都是维护少量的视图来显示大量的数据，它们的视图是根据屏幕大小生成的，在视图层次结构中的视图并没有比屏幕上显示的视图更多。ListView 本身并不知道它所包含的视图的细节，例如类型和内容，当用户上下滚动，需要显示新的视图的时候，它从 ListAdapter 请求视图，通过调用 setAdapter(ListAdapter) 和带有列表项的 ListAdapter 关联起来。RecyclerView 或者 ListView 都不能被添加到 ScrollView 中，因为它们负责自己的垂直滚动操作。最重要的是，这样做会破坏它们针对大型列表的优化，因为 ScrollView 迫使 ListView 显示它的整个列表，以填充由 ScrollView 提供的无限容器，这样会导致很差的用户接口体验。

TextView 类也能处理它自己的滚动，因此不需要使用 ScrollView，但是同时使用，可以实现在较大容器中的显示文本视图的效果。

在一个可以滚动的视图中添加另一个可以滚动的视图，叫嵌套滚动。在材料设计中，应该使用 NestedScrollView 代替 ScrollView 进行垂直滚动，因为 NestedScrollView 支持嵌套的滚动，但它既实现了 NestedScrollingParent 接口，又实现了 NestedScrollingChild2 接口。

CoordinatorLayout 是实现了 NestedScrollingParent2 接口的 ViewGroup，它是超级的 FrameLayout。CoordinatorLayout 有两种应用场景：作为顶层应用的装饰或者 chrome 布局；作为和一个和多个子视图进行交互的容器。通过给 CoordinatorLayout 的子视图指定 Behavior，允许子视图和父视图以及子视图之间进行不同的交互。当视图类用作 CoordinatorLayout 的子视图时，可以在视图类前使用 DefaultBehavior 注解(@DefaultBehavior)

指定默认的 Behavior。

AppBarLayout 是一个垂直的 LinearLayout,实现了材料设计的 App bar 的很多特征,它的子视图通过在布局文件中的属性 layout_scrollFlags 设置期望的滚动特征,也可以在代码中调用 setScrollFlags(int)设置滚动特征。AppBarLayout 仅仅用作 CoordinatorLayout 的直接子视图时才能发挥其性能,如果把 AppBarLayout 放入其他的 ViewGroup,它的大部分功能将不能工作。

11.1 Android 触摸事件的消息传递机制

Android 系统中,触摸事件的消息处理机制是比较偏底层的知识,在简单的应用开发中,一般不会涉及这方面的知识,初次阅读时,这一小节可以直接跳过去。

Activity 的窗口与视图的层次关系如图 11-1 所示,Activity 包含一个 Window 类型的对象 mWindow,在 Activity 的 attach 方法中通过 mWindow = new PhoneWindow(this, window)初始化 mWindow。Window 是一个抽象的类,PhoneWindow 是 Window 的一个实现类,PhoneWindow 中包含一个 DecorView 类型的对象 mDecor 和一个 ViewGroup 类型 mContentParent。DecorView 代表了当前 Window 最顶级的视图,DecorView 中包含一个 ViewGroup 类型对象 mContentRoot 和一个 DecorCaptionView 类型对象 mDecorCaptionView,其中 mDecorCaptionView 包含标题和窗口控制按钮,这个视图是否可见取决于窗口类型,如果窗口类型不需要这个标题,mDecorCaptionView 就为 null; mContentRoot 对应 Android 系统定义的一个屏幕的基本布局,例如 sdk\platforms\android-24\data\res\layout\screen_title.xml。PhoneWindow 中的 ViewGroup 类型的对象 mContentParent,它要么就是 mDecor,要么就是 mDecor 中的屏幕的基本布局中的 id 为"@android:id/content"的 FrameLayout,mContentParent 就是 Activity 中通过 setContentView 方法设置的视图。

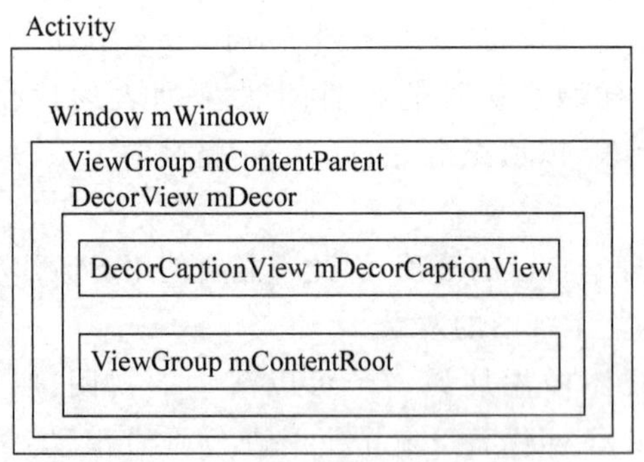

图 11-1 Activity 的窗口与视图的层次关系

一次触摸事件是由 ACTION_DOWN,0 次或者多次 ACTION_MOVE、ACTION_UP 组成的。触摸事件的处理是从调用 Activity 的 dispatchTouchEvent 方法开始的。图 11-2 是 dispatchTouchEvent 方法的调用的层次关系。在 Activity 的 dispatchTouchEvent 方法中,调用与 Activity 关联的 PhoneWindow 的 superDispatchTouchEvent 方法。在 PhoneWindow 的 superDispatchTouchEvent 方法中,调用 mDecor 的 superDispatchTouchEvent 方法,

DecorView 类型的对象 mDecor 是 PhoneWindow 窗口的顶层视图。在 DectorView 的 superDispatchTouchEvent 方法,调用它的父类的 dispatchTouchEvent 方法,DecorView 继承于 FrameLayout,FrameLayout 并没有重写它的父类 ViewGroup 的 dispatchTouchEvent 方法,所以调用 DecorView 的 superDispatchTouchEvent 方法就是调用 ViewGroup 的 dispatchTouchEvent 方法。在 ViewGroup 的 dispatchTouchEvent 方法,会调用它的子视图的 dispatchTouchEvent 方法。

图 11-2 dispatchTouchEvent 方法的调用的层次关系

图 11-3 是一个用户定义的布局文件的层次图,以它为例,触摸事件从 Activity 依次传递到 PhoneWindow、DecorView、mContentParent,mContentParent 就是在 Activity 中通过 setContentView 方法设置的视图,在这个图中就是 ViewGroup0,再继续按照先后顺序往 ViewGroup0 的各个子视图传递,直到找到一个视图消费掉这个事件为止。在这个图中,ViewGroup0 有两个子视图:ViewGroup1 和 ViewGroup2,触摸事件会依次传递到这两个子视图。当把触摸事件传到了 ViewGroup1 的时候,就会遍历 ViewGroup1 的所有子视图,当 ViewGroup1 的所有子视图都没有消费这个事件的时候,就会调用 ViewGruop1 的父类去试着消费这个事件,要是还是没有被消费,则触摸事件就会传递到 ViewGroup2;与 ViewGroup1 的传递过程一样,也会遍历 ViewGroup2 的所有子视图,当 ViewGroup2 的所有子视图都没有消费这个事件的时候,就会调用 ViewGruop2 的父类 View 去试着消费这个事件。要是 ViewGroup0 的所有子视图 ViewGroup1 和 ViewGroup2 都没有消费掉这个触摸事件,又会传到 ViewGroup0 的父类去试着消费,如果仍然没有消费掉,就会传到 mDecor 的父类。仍然没有消费、最后由回到 Activity 中。

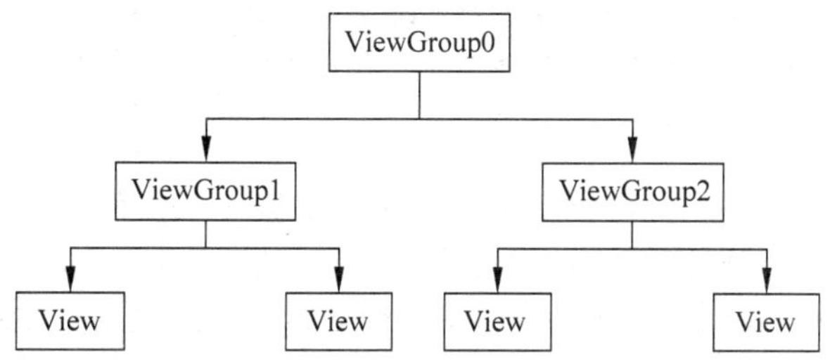

图 11-3 用户定义的布局层次关系图

再从源代码的角度,分析触摸事件传递的流程。触摸事件的处理是从调用 Activity 的 dispatchTouchEvent 方法开始的。下面是 Activity 的 dispatchTouchEven 方法的代码:

```
public boolean dispatchTouchEvent(MotionEvent ev) {
    if (ev.getAction() == MotionEvent.ACTION_DOWN) {
        onUserInteraction();
    }
    if (getWindow().superDispatchTouchEvent(ev)) {
        return true;
    }
    return onTouchEvent(ev);
}
```

Activity 的 dispatchTouchEven 方法的参数是一个 MotionEvent，在这个方法中，通过 getWindow(). superDispatchTouchEvent(ev) 调用了 Activity 所依附的 Window 的 superDispatchTouchEvent(ev)方法来进行触摸事件的分发。如果 Window 消耗了此事件，则返回值为 true，进而整个方法返回 true，就不会执行后面的 Activity 的 onTouchEvent 方法了。只有 Window 没有处理触摸事件的情况下，才会调用 Activity 的 onTouchEvent 方法去处理事件。也就是说，只有当触摸事件没有被任何的 View 或 ViewGroup 处理过的时候，Activity 才会执行自己的 onTouchEvent 去处理触摸事件。

实际上，getWindow()返回的是一个 PhoneWindow 类型的实例，这样 Activity 方法中调用 getWindow(). superDispatchTouchEvent(ev)实际就是调用 PhoneWindow 的 superDispatchTouchEvent 方法。PhoneWindow 的 superDispatchTouchEvent 方法的源码如下所示：

```
@Override
public boolean superDispatchTouchEvent(MotionEvent event) {
    return mDecor.superDispatchTouchEvent(event);
}
```

mDecor 是 DectorView 类型的对象，下面是 DectorView 的 superDispatchTouchEvent 方法的源码：

```
@Override
public boolean dispatchTouchEvent(MotionEvent ev) {
    final Window.Callback cb = mWindow.getCallback();
    return cb != null && !mWindow.isDestroyed() && mFeatureId < 0
            ? cb.dispatchTouchEvent(ev) : super.dispatchTouchEvent(ev);
}
```

在 DecorView 的 superDispatchTouchEvent 方法中，或者调用它关联 Window 的 Callback 的 dispatchTouchEvent 方法或者调用其父类的 dispatchTouchEvent 方法。DecorView 的父类是 FrameLayout，FrameLayout 的父类是 ViewGroup，FrameLayout 并没有重写 ViewGroup 的 dispatchTouchEvent，所以执行其父类 ViewGroup 对应的 dispatchTouchEvent 方法。

ViewGroup 中定义了一个 TouchTarget 类型的成员变量 mFirstTouchTarget，用于保存当前 ViewGroup 中处理了触摸事件的子视图。当前面的 ACTION_DOWN 事件找到了消费此触摸事件的视图时，则 mFirstTouchTarget 不为空。

在 ViewGroup 的 dispatchTouchEvent 方法中，首先会判断是否需要调用 onInterceptTouchEvent 方法拦截触摸事件。当触摸事件为 ACTION_DOWN 或者 mFirstTouchTarget 不为空，才有可能调用该对象的 onInterceptTouchEvent 方法。如果该方法返回 true，就表示 ViewGroup 拦截了触摸事件；如果返回 false，表示 ViewGroup 没有拦截触摸事件，应该将触摸事件传递给该对象的子视图。dispatchTouchEvent 方法的这部分代码如下：

```
@Override
  public boolean dispatchTouchEvent(MotionEvent ev) {
     …
     // Check for interception.
     final boolean intercepted;
     if (actionMasked == MotionEvent.ACTION_DOWN
                  || mFirstTouchTarget != null) {
       final boolean disallowIntercept = (mGroupFlags & FLAG_DISALLOW_INTERCEPT) != 0;
             if (!disallowIntercept) {
                 intercepted = onInterceptTouchEvent(ev);
                 ev.setAction(action);           // restore action in case changed
             } else {
                 intercepted = false;
             }
         } else {
        // There are no touch targets and this action is not an initial down
             // so this view group continues to intercept touches.
             intercepted = true;
         }
…
}
```

如果当前 ViewGroup 没有拦截，即 intercepted 为 false 时，ViewGroup 才会遍历所有子视图，通过调用 isTransformedTouchPointInView 方法判断 MotionEvent 所携带的触摸事件的坐标是否落在子视图的范围内，如果触摸事件的坐标恰好落在了该子视图范围内，说明当前 ViewGroup 内的该子视图被触摸了，这样 ViewGroup 就会把触摸事件的坐标以及该子视图传递给 dispatchTransformedTouchEvent 方法。即如果该子视图能够接受触摸事件，就调用 dispatchTransformedTouchEvent (ev, false, child, idBitsToAssign) 把 MotionEvent 传递给子视图，dispatchTouchEvent 方法的对应代码如下：

```
final View[] children = mChildren;
for (int i = childrenCount - 1; i >= 0; i--) {
   if (dispatchTransformedTouchEvent(ev, false, child, idBitsToAssign)) {
         // Child wants to receive touch within its bounds.
             mLastTouchDownTime = ev.getDownTime();
                mLastTouchDownX = ev.getX();
                mLastTouchDownY = ev.getY();
                 newTouchTarget = addTouchTarget(child, idBitsToAssign);
```

```
                    alreadyDispatchedToNewTouchTarget = true;
                    break;
        }
            /* The accessibility focus didn't handle the event, so clear
             the flag and do a normal dispatch to all children. */
                    ev.setTargetAccessibilityFocus(false);
    }
```

在 dispatchTransformedTouchEvent 方法中，会让子视图重复父容器类似的分发方式，这个方法的返回值表示是否消费了该事件。如果 dispatchTransformedTouchEvent 方法返回 true，表示子视图处理了触摸事件，这样 ViewGroup 会通过调用 addTouchTarget 方法将 newTouchTarget 绑定该子视图，并且变量 alreadyDispatchedToNewTouchTarget 也会设置为 true，表示已经有子视图处理了触摸事件。一旦有子视图处理了触摸事件，ViewGroup 就会通过 break 跳出 for 循环，不再对其他子视图进行遍历。

若在遍历了所有的子视图后，mFirstTouchTarget 还是 null，说明没有任何一个子视图消费该事件，就会调用 dispatchTransformedTouchEvent (ev, canceled, null, TouchTarget . ALL_POINTER_IDS)，第三个参数传递的是 null，所以就会调用 super. dispatchTouchEvent (event)方法。需要注意的是，这里的 super 在 Java 语法中指父类，如果子类重写了父类的方法，要访问父类的方法，就需要使用 super 指定。也就是说，如果遍历了该对象的所有的子视图后，还是没有一个子视图能够消费此触摸事件，就会调用该对象父类的 dispatchTouchEvent。由于 ViewGroup 继承自 View，就相当于执行了从 View 中继承来的 dispatchTouchEvent (event)方法。下面是 ViewGroup 类的 dispatchTransformedTouchEvent 的源代码。

```
private boolean dispatchTransformedTouchEvent(MotionEvent event, boolean cancel,
        View child, int desiredPointerIdBits) {
    final boolean handled;
    // Canceling motions is a special case. We don't need to perform any transformations
    // or filtering. The important part is the action, not the contents.
    final int oldAction = event.getAction();
    if (cancel || oldAction == MotionEvent.ACTION_CANCEL) {
        event.setAction(MotionEvent.ACTION_CANCEL);
        if (child == null) {
            handled = super.dispatchTouchEvent(event);
        } else {
            handled = child.dispatchTouchEvent(event);
        }
        …
```

在 View 类的 dispatchTouchEvent (event)的方法中，View 首先会查看其有没有设置过 OnTouchListener，如果设置过就调用 OnTouchListener 的 onTouch 方法，如果其返回了 true，就表明触摸事件被处理了，result 就会设置为 true。如果 result 不为 true，触摸事件没有被 OnTouchListener 处理，那么就会执行 View 的 onTouchEvent 方法，如果 onTouchEvent 返回了 true，就表示触摸事件被 View 处理了，result 就被设置为了 true。

在 View 类的 onTouchEvent()方法中，会根据 ACTION 的不同，执行不同的处理，例如，如果是 ACTION_UP，会执行 post(mPerformClick)，mPerformClick 是 PerformClick 类型的对象，PerformClick 实现了 Runnable 接口，在 PerformClick 的 run 方法中，调用了 performClick()方法，该方法会触发 OnClickListener.onClick()的执行。如果 View 没有注册任何的 CLICK 或 LONG_CLICK 等类型的事件监听器，那么 onTouchEvent 就返回 false，表示 onTouchEvent 没有对传入的触摸事件做任何处理。

最后，总结一下 Android 触摸事件的相关知识。

dispatchTouchEvent 方法：用来分发 TouchEvent。

onInterceptTouchEvent 方法：用来拦截 TouchEvent，改变事件的传递方向。决定传递方向的是返回值，返回为 false 时，表示不拦截，事件会传递给子控件；返回值为 true 时，事件会传递给当前控件的 onTouchEvent()。在 Activity 中和 View 中没有 onInterceptTouchEvent，因为 activity 本身只是一个容器，不存在对触摸事件进行拦截。而如果一个 View 不是 ViewGroup，内部就不可能有其他子视图，因而也就没有 onInterceptTouchEvent 方法。View 只有 dispatchTouchEvent 和 onTouchEvent 方法。如果 onInterceptTouchEvent 方法的返回值是 true，当前触摸事件就不会传递给子视图，即被当前 ViewGroup 拦截并由当前 ViewGroup 控件的 onTouchEvent(MotionEvent event)方法进行处理；如果这个方法的返回值是 false，那么这个触摸事件就会传递给子视图，由子视图继续处理。

onTouchEvent 方法：用来处理 TouchEvent 事件，返回值决定当前控件是否消费了这个事件。尤其对于 ACTION_DOWN 事件，返回 true，表示该控件消费了此事件，想要处理后续事件；返回 false，表示该控件没有消费此事件，然后交给由父控件进行处理。

在没有重写 onInterceptTouchEvent()和 onTouchEvent()的情况下，这两个方法的返回值都是 false。onInterceptTouchEvent 是从根视图向子视图传递，onTouchEvent 正好相反，从子视图向父视图传递。ACTION_MOVE 或者 ACTION_UP 发生的前提是一定曾经发生了 ACTION_DOWN，如果没有消费 ACTION_DOWN，那么系统会认为 ACTION_DOWN 没有发生过，所以 ACTION_MOVE 或者 ACTION_UP 就不能被捕获。

11.2　嵌套滚动

如果视图实现了 NestedScrollingChild 接口，就支持分发嵌套滚动事件给协同操作的父视图。实现了 NestedScrollingChild 接口的类通常创建一个 NestedScrollingChildHelper 类型的一个 final 实例作为它的一个域，所有的视图方法的实现基本上就是调用 NestedScrollingChildHelper 实例的相应方法，实现 Child 和 Parent 交互的逻辑。

在一个可以滚动的视图中添加另一个可以滚动的视图，叫嵌套滚动。嵌套滚动的主要方法的调用过程，如表 11-1 所示。滚动动作是 Child 主动发起，在子控件接收到滚动一段距离的请求时，先调用 dispatchNestedPreScroll 询问 Parent 是否需要滚动，在这个方法里回调 Parent 的 onNestedPreScroll()。如果父控件滚动了，它就通知子控件它消耗了一部分滚动距离，子控件就只处理剩下的滚动距离。然后子控件滚动完毕后再把剩余的滚动距离传给父控件。在一次滚动操作过程中，父控件和子控件都有机会对滚动操作出响应，尤其父控件能够分别在子控件处理滚动距离之前和之后对滚动距离进行响应。在 API 级别 21 以后，嵌套滚动的相关逻辑作为普通方法直接写进了 View 和 ViewGroup 类。为了兼容 API 级别 21 之前的版本，android. support. v4 兼容包提供了两个接口 NestedScrollingChild 和 NestedScrollingParent。还有两个辅助类 NestedScrollingChildHelper 和 NestedScrollingParentHelper 来帮助控件实现嵌套滚动。表 11-2 是 NestedScrollingChild 接口定义的主要方法。因而控件能够进行嵌套滚动的前提条件是控件要么是继承于 API 21 之后的 View 或 ViewGroup，要么实现了这两个接口。

表 11-1　嵌套滚动的主要方法的调用过程

子视图发起调用	父视图接受回调
startNestedScroll：作用是找到接收滚动距离信息的外控件。在子视图的 onInterceptTouchEvent 或者 onTouch 中（一般在 MotionEvent. ACTION_DOWN 事件里），调用 startNestedScroll(int axes)方法开启嵌套滚动流程（实际上是进行了一些嵌套滚动前准备工作），通知 Parent，准备进入滚动状态	onStartNestedScroll：内控件通过调用外控件的这个方法来确定外控件是否接收滚动信息 onNestedScrollAccepted：当外控件确定接收滚动信息后该方法被回调，可以让外控件针对嵌套滚动做一些前期工作
dispatchNestedPreScroll：在内控件处理滚动前调用这个方法把滚动信息分发给外控件	onNestedPreScroll：接收内控件在处理滚动前发出的滚动距离信息，在这里外控件可以优先响应滚动操作，消耗部分或者全部滚动距离
dispatchNestedScroll：在内控件处理完滚动后调用该方法把剩下的滚动距离信息分发给外控件。在子视图的 onInterceptTouchEvent 或者 onTouch 中（一般在 MontionEvent. ACTION_MOVE 事件里），如果父视图滚动了一段距离，子视图需要重新计算一下父视图滚动后剩下的滚动距离余量。然后，子视图进行余下的滚动。最后，如果滚动距离还有剩余，就调用 dispatchNestedScroll()，向父视图汇报滚动情况，包括子视图消费的部分和子视图没有消费的部分，询问父视图是否需要继续滚动剩下的距离。如果父视图接受了它的滚动参数，消耗掉了部分偏移，则这个函数返回 true，否则为 false	onNestedScroll：接收内控件处理完滚动后的滚动距离信息，在这里外控件可以选择是否处理剩余的滚动距离
stopNestedScroll：结束方法，主要作用就是清空嵌套滚动的相关状态。子视图在 onInterceptTouchEvent（一般在 MontionEvent. ACTION_UP 事件里），调用 stopNestedScroll 结束整个流程	onStopNestedScroll：用来做一些收尾工作

表 11-2　NestedScrollingChild 接口的 Public 方法

返回值类型	方法及解释
abstract boolean	dispatchNestedFling(float velocityX, float velocityY, boolean consumed) 在这个视图处理之后，发出快速拨动操作给嵌套滚动的父
abstract boolean	dispatchNestedPreFling(float velocityX, float velocityY) 在这个视图处理之前，发出快速拨动操作给嵌套滚动的父

续表

返回值类型	方法及解释
abstract boolean	dispatchNestedPreScroll(int dx,int dy,int[] consumed,int[] offsetInWindow) 在这个视图消费任何滚动之前调用此方法,发出正在进行中的嵌套滚动的其中一步操作给父视图,询问父视图是否要在子 view 之前进行滚动。dispatchNestedPreScroll 给父视图提供了一个在子视图消费任何滚动之前消费部分或者全部滚动操作的机会。该方法告诉父 View 需要滚动的距离。如果父视图滚动了一定距离,该函数返回父视图是否消耗了部分或者全部距离,以及消耗的距离的值(水平距离,垂直距离)。Parent 回调 onNestedPreScroll(),Parent 可以在这个回调中“劫持”掉 Child 的滚动,也就是先于 Child 滚动 dx,dy: 输入参数,即常规的函数参数,调用函数的时候需要为其传递确切的值,分别告诉父 View 要滚动的水平距离和垂直距离,单位像素 consumed: 输出参数,用于子视图获取父视图消费掉的距离,调用函数时只需要传递容器(在这里就是数组),在调用结束后,就可以从容器中获取函数输出的值。如果不为 null,consumed[0]返回父 View 消耗掉的 dx 分量,consumed[1]返回父 View 消耗掉的 dx 分量 offsetInWindow: 输出参数,用于获取子视图的窗口偏移量,调用函数时只需要传递容器(在这里就是数组),在调用结束后,就可以从容器中获取函数输出的值。如果不为 null,将返回父视图在这次操作前后的水平偏移和垂直偏移(本地视图坐标)。 offsetInWindow[0]返回子 View 窗口的水平偏移,offsetInWindow[1]返回子 View 窗口的垂直偏移。由于窗体进行了移动,如果记录了手指最后的位置,需要根据第四个参数 offsetInWindow 计算偏移量,才能保证下一次的 touch 事件的计算是正确的 返回值: 如果 parent 消费了一部分或全部滚动距离,则此方法返回 true,否则返回 false
abstract boolean	dispatchNestedScroll(int dxConsumed, int dyConsumed, int dxUnconsumed, int dyUnconsumed, int[] offsetInWindow) 在这个视图(支持嵌套滚动的子视图)处理滚动后、如果滚动距离还有剩余,应该调用这个方法,发出正在进行中的嵌套滚动的其中一步操作,向嵌套滚动的父视图汇报滚动情况,包括子视图消费的部分和没有消费的部分。如果嵌套的滚动不在进行中,或者这个视图的嵌套滚动没有被使能,则这个方法什么也不做。参数 dxConsumed,dyConsumed 分别表示在这一步滚动中被这个视图消耗掉的水平距离和垂直距离,单位是像素。dxUnconsumed,dyUnconsumed 分别表示在这一步滚动中被这个视图消耗掉的水平距离和垂直距离,单位是像素 offsetInWindow: 输出参数,用于获取子 View 的窗口偏移量,调用函数时只需要传递容器(在这里就是数组),在调用结束后,就可以从容器中获取函数输出的值。如果不为 null,将返回父视图在这次操作前后的水平偏移和垂直偏移(本地视图坐标)。 offsetInWindow[0]返回子 View 窗口的水平偏移,offsetInWindow[1]返回子 View 窗口的垂直偏移。由于窗体进行了移动,如果记录了手指最后的位置,需要根据第四个参数 offsetInWindow 计算偏移量,才能保证下一次的 touch 事件的计算是正确的 返回值: 如果事件被派发出去了,父 View 进行了部分消耗,就返回 true,否则返回 flase

续表

返回值类型	方法及解释
abstract boolean	hasNestedScrollingParent()如果有一个嵌套滚动的父视图，就返回 true
abstract boolean	isNestedScrollingEnabled()如果这个视图的嵌套滚动操作被允许，就返回 true
abstract void	setNestedScrollingEnabled(boolean enabled)使能或者禁止这个视图的嵌套滚动操作
abstract boolean	startNestedScroll(int axes) 参数 axes 取值为 SCROLL_AXIS_HORIZONTAL 和/或者 SCROLL_AXIS_VERTICAL。返回值：当找到了能够配合当前子 View 进行嵌套滚动的父 View，并且用于给定手势的嵌套滚动已经使能时，返回值为 true 开始沿着给定的轴方向的嵌套的滚动操作。在触摸滚动的情境下，这对应于初始的 ACTION_DOWN。如果子视图准备开始滚动了，调用 startNestedScroll()开启嵌套滚动流程(实际上是进行了一些嵌套滚动前准备工作)，通知 Parent，准备进入滚动状态。Parent 收到后回调 onStartNestedScroll()，决定是否需要配合 Child 一起进行处理滚动，如果需要配合，还会回调 onNestedScrollAccepted() 如果 startNestedScroll 返回 true，说明已经找到了协同操作的父；如果 startNestedScroll 返回 false，调用者可以忽略这次协议的剩下部分，直到下一次滚动。当嵌套滚动已经在进行中，调用 startNestedScroll 将返回 true
abstract void	stopNestedScroll( )方法用于停止正在进行中的嵌套的滚动，回调 onStopNestedScroll()表示本次处理结束

11.3 RecyclerView

RecyclerView 和 ListView 不同的是，RecyclerView 不再负责 Item 内的布局和显示相关的功能，ViewHolder 负责 Item 内的布局，LayoutManager 用来确定各个 item 如何进行排列摆放、何时展示和隐藏。RecyclerView 结构，如图 11-4 所示。

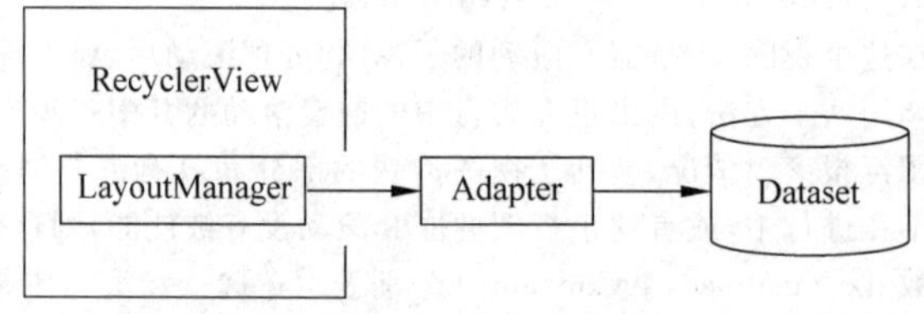

图 11-4 RecyclerView 的结构

为了使用 RecyclerView 控件，还需要创建一个 LayoutManager 和一个 Adapter。RecyclerView. LayoutManager 主要用来设置列表的每一项的视图在 RecyclerView 中的位置布局以及控制视图的显示或者隐藏。当重用或者回收一个 item 的视图的时候，LayoutManger 都会向 Adapter 来请求新的数据来进行替换原来的数据。这种回收重用的机制可以提高性能，避免创建很多的视图或者是频繁的调用 findViewById 方法。Android 系统提供了三种 RecyclerView. LayoutManager：①LinearLayoutManager：横向或者纵向滚动列表，实现 ListView 效果；②GridLayoutManager：网格布局管理器，实现 GridView 效果；③StaggeredGridLayoutManager：交错网格布局管理器，呈现交错网格布局的 RecyclerView。StaggeredGridLayoutManager 是一种特殊的 GridLayoutManager，与 GridLayoutManager 的区别在于，它允许每个 Item 的宽度或高度不一致，实现瀑布流效果。可以这样理解：每个 item 的尺寸都是相同的 StaggeredGridLayoutManager，则是 GridLayoutManager。

RecyclerView. Adapter 的子类，主要用来将数据和 item 进行绑定。一个以前用于显示适配器某个特定位置数据的视图可以被放置在缓存中，以便后来再次显示相同类型的数据，这可以跳过初始的布局扩展，从而大大地改善了性能。在 RecyclerView 中，有两类位置：布局位置和适配器位置。布局位置是从 LayoutManager 的角度看的位置，当编写 RecyclerView. LayoutManager 程序时，总是期望使用布局位置，例如 getLayoutPosition()，findViewHolderForLayoutPosition(int)。适配器位置是从适配器的角度看的位置，当编写 RecyclerView. Adapter 程序时，总是期望使用适配器位置，例如 getAdapterPosition()，findViewHolderForAdapterPosition(int)。这两个位置是相同的，除了分发 adapter. notify* 时和计算更新的布局时。

android. support. v7. widget. RecyclerView. ViewHolder 用于描述在 RecyclerView 中的一个 Item 的视图和关于它的位置的元数据。在 RecyclerView. Adapter 的实现中，首先应创建 RecyclerView. ViewHolder 的子类，在 ViewHolder 中添加域用于缓存 findViewById(int)的结果。RecyclerView. Adapter 应该使用自定义的 ViewHolder 存储数据，每个 item 的视图应该持有对 ViewHolder 对象的引用。在 RecycleView 里，ViewHolder 是适配器的默认方式。在 listView 的适配器里，可以自己选择实现 ViewHolder。

android. support. v7. widget. RecyclerView. Adapter < VH extendsandroid. support. v7. widget. RecyclerView. ViewHolder >用作 RecyclerView 的适配器的基类。适配器提供一种应用特定的数据集与 RecyclerView 中显示的视图的绑定方法。表 11-3 列出了 RecyclerView. Adapter 的抽象方法，在创建 RecyclerView. Adapter 的子类时，必须重写这些抽象方法。

表 11-3　RecyclerView. Adapter 的抽象方法

方　法	解　释
abstract int getItemCount()	返回适配器持有的数据集的 item 的总数量
abstract void onBindViewHolder(VH holder, int position)	被 RecyclerView 调用以便显示特定位置的数据
abstract VH onCreateViewHolder (ViewGroup parent, int viewType)	当 RecyclerView 需要一个新的 RecyclerView. ViewHolder 时，调用这个方法

11.3.1　ItemToucherHelper 基本使用

ItemTouchHelper 是一个支持 RecyclerView 滑动删除、长按拖拽的一个工具类。它需要和 ItemTouchHelper. Callback 一起工作，Callback 用于配置允许哪些交互。ItemTouchHelper 的使用步骤如下

1. 新建接口 IOperationData，Adapter 实现接口

ItemTouchHelper 在完成触摸的各种动作后，就要对 Adapter 的数据进行操作，比如侧滑删除操作，最后需要调用 Adapter 的 notifyItemRemove()方法来移除该数据。因此可以把 Adapter 中数据操作的部分抽象成一个接口方法，Adapter 实现接口，让 ItemTouchHelper.

Callback 的方法调用 Adapter 的对应方法即可。例如：

接口定义如下：

```
public interface IOperationData {
    void onItemMove(int fromPosition, int toPosition);
    void onItemDissmiss(int position);
}
```

Adapter 实现接口：

```
public class MyAdapter extends RecyclerView.Adapter<MyAdapter.MyViewHolder> implements
IOperationData{
    private static final String TAG = "MyAdapter";
    private int mPosition;
    private List<String> mData = new ArrayList<>();

    @Override
    public void onItemMove(int fromPosition, int toPosition) {
        Collections.swap(mData,fromPosition,toPosition);
        notifyItemMoved(fromPosition,toPosition);
    }
    @Override
    public void onItemDissmiss(int position) {
        mData.remove(position);
        notifyItemRemoved(position);
    }
    public int getmPosition() {
        return mPosition;
    }

    public void setmPosition(int mPosition) {
        this.mPosition = mPosition;
    }

    public MyAdapter() {
        for(int i=0;i<20;i++){
            mData.add("item_" + i);
        }
    }
    @NonNull
    @Override
    public MyViewHolder onCreateViewHolder(@NonNull ViewGroup viewGroup, int i) {
        View view = LayoutInflater.from( viewGroup.getContext()).inflate(R.layout.item,
viewGroup,false);

        return new MyViewHolder(view);
    }

    @Override
```

```
    public void onBindViewHolder(@NonNull MyViewHolder myViewHolder, final int position) {
        myViewHolder.testTxt.setText(mData.get(position));
    }

    @Override
    public int getItemCount() {
        return mData.size();
    }

      class    MyViewHolder    extends    RecyclerView.    ViewHolder    implements    View.
OnCreateContextMenuListener {
        TextView testTxt;
      public MyViewHolder(@NonNull View itemView) {
          super(itemView);
          testTxt = (TextView) itemView;
//        itemView.setOnCreateContextMenuListener(this);
      }
}
```

2. 新建 ItemTouchHelper. Callback 的子类，调用接口里声明的方法

需要重写的几个常用的方法：

public int getMovementFlags(RecyclerView, RecyclerView. ViewHolder)：该方法在基类中是抽象方法，必须重写。该方法用于返回可以滑动的方向，比如说允许从右到左侧滑，允许上下拖曳等。一般使用 makeMovementFlags(int,int)或 makeFlag(int,int)来构造我们的返回值。

public abstract boolean onMove(RecyclerView recyclerView, ViewHolder viewHolder, ViewHolder target)：该方法在基类中是抽象方法，必须重写。当用户拖曳一个 Item 从旧的位置到新的位置的时候会调用该方法，在该方法内，可以调用 Adapter 的 notifyItemMoved 方法来交换两个 ViewHolder 的位置，最后返回 true，表示被拖动的 ViewHolder 已经移动到了目的位置。所以，如果要实现拖动交换位置，可以重写该方法（前提是支持上下拖动）。

public abstract void onSwiped(ViewHolder viewHolder,int direction)：该方法在基类中是抽象方法，必须重写。当用户左右滑动 Item 达到删除条件时，会调用该方法，一般手指触摸滑动的距离达到 RecyclerView 宽度的一半时，再松开手指，此时该 Item 会继续向原先滑动方向滑过去并且调用 onSwiped 方法进行删除，否则会反向滑回原来的位置。

public void onSelectedChanged(ViewHolder viewHolder, int actionState)：当 ViewHolder 通过 ItemTouchHelper 滑动或拖动时被调用。

public void clearView(RecyclerView recyclerView, ViewHolder viewHolder)：当用户与一个元素的交互结束、动画完成后，由 ItemTouchHelper 调用。一般在该方法内恢复 ItemView 的初始状态。

public boolean canDropOver(RecyclerView recyclerView, ViewHolder current, ViewHolder target)：如果正在拖动的 ViewHolder 可以放在 target 位置，返回 true。

public boolean isLongPressDragEnabled()：该方法返回 true 时，表示支持长按拖动，即长按 ItemView 后才可以拖动。默认是返回 true。

public boolean isItemViewSwipeEnabled() ：该方法返回 true 时，表示用户可以触摸滑动 item。

例如：下面的代码允许 RecyclerView 的 Item 可以上下拖动，同时允许从右到左侧滑，但不许允许从左到右的侧滑。

```
public class ItemTouchHelperCallback extends ItemTouchHelper.Callback {
    private MyAdapter myAdapter;
    public ItemTouchHelperCallback(MyAdapter myAdapter) {
        this.myAdapter = myAdapter;
    }

    @Override
    public void onSelectedChanged (@ Nullable RecyclerView. ViewHolder viewHolder, int actionState) {
        super.onSelectedChanged(viewHolder, actionState);
        if (viewHolder != null && actionState == ItemTouchHelper.ACTION_STATE_DRAG){
            viewHolder.itemView.setBackgroundColor(Color.RED);
        }
    }

    @Override
    public void clearView (@NonNull RecyclerView recyclerView, @NonNull RecyclerView.ViewHolder viewHolder) {
        super.clearView(recyclerView, viewHolder);
            viewHolder.itemView.setBackgroundColor(0);
    }

    @Override
    public int getMovementFlags (@ NonNull RecyclerView recyclerView, @ NonNull RecyclerView.ViewHolder viewHolder) {
        int dragFlags = ItemTouchHelper.UP|ItemTouchHelper.DOWN;
        int swipeFlags = ItemTouchHelper.LEFT;
        return makeMovementFlags(dragFlags,swipeFlags);
    }

    @Override
    public boolean onMove (@ NonNull RecyclerView recyclerView, @ NonNull RecyclerView.ViewHolder viewHolder,
                          @NonNull RecyclerView.ViewHolder viewHolder1) {
            myAdapter. onItemMove ( viewHolder. getAdapterPosition ( ), viewHolder1.getAdapterPosition());
        return false;
    }
    @Override
    public void onSwiped(@NonNull RecyclerView.ViewHolder viewHolder, int direction) {
        myAdapter.onItemDismiss(viewHolder.getAdapterPosition());
    }
}
```

3. 创建 ItemTouchHelper，并与 RecyclerView 关联

```
recyclerView.setAdapter(myAdapter);
recyclerView.setLayoutManager(new LinearLayoutManager(this));
recyclerView.addItemDecoration(
      new DividerItemDecoration(this,DividerItemDecoration.VERTICAL));
ItemTouchHelper helper = new ItemTouchHelper(new ItemTouchHelperCallback(myAdapter);
helper.attachToRecyclerView(recyclerView);
```

11.4　CoordinatorLayout

android.support.design.widget.CoordinatorLayout 是 android.view.ViewGroup 的直接子类，实现了 NestedScrollingParent2 接口，用作最上层的应用布局（嵌套滚动的父布局），与它的一个或者多个直接子视图进行交互。子视图只要实现了 NestedScrollingChild 接口，就可以将滚动事件传递给它的父视图 CoordinatorLayout。CoordinatorLayout 默认情况下可理解是一个 FrameLayout，布局方式默认是一层一层叠上去。当 CoordinatorLayout 用作和它的一个或者多个直接子视图进行交互的容器时，通过给 CoordinatorLayout 的子视图指定 Behavior，可以提供在同一个父视图中的子视图彼此之间的交互功能。可以通过在视图的类定义前添加@CoordinatorLayout.DefaultBehavior 注解修饰符来指定默认的 behavior。

android.support.design.widget.CoordinatorLayout.Behavior < V extendsandroid.view.View > 封装了 CoordinatorLayout 的子视图随着其他子视图的变化而变化的规则，从而实现子视图之间的一种和多种交互行为，这包括 drags，swipes，flings 或者其他手势。参数 V 代表了 Behavior 操作的视图类型。表 11-4 列出了 CoordinatorLayout.Behavior 的常用的方法。

CoordinatorLayout 的工作原理是搜索定义了 CoordinatorLayout.Behavior 的子视图，当 CoordinatorLayout 收到某个子视图发来的滚动事件的时候，CoordinatorLayout 会尝试触发那些声明了 Behavior 的子视图。阅读 AppBarLayout.Behavior 和 FloatingActionButtion.Behavior 的代码，可以帮助理解如何自定义 behavior。

表 11-4　CoordinatorLayout.Behavior 的常用的方法

返回值类型	方法及解释
boolean	blocksInteractionBelow(CoordinatorLayout parent，V child) 决定是否阻止和给定的子视图之后的视图交互
static Object	getTag(View child) 得到与给定的子视图关联的 tag
boolean	layoutDependsOn(CoordinatorLayout parent，V child，View dependency) 决定 child 是否依赖 dependency，这个方法在响应布局请求时至少调用一次。当调用这个方法返回 true 时，CoordinatorLayout 总是先布局 dependency，再布局 child，在 dependency 的位置或者布局改变时，调用 onDependentViewChanged(CoordinatorLayout，V，View) onDependentViewChanged(CoordinatorLayout parent，V child，View dependency) 响应依赖视图的变化。当依赖视图在标准的布局流程外大小或者位置有改变时，这个方法被调用
void	onDependentViewRemoved(CoordinatorLayout parent，V child，View dependency) 响应依赖视图的移除。当依赖视图从父视图中移除后，这个方法被调用

续表

返回值类型	方法及解释
boolean	onInterceptTouchEvent(CoordinatorLayout parent,V child,MotionEvent ev) 在触摸事件被分发到子视图前,响应 CoordinatorLayout 的触摸事件 onLayoutChild(CoordinatorLayout parent,V child,int layoutDirection) 当父 CoordinatorLayout 即将布局给定的子视图时被调用 onMeasureChild(CoordinatorLayout parent,V child,int parentWidthMeasureSpec,int widthUsed,int parentHeightMeasureSpec,int heightUsed) 当父 CoordinatorLayout 即将测量给定的子视图时被调用 onNestedFling(CoordinatorLayout coordinatorLayout, V child, View target, float velocityX,float velocityY,boolean consumed) 当嵌套滚动子视图正在开始快速拨动或者动作将是快速拨动时被调用 onNestedPreFling(CoordinatorLayout coordinatorLayout, V child, View target, float velocityX,float velocityY)当嵌套滚动子视图即将快速拨动时被调用
void	onNestedPreScroll(CoordinatorLayout coordinatorLayout, V child, View target, int dx,int dy,int[] consumed) 这个方法在 api 级别 26.1.0 已废弃,用 onNestedPreScroll(CoordinatorLayout,View,View,int,int,int[],int)代替。如果类型是 TYPE_TOUCH,这个方法将继续被调用 onNestedPreScroll(CoordinatorLayout coordinatorLayout, V child, View target, int dx,int dy,int[] consumed,int type)　在目标(发起嵌套滚动的视图)消耗任何滚动距离前,当正在进行中的嵌套滚动即将更新时被调用 onNestedScroll(CoordinatorLayout coordinatorLayout, V child, View target, int dxConsumed,int dyConsumed,int dxUnconsumed,int dyUnconsumed) 这个方法在 api 级别 26.1.0 已废弃,用 onNestedScroll(CoordinatorLayout,View,View,int,int,int,int,int)代替　如果类型是 TYPE_TOUCH,这个方法将继续被调用 onNestedScroll(CoordinatorLayout coordinatorLayout, V child, View target, int dxConsumed,int dyConsumed,int dxUnconsumed,int dyUnconsumed,int type) 当正在进行中的嵌套的滚动已经更新、目标已经滚动时被调用。dxConsumed 和 dyConsumed 为目标自己的滚动操作所消耗的水平像素数和垂直水平像素数。dxUnconsumed 和 dyUnconsumed 指目标自己的滚动操作没有消耗的水平像素数和垂直水平像素数 onNestedScrollAccepted(CoordinatorLayout coordinatorLayout, V child, View directTargetChild,View target,int axes,int type) 当嵌套的滚动已经被 CoordinatorLayout 接受时被调用 onNestedScrollAccepted(CoordinatorLayout coordinatorLayout, V child, View directTargetChild,View target,int axes)这个方法在 api 级别 26.1.0 已废弃,用 onNestedScrollAccepted(CoordinatorLayout,View,View,View,int,int)代替。如果类型是 TYPE_TOUCH,这个方法将继续被调用
boolean	onStartNestedScroll(CoordinatorLayout coordinatorLayout,V child,View directTargetChild,View target,int axes)这个方法在 api 级别 26.1.0 已废弃,用 onStartNestedScroll(CoordinatorLayout,View,View,View,int,int)代替。如果类型是 TYPE_TOUCH,这个方法将继续被调用

续表

返回值类型	方法及解释
boolean	onStartNestedScroll(CoordinatorLayout coordinatorLayout, V child, View directTargetChild, View target, int axes, int type)当 CoordinatorLayout 的子视图发起嵌套的滚动时被调用。与 CoordinatorLayout 的直接子视图相关联的 Behavior 都可以响应这个事件，如果这个 Behavior 期望接收嵌套滚动，就返回 true；否则返回 false。只有返回 true 的 Behavior 才会响应随后的嵌套滚动事件
void	onStopNestedScroll(CoordinatorLayout coordinatorLayout, V child, View target) 这个方法在 api 级别 26.1.0 已废弃，用 onStopNestedScroll(CoordinatorLayout, View, View, int)代替。如果类型是 TYPE_TOUCH，这个方法将继续被调用 onStopNestedScroll(CoordinatorLayout coordinatorLayout, V child, View target, int type)当嵌套的滚动结束时被调用，target 指正在发起嵌套滚动的 CoordinatorLayout 的子视图
boolean	onTouchEvent(CoordinatorLayout parent, V child, MotionEvent ev) 在 Behavior 已经开始拦截触摸事件后，响应 CoordinatorLayout 的触摸事件
static void	setTag(View child, Object tag)　设置给定的 View 和 Tag 对象关联

11.4.1　设置为子视图的 Behavior

当 CoordinatorLayout 用作和一个或者多个子视图进行交互的容器时，通过给 CoordinatorLayout 的子视图设置 Behavior，可以提供在同一个父视图中的子视图彼此之间的交互功能。为子视图设置 Behavior 主要有以下三种方式：

在视图的类的定义前添加@CoordinatorLayout. DefaultBehavior 修饰符指定视图的默认 Behavior，这样在使用的过程中就不用每次再定义了。一个例子如下：

```
@CoordinatorLayout.DefaultBehavior(AppBarLayout.Behavior.class)
public class AppBarLayout extends LinearLayout {}
```

如果该默认的 Behavior 不满足要求，则还可以通过在 Java 代码中设置或者在 XML 布局文件中设置来覆盖默认的 Behavior。

在 XML 布局文件中，可以通过 app：layout_behavior 属性指定 behavior。

```
app:layout_behavior = "Behavior 的全限定类名"
```

app：layout_behavior 是 CoordinatorLayout. LayoutParams 的 XML 属性，属性的取值必须使用 Behavior 的全限定类名，全限定类名由包的名字和类的名字两部分组成。只有直接子视图才存有 CoordinatorLayout. LayoutParams 定义的属性，因此只能给 CoordinatorLayout 的直接子视图添加 app：layout_behavior 属性。在 CoordinatorLayout 的内部类 LayoutParams 的构造函数中，首先检查了是不是定义了 layout_behavior 属性，如果定义了 layout_behavior 属性，就调用了 parseBehavior 方法，返回 Behavior 的实例，在自定义 Behavior 的时候，必须重写 Behavior(Context context, AttributeSet attrs)这个构造方法。

(3) 在 Java 代码中,调用 setBehavior 方法为某个视图指定 Behavior,例如:

```
YourBehavior yourBehavior = new YourBehavior();
(CoordinatorLayout.LayoutParams)childView.getLayoutParams().setBehavior(yourBehavior);
```

11.4.2 实现自定义 Behavior

当 CoordinatorLayout 收到某个视图的大小、位置、显示状态变化或者嵌套滚动事件时,CoordinatorLayout 就会尝试把事件分发给 Behavior,绑定了该 Behavior 的视图就会对事件做出响应。在自定义 Behavior 的时候,监听的内容通常分为两类:

第一类:一个视图监听另一个视图的状态变化,例如大小、位置、显示状态等,这时需要实现 layoutDependsOn 和 onDependentViewChanged 方法。

layoutDependsOn 方法可以根据 id 值来判断依赖关系:

```
public boolean layoutDependsOn(CoordinatorLayout parent, View child, View dependency) {
  return dependency.getId() == R.id.xxx;
}
```

layoutDependsOn 方法可以根据类型来判断依赖关系:

```
public boolean layoutDependsOn(CoordinatorLayout parent, View child, View dependency) {
return dependency instanceof CustomView;
}
```

第二类:某个视图监听 CoordinatorLayout 里的滚动状态,对滚动事件的处理主要实现这 3 个方法:onStartNestedScroll、onNestedPreScroll、onNestedScroll、onNestedFling 方法。

android.support.design.widget.CoordinatorLayout.LayoutParams 是 android.view.ViewGroup.MarginLayoutParams 的直接子类,主要属性如表 11-5 所示。

表 11-5 CoordinatorLayout.LayoutParams 的 XML 属性

属性	解释
android:gravity	Gravity 值,设置这个子视图的放置位置
app:layout_anchor	设置子视图以哪个视图作参考
app:layout_anchorGravity	Gravity 值,设置这个子视与它的锚视图的哪个边沿对齐
app:layout_behavior	设置 CoordinatorLayout 的子视图的交互行为
app:layout_dodgeInsetEdges	Gravity 值,设置这个子视图怎样避让 CoordinatorLayout 的内嵌子视图
app:layout_insetEdge	Gravity 值,设置这个子视图怎样内嵌在 CoordinatorLayout 中,其他被设置成避开同一内嵌沿的子视图将被适当移动以便视图之间不重叠
app:layout_keyline	Keyline 允许把元素放置在布局栅格的外边。Keyline 是垂直线,用于表示把不能对齐到栅格的元素放置在哪里。Keyline 是由元素到屏幕的左边沿的距离决定的,增量是 8dp。Keyline 和响应式布局组合使用

11.5 材料设计中的 AppBar

当设计滚动行为时，App Bar 从上到下包括四个主要区域：状态条、工具条、Tab bar/search bar，灵活空间。android. support. design. widget. AppBarLayout 是 android. widget. LinearLayout 的直接子类，它是一个垂直方向的 LinerLayout，它的作用是把 AppBarLayout 包括的内容都作为 App Bar。这个类需要跟 CoordinatorLayout 配合使用，即 AppBarLayout 必须是 CoordinatorLayout 的直接子视图。如果 AppBarLayout 不是 CoordinatorLayout 的直接子视图，而是在别的 ViewGroup 内，它的功能将不起作用。

AppBarLayout 为了知道何时滚动，必须有一个分离的可以滚动的兄弟视图，必须在这个兄弟视图中指定 behavior 属性为 AppBarLayout. ScrollingViewBehavior 类的一个实例，可以使用内置的 @ string/appbar _ scrolling _ view _ behavior 表示这个默认的实例。AppBarLayout 的子视图必须通过调用 setScrollFlags(int)来提供它们期望的滚动行为，或者通过相关联的 XML 属性 app：layout_scrollFlags 来设置。

android. support. design. widget. AppBarLayout. LayoutParams 是 android. widget. LinearLayout. LayoutParams 的直接子类，除了继承的 LinearLayout. LayoutParams 的 XML 属性外，还包含的 XML 属性如表 11-6 所示。

表 11-6 AppBarLayout. LayoutParams 的 XML 属性

属　　性	解　　释
android. support. design：layout_scrollFlags	取值为：scroll，exitUntilCollapsed，enterAlways，enterAlwaysCollapsed，snap，这些分别对应比特形式的整数，还可以将它们的一个或者多个用竖线隔开，例如设置 app：layout _ scrollFlags = " scroll | enterAlways "。enterAlwaysCollapsed 要和 scroll 和 enterAlways 一起使用才有效果。exitUntilCollapsed 只有和 scroll 一起组合才会有效果 scroll 表示向上滚动时，这个 View 会被滑出屏幕范围直到隐藏。所有想滚动出屏幕的 View 都需要 app：layout_scrollFlags="scroll"，没有设置 Scroll 的视图将不能滑出屏幕顶部。enterAlways 表示向下滚动时，这个 View 会随着滚动手势向下滚动(无论滚动视图是否正在滚动)，直到恢复原来的位置，也被称为"快速返回模式"。enterAlwaysCollapsed：当视图已经设置 minHeight 属性又使用此标志时，该视图只能以最小高度进入，只有当滚动视图到达顶部时才扩大到完整高度。exitUntilCollapsed：滚动退出屏幕，最后折叠在顶端
android. support. design：layout_scrollInterpolator	设置滚动视图的时候要使用的 Interpolator

11.5.1 在布局文件中使用 AppBarLayout

首先需要使用 CoordinatorLayout 作为布局文件的顶层布局，在< android. support. design. widget. CoordinatorLayout >内部放置< android. support. design. widget. AppBarLayout >元素。

在 CoordinatorLayout 里面放置一个实现了 NestedScrollingChild 接口的视图，也就是

主内容视图，作为 AppBarLayout 的兄弟视图。

AppBar 中的视图都放在 AppBarLayout 里面，并且通过 app：layout_scrollFlags 属性来控制滚动时候的表现。通常带有 scroll 标志的视图放在前面(即屏幕上面)，不带 Scroll 标志的视图放在带有 Scroll 标志的 View 的后面(即屏幕下面)，这样带有 Scroll 标志的视图才能正常滑出屏幕，而不带 Scroll 标志的视图继续留在屏幕里，这是因为 AppBarLayout 是一个垂直方向的 LinearLayout 布局。

在 CoordinatorLayout 中使用 AppBarLayout，如果 AppBarLayout 的子视图(如 ToolBar、TabLayout)使用 app：layout_scrollFlags 设置了滚动事件，那么在 CoordinatorLayout 布局里其他设置了 layout_behavior 的子视图(LinearLayout、RecyclerView、NestedScrollView 等)就能够响应 AppBarLayout 中控件被标记的滚动事件。

实例 1：AppBarLayout。

```
<android.support.design.widget.CoordinatorLayout
        xmlns:android = "http://schemas.android.com/apk/res/android"
        xmlns:app = "http://schemas.android.com/apk/res - auto"
        android:layout_width = "match_parent"
        android:layout_height = "match_parent">
    <android.support.v4.widget.NestedScrollView
        android:layout_width = "match_parent"?android:layout_height = "match_parent"
        app:layout_behavior = "@string/appbar_scrolling_view_behavior">
      <! -- Your scrolling content -->
    </android.support.v4.widget.NestedScrollView>

    <android.support.design.widget.AppBarLayout
                                        android:layout_height = "wrap_content" android:
layout_width = "match_parent">
        <android.support.v7.widget.Toolbar…
          app:layout_scrollFlags = "scroll|enterAlways"/>
        <android.support.design.widget.TabLayout?…
          app:layout_scrollFlags = "scroll|enterAlways"/>
    </android.support.design.widget.AppBarLayout>

</android.support.design.widget.CoordinatorLayout>
```

11.5.2 ToolBar

android. widget. toolbar 是 android. view. ViewGroup 的直接子类，是 Android 5.0(API21)增加的新特性、用于替代 actionbar。若需要在 Android 5.0(API21)以前的平台上使用，可以使用 android. support. v7. widget. toolbar。ActionBar 是由框架控制的不透明的 window décor(视图层级的顶层)的一部分，ToolBar 可以放置在视图层级图的任何一层，可以被放置在界面的任何地方，也提供了可定制化的空间。ToolBar 的 XML 属性，如表 11-7 所示。应用程序可以使用 setSupportActionBar()或者 setActionBar()方法选择指定的 ToolBar 作为 ActionBar。

表 11-7 android. support. v7. widget. Toolbar 的 xml 属性

属 性	解 释
app：buttonGravity	导航按钮的位置，top 或者 bottom，仅在 API 24 以上有效
app：collapseContentDescription	用于折叠按钮的内容描述的文本
app：collapseIcon	用于折叠按钮的可绘制图像
app：contentInsetEnd	设置在 ToolBar 的结束位置插入的空间的最小宽度（离 ToolBar 内容视图的距离）。ToolBar 的左右两侧默认都有 16dp 的 padding 的，如果需要让 ToolBar 上的内容与左右两侧的距离有变化，便可以通过 contentInsetStart、contentInsetLeft、contentInsetEnd、contentInsetRight 属性来进行相应的设置。比如要让内容紧贴左侧或起始侧便可以将 contentInsetLeft 或 contentInsetStart 设为 0
app：contentInsetEndWithActions	如果 Action 存在，设置 Action 与其左侧内容的距离，默认为 16dp
app：contentInsetLeft	设置在 ToolBar 的左边位置插入的空间的最小宽度（离 ToolBar 内容视图的距离）
app：contentInsetRight	设置在 ToolBar 的右边位置插入的空间的最小宽度（离 ToolBar 内容视图的距离）
app：contentInsetStart	设置在 ToolBar 的开始位置插入的空间的最小宽度（离 ToolBar 内容视图的距离）
app：contentInsetStartWithNavigation	如果该 ToolBar 有 Navigation Button 的话，通过该属性设置 Navigation Button 与它右侧的内容之间的距离，默认是 16dp
app：logo	取值为可绘制图像，用于设置 logo 图标，它出现在 ToolBar 的开始位置，仅仅在导航按钮的后面
app：logoDescription	Logo 的内容描述字符串
app：maxButtonHeight	设置最大的按钮高度
app：navigationContentDescription	文本串，对导航按钮的内容描述
app：navigationIcon	取值为可绘制图标，用作在 ToolBar 开始位置的导航按钮
app：popupTheme	对 theme 的引用，用于填充 ToolBar 上某个控件的弹出式菜单
app：subtitle	指定了当导航模式为普通时的副标题文本串
app：subtitleTextAppearance	设置副标题的文本样式
app：subtitleTextColor	设置副标题字符串的颜色
app：title	设置 ToolBar 的标题
app：titleMargin	指定了 ToolBar 标题的左边、右边、开始以及结束位置保留的额外空间，如果同时设置 titleMargin 和 titleMarginStart（End，Top，Bottom），则优先取后面的属性值
app：titleMarginBottom	指定 ToolBar 的标题的底部保留的额外空间
app：titleMarginEnd	指定标题的结束位置一侧保留的额外空间
app：titleMarginStart	指定标题的开始位置一侧保留的额外空间
app：titleMarginTop	指定标题的顶部位置保留的额外空间
app：titleTextAppearance	设置标题的文本样式
app：titleTextColor	设置标题字符串的文本颜色

一个 ToolBar 从左到右包括了一个 navigation button、一个 logo、一个 title 和 subtitle、一个或多个自定义的视图和一个 action menu 这 5 部分，如图 11-5 所示。导航按钮可能是一个箭头用于导航菜单切换，关闭。logo 图标的高度可以扩展到 ToolBar 的高度，宽度任意。标题应该是导航层次结构和其所包含的内容的指示。如果有副标题，应该暗示当前内容的扩展信息。如果一个应用程序使用 logo 图标，应该强烈建议省略标题和副标题。在 API 21 后，使用应用程序图标＋标题作为标准布局不再被鼓励。应用程序可以将任意多个自定义的子视图添加到 ToolBar 中。如果一个子视图的 Toolbar. LayoutParams 中 Gravity (对齐方式)的值是 CENTER_HORIZONTAL(水平居中)，视图将定位在其他元素获取到空间后的 ToolBar 的剩余空间的中心。操作菜单及溢出菜单将绑定到 ToolBar 的结束位置，操作菜单提供频繁的、重要的或是典型的操作。

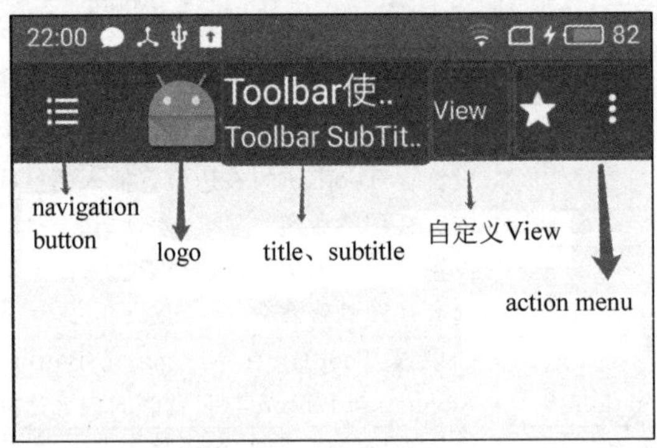

图 11-5　ToolBar 的组成

给 Activity 添加 ToolBar 的步骤：

(1) 在 build. gradle(module：app)文件中中添加 appcompat v7 支持库。例如：

```
compile 'com.android.support:appcompat-v7:24.2.1'
```

(2) 确保要使用 Toolbar 的 Activity 继承了 AppCompatActivity。

```
public class MainActivity extends AppCompatActivity {//…
}
```

Android Studio 在创建 Basic Activity 的时候，就可以保证以上两点。

(3) 隐藏 ActionBar. 可以通过在 Java 中编程隐藏，也可以通过配置文件的方式。下面介绍配置文件，可以在 styles. xml 文件中的 AppTheme 标签中加入如下两行就能禁止 Actionbar：

```
<item name="windowActionBar">false</item>
<item name="windowNoTitle">true</item>
```

也可以新建一个 style 标签，将上面两行代码加入，并且将这个新建的标签作为 application 或者 activity 的 theme；也可以将 AppTheme 的 parent 设置为 Theme. AppCompat. Light. NoActionBar；也可以将 application 元素的主题设置为使用 appcompat 的任何一个

NoActionBar 主题，使用 NoActionBar 的主题可以防止应用使用原生 ActionBar 类。如果所有的 activity 都是一样的主题，可以设置在 application 结点。例如：

```
<application
    android:theme = "@style/Theme.AppCompat.Light.NoActionBar">
...
</ application>
```

当然，也可以不在 application 结点，而是在使用 ToolBar 的 activity 结点设置任意一个 appCompat 的其中一个 NoActionBar 主题或者它们的子主题（比如自定义的主题，parent 为 appCompat 的 NoActionBar）

例如 AndroidManifest.xml 内容如下：

```
<?xml version = "1.0" encoding = "utf - 8"?>
<manifest xmlns:android = "http://schemas.android.com/apk/res/android"
    package = "cuc.menu">
    <application
        android:allowBackup = "true"
        android:icon = "@mipmap/ic_launcher"
        android:label = "@string/app_name"
        android:supportsRtl = "true"
        android:theme = "@style/AppTheme">
        <activity
            android:name = ".MainActivity" android:label = "@string/app_name"
            android:theme = "@style/AppTheme.NoActionBar" >
            <intent - filter>
                    <action android:name = "android.intent.action.MAIN" />
                <category android:name = "android.intent.category.LAUNCHER"/>
            </intent - filter>
        </activity>
    </application>
</manifest>
```

上述清单文件中，application 使用 AppTheme，MainActivity 使用了 AppTheme.NoActionBar，这两个主题都定义在 res/values/styles.xml 中。res/values/styles.xml 内容如下：

```
<resources>
    <!-- Base application theme. -->
    <style name = "AppTheme" parent = "Theme.AppCompat.Light.DarkActionBar">
        <!-- Customize your theme here. -->
        <item name = "colorPrimary">@color/colorPrimary</item>
        <item name = "colorPrimaryDark">@color/colorPrimaryDark</item>
        <item name = "colorAccent">@color/colorAccent</item>
    </style>
    <style name = "AppTheme.NoActionBar">
        <item name = "windowActionBar">false</item>
```

```
        <item name="windowNoTitle">true</item>
    </style>
    <style name="AppTheme.AppBarOverlay" parent="ThemeOverlay.AppCompat.Dark.ActionBar" />
<style name="AppTheme.PopupOverlay" parent="ThemeOverlay.AppCompat.Light"/>
</resources>
```

(4) 在 Activity 的布局文件中添加 ToolBar。

```
<?xml version="1.0" encoding="utf-8"?>
<android.support.design.widget.CoordinatorLayout
xmlns:android="http://schemas.android.com/apk/res/android"
    xmlns:app="http://schemas.android.com/apk/res-auto"
    xmlns:tools="http://schemas.android.com/tools"
    android:layout_width="match_parent"
    android:layout_height="match_parent"
    android:fitsSystemWindows="true"
    tools:context="cuc.menu.MainActivity">
    <android.support.design.widget.AppBarLayout
        android:layout_width="match_parent"
        android:layout_height="wrap_content"
        android:theme="@style/AppTheme.AppBarOverlay">
        <android.support.v7.widget.Toolbar
            android:id="@+id/toolbar"
            android:layout_width="match_parent"
            android:layout_height="?attr/actionBarSize"
            android:background="?attr/colorPrimary"
            app:popupTheme="@style/AppTheme.PopupOverlay" />
    </android.support.design.widget.AppBarLayout>
    <include layout="@layout/content_main" />
</android.support.design.widget.CoordinatorLayout>
```

在使用 Toolbar 的布局文件中，应当使用自定义的命名空间（xmlns: app="http://schemas.android.com/apk/res-auto"）。ToolBar 的常用属性采用自定义命名空间 App，而不是 Android，如果使用 android: navigationIcon 必须是 API 21 以上。

(5) 在 Activity 中使用 ToolBar。在 Activity 中使用 ToolBar 有两种方式：把 ToolBar 当 ActionBar 使用和当普通的 widget 使用。第一种方法，调用 setSupportActionBar 方法设置 ToolBar 为 ActionBar，要显示菜单只需要重写 Activity 的 onCreateOptionsMenu 方法，而监听菜单项的事件也只需重写 Activity 的 onOptionsItemSelected 方法即可。第二种方法，ToolBar 当普通的 widget 使用，通过 inflateMenu 方法来加载菜单文件并通过 setOnMenuItemClickListener 方法为菜单的每一项设置监听，在这种情况下 Activity 不必继承于 AppCompatActivity。

在 Activity 中把 ToolBar 当 ActionBar 使用的例子：

```
public class MainActivity extends AppCompatActivity {

    @Override
```

```
    protected void onCreate(Bundle savedInstanceState) {
        super.onCreate(savedInstanceState);
        setContentView(R.layout.activity_main);
        Toolbar toolbar = (Toolbar) findViewById(R.id.toolbar);
        setSupportActionBar(toolbar);
    }
    @Override
    public boolean onCreateOptionsMenu(Menu menu) {
        getMenuInflater().inflate(R.menu.abc, menu);
        return super.onCreateOptionsMenu(menu);
    }
    @Override
    public boolean onOptionsItemSelected(MenuItem item) {
        switch (item.getItemId()) {
            case R.id.item1:
                Toast.makeText(this, "R.id.item1", Toast.LENGTH_LONG).show();
                return true;
            case R.id.group:
                Toast.makeText(this, "R.id.group", Toast.LENGTH_LONG).show();
                return true;
          }
        return super.onOptionsItemSelected(item);
    }
}
```

在 Activity 中把 ToolBar 当普通 widget 使用的例子：

```
public class MainActivity extends AppCompatActivity {
    @Override
    protected void onCreate(Bundle savedInstanceState) {
        super.onCreate(savedInstanceState);
        setContentView(R.layout.activity_main);
        Toolbar toolbar = (Toolbar) findViewById(R.id.toolbar);
        toolbar.inflateMenu(R.menu.menu_main);
toolbar.setNavigationOnClickListener(new View.OnClickListener() {
       @Override
        public void onClick(View v) {
         finish();
        }
      });
        toolbar.setOnMenuItemClickListener(new Toolbar.OnMenuItemClickListener() {
            @Override
            public boolean onMenuItemClick(MenuItem item) {
                int id = item.getItemId();
                String msg = "";
                switch (id){
                    case R.id.ic_delete:
                        msg += "ic_delete";
                        break;
```

```
                    case R.id.ic_input_add:
                        msg += "ic_input_add";
                        break;
                }
              Toast.makeText(MainActivity.this,msg,Toast.LENGTH_SHORT).show();
                return true;
            }
        });
    }
}
```

11.6 NestedScrollView

android.support.v4.widget.NestedScrollView 是 android.widget.FrameLayout 的直接子类，实现了 NestedScrollingParent、NestedScrollingChild 和 ScrollingView 接口，它支持传递滚动事件到实现了 NestedScrollingParent 接口的 CoordinatorLayout 也可以接收滚动事件，并且把滚动事件发送给绑定了 Behavior 的视图。NestedScrollView 只能接受一个子视图(通常 LinearLayout)，并且使子视图拥有滚动效果。NestedScrollView 是一个增强型的 ScrollView。

11.7 侧滑抽屉

android.support.v4.widget.DrawerLayout 是 android.view.ViewGroup 的直接子类，用作窗口内容的顶层容器，它通常包含侧边抽屉和主内容区两部分，侧边抽屉可以根据手势显示与隐藏(DrawerLayout 自身特性)，就好像抽屉被拉出了屏幕一样，主内容区的内容可以随着菜单的单击而变化(这需要使用者自己实现)。根据 Android 设计指南，任何位置在屏幕左边/开始边沿的抽屉应该总是包含导航，任何位置在屏幕右边/结束边沿的抽屉应该总是包含对现在的内容可以采用的动作，这和 ActionBar 的导航在左边、动作在右边的结构一致。抽屉式导航栏将应用的主要导航选项显示在屏幕左边缘。大多数情况下，它处于隐藏状态，但是如果用户从屏幕左边缘向屏幕中心滑动手指，或者在应用的顶层时，用户触摸 ActionBar 中的应用图标，它就会显示出来。在窗口的每个垂直边沿上都允许有唯一的一个抽屉视图。如果在视图的某个垂直边沿上配置了超过一个抽屉视图，运行时将抛出例外。

为了使用 DrawerLayout，DrawerLayout 最好为界面的根布局，否则可能会出现触摸事件被屏蔽的问题。在 DrawerLayout 内，显示屏幕主要内容的主视图(当抽屉式导航栏处于隐藏状态时为主要布局)作为 DrawerLayout 的第一个子视图，并且宽度和高度都设置为 match_parent，并且不能设置 layout_gravity 属性，因为 XML 顺序意味着按 z 序(层叠顺序)排序，并且抽屉式导航栏必须位于内容顶部，在抽屉式导航栏处于隐藏状态时，主视图代表整个 UI。主要的内容视图后面紧跟着一个或者两个抽屉视图，抽屉的摆放位置通过

android: layout_gravity 属性来控制，可选值为 left、right 或 start、end，对应于抽屉从视图的左边还是右边出现(或者在支持布局方向的平台上 start/end)。抽屉视图的高度通常设置成 match_parent，宽度设置成固定值，宽度不应超过 320dp，以便用户把内容展示区和抽屉菜单区区分开，始终可以看到部分主内容。左侧抽屉用于导航，右侧抽屉显示与主内容区相关的动作。左侧导航可以使用 NavigationView，也可以使用普通的 View，例如 RecyclerView 或者 ListView，也可以是一个 Button，侧边菜单跟 Activity 的选项菜单完全不同，Activity 的菜单只需要在资源文件中定义好，就能按照固定的形式显示出来，而 drawerLayout 的侧边菜单显示成什么样完全取决于自己的设计。在 Activity 中，需要初始化抽屉式导航栏的项目列表，具体怎么初始化 Item 列表、怎样监听选项变化取决于抽屉视图的类型。

android. support. v4. widget. DrawerLayout. LayoutParams 是 android. view. ViewGroup. MarginLayoutParams 的直接子类，描述了 DrawerLayout 的子视图中跟 DrawerLayout 有关联的属性。DrawerLayout 布局的子视图的有些属性跟 DrawerLayout 有关联，这些属性定义在 DrawerLayout. LayoutParams 中。DrawerLayout. LayoutParams 包含 android: layout_gravity 属性，用于设置 DrawerLayout 的抽屉的位置：在左边还是在右边。抽屉视图必须使用 android: layout_gravity 属性指定其水平 gravity。要支持从右到左(RTL)的语言，请使用 start(而非 left)指定该值，这样当布局为 RTL 时，抽屉式导航栏会显示在右侧。

下面是一个 DrawerLayout 布局资源文件，它使用 DrawerLayout 作为父元素，包含一个 FrameLayout(主内容、在运行时由 Fragment 填充)和一个 ListView(用作抽屉式导航栏)。

```
<android.support.v4.widget.DrawerLayout
  xmlns:android = "http://schemas.android.com/apk/res/android"
  android:id = "@ + id/drawer_layout"
  android:layout_width = "match_parent"
  android:layout_height = "match_parent">
  <!-- The main content view -->
  <FrameLayout
    android:id = "@ + id/content_frame"
    android:layout_width = "match_parent"
    android:layout_height = "match_parent" />
  <!-- The navigation drawer -->
  <ListView android:id = "@ + id/left_drawer"
    android:layout_width = "240dp"
    android:layout_height = "match_parent"
    android:layout_gravity = "start"
    android:choiceMode = "singleChoice"
    android:divider = "@android:color/transparent"
    android:dividerHeight = "0dp"
    android:background = "#111"/>
</android.support.v4.widget.DrawerLayout>
```

11.7.1 侧边菜单的显示与隐藏

用户可以通过朝向或者远离屏幕中心的滑动手势显示和隐藏导航抽屉，DrawerLayout.DrawerListener接口可以用来监听抽屉的显示与隐藏事件，此接口为抽屉事件提供了回调，从而在菜单显示与隐藏发生的时刻做一些希望做的事情，比如更新actionbar菜单。要侦听抽屉的打开和关闭事件，调用DrawerLayout对象的setDrawerListener()并向其传递DrawerLayout.DrawerListener的实现。但是，如果Activity中包括Actionbar，则可用ActionBarDrawerToggle子类来监听。ActionBarDrawerToggle实现了DrawerLayout.DrawerListener接口，所以它能做DrawerLayout.DrawerListener可以做的任何事情，同时它还能将抽屉的显示和隐藏与ActionBar的App图标关联起来，单击App图标还能显示或者隐藏抽屉。ActionBarDrawerToggle的onOptionsItemSelected方法，该方法判断单击事件是否来自于app图标，然后调用DrawerLayout对象的closeDrawer或者DrawerLayout对象的openDrawer来隐藏或者显示抽屉。在单击侧边菜单选项后，往往需要隐藏侧边菜单以便全屏显示菜单对应的内容，DrawerLayout对象的closeDrawer方法用于隐藏侧边菜单，DrawerLayout对象的openDrawer方法用于显示侧边菜单。可以使用下面的构造函数创建ActionBarDrawerToggle的实例。

```
public ActionBarDrawerToggle(Activity activity, DrawerLayout drawerLayout,
            Toolbar toolbar, @StringRes int openDrawerContentDescRes,
            @StringRes int closeDrawerContentDescRes)
```

函数参数依次为：托管抽屉的Activity、与ToolBar联系的DrawerLayout、独立的ToolBar、描述“打开抽屉”操作的字符串资源、描述“关闭抽屉”操作的字符串资源。

如果ToolBar用作Activity的ActionBar，可以直接使用下面的构造函数：

```
ActionBarDrawerToggle(Activity activity, DrawerLayout drawerLayout,
            @StringRes int openDrawerContentDescRes,
            @StringRes int closeDrawerContentDescRes)
```

使用ActionBarDrawerToggle时，ActionBar左边的up按钮将被设置成固定的图标。当抽屉关闭时显示汉堡包的图标，抽屉打开时显示箭头的图标。

11.7.2 NavitationView

android.support.design.widget.NavigationView是android.widget.FrameLayout的直接子类，它是一个简单高效的导航菜单框架，它提供了默认样式、选中项高亮、分组单选、分组子标题，以及可选的Header，菜单内容由菜单资源文件提供。使用NavigationView比使用ListView作为抽屉菜单更加简单高效。

NavigationView的典型应用是放置在DrawerLayout里，作为DrawerLayout的Drawer，即导航菜单的本体部分。NavigationView的xml属性，如表11-8所示，在xml文件中使用这些属性，并没有必要使用硬编码的包名，只需要输入App代替相应的包名android.support.design。

表 11-8 NavigationView 的 XML 属性

属　性	解　释
android. support. design：itemBackground	设置菜单项的背景为给定的 Drawable 资源
android. support. design：itemIconTint	设置菜单项图标的着色
android. support. design：itemTextAppearance	设置菜单项的背景为给定的资源
android. support. design：itemTextColor	设置菜单项的文本颜色为给定的资源

下面给出了一个示例，其中 NavigationView 的两个自定义属性：app：headerLayout 和 app：menu。示例中的 app：menu＝"@menu/my_navigation_menu"指定了导航所用的菜单，几乎是必选项，不然这个控件就失去意义了，但也可以在运行时动态改变 menu 属性。app：headerLayout 指定导航菜单顶部的布局，可选项。

```
< android.support.v4.widget.DrawerLayout
xmlns:android = "http://schemas.android.com/apk/res/android"
  xmlns:app = "http://schemas.android.com/apk/res - auto"
  android:id = "@ + id/drawer_layout"
  android:layout_width = "match_parent" ?
android:layout_height = "match_parent"
  android:fitsSystemWindows = "true">
  <! -- Your contents -- >
  < android.support.design.widget.NavigationView ?
android:id = "@ + id/navigation"
    android:layout_width = "wrap_content" ?
android:layout_height = "match_parent"
    android:layout_gravity = "start" ?app:menu = "@menu/my_navigation_menu"
app:headerLayout = "@layout/my_drawer_header"/>
</android.support.v4.widget.DrawerLayout >
```

用于 NavigationView 的典型 menu 文件，应该是一个可选中菜单项的集合，菜单中 checked 为 true 的 item 将会高亮显示，以便用户知道当前选中的菜单项是哪个。item 的选中状态可以在代码中设置。

调用 NavigationView 对象的 setNavigationItemSelectedListener 方法来设置当导航项被单击时的回调。NavigationView. OnNavigationItemSelectedListener(MenuItem menuItem)会提供被选中的 MenuItem，这与 Activity 的 onOptionsItemSelected 非常类似。通过这个回调方法处理单击事件，例如改变 item 的选中状态、更新页面内容、关闭导航菜单等。

11.8 水平翻页

android. support. v4. view. ViewPager 是 android. view. ViewGroup 的直接子类，是一个允许用户左右滑动数据页面的布局管理器。可以实现 PagerAdapter 来生成视图要显示的页面，给 ViewPager 提供数据。ViewPager 更多的时候会与 Fragment 一起使用，这是管理各个页面的生命周期的一种很方便的方法。Android 提供了一些标准的适配器：

FragmentPagerAdapter 与 FragmentStatePagerAdapter，来让 ViewPager 与 Fragment 一起工作。FragmentPagerAdapter 适用于面数量比较少且相对静态的页的情形，例如几个 Tab 页。如果需要处理较多的页面集合，并且数据动态性较大、占用内存较多的情况，应该使用 FragmentStatePagerAdapter。

PagerTitleStrip、PagerTabStrip 都是 ViewPager 的一个关于当前页面、上一个页面和下一个页面的指示器。pagerTitleStrip 是非交互的指示器，而 PagerTabStrip 是交互的指示器。PagerTabStrip 的 Tab 是可以单击的，当用户单击某一个 Tab 时，就会跳转到这个页面，并在标题下面画线来提示当前页面的 Tab 是哪个，而 PagerTitleStrip 则没有 Tab 的单击和画线功能。图 11-6 显示了 PagerTabStrip 类的继承关系。

java.lang.Object
android.view.View
android.view.ViewGroup
android.support.v4.view.PagerTitleStrip
↳ android.support.v4.view.PagerTabStrip

图 11-6 PagerTabStrip 类的继承关系

在 XML 布局文件中，PagerTitleStrip、PagerTabStrip 都必须用作 ViewPager 的子元素，并且要把它的 layout_gravity 属性设置为 TOP 或 BOTTOM 以便把它显示在 ViewPager 的顶部或底部。每个页面的标题是通过 ViewPager 的适配器的 getPageTitle(int)方法提供的。PagerTitleStrip 的定义如下：

```
@ViewPager.DecorView
public class PagerTitleStrip extends ViewGroup {…}
```

PagerTitleStrip 的定义前有注解@ViewPager.DecorView，带有@ViewPager.DecorView 注解的视图都可以看成是 ViewPager 的装饰的一部分，每个装饰视图的位置通过它的 layout_gravity 属性控制。

TabLayout 可以和任何种类的子视图一起使用，作为 Activity/Fragment 的一部分；TabLayout 提供了 setupWithViewPager(ViewPager viewPager)方法用于附着到 ViewPager，能够代替 PagerTabStrip 作为 ViewPager 的一部分。ViewPager 的示例如下：

```
<android.support.v4.view.ViewPager
  android:layout_width = "match_parent" ?android:layout_height = "match_parent">
  <android.support.v4.view.PagerTitleStrip android:id = "@ + id/pagerTitleStrip"
    android:layout_width = "match_parent" ?android:layout_height = "wrap_content"
    android:layout_gravity = "top" />
</android.support.v4.view.ViewPager>
```

PagerAdapter 是用于将多个页面填充到 ViewPager 的适配器的一个抽象类，大多数情况下，可能更倾向于使用 PagerAdapter 类的更加特定的实现，例如 FragmentPagerAdapter 或者 FragmentStatePagerAdapter。ViewPager 通过 setAdapter()方法来建立与 PagerAdapter 的联系。这个联系是双向的，一方面，ViewPager 会拥有 PagerAdapter 对象，从而可以在需要时调用 PagerAdapter 的方法来取得所需显示的页；另一方面，ViewPager 会在 setAdapter()中调用 PagerAdapter 的 registerDataSetObserver()方法，注册 PagerObserver 对象，从而在 PagerAdapter 的数据变化时(如 notifyDataSetChanged()时)，可以调用 Observer 的 onChanged()方法，从而实现 PagerAdapter 向 ViewPager 方向发送信息。实现 PagerAdapter 通常需要重写的方法，如表 11-9 所示。

表 11-9 实现 PagerAdapter 通常需要重写的方法

返回值类型	方法及解释
public Object	instantiateItem(ViewGroup container,int position) 参数 container：容器视图，在容器视图里显示页面 参数 position：要被初始化的页的位置 返回值：返回一个代表新增加的视图页面的 Object，可以返回视图本身，也可以返回这个页面的其他容器，即代表当前页面的任意值，只要与新增加的视图一一对应即可。 该方法实现的功能是创建 position 位置的页面视图。在每次 ViewPager 需要显示新的 Object 关联的页面的时候，该函数都会被 ViewPager. addNewItem()调用。适配器负责把新创建的 View 视图添加到 container 容器中，确保在 finishUpdate(viewGroup)返回前创建的页面视图已经添加到 container 容器中
abstract Boolean	isViewFromObject(View,Object) 该方法用来判断 View 是否和给定的 Key 对象(instantiateItem 返回的 key 对象)关联。如果 view 和 object 是关联的，返回 True，否则返回 false
abstractint	getCount()返回可得到的视图的数量
void	destroyItem(ViewGroup container,int position,Object object) 该方法用于移除 position 位置的页面，从 container 中移除 object 关联的页面。适配器负责从容器中移除视图页面，必须确保从 finishUpdate(ViewGroup)返回前就完成页面的移除工作

PagerAdapter 的方法 getPageTitle(int position)可以被 ViewPager 调用，获得 position 位置的页面的标题字符串。这个方法可以返回 null，表示指定位置的页面没有标题。默认的实现就是返回 null，表示指定位置的页面没有标题。

在数据集发生变化、视图需要更新的时候，Activity 会调用 PagerAdapter 的方法 notifyDataSetChanged()，通知注册过的所有 DataSetObserver。其中之一就是在 ViewPager. setAdapter()中注册过的 PageObserver。PageObserver 则进而调用 ViewPager. dataSetChanged()，从而导致 ViewPager 开始触发更新其内部 View 的操作。

ViewPager 使用回调函数来指示视图更新的步骤，而不是直接提供视图的回收机制，如果需要，PagerAdapter 也可以实现视图的回收机制，或者采用更复杂的方法(例如 Fragment 事务，每一页表示成一个 Fragment)来管理页面视图。

ViewPager 并不直接管理页面视图，而是通过一个 key 把每个页面关联起来。这个 key 用来跟踪和唯一标识一个给定的页面，但与该页面在 adapter 里的位置独立。PagerAdapter 的 startUpdate(ViewGroup)方法一旦被调用，就标志着 ViewPager 的内容即将开始改变。紧接着，将会一次或者多次调用 instantiateItem(ViewGroup, int)和/或

destroyItem(ViewGroup,int,Object)方法,然后调用 finishUpdate(ViewGroup)就意味着这一次更新的结束。在 finishUpdate(ViewGroup)方法返回前,与 instantiateItem(ViewGroup, int)方法返回的 key 对象相关联的视图将会被加入到父 ViewGroup 中,而与传递给 destroyItem(ViewGroup, int, Object)方法的 key 对象相关联的视图将会被移除。方法 isViewFromObject(View,Object)则判断一个视图是否与一个给定的 key 对象相关联。

一个简单的 PagerAdapter 的实现可以选择将页面视图本身作为 key 对象,在创建视图并加入父 ViewGroup 之后通过 instantiateItem(ViewGroup,int)返回。这种情况下,一个匹配的 destroyItem(ViewGroup,int,Object)的实现方法只需要把 View 从父 ViewGroup 中移除即可,而 isViewFromObject(View,Object)的实现方法可以直接写成"return view == object"。

PagerAdapter 支持数据集的改变。数据集的改变必须放在主线程中,并且在结束时调用 notifyDataSetChanged()方法,这与 AdapterView 类似。一个数据集的改变包含了页面的添加、移除或者位置改变。只要适配器实现 getItemPosition(Object)方法,ViewPager 就会保持当前页面处于活跃状态。

android. support. v4. app. FragmentPagerAdapter 和 android. support. v4. app. FragmentStatePagerAdapter 都是 PagerAdapter 的直接子类,这两个类都是抽象类,都采用 Fragment 来管理页面,每个 Fragment 代表一个页面。相比通用的 PagerAdapter,这两个类更专注于每一页均为 Fragment 的情况。FragmentPagerAdapter 适合只有少量的静态的 Fragment 的情况。FragmentPagerAdapter 的视图层级可能会在不可见时被销毁,在需要时重新装载,即当 Fragment 不再需要时,FragmentPagerAdapter 不是调用 remove (Fragment)而是调用 detach(Fragment),销毁了 Fragment 的视图层级。只要用户可能返回到这个页面,这个 Fragment 页面就会持续保存在 FragmentManager 中。因而尽管它的视图层级可能在不可看见时被销毁,FragmentPagerAdapter 创建过的 Fragment 都会被保存在内存中,这可能消耗大量的内存,因此只适用于面数量比较少且相对静态的页的情形。如果需要处理较多的页面集合,并且数据动态性较大、占用内存较多的情况,应该使用 FragmentStatePagerAdapter。

与 FragmentPagerAdapter 不同的是,FragmentStatePagerAdapter 工作更加类似于 ListView,当页面对用户不可见时,Fragment 被完全销毁,进行一次事务提交,销毁时把状态保存到 Bundle 中。如果用户需要返回那一页,就需要获得保存过的状态以创建新的 Fragment。例如,书本阅读器应用并不需要把所有的 Fragment 一起装入内存,这时应该选用 FragmentStatePagerAdapter。

FragmentPagerAdapter 或者 FragmentStatePagerAdapter 重载实现了 PagerAdapter 几个必需的方法,PagerAdapter 的方法只有 getCount()需要实现。另外 FragmentPagerAdapter. instantiateItem()的实现中,调用了一个新增的抽象方法 getItem(int position),返回与 position 位置相关联的 Fragment。因此,FragmentPagerAdapter 或者 FragmentStatePagerAdapter 派生类都只需要实现 getCount()和 getItem(int)即可。

android. support. design. widget. TabLayout 提供了一个显示 Tab 的水平布局。TabLayout 类的继承关系如图 11-7 所示。TabLayout 是 HorizontalScrollView 的子类, HorizontalScrollView 是一个仅仅支持水平滚动的布局容器,允许它比实际的物理显示更大一些。HorizontalScrollView 只能有一个直接子视图,这个子视图可以是一个布局管理器，拥有复杂的对象层级关系，经常使用的子视图是水平方向的 LinearLayout。

java.lang.Object					
	android.view.View				
		android.view.ViewGroup			
			android.widget.FrameLayout		
				android.widget.HorizontalScrollView	
					android.support.design.widget.TabLayout

图 11-7 TabLayout 类的继承关系

下面是采用 HorizontalScrollView 的一个布局文件，以这个文件作为视图的 Activity，这个视图可以水平滚动显示。

```
<?xml version = "1.0" encoding = "utf - 8"?>
< HorizontalScrollView
xmlns:android = "http://schemas.android.com/apk/res/android"
    android:layout_width = "match_parent" android:layout_height = "match_parent">
    < LinearLayout android:orientation = "horizontal"
        android:layout_width = "match_parent"
anroid:layout_height = "wrap_content">
    < Button android:text = "1"
        android:layout_width = "wrap_content"
android:layout_height = "wrap_content" />
    < Button android:text = "2"
        android:layout_width = "wrap_content"
android:layout_height = "wrap_content" />
    < Button android:text = "3"
        android:layout_width = "wrap_content"
android:layout_height = "wrap_content" />
    < Button android:text = "4"
        android:layout_width = "wrap_content"
android:layout_height = "wrap_content" />
    < Button android:text = "5"
        android:layout_width = "wrap_content"
android:layout_height = "wrap_content" />
    < Button android:text = "6"
        android:layout_width = "wrap_content"
android:layout_height = "wrap_content" />
    < Button android:text = "7"
        android:layout_width = "wrap_content"
android:layout_height = "wrap_content" />
    < Button android:text = "8"
        android:layout_width = "wrap_content"
android:layout_height = "wrap_content" />
    < Button android:text = "9"
        android:layout_width = "wrap_content"
android:layout_height = "wrap_content" />
```

```
</LinearLayout>
</HorizontalScrollView>
```

TextView 也能管理自己的滚动，所以不需要 HorizontalScrollView，把 TextView 放在 HorizontalScrollView 中可以获得一个较大的容器中文本显示的效果。

TabLayout 是材料设计的概念，用于显示可切换的标签效果，用来代替已经废弃的 Actionbar. Tab 导航效果，可以在水平方向上添加 Tab，Tab 可以包含文字、图标或者自定义的视图。TabLayout 通常用作 ViewPager 的页面指示器。当同时使用 ViewPager 和 TabLayout 时，应该先为 viewPager 设置适配器，再调用 TabLayout 对象的方法 setupWithViewPager(ViewPager)把 ViewPager 和 TabLayout 联系起来，TabLayout 将从 PagerAdapter 的页标题中自动获取标题，代替 PagerTabStrip 作为 ViewPager 的标签指示。

TabLayout 元素可以放在 ViewPager 元素的外面，它的滚动是和页面分开的。TabLayout 也支持作为 ViewPager 的装饰视图的一部分，在布局文件中把 TabLayout 作为 ViewPager 的子元素。示例代码如下：

```
<android.support.v4.view.ViewPager
    android:layout_width="match_parent"
    android:layout_height="match_parent">
    <android.support.design.widget.TabLayout
app:tabMode="scrollable"
        android:layout_width="match_parent"
        android:layout_height="wrap_content"
        android:layout_gravity="top" />
</android.support.v4.view.ViewPager>
```

需要把 TabLayout 的 tabMode 属性设置为 scrollable，tabMode 的默认值是 fixed，不能滑动，当标签超过屏幕实际宽度时被挤压。

TabLayout 提供了一个水平的布局用来展示选项卡 Tab。可以通过调用 TabLayout 对象的 newTab()方法来创建 Tab，然后调用 setText(int)和 setIcon(int)设置 Tab 的标签和图标，最后调用 TabLayout 对象的 addTab(Tab)方法添加到布局中。例如：

```
tabLayout = (TabLayout) findViewById(R.id.tabLayout);
tabLayout.addTab(tabLayout.newTab().setText("tab1").setIcon(R.mipmap.ic_launcher));
tabLayout.addTab(tabLayout.newTab().setText("tab2").setIcon(android.R.drawable.ic_
delete));
```

调用 TabLayout 对象的 setOnTabSelectedListener(OnTabSelectedListener)方法可以为 TabLayout 对象设置监听器，以便当选项卡的选择状态改变时得到通知。

11.9 实验：一个 View 跟着另一个 View 移动

实验目的：实现在 CoordinatorLayout 中一个 View 监听 view 的大小、位置、显示状态变化。

1. 新建工程 CoordinatorLayout1

具体步骤：运行 Android Studio，单击 Start a new Android Studio project，在 Create New Project 页中，Application name 为 CoordinatorLayout1，Company domain 为 cuc，Project location 为 C:\android\myproject\CoordinatorLayout1，单击 Next 按钮，选中 Phone and Tablet，minimum sdk 设置为 API 19：Android 4.4(minSdkVersion)，单击 Next 按钮，选择 Empty Activity，再单击 Next 按钮，勾选 Generate Layout File 选项，勾选 Backwards Compatibility(AppCompat)选项，单击 Finish 按钮。

2. 添加 com.android.support：design 库

因为在布局文件中需要使用 android.support.design.widget.CoordinatorLayout 作为根结点，所以需要 design 库的支持。可以通过菜单添加需要的库，库的版本应该和 compileSdkVersion 版本一致。Android Studio 新建工程的时候要求选择最低支持的 SDK 版本，默认的目标编译 SDK 版本会以系统当前 SDK 中最新 SDK platform 作为目标的 API Level。具体步骤为：

单击 Android Studio 的菜单 File|Project Structure|app|Dependencies，单击右边的“+”，选择 1 Library dependency 选项，在输入栏中输入 design，并且按下回车键进行搜索，搜索到 cardview 库后，如果没有版本信息，就添加版本信息，例如 28.+，单击 OK 按钮完成库的添加，再单击 OK 按钮，关闭 Project Structure 界面。

也可以在 build.gradle(module：app)文件的 dependencies 中直接添加依赖：

```
implementation 'com.android.support:design:28.+'
```

由于 build.gradle 文件内容已被修改，这个文件窗口的上面就出现提示：

“Gradle files have changed since last project sync. A project sync may be necessary for the IDE to work properly. sync now”。单击 sync now 按钮进行同步。

在 sdk 的安装文件夹里有 design 库的各种版本，具体位置为 sdk\extras\android\m2repository\com\android\support\design\。

3. 创建一个名叫 DependentView 的 java 类

在工程结构视图中 main 源集 java 文件夹里右击 cuc.coordinatorlayout1 包，选择 New|Java Class，Name 输入框中输入 DependentView，Superclass 输入框中输入 android.view.View，单击 OK 按钮。DependentView.java 文件内容如下：

```
package cuc.coordinatorlayout1;
import android.content.Context;
import android.support.annotation.Nullable;
import android.support.design.widget.CoordinatorLayout;
import android.util.AttributeSet;
import android.view.MotionEvent;
import android.view.View;

public class DependentView extends View {
    private int lastX;
    private int lastY;
```

```
    private int mWidth;
    private int mHeight;
    private int screenWidth;
    private int screenHeight;
    public DependentView(Context context, @Nullable AttributeSet attrs) {
        super(context, attrs);
        mWidth = getMeasuredWidth();
        mHeight = getMeasuredHeight();
    }

    @Override
    public boolean onTouchEvent(MotionEvent event) {
        int x = (int) event.getRawX();
        int y = (int)event.getRawY();
        switch (event.getAction()){
            case MotionEvent.ACTION_DOWN:
                lastX = x;
                lastY = y;
                break;
            case MotionEvent.ACTION_UP:
                break;
            case MotionEvent.ACTION_MOVE:
               CoordinatorLayout.MarginLayoutParams
marginLayoutParams = (CoordinatorLayout.MarginLayoutParams)
                     getLayoutParams();
marginLayoutParams.leftMargin = marginLayoutParams.leftMargin + (x - lastX);
marginLayoutParams.topMargin = marginLayoutParams.topMargin + (y - lastY);
               System.out.println("leftmargin:" + marginLayoutParams.leftMargin);
                System.out.println("topMargin:" + marginLayoutParams.topMargin);
                setLayoutParams(marginLayoutParams);
                requestLayout();
                break;
        }
        lastX = x;
        lastY = y;
        return true;
    }
}
```

上面代码中，context.getResources().getDisplayMetrics()是获取手机屏幕参数。

4. 创建 Behavior 类型的类 MyBehavior

在工程结构视图中 main 源集 Java 文件夹里右击 cuc.coordinatorlayout1 包，选择 New|Java Class，Name 输入框中输入 MyBehavior1，Superclass 输入框中输入 android.support.design.widget.CoordinatorLayout.Behavior，单击 OK 按钮。MyBehavior1.java 文件内容如下：

```
package cuc.coordinatorlayout1;
import android.content.Context;
import android.support.design.widget.CoordinatorLayout;
```

```
import android.util.AttributeSet;
import android.view.View;
import android.widget.Button;

public class MyBehavior1 extends CoordinatorLayout.Behavior<Button> {
    private int width;
    public MyBehavior1(Context context, AttributeSet attrs) {
        super(context, attrs);
        width = context.getResources().getDisplayMetrics().widthPixels;
    }

    @Override
    public boolean layoutDependsOn(CoordinatorLayout parent, Button child, View
dependency) {
        return dependency instanceof DependentView;
    }
    private void setPosition(View view, int leftMargin, int topMargin){
          CoordinatorLayout.MarginLayoutParams marginLayoutParams = (CoordinatorLayout.
MarginLayoutParams)
                view.getLayoutParams();
        marginLayoutParams.leftMargin = leftMargin;
        marginLayoutParams.topMargin = topMargin;
        view.setLayoutParams(marginLayoutParams);
    }
    @Override
     public boolean onDependentViewChanged(CoordinatorLayout parent, Button child, View
dependency) {
        int leftMargin =  width - (dependency.getLeft() + child.getWidth());
        int topMargin = dependency.getTop();
        setPosition(child, leftMargin, topMargin);
        return true;
    }
}
```

5. 新建布局文件 activity_main1.xml

新建布局文件 activity_main1.xml 的具体步骤：在工程结构视图中(单击 Project 按钮可以显示或者隐藏工程结构图，使用工程结构图中的下拉列表框可以选择 Android 以显示工程逻辑结构，或者选择 Project 显示工程实际的目录结构)，右击 layout，依次选择 New|Layout resource file，File name 输入框输入 activity_main1，Root element 输入框输入 android.support.design.widget.CoordinatorLayout，Source set 为 main，单击 OK 按钮。

在布局文件 activity_main1.xml 的根结点，添加 App 名称空间，CoordinatorLayout 内部放在一个 DependentView 和一个 Button，Button 定义了 app: layout_behavior = "cuc.coordinatorlayout1.MyBehavior1"，Button 能够随着 DependentView 的移动而移动。编辑后布局文件 activity_main1.xml 内容如下：

```
<?xml version = "1.0" encoding = "utf - 8"?>
```

```
<android.support.design.widget.CoordinatorLayout
    xmlns:android="http://schemas.android.com/apk/res/android"
    xmlns:app="http://schemas.android.com/apk/res-auto"
    android:orientation="vertical" android:layout_width="match_parent"
    android:layout_height="match_parent">
    <cuc.coordinatorlayout1.DependentView
        android:layout_width="100dp"
        android:layout_height="100dp"
        android:layout_marginLeft="50dp"
        android:layout_marginTop="50dp"
        android:background="@color/colorPrimary"
        android:tag="10"/>
    <Button
        android:text="this button is Child "
        android:layout_width="wrap_content"
        android:layout_height="wrap_content"
        android:layout_marginLeft="200dp"
        android:layout_marginTop="200dp"
        android:background="@android:color/holo_orange_light"
        app:layout_behavior="cuc.coordinatorlayout1.MyBehavior1"/>
</android.support.design.widget.CoordinatorLayout>
```

6. 编辑 MainActivity.java，显示布局 activity_main1

编辑后，MainActivity.java 内容如下：

```
package cuc.coordinatorlayout1;
import android.support.v7.app.AppCompatActivity;
import android.os.Bundle;
public class MainActivity extends AppCompatActivity {
    @Override
    protected void onCreate(Bundle savedInstanceState) {
        super.onCreate(savedInstanceState);
        setContentView(R.layout.activity_main1);
    }
}
```

7. 编译运行

在手机上运行后，可以看到自定义的 DependentView 能够跟着手指移动，而定义了 app：layout_behavior ="cuc.coordinatorlayout1.MyBehavior" 的 Button 能够随着 DependentView 的移动而移动。

8. 创建 Behavior 类 ScrollToTopBehavior

具体步骤为：在工程结构视图中 main 源集 java 文件夹里右击 cuc.coordinatorlayout1 包，选择 New|Java Class，Name 输入框中输入 ScrollToTopBehavior，Superclass 输入框中输入 android.support.design.widget.CoordinatorLayout.Behavior，单击 OK 按钮。ScrollToTopBehavior.java 文件内容如下：

```
package cuc.coordinatorlayout1;
import android.content.Context;
import android.support.annotation.NonNull;
import android.support.design.widget.CoordinatorLayout;
import android.util.AttributeSet;
import android.view.View;

public class ScrollToTopBehavior extends CoordinatorLayout.Behavior {
    int offsetY = 0;
    boolean scrolling = false;
    public ScrollToTopBehavior(Context context, AttributeSet attrs) {
        super(context, attrs);
    }
    @Override
    public boolean onStartNestedScroll(@NonNull CoordinatorLayout coordinatorLayout, @
NonNull View child,
                                       @NonNull View directTargetChild, @NonNull View
target, int axes, int type) {
        return axes == View.SCROLL_AXIS_VERTICAL;
    }

    @Override
    public void onNestedScroll(@NonNull CoordinatorLayout coordinatorLayout, @NonNull
View child, @NonNull View target,
                               int dxConsumed, int dyConsumed, int dxUnconsumed, int
dyUnconsumed, int type) {
         super.onNestedScroll(coordinatorLayout, child, target, dxConsumed, dyConsumed,
dxUnconsumed, dyUnconsumed, type);
        int oldOffsetY = offsetY;
        int newOffsetY = oldOffsetY - dyConsumed;
        newOffsetY = Math.max(newOffsetY, - child.getHeight());
        newOffsetY = Math.min(0,newOffsetY);
        offsetY = newOffsetY;
        if(oldOffsetY == newOffsetY){
            scrolling = false;
            return;
        }
        child.offsetTopAndBottom(newOffsetY - oldOffsetY);
        scrolling = true;
        return;
    }
}
```

9. 新建布局文件 activity_main2.xml

具体步骤为：在工程结构视图中(单击 Project 按钮可以显示或者隐藏工程结构图，使用工程结构图中的下拉列表框可以选择 Android 以显示工程逻辑结构，或者选择 Project 显示工程实际的目录结构)，右击 Layout，依次选择 New|Layout resource file，File name 输入框输入 activity_main2，Root element 输入框输入 android.support.design.widget.CoordinatorLayout，

Source set 为 main，单击 OK 按钮。并编辑 activity_main2. xml，编辑后，activity_main2. xml 文件内容如下：

```
<?xml version="1.0" encoding="utf-8"?>
<android.support.design.widget.CoordinatorLayout
xmlns:app="http://schemas.android.com/apk/res-auto"
    xmlns:android="http://schemas.android.com/apk/res/android" android:layout_width=
"match_parent"
    android:layout_height="match_parent">
    <android.support.v4.widget.NestedScrollView
        android:id="@+id/second"
        android:layout_width="match_parent"
        android:layout_height="match_parent">
        <TextView
            android:layout_width="match_parent"
            android:layout_height="wrap_content"
            android:layout_marginTop="128dp"
            style="@style/TextAppearance.AppCompat.Display3"
            android:text="1 A\n 2 A\n 3 A\n 4 A\n 5 A\n 6 A\n 7 A\n 8 A\n 9 A\n 10 A\n
11 A\n 12 A\n 13 A\n 14 A\n 15 A\n" />
    </android.support.v4.widget.NestedScrollView>
    <View
        android:layout_width="match_parent"
        android:layout_height="128dp"
        app:layout_behavior="cuc.coordinatorlayout1.ScrollToTopBehavior"
        android:background="@android:color/holo_blue_light"/>
</android.support.design.widget.CoordinatorLayout>
```

10. 编辑 MainActivity. java，显示 activity_main2，编译运行

把 onCreate 方法中的 setContentView(R. layout. activity_main1)更改为 setContentView(R. layout. activity_main2)，并编译运行。

11.10 实验：ToolBar 当 ActionBar 使用

实验目的：

(1) 熟悉 ToolBar 的使用和 app：layout_scrollFlags 属性。

(2) 熟悉 OptionMenu 和 ContextMenu 的使用。

1. 新建工程 ToolBarAsActionBar1

具体步骤：运行 Android Studio，单击 Start a new Android Studio project，在 Create New Project 页中，Application name 为 ToolBarAsActionBar1，Company domain 为 cuc，Project location 为 C:\android\myproject\ToolBarAsActionBar1，单击 Next 按钮，选中 Phone and Tablet，minimum sdk 设置为 API 19：Android 4.4(minSdkVersion)，单击 Next 按钮，选择 Basic Activity，再单击 Next 按钮，勾选 Generate Layout File 选项，勾选 Backwards Compatibility (AppCompat)选项，不用勾选 Use a Fragment 选项，单击 Finish 按钮。这时候，Android Studio 已经根据 compileSdkVersion 版本 28 在 build. gradle(module：app)添加了同样版本的依赖

com. android. support：design：28. +，并同步了相应的库。build. gradle(module：app)内容如下：

```
apply plugin: 'com.android.application'
android {
    compileSdkVersion 28
    defaultConfig {
        applicationId "cuc.toolbarasactionbar1"
        minSdkVersion 19
        targetSdkVersion 28
        versionCode 1
        versionName "1.0"
        testInstrumentationRunner "android.support.test.runner.AndroidJUnitRunner"
    }
    buildTypes {
        release {
            minifyEnabled false
             proguardFiles getDefaultProguardFile('proguard-android.txt'), 'proguard-
rules.pro'
        }
    }
}
dependencies {
    implementation fileTree(dir: 'libs', include: ['*.jar'])
    implementation 'com.android.support:appcompat-v7:28.+'
    implementation 'com.android.support.constraint:constraint-layout:1.0.2'
    implementation 'com.android.support:design:28.+'
    testImplementation 'junit:junit:4.12'
    androidTestImplementation 'com.android.support.test:runner:1.0.1'
    androidTestImplementation 'com.android.support.test.espresso:espresso-core:3.0.1'
}
```

2. 修改布局文件 activity_main. xml

编辑后，activity_main. xml 内容如下：

```
<?xml version="1.0" encoding="utf-8"?>
<android.support.design.widget.CoordinatorLayout
xmlns:android="http://schemas.android.com/apk/res/android"
    xmlns:app="http://schemas.android.com/apk/res-auto"
    xmlns:tools="http://schemas.android.com/tools"
    android:layout_width="match_parent"
    android:layout_height="match_parent"
    tools:context="cuc.toolbarasactionbar1.MainActivity">

    <android.support.design.widget.AppBarLayout
        android:layout_width="match_parent"
        android:layout_height="wrap_content"
        android:theme="@style/AppTheme.AppBarOverlay">
```

```
        <android.support.v7.widget.Toolbar
            app:layout_scrollFlags = "scroll|enterAlways"
            android:id = "@ + id/toolbar"
            android:layout_width = "match_parent"
            android:layout_height = "?attr/actionBarSize"
            android:background = "?attr/colorPrimary"
            app:popupTheme = "@style/AppTheme.PopupOverlay" />
        <TextView
            android:layout_width = "match_parent"
            android:layout_height = "wrap_content"
            android:text = "donot scroll"
            android:textSize = "30sp" />

    </android.support.design.widget.AppBarLayout>
    <TextView
        android:id = "@ + id/leftTxt"
        android:layout_gravity = "top|left"
        android:layout_width = "wrap_content"
        android:layout_height = "wrap_content"
        android:textSize = "40sp"

    android:text = "0\n1\n2\n3\n4\n5\n6\n7\n8\n9\na\nb\nc\nd\ne\nf\ng\nh
        \ni\nj\nk\nl\nm\nn\no\np\nq\nr\ns\nt\nu\nv\nw\nx\ny\nz"/>
    <TextView
        android:id = "@ + id/centerTxt"
        android:layout_gravity = "top|center_horizontal"
        app:layout_behavior = "@string/appbar_scrolling_view_behavior"
        android:layout_width = "wrap_content"
        android:layout_height = "wrap_content"
        android:textSize = "40sp"

    android:text = "0\n1\n2\n3\n4\n5\n6\n7\n8\n9\na\nb\nc\nd\ne\nf\ng\nh
        \ni\nj\nk\nl\nm\nn\no\np\nq\nr\ns\nt\nu\nv\nw\nx\ny\nz" />
    <android.support.v4.widget.NestedScrollView
        app:layout_behavior = "AppBarLayout $ ScrollingViewBehavior"
        android:layout_width = "wrap_content"
        android:layout_height = "wrap_content"
        android:layout_gravity = "top|right">
        <TextView
            android:id = "@ + id/rightTxt"
            android:layout_width = "wrap_content"
            android:layout_height = "wrap_content"
            android:textSize = "40sp"

    android:text = "0\n1\n2\n3\n4\n5\n6\n7\n8\n9\na\nb\nc\nd\ne\nf\ng\nh
        \ni\nj\nk\nl\nm\nn\no\np\nq\nr\ns\nt\nu\nv\nw\nx\ny\nz"/>
    </android.support.v4.widget.NestedScrollView>

    <android.support.design.widget.FloatingActionButton
        android:id = "@ + id/fab"
        android:layout_width = "wrap_content"
        android:layout_height = "wrap_content"
```

```
        android:layout_gravity = "bottom|end"
        android:layout_margin = "@dimen/fab_margin"
        app:srcCompat = "@android:drawable/ic_dialog_email" />
</android.support.design.widget.CoordinatorLayout>
```

在上面的布局文件中，AppBarLayout 里面，有一个 Toolbar 和一个 TextView，Toolbar 设置了 app: layout_scrollFlags = "scroll | enterAlways"，是可以向上滚动出屏幕的，TextView 没有设置 app: layout_scrollFlags 标志，因而是不能向上滚动出屏幕的，不能向上滚动出屏幕的子视图应该放在可以滚动出屏幕的下方。AppBarLayout 是一个垂直方向的线性布局，所以需要向上滚动出屏幕的子视图在布局文件中先出现。

在上面的布局文件中，AppBarLayout 后面，有两个 TextView 和一个嵌入了 TextView 的 NestedScrollView，分别设置了 android: layout_gravity = "top | left"、"top | center_horizontal"和"top | right"，所以位置分别在父视图的左边、水平中间和右边。左边的 TextView 没有设置 app: layout_behavior，所以如果用手滚动 AppBarLayout，左边的这个 TextView 不会随着 AppBarLayout 的滚动而滚动。中间的 TextView 和右边的 NestedScrollView 都设置了 app: layout_behavior = "android.support.design.widget.AppBarLayout$ScrollingViewBehavior"，所以如果用手滚动 AppBarLayout，都能够随着 AppBarLayout 的滚动而滚动。另外由于右边的 TextView 嵌入在 NestedScrollView 内部，所以 NestedScrollView 内部的 TextView 是可以用手指滚动的，向上滚动 NestedScrollView 时，AppBarLayout 后里面设置了 app: layout_scrollFlags="scroll"的子视图可以向上滚动出屏幕。向下滚动 NestedScrollView 时，AppBarLayout 后里面滚动出屏幕外面的子视图可以滚动进入屏幕。当然实际的图形界面，应该使用嵌入滚动的视图，左边和中间的两个 TextView 没有放在 NestedScrollView，不能滚动是没有实际意义的。

3. 修改菜单资源文件 menu_main.xml

系统创建的文件只有一个菜单项，另外添加几个菜单项。修改后，菜单资源文件 menu_main.xml 内容如下：

```
<menu xmlns:android = "http://schemas.android.com/apk/res/android"
    xmlns:app = "http://schemas.android.com/apk/res-auto"
    xmlns:tools = "http://schemas.android.com/tools"
    tools:context = "cuc.toolbarasactionbar1.MainActivity">
    <item android:id = "@+id/item1"
        android:title = "item1"
        android:icon = "@android:drawable/ic_dialog_email"
        app:showAsAction = "ifRoom|withText"/>
    <item
        android:id = "@+id/action_settings"
        android:orderInCategory = "120"
        android:title = "@string/action_settings"
        app:showAsAction = "never" />
    <group android:id = "@+id/group">
        <item android:id = "@+id/group_item1"
            android:onClick = "onClickMy"
```

```
            android:title = "group_item1"
            android:icon = "@android:drawable/ic_input_add" />
        < item android:id = "@ + id/group_item2"
            android:onClick = "onClickMy"
            android:title = "group_item2"
            android:icon = "@android:drawable/ic_input_delete" />
    </group >
    < item android:id = "@ + id/submenu"
        android:title = "submenu_title"
        app:showAsAction = "ifRoom|withText" >
        < menu >
            < item android:id = "@ + id/submenu_item1"
                android:title = "submenu_item1" />
            < item android:id = "@ + id/submenu_item2"
                android:title = "submenu_item1" />
        </menu >
    </item >
</menu >
```

其中 app：showAsAction＝"ifRoom"表示如果工具栏有空间，就在工具栏显示。app：showAsAction＝"never"表示永远不会显示在工具栏里，而是隐藏起来，单击工具栏最右边的扩展，才能显示相应的图标。

4. 新建两个菜单资源文件，用作 ContextMenu

新建两个菜单资源文件 contextmenu. xml 和 contextmenu2. xml，分别用作中间的文本框 centerTxt 和右边的文本框 rightTxt 的上下文菜单。

新建菜单文件 contextmenu. xml 的具体步骤：在工程结构视图中(单击 Project 按钮可以显示或者隐藏工程结构图，使用工程结构图中的下拉列表框可以选择 Android 以显示工程逻辑结构，或者选择 Project 显示工程实际的目录结构)，右击 menu 文件夹，依次选择 New|Menu resource file，File name 输入框输入 contextmenu，Source set 为 main，单击 OK 按钮。contextmenu. xml 内容如下：

```
<?xml version = "1.0" encoding = "utf - 8"?>
< menu xmlns:android = "http://schemas.android.com/apk/res/android">
    < item android:id = "@ + id/context_item_textColorRed"
        android:title = " textColorRed"/>
    < item android:id = "@ + id/context_item_textColorGreen"
        android:title = "textColorGreen"/>
</menu >
```

同样的方法，创建 contextmenu2. xml。contextmenu2. xml 文件内容如下：

```
<?xml version = "1.0" encoding = "utf - 8"?>
< menu xmlns:android = "http://schemas.android.com/apk/res/android">
    < item android:id = "@ + id/context2_item_backgroundColorRed"
        android:title = " backgroundColorRed"/>
```

```
    <item android:id="@+id/context2_item_backgroundColorGreen"
        android:title=" backgroundColorGreen"/>
</menu>
```

5. 编辑 mainActivity. java

首先使用插件从菜单资源文件 menu_main. xml 生成菜单，并拷贝到 mainActivity. java 文件中。具体步骤如下：在工程结构视图中选中 menu_main. xml，右击，选择 Generate Android code|Menu，单击 Copy code to Clipboard 按钮，然后选中文件 MainActivity. java 中的对应部分，右击，选择 paste 选项以便用 Clipboard 中内容替换掉 MainActivity. java 中的对应部分。并添加 menu_main. xml 文件中由 android：onClick="onClickMy"声明的菜单响应函数 onClickMy，然后在 MainActivity 类的 onCreate(Bundle savedInstanceState)方法中调用 registerContextMenu1()实现上下文菜单的注册。最后 MainActivity. java 内容如下：

```
package cuc.toolbarasactionbar1;
import android.graphics.Color;
import android.os.Bundle;
import android.support.design.widget.FloatingActionButton;
import android.support.design.widget.Snackbar;
import android.support.v7.app.AppCompatActivity;
import android.support.v7.widget.Toolbar;
import android.view.ContextMenu;
import android.view.Menu;
import android.view.MenuItem;
import android.view.View;
import android.widget.TextView;
import android.widget.Toast;

public class MainActivity extends AppCompatActivity {
    private Toolbar toolbar;
    private FloatingActionButton fab;
    private TextView leftTxt;
    private TextView centerTxt;
    private TextView rightTxt;
    private void assignViews() {
        toolbar = (Toolbar) findViewById(R.id.toolbar);
        fab = (FloatingActionButton) findViewById(R.id.fab);
        leftTxt = (TextView) findViewById(R.id.leftTxt);
        centerTxt = (TextView) findViewById(R.id.centerTxt);
        rightTxt = (TextView) findViewById(R.id.rightTxt);
    }
    void registerContextMenu1(){
        centerTxt.setOnCreateContextMenuListener(new View.OnCreateContextMenuListener() {
            @Override
             public void onCreateContextMenu(ContextMenu contextMenu, View view, ContextMenu.
ContextMenuInfo contextMenuInfo) {
```

```
                getMenuInflater().inflate(R.menu.contextmenu1,contextMenu);
            }
        });
        rightTxt.setOnCreateContextMenuListener(new View.OnCreateContextMenuListener() {
            @Override
            public void onCreateContextMenu(ContextMenu contextMenu, View view, ContextMenu.
ContextMenuInfo contextMenuInfo) {
                getMenuInflater().inflate(R.menu.contextmenu2,contextMenu);
            }
        });
    }

    @Override
    public boolean onContextItemSelected(MenuItem menuItem) {
        switch (menuItem.getItemId()) {
            case R.id.context_item_textColorRed:
             Toast.makeText(this, "context_item1", Toast.LENGTH_SHORT).show();
                centerTxt.setTextColor(Color.RED);
                break;
            case R.id.context_item_textColorGreen:
             Toast.makeText(this, "context_item2", Toast.LENGTH_SHORT).show();
                centerTxt.setTextColor(Color.GREEN);
                break;
            case R.id.context2_item_backgroundColorRed:
            Toast.makeText(this, "context2_item1", Toast.LENGTH_SHORT).show();
                rightTxt.setBackgroundColor(Color.RED);
                break;
            case R.id.context2_item_backgroundColorGreen:
            Toast.makeText(this, "context2_item2", Toast.LENGTH_SHORT).show();
                rightTxt.setBackgroundColor(Color.GREEN);
                break;
        }
        return true;
    }
    @Override
    protected void onCreate(Bundle savedInstanceState) {
        super.onCreate(savedInstanceState);
        setContentView(R.layout.activity_main);
        Toolbar toolbar = (Toolbar) findViewById(R.id.toolbar);
        setSupportActionBar(toolbar);
        assignViews();
        registerContextMenu1();

        fab = (FloatingActionButton) findViewById(R.id.fab);
        fab.setOnClickListener(new View.OnClickListener() {
            @Override
            public void onClick(View view) {
                Snackbar.make(view, "Replace with your own action", Snackbar.LENGTH_LONG)
                        .setAction("Action", null).show();
```

```
            }
        });
    }

    @Override
    public boolean onCreateOptionsMenu(Menu menu) {
      // Inflate the menu; this adds items to the action bar if it is present.
        getMenuInflater().inflate(R.menu.menu_main, menu);
        return true;
    }
    public void onClickMy(MenuItem item) {
        switch (item.getItemId()){
            case R.id.group_item1:
                Toast.makeText(this, "R.id.group_item1,onClickMy", Toast.LENGTH_LONG).
show();
                break;
            case R.id.group_item2:
                Toast.makeText(this, "R.id.group_item2,onClickMy", Toast.LENGTH_LONG).
show();
                break;
        }
    }
    @Override
    public boolean onOptionsItemSelected(MenuItem item) {
        switch (item.getItemId()) {
            case R.id.item1:
                Toast.makeText(this, "item1", Toast.LENGTH_LONG).show();
                return true;
            case R.id.action_settings:
            Toast.makeText(this, "action_settings", Toast.LENGTH_LONG).show();
                return true;
            case R.id.group:
                Toast.makeText(this, "group", Toast.LENGTH_LONG).show();
                return true;
            case R.id.group_item1:
                Toast.makeText(this, "group_item1", Toast.LENGTH_LONG).show();
                return true;
            case R.id.group_item2:
                Toast.makeText(this, "group_item2", Toast.LENGTH_LONG).show();
                return true;
            case R.id.submenu:
                Toast.makeText(this, "submenu", Toast.LENGTH_LONG).show();
                return true;
            case R.id.submenu_item1:
              Toast.makeText(this, "submenu_item1", Toast.LENGTH_LONG).show();
                return true;
            case R.id.submenu_item2:
              Toast.makeText(this, "submenu_item2", Toast.LENGTH_LONG).show();
                return true;
```

```
            }
            return super.onOptionsItemSelected(item);
        }
    }
```

在上面的 registerContextmenu1 方法中，调用 View.setOnCreateContextMenuListener 方法的参数使用了匿名内部类的方法。也可以让最外层的 MainActivity 类实现 View.OnCreateContextMenuListener 接口，重写 onCreateContextMenu 方法，并采用这个对象作为监听器。在 MainActivity 类中添加以下代码，并在 onCreate 方法中调用 registerContextMenu2 方法（替代 registerContextMenu1）也能达到注册 ContextMenu 的效果。

```
@Override
public void onCreateContextMenu(ContextMenu menu, View view, ContextMenu.ContextMenuInfo
menuInfo) {
    super.onCreateContextMenu(menu, view, menuInfo);
    switch (view.getId()){
        case R.id.centerTxt:
            getMenuInflater().inflate(R.menu.contextmenu1,menu);
            break;
        case R.id.rightTxt:
            getMenuInflater().inflate(R.menu.contextmenu2,menu);
            break;
    }
}
void registerContextMenu2(){
    centerTxt.setOnCreateContextMenuListener(this);
    rightTxt.setOnCreateContextMenuListener(this);
}
```

6. 编译运行

运行后，可以发现，用手指按住 AppBarLayout 并向上滑动，AppBarLayout 中的 ToolBar 可以向上滑出屏幕，AppBarLayout 中的文本框却不能向上滑出屏幕。这是由于 ToolBar 设置了 app：layout_scrollFlags＝"scroll|enterAlways"，而 AppBarLayout 中的文本框没有设置 app：layout_scrollFlags 属性。

手指按住 AppBarLayout 向上滑动时，AppBarLayout 下面的左边的文本框 leftTxt 不能跟随 AppBarLayout 的滑动而滑动，而中间的文本框 centerTxt 和右边的 NestedScrollView 都能跟随 AppBarLayout 的滑动而滑动，这是由于左边的文本框 leftTxt 没有设置 app：layout_behavior 属性，而中间的文本框 centerTxt 和右边的 NestedScrollView 都设置了 app：layout_behavior＝"@string/appbar_scrolling_view_behavior"。

手指按住 NestedScrollView 向上滑动时，AppBarLayout 能跟着向上滑动，这是由于 NestedScrollView 实现了 NestedScrollingChild 接口，CoordinatorLayout 实现了 NestedScrollingParent2 接口，AppBarLayout 类通过注解@CoordinatorLayout.DefaultBehavior(AppBarLayout.Behavior.class)定义了默认的 Behavior。查看 sdk 中的源代码，可以看到 AppBarLayout 类的定义如下：

```
@CoordinatorLayout.DefaultBehavior(AppBarLayout.Behavior.class)
public class AppBarLayout extends LinearLayout {
...
}
```

11.11 实验：一个 NestedScrollView 跟随另一个垂直滚动

实验目的：学习自己重写 CoordinatorLayout.Behavior。

1. 新建工程 NestedScrollView1

具体步骤：运行 Android Studio，单击 Start a new Android Studio project，在 Create New Project 页中，Application name 为 NestedScrollView1，Company domain 为 cuc，Project location 为 C:\android\myproject\NestedScrollView1，单击 Next 按钮，选中 Phone and Tablet，minimum sdk 设置为 API 19：Android 4.4(minSdkVersion)，单击 Next 按钮，选择 Empty Activity，再单击 Next 按钮，不勾选 Generate Layout File 选项，勾选 Backwards Compatibility(AppCompat)选项，单击 Finish 按钮。

2. 添加 com.android.support：design 库

因为在布局文件中需要使用 android.support.design.widget.CoordinatorLayout 作为根结点，所以需要 design 库的支持。可以通过菜单添加需要的库，库的版本应该和 build.gradle(module：app)文件中 compileSdkVersion 版本一致。具体步骤为：

单击 Android Studio 的菜单 File|Project Structure|app|Dependencies，单击右边的“+”，选择 1 Library dependency 选项，在输入栏中输入 design，并且按下回车键进行搜索，搜索到 cardview 库后，如果没有版本信息，就添加版本信息，例如 28.+，单击 OK 按钮完成库的添加，再单击 OK 按钮，关闭 Project Structure 界面。

也可以在 build.gradle(module：app)文件的 dependencies 中直接添加依赖：

```
implementation 'com.android.support:design:28.+'
```

由于 build.gradle 文件内容已被修改，这个文件窗口的上面就出现提示：

“Gradle files have changed since last project sync. A project sync may be necessary for the IDE to work properly. sync now”。单击 sync now 按钮进行同步。

build.gradle(Module：app)内容如下：

```
apply plugin: 'com.android.application'
android {
    compileSdkVersion 28
    defaultConfig {
        applicationId "cuc.nestedscrollview1"
        minSdkVersion 19
        targetSdkVersion 28
```

```
        versionCode 1
        versionName "1.0"
        testInstrumentationRunner "android.support.test.runner.AndroidJUnitRunner"
    }
    buildTypes {
        release {
            minifyEnabled false
            proguardFiles getDefaultProguardFile('proguard-android.txt'), 'proguard-
rules.pro'
        }
    }
}
dependencies {
    implementation fileTree(include: ['*.jar'], dir: 'libs')
    implementation 'com.android.support:appcompat-v7:28.+'
    implementation 'com.android.support.constraint:constraint-layout:1.0.2'
    testImplementation 'junit:junit:4.12'
    androidTestImplementation 'com.android.support.test:runner:1.0.1'
    androidTestImplementation 'com.android.support.test.espresso:espresso-core:3.0.1'
    implementation 'com.android.support:design:28.+'
}
```

3. 创建 Behavior 类型的类 MyBehavior

具体步骤：在工程结构视图中 main 源集 Java 文件夹里右击 cuc.nestedscrollview1 包，依次选择 New | Java Class，Name 输入框中输入 MyBehavior，Superclass 输入框中输入 android.support.design.widget.CoordinatorLayout.Behavior，单击 OK 按钮。编辑后，MyBehavior.java 文件内容如下：

```
package cuc.nestedscrollview1;
import android.content.Context;
import android.support.annotation.NonNull;
import android.support.design.widget.CoordinatorLayout;
import android.support.v4.view.ViewCompat;
import android.support.v4.widget.NestedScrollView;
import android.util.AttributeSet;
import android.view.View;

public class MyBehavior extends CoordinatorLayout.Behavior {
    public MyBehavior(Context context, AttributeSet attrs) {
        super(context, attrs);
    }
    @Override
    public boolean onStartNestedScroll(@NonNull CoordinatorLayout coordinatorLayout, @
NonNull View child, @NonNull View directTargetChild, @NonNull View target, int axes, int
type) {
        return axes == ViewCompat.SCROLL_AXIS_VERTICAL;
    }
    @Override
```

```
    public void onNestedScroll(@NonNull CoordinatorLayout coordinatorLayout, @NonNull
View child, @NonNull View target, int dxConsumed, int dyConsumed, int dxUnconsumed, int
dyUnconsumed, int type) {
        super.onNestedScroll(coordinatorLayout, child, target, dxConsumed, dyConsumed,
dxUnconsumed, dyUnconsumed, type);
        child.setScrollY(target.getScrollY());
    }
    @Override
    public boolean onNestedFling(CoordinatorLayout coordinatorLayout, View child, View
target, float velocityX, float velocityY, boolean consumed) {
        if (child instanceof NestedScrollView){
            ((NestedScrollView)child).fling((int)velocityY);
        }
        return true;
    }
}
```

该类首先实现了带两个参数的构造函数，这是在布局文件中通过 app：layout_behavior 指定 Behavior，通过 Java 反射机制生成 Behavior 所必需的。该类还实现了 onStartNestedScroll，用以判断是否接收后续的 onNestedScroll 事件，仅仅监听垂直方向的滚动。还实现了 onNestedFling 方法，在 fling（快速滚动）事件触发的时候调用，设置与 behavior 联系的子视图和发起滚动的目标视图的垂直方向的速度一样。

4. 新建 layout 文件夹

由于在新建工程时，没有选择 Generate Layout File 选项，所以 Android Studio 没有在 res 中创建类型为 layout 的 Android Resource Directory，所以需要创建一个类型为 layout 的 Android 资源文件夹。如果已经存在 layout 文件夹，就跳过此步骤。具体步骤为：在工程结构视图中，右击 res 文件夹，选择 New|Android Resource Directory，从 Resource type 右边的下拉列表框中，选择 layout 选项，单击 OK 按钮，这时系统已经创建了 layout 资源文件夹。

5. 新建布局文件 activity_main1.xml

具体步骤：在工程结构视图中（单击 Project 按钮可以显示或者隐藏工程结构图，使用工程结构图中的下拉列表框可以选择 Android 以显示工程逻辑结构，或者选择 Project 显示工程实际的目录结构），右击 layout，依次选择 New|Layout resource file，File name 输入框输入 activity_main1，Root element 输入框输入 android.support.design.widget.CoordinatorLayout，Source set 为 main，单击 OK 按钮。编辑后，layout/activity_main1.xml 内容如下：

```
<?xml version="1.0" encoding="utf-8"?>
<android.support.design.widget.CoordinatorLayout
    xmlns:android="http://schemas.android.com/apk/res/android"
    xmlns:app="http://schemas.android.com/apk/res-auto"
    android:layout_width="match_parent"
    android:layout_height="match_parent">
    <android.support.v4.widget.NestedScrollView
        app:layout_behavior="cuc.nestedscrollview1.MyBehavior"
```

```
        android:layout_width = "wrap_content"
        android:layout_height = "wrap_content"
        android:layout_gravity = "top|left" >
        <TextView
            android:layout_width = "wrap_content"
            android:layout_height = "wrap_content"
            android:textSize = "40sp"

    android:text = "0\n1\n2\n3\n4\n5\n6\n7\n8\n9\na\nb\nc\nd\ne\nf\ng\nh
        \ni\nj\nk\nl\nm\nn\no\np\nq\nr\ns\nt\nu\nv\nw\nx\ny\nz"/>
    </android.support.v4.widget.NestedScrollView>
    <android.support.v4.widget.NestedScrollView
        app:layout_behavior = "cuc.nestedscrollview1.MyBehavior"
        android:layout_width = "wrap_content"
        android:layout_height = "wrap_content"
        android:layout_gravity = "top|center_horizontal" >
        <TextView
            android:layout_width = "wrap_content"
            android:layout_height = "wrap_content"
            android:textSize = "40sp"

    android:text = "0\n1\n2\n3\n4\n5\n6\n7\n8\n9\na\nb\nc\nd\ne\nf\ng\nh
        \ni\nj\nk\nl\nm\nn\no\np\nq\nr\ns\nt\nu\nv\nw\nx\ny\nz" />
    </android.support.v4.widget.NestedScrollView>
    <android.support.v4.widget.NestedScrollView
        android:layout_width = "wrap_content"
        android:layout_height = "wrap_content"
        android:layout_gravity = "top|right" >
        <TextView
            android:layout_width = "wrap_content"
            android:layout_height = "wrap_content"
            android:textSize = "40sp"

    android:text = "0\n1\n2\n3\n4\n5\n6\n7\n8\n9\na\nb\nc\nd\ne\nf\ng\nh
        \ni\nj\nk\nl\nm\nn\no\np\nq\nr\ns\nt\nu\nv\nw\nx\ny\nz"/>
    </android.support.v4.widget.NestedScrollView>
</android.support.design.widget.CoordinatorLayout>
```

在上面的布局文件中，有三个嵌入了 TextView 的 NestedScrollView，分别设置了 android：layout_gravity＝"top|left"、android：layout_gravity＝"top|cener_horizontal"、android：layout_gravity＝"top|right"，所以位置分别在父视图 CoordinatorLayout 的左边、中间、右边位置。三个 NestedScrollView 都能随着手指单击自己的内部滚动而滚动，这是由于 NestedScrollView 实现了 ScrollingView 接口。由于位于左边、中间的两个 NestedScrollView 都设置了 app：layout_behavior＝"cuc.nestedscrollview1.MyBehavior" 属性，所以这两个 NestedScrollView 都能随着其他的 NestedScrollView 的滚动而滚动。而位于右边的 NestedScrollView 没有设置 app：layout_behavior 属性，所以不能随着其他的 NestedScrollView 的滚动而滚动。

6. 编辑 MainActivity.java，并编译运行

编辑后，MainActivity.java 内容如下：

```
package cuc.nestedscrollview1;

import android.support.v7.app.AppCompatActivity;
import android.os.Bundle;
public class MainActivity extends AppCompatActivity {
    @Override
    protected void onCreate(Bundle savedInstanceState) {
        super.onCreate(savedInstanceState);
        setContentView(R.layout.activity_main1);
    }
}
```

最后编译运行。

11.12　实验：RecyclerView 实验

实验目的：

(1) 熟悉 RecyclerView 的用法。

(2) 实现 item 的上下拖动和向左滑动删除的功能。

(3) 给 RecyclerView 添加上下文菜单的方法。

1. 新建工程 RecyclerView1

具体步骤：运行 Android Studio，单击 Start a new Android Studio project，在 Create New Project 页中，Application name 为 RecyclerView1，Company domain 为 cuc，Project location 为 C:\android\myproject\RecyclerView1，单击 Next 按钮，选中 Phone and Tablet，minimum sdk 设置为 API 19：Android 4.4(minSdkVersion)，单击 Next 按钮，选择 Empty Activity，再单击 Next 按钮，勾选 Generate Layout File 选项，勾选 Backwards Compatibility(AppCompat)选项，单击 Finish 按钮。

2. 添加 RecyclerView-v7 库

因为要使用 RecyclerView，RecyclerView 位于 RecyclerView-v7 之中，所以需要添加 RecyclerView-v7 库。可以通过菜单添加需要的库，库的版本应该和 build.gradle(module：app)文件中 compileSdkVersion 版本一致。具体步骤为：

单击 Android Studio 的菜单 File|Project Structure|app|Dependencies，单击右边的"+"，选择 1 Library dependency 选项，在输入栏中输入 recyclerview，并且按下回车键进行搜索，搜索到 cardview 库后，如果没有版本信息，就添加版本信息，例如 28.+，单击 OK 按钮完成库的添加，再单击 OK 按钮，关闭 Project Structure 界面。

也可以在 build.gradle(module：app)文件的 dependencies 中直接添加依赖：

```
implementation 'com.android.support:recyclerview-v7:28.+'
```

由于 build.gradle 文件内容已被修改，这个文件窗口的上面就出现提示：

"Gradle files have changed since last project sync. A project sync may be necessary for the IDE to work properly. sync now"。单击 sync now 按钮进行同步。

compileSdkVersion 应该尽量设置为最新版本，compileSdkVersion 可以这样设置：单击 Android Studio 的菜单 File | Project Structure | app | Properties，再单击 Comple Sdk Version 右侧的列表框，选择最新的 SDK 版本。最新的 SDK 版本可以通过 SDK manager 下载。

3. 新建一个 Java 类

在工程结构视图中 main 源集 java 文件夹里，右击 cuc. recyclerview1 包，选择 New | Java Class，Name 输入框中输入 Student，单击 OK 按钮。Student. java 内容如下：

```
package cuc.recyclerview1;
public class Student {
    private String name;
    private int id;
    private int age;
    public Student(String name, int id, int age) {
        this.name = name;
        this.id = id;
        this.age = age;
    }
    public String getName() {
        return name;
    }
    public void setName(String name) {
        this.name = name;
    }
    public int getId() {
        return id;
    }
    public void setId(int id) {
        this.id = id;
    }
    public int getAge() {
        return age;
    }
    public void setAge(int age) {
        this.age = age;
    }
}
```

4. 新建 Item 的布局文件 item. xml

item. xml 文件内容如下：

```
<?xml version="1.0" encoding="utf-8"?>
<LinearLayout xmlns:android="http://schemas.android.com/apk/res/android"
    android:orientation="horizontal" android:layout_width="match_parent"
    android:layout_height="wrap_content">
    <TextView
        android:layout_marginStart="20dp"
```

```
            android:id = "@ + id/nameTxt"
            android:layout_width = "wrap_content"
            android:layout_height = "wrap_content" />
        <TextView
            android:layout_marginStart = "20dp"
            android:id = "@ + id/idTxt"
            android:layout_width = "wrap_content"
            android:layout_height = "wrap_content" />
        <TextView
            android:layout_marginStart = "20dp"
            android:id = "@ + id/ageTxt"
            android:layout_width = "wrap_content"
            android:layout_height = "wrap_content" />
</LinearLayout>
```

5. 新建一个 Java 类 ItemHolder

具体步骤：在工程结构视图中 main 源集 Java 文件夹里，右击 cuc. recyclerview1 包，选择 New|Java Class，Name 输入框中输入 ItemHolder，Superclass 输入框中输入 android. support. v7. widget. RecyclerView. ViewHolder，单击 OK 按钮。ItemHolder. java 文件内容如下：

```
package cuc.recyclerview1;
import android.support.v7.widget.RecyclerView;
import android.view.View;
import android.widget.TextView;
public class ItemHolder extends RecyclerView.ViewHolder {
    public TextView nameTxt;
    public TextView idTxt;
    public TextView ageTxt;
    public ItemHolder(View itemView) {
        super(itemView);
        nameTxt  =  (TextView) itemView.findViewById(R.id.nameTxt);
        idTxt  =  (TextView) itemView.findViewById(R.id.idTxt);
        ageTxt  =  (TextView) itemView.findViewById(R.id.ageTxt);
    }
}
```

6. 新建一个 Java 类 StudentAdapter

具体步骤：在工程结构视图中 main 源集 Java 文件夹里，右击 cuc. recyclerview1 包，选择 New | Java Class，Name 输入框中输入 StudentAdapter，Superclass 输入框中输入 android. support. v7. widget. RecyclerView. Adapter，单击 OK 按钮。这时系统生成的代码如下：

```
package cuc.recyclerview1;
import android.support.v7.widget.RecyclerView;
public class StudentAdapter extends RecyclerView.Adapter {
}
```

在 RecyclerView. Adapter 后加上 Java 泛型，这时 StudentAdapter. java 内容如下：

```
package cuc.recyclerview1;
import android.support.v7.widget.RecyclerView;
public class StudentAdapter extends RecyclerView.Adapter<ItemHolder>{
}
```

由于 RecyclerView. Adapter 是抽象类，有一些方法需要实现，把光标移动到 StudentAdapter 位置，同时按住 Alt+Enter 这两个键，就会出现提示，选择 Implement methods 选项，单击 OK 按钮，Android Studio 就会在 UserAdapter 中生成 onCreateViewHolder()、onBindViewHolder 和 getItemCount 方法，编辑 UserAdapter. java，编辑后 StudentAdapter. java 内容如下：

```
package cuc.recyclerview1;
import android.support.v7.widget.RecyclerView;
import android.view.LayoutInflater;
import android.view.View;
import android.view.ViewGroup;
import java.util.ArrayList;
import java.util.List;

public class StudentAdapter extends RecyclerView.Adapter<ItemHolder>{
    private List<Student> studentList;
    public StudentAdapter(){
        studentList = new ArrayList<>();
        for (int i = 0;i<100;i++){
            studentList.add(new Student("student" + i,i,25));
        }
    }
    @Override
    public ItemHolder onCreateViewHolder(ViewGroup parent, int viewType) {
        View itemView = LayoutInflater.from(parent.getContext())
                .inflate(R.layout.item,parent,false);
        return new ItemHolder(itemView);
    }
    @Override
    public void onBindViewHolder(ItemHolder holder, int position) {
        Student student = studentList.get(position);
        holder.nameTxt.setText(studentList.get(position).getName());
        holder.ageTxt.setText(String.valueOf(student.getAge()));
        holder.idTxt.setText(String.valueOf(student.getId()));
    }
    @Override
    public int getItemCount() {
        return studentList.size();
    }
}
```

在 StudentAdapter 的 onCreateViewHolder 方法里调用了 LayoutInflater. inflate(int resource,ViewGroup root,boolean attachToRoot)方法，需要注意的是，attachToRoot 传入的是 false，即以后由适配器来管理每个 item 对应的视图。

下面简单介绍 LayoutInflater. inflate(int resource,ViewGroup root,boolean attachToRoot)方法的使用。第一个参数传入布局的资源 ID，用于生成视图。第二个参数是资源 ID 对应视图的父视图。第三个参数告知布局生成器是否将生成的视图添加给父视图 root 中。如果提供了 root，当 attachToRoot 是 true 时，就自动将布局 resource 添加到 root 父布局，不需要手动调用 root. addView 方法，返回的是 root 视图；否则返回的是由 resource 扩展的视图，不会添加这个布局到 root 父布局，但会根据布局文件最外层的所有 layout 属性进行设置，当以后采用 root. addView 方法该 view 添加到父 view 当中时，这些 layout 属性会自动生效。如果 root 为 null，attachToRoot 将失去作用，设置任何值都没有意义，同时这个布局的最外层参数就无效了。

7. 在 build. gradle(module: app)中，添加 DataBinding 支持，并同步

在文件 build. gradle(module: app)的 android 语句的花括号内添加：

```
dataBinding.enabled = true
```

由于修改了 build. gradle 文件内容，这个文件窗口的上面就出现提示："Gradle files have changed since last project sync. A project sync may be necessary for the IDE to work properly. sync now"，单击 sync now 按钮进行同步。

8. 新建布局文件 activity_main1. xml

activity_main1. xml 文件内容如下：

```
<?xml version = "1.0" encoding = "utf - 8"?>
<layout>
    <data/>
    <android.support.v7.widget.RecyclerView xmlns:android = "http://schemas.android.com/
apk/res/android"
        android:id = "@ + id/recyclerView"
        android:layout_width = "match_parent" android:layout_height = "match_parent"/>
</layout>
```

9. 编辑 MainActivity. java，并编译运行

编辑后，MainActivity. java 内容如下：

```
package cuc.recyclerview1;

import android.databinding.DataBindingUtil;
import android.os.Bundle;
import android.support.v7.app.AppCompatActivity;
import android.support.v7.widget.DefaultItemAnimator;
import android.support.v7.widget.DividerItemDecoration;
import android.support.v7.widget.LinearLayoutManager;

import cuc.recyclerview1.databinding.ActivityMain1Binding;
```

```
public class MainActivity extends AppCompatActivity {
        private StudentAdapter adapter;
        private ActivityMain1Binding binding;
        @Override
        protected void onCreate(Bundle savedInstanceState) {
            super.onCreate(savedInstanceState);
            binding = DataBindingUtil.setContentView(this,R.layout.activity_main1);
            binding.recyclerView.setLayoutManager(new LinearLayoutManager(this));
            adapter = new StudentAdapter();
            binding.recyclerView.setAdapter( adapter);
            binding.recyclerView.setItemAnimator(new DefaultItemAnimator());
            binding.recyclerView.addItemDecoration(new
DividerItemDecoration(this,DividerItemDecoration.VERTICAL));
        }
}
```

下面实现 item 的拖拽和左滑删除功能。

10. 新建接口 IOperationData，Adapter 实现接口

接口定义如下：

```
public interface IOperationData {
    void onItemMove(int fromPosition,int toPosition);
    void onItemDismiss(int position);
}
```

在 StudentAdapter 定义的大括号前添加 implements IOperationData，同时添加 IOperationData 接口中定义的两个方法，这时 StudentAdapter.java 内容如下：

```
public class StudentAdapter extends RecyclerView.Adapter<ItemHolder> implements
IOperationData {
    private List<Student> studentList;
    public StudentAdapter() {
        studentList = new ArrayList<>();
        for (int i = 0;i<100;i++){
            studentList.add(new Student("student" + i,i,25));
        }
    }
    @Override
    public ItemHolder onCreateViewHolder(ViewGroup parent, int viewType) {

        View itemView = LayoutInflater.from(parent.getContext())
                .inflate(R.layout.item,parent,false);
        return new ItemHolder(itemView);
    }

    @Override
    public void onBindViewHolder(ItemHolder holder, int position) {
        Student student = studentList.get(position);
```

```
            holder.nameTxt.setText(studentList.get(position).getName());
            holder.ageTxt.setText(String.valueOf(student.getAge()));
            holder.idTxt.setText(String.valueOf(student.getId()));
        }
        @Override
        public int getItemCount() {
            return studentList.size();
        }

        @Override
        public void onItemMove(int fromPosition, int toPosition) {
            Collections.swap(studentList,fromPosition,toPosition);
            notifyItemMoved(fromPosition,toPosition);
        }
        @Override
        public void onItemDissmiss(int position) {
            studentList.remove(position);
            notifyItemRemoved(position);
        }
    }
```

11. 新建 ItemTouchHelper.Callback 的子类

ItemTouchHelperCallback.java 内容如下：

```
package cuc.recyclerview1;
import android.graphics.Color;
import android.support.annotation.NonNull;
import android.support.annotation.Nullable;
import android.support.v7.widget.RecyclerView;
import android.support.v7.widget.helper.ItemTouchHelper;

public class ItemTouchHelperCallback extends ItemTouchHelper.Callback {
    StudentAdapter studentAdapter;

    public ItemTouchHelperCallback(StudentAdapter studentAdapter) {
        this.studentAdapter = studentAdapter;
    }

    @Override
     public int getMovementFlags (@ NonNull RecyclerView recyclerView, @ NonNull
RecyclerView.ViewHolder viewHolder) {
        int dragFlag = ItemTouchHelper.UP | ItemTouchHelper.DOWN;
        int swipeFlag = ItemTouchHelper.LEFT;
        return makeMovementFlags(dragFlag,swipeFlag);
    }
    @Override
    public boolean onMove (@ NonNull RecyclerView recyclerView, @ NonNull RecyclerView.
ViewHolder viewHolder,
                            @NonNull RecyclerView.ViewHolder viewHolder1) {
```

```
            studentAdapter. onItemMove ( viewHolder. getAdapterPosition ( ), viewHolder1.
getAdapterPosition());
            return true;
        }
        @Override
        public void onSwiped(@NonNull RecyclerView.ViewHolder viewHolder, int i){
            studentAdapter.onItemDissmiss(viewHolder.getAdapterPosition());
        }

        @Override
        public void onSelectedChanged ( @ Nullable RecyclerView. ViewHolder viewHolder, int
actionState) {
            super.onSelectedChanged(viewHolder, actionState);

if((viewHolder!= null)&&(actionState == ItemTouchHelper.ACTION_STATE_DRAG)){
                viewHolder.itemView.setBackgroundColor(Color.RED);
            }
        }
        @Override
        public void clearView(@NonNull RecyclerView recyclerView, @NonNull RecyclerView.
ViewHolder viewHolder) {
            super.clearView(recyclerView, viewHolder);
            viewHolder.itemView.setBackgroundColor(0);
        }
    }
```

12. 编辑 MainActivity.java，并编译运行

在 MainActivity 类的 onCreate 方法的最后添加以下内容，实现上下拖动和向左滑动删除 item 的功能。

```
ItemTouchHelper helper = new ItemTouchHelper(new ItemTouchHelperCallback(adapter));
  helper.attachToRecyclerView(binding.recyclerView);
```

编辑后，编译运行，可以发现上下拖动和向左滑动删除 item 的功能都实现了。

13. 给 RecyclerView 添加右键菜单，实现删除功能

首先编辑 ItemHolder，给 itemView 设置 View.OnCreateContextMenuListener。ItemHolder.java 内容如下：

```
package cuc.recyclerview1;
import android.support.v7.widget.RecyclerView.ViewHolder;
import android.view.ContextMenu;
import android.view.Menu;
import android.view.View;
import android.widget.TextView;

public class ItemHolder extends ViewHolder implements View.OnCreateContextMenuListener {
    public TextView nameTxt;
```

```
        public TextView idTxt;
        public TextView ageTxt;
        public ItemHolder(View itemView) {
            super(itemView);
            nameTxt = (TextView) itemView.findViewById(R.id.nameTxt);
            idTxt = (TextView) itemView.findViewById(R.id.idTxt);
            ageTxt = (TextView) itemView.findViewById(R.id.ageTxt);
            itemView.setOnCreateContextMenuListener(this);
        }

        @Override
        public void onCreateContextMenu(ContextMenu menu, View v, ContextMenu.ContextMenuInfo
menuInfo) {
            menu.add(Menu.NONE,Menu.NONE,0,"delete");
            menu.add(Menu.NONE,Menu.FIRST,1,"other");
        }
    }
```

在StudentAdapter类中，给ItemHolder的itemView设置监听器View.OnLongClickListener()，监听器返回值必须为false，以便实现上下文菜单。并添加removeItem()方法用于删除被长按的item。编辑后StudentAdapter.java内容如下：

```
package cuc.recyclerview1;
import android.support.v7.widget.RecyclerView;
import android.view.LayoutInflater;
import android.view.View;
import android.view.ViewGroup;

import java.util.ArrayList;
import java.util.Collections;
import java.util.List;

public class StudentAdapter extends RecyclerView.Adapter<ItemHolder> implements
IOperationData {
    private List<Student> studentList;
    public int mPosition;
    public StudentAdapter() {
        studentList = new ArrayList<>();
        for (int i = 0;i < 100;i++){
            studentList.add(new Student("student" + i,i,25));
        }
    }
    public void removeItem(){
        onItemDismiss(mPosition);
    }

    @Override
```

```
    public ItemHolder onCreateViewHolder(ViewGroup parent, int viewType) {

        View itemView = LayoutInflater.from(parent.getContext())
                .inflate(R.layout.item,parent,false);
        return new ItemHolder(itemView);
    }

    @Override
    public void onBindViewHolder(ItemHolder holder, final int position) {
        Student student = studentList.get(position);
        holder.nameTxt.setText(studentList.get(position).getName());
        holder.ageTxt.setText(String.valueOf(student.getAge()));
        holder.idTxt.setText(String.valueOf(student.getId()));
        holder.itemView.setOnLongClickListener(new View.OnLongClickListener() {
            @Override
            public boolean onLongClick(View v) {
                mPosition = position;
                return false;
            }
        });
    }
    @Override
    public int getItemCount() {
        return studentList.size();
    }

    @Override
    public void onItemMove(int fromPosition, int toPosition) {
        Collections.swap(studentList,fromPosition,toPosition);
        notifyItemMoved(fromPosition,toPosition);
    }

    @Override
    public void onItemDissmiss(int position) {
        studentList.remove(position);
        notifyItemRemoved(position);
    }
}
```

在 MainActivity 类中，重写 onContextItemSelected(MenuItem item)，内容如下：

```
@Override
public boolean onContextItemSelected(MenuItem item) {
    switch (item.getItemId()){
        case Menu.NONE:
            adapter.removeItem();
            break;
        case Menu.FIRST:
            break;
    }
    return super.onContextItemSelected(item);
}
```

编译运行，可以看到长按 RecyclerView 的某个 item，出现上下文菜单，选择 delete 可以删除这个 item。

11.13　侧滑菜单实验

实验目的：

(1) 熟悉 DrawerLayout 的用法。

(2) 熟悉 NavitationView 实现侧滑菜单的方法。

1. 新建工程 DrawerLayout1

具体步骤：运行 Android Studio，单击 Start a new Android Studio project，在 Create New Project 页中，Application name 为 DrawerLayout1，Company domain 为 cuc，Project location 为 C:\android\myproject\DrawerLayout1，单击 Next 按钮，选中 Phone and Tablet，minimum sdk 设置为 API 19：Android 4.4(minSdkVersion)，单击 Next 按钮，选择 Empty Activity，再单击 Next 按钮，勾选 Generate Layout File 选项，勾选 Backwards Compatibility(AppCompat)选项，单击 Finish 按钮。

2. 新建三个布局文件

新建布局文件 recentcase.xml 的具体步骤：在工程结构视图中(单击 Project 按钮可以显示或者隐藏工程结构图，使用工程结构图中的下拉列表框可以选择 Android 以显示工程逻辑结构，或者选择 Project 显示工程实际的目录结构)，右击 layout，依次选择 New|Layout resource file，File name 输入框输入 activity_main1，Root element 输入框输入 RelativeLayout，Source set 为 main，单击 OK 按钮。layout/recentcase.xml 内容如下：

```
<?xml version = "1.0" encoding = "utf - 8"?>

<RelativeLayout xmlns:android = "http://schemas.android.com/apk/res/android"
    android:layout_width = "match_parent"
    android:layout_height = "match_parent"
    android:background = "@android:color/holo_blue_bright">
    <TextView
        android:layout_width = "wrap_content"
        android:layout_height = "wrap_content"
        android:text = "recent case"
        android:textSize = "25sp"
        android:layout_centerInParent = "true" />
</RelativeLayout>
```

同样的方法创建另外两个布局文件。wordspeak.xml 内容如下：

```
<?xml version = "1.0" encoding = "utf - 8"?>
<RelativeLayout xmlns:android = "http://schemas.android.com/apk/res/android"
    android:layout_width = "match_parent"
    android:layout_height = "match_parent"
    android:background = "@android:color/holo_blue_bright">
    <TextView
```

```
        android:layout_width = "wrap_content"
        android:layout_height = "wrap_content"
        android:text = "words to speak"
        android:textSize = "25sp"
        android:layout_centerInParent = "true" />
</RelativeLayout>
```

peoplenearby. xml 内容如下：

```
<?xml version = "1.0" encoding = "utf - 8"?>
<RelativeLayout xmlns:android = "http://schemas.android.com/apk/res/android"
    android:layout_width = "match_parent"
    android:layout_height = "match_parent"
    android:background = "@android:color/holo_orange_light">
    <TextView
        android:layout_width = "wrap_content"
        android:layout_height = "wrap_content"
        android:text = "people nearby"
        android:textSize = "25sp"
        android:layout_centerInParent = "true" />
</RelativeLayout>
```

3. 由布局文件生成 Fragment 类

从布局文件生成 Fragment 类使用了 Android code Generator 插件，可以通过 Android Studio 菜单中的 File|settings|plugins 查看该插件是否已经安装，若没有安装，请参看 1.5 节 Android Studio 常用插件的安装。

具体步骤：在 Android 工程结构视图中，选中布局文件 recentcase. xml，右击，在弹出菜单中选中 generate Android Code|Fragment，Source path 选择\app\src\main\java，包名采用默认的 cuc. fragments1，单击 Create File 按钮，就生成了文件 RecentcaseFragment. java。

同样的方法，为另外两个布局文件 wordspeak. xml、peoplenearby. xml 生成相应的 Fragment 类。

4. 新建布局文件 navigation_headerlayout. xml，用作侧拉菜单的顶部布局

新建布局文件 navigation_headerlayout. xml 的具体步骤：在工程结构视图中（单击 Project 按钮可以显示或者隐藏工程结构图，使用工程结构图中的下拉列表框可以选择 Android 以显示工程逻辑结构，或者选择 Project 显示工程实际的目录结构），右击 layout，依次选择 New|Layout resource file，File name 输入框输入 navigation_headerlayout，Root element 输入框输入 RelativeLayout，Source set 为 main，单击 OK 按钮。layout/navigation_headerlayout. xml 内容如下：

```
<?xml version = "1.0" encoding = "utf - 8"?>
<RelativeLayout xmlns:android = "http://schemas.android.com/apk/res/android"
    android:layout_width = "match_parent"
    android:layout_height = "wrap_content"
```

```
        android:background = "@color/colorPrimary">
        <TextView
            android:layout_width = "wrap_content"
            android:layout_height = "wrap_content"
            android:text = "left side menu"
            android:textColor = "@android:color/white"
            android:layout_centerInParent = "true"/>
    </RelativeLayout>
```

5. 新建一个菜单文件,用作侧拉菜单

具体步骤:如果 res 目录里有 menu 文件夹,就跳过创建 menu 文件夹这一步。如果 res 目录里没有 menu 文件夹,就需要新建一个 menu 文件夹。创建 menu 文件夹的步骤为:在工程结构视图中,右击 res 文件夹,选择 New | Android Resource Directory,从 Resource type 右边的下拉列表框中,选择 menu 选项,单击 OK 按钮,这时系统已经创建了 menu 资源文件夹。

再创建 menu 文件 drawer. xml。具体步骤为:在工程结构视图中(单击 Project 按钮可以显示或者隐藏工程结构图,使用工程结构图中的下拉列表框可以选择 Android 以显示工程逻辑结构,或者选择 Project 显示工程实际的目录结构),右击 menu,依次选择 New | Menu resource file,File name 输入框输入 drawer,Source set 为 main,单击 OK 按钮。在 drawer. xml 添加菜单项,编辑后 drawer. xml 内容如下:

```
<?xml version = "1.0" encoding = "utf - 8"?>
<menu xmlns:android = "http://schemas.android.com/apk/res/android">
    <group android:checkableBehavior = "single">
        <item android:id = "@ + id/recentcase"
            android:icon = "@mipmap/ic_launcher"
            android:title = "recent case"/>
        <item android:id = "@ + id/wordspeak"
            android:icon = "@android:drawable/ic_input_get"
            android:title = "word speak"/>
        <item android:id = "@ + id/peoplenearby"
            android:icon = "@android:drawable/ic_input_add"
            android:title = "people nearby"/>
    </group>
</menu>
```

6. 添加 com. android. support:design 库

因为在布局文件中需要使用 android. support. design. widget. NavigationView,所以需要 design 库的支持。可以通过菜单添加需要的库,库的版本应该和 build. gradle(module: app)文件中 compileSdkVersion 版本一致。具体步骤为:

单击 Android Studio 的菜单 File | Project Structure | app | Dependencies,单击右边的"+",选择 1 Library dependency 选项,在输入栏中输入 design,并且按下回车键进行搜索,搜索到 cardview 库后,如果没有版本信息,就添加版本信息,例如 28. +,单击 OK 按钮完成库的添加,再单击 OK 按钮,关闭 Project Structure 界面。

也可以在 build. gradle(module：app)文件的 dependencies 中直接添加依赖：

```
implementation 'com.android.support:design:28.+'
```

由于 build. gradle 文件内容已被修改，这个文件窗口的上面就出现提示：

“Gradle files have changed since last project sync. A project sync may be necessary for the IDE to work properly. sync now”。单击 sync now 按钮进行同步。

7. 在 values/styles. xml 中把 AppTheme 的 parent 修改为 Theme. AppCompat. Light. NoActionBar

具体步骤：在 AndroidManifest. xml 文件中，可以看到< application >的属性 android: theme="@style/AppTheme"，名字为"AppTheme"的 style 定义在 res\values\styles. xml 文件中，其内容如下：

```
<resources>
    <!-- Base application theme. -->
    <style name="AppTheme" parent="Theme.AppCompat.Light.DarkActionBar">
        <!-- Customize your theme here. -->
        <item name="colorPrimary">@color/colorPrimary</item>
        <item name="colorPrimaryDark">@color/colorPrimaryDark</item>
        <item name="colorAccent">@color/colorAccent</item>
    </style>
</resources>
```

AppTheme 的 parent 为 Theme. AppCompat. Light. DarkActionBar，采用了 DarkActionBar。为了使用自定义的 Toolbar 来代替 DarkActionBar，所以需要取消 DarkActionBar，将 AppTheme 的 parent 修改为 Theme. AppCompat. Light. NoActionBar，以便在布局文件中使用 ToolBar。修改后 styles. xml 文件内容如下：

```
<resources>
    <!-- Base application theme. -->
    <style name="AppTheme" parent="Theme.AppCompat.Light.NoActionBar">
        <!-- Customize your theme here. -->
        <item name="colorPrimary">@color/colorPrimary</item>
        <item name="colorPrimaryDark">@color/colorPrimaryDark</item>
        <item name="colorAccent">@color/colorAccent</item>
    </style>
</resources>
```

8. 新建一个布局文件 activity_main1. xml

activity_main1. xml 内容如下：

```
<?xml version="1.0" encoding="utf-8"?>
<android.support.v4.widget.DrawerLayout xmlns:android="http://schemas.android.com/apk/
res/android"
    xmlns:app="http://schemas.android.com/apk/res-auto"
    android:id="@+id/drawerLayout"
```

```
    android:layout_width = "match_parent" android:layout_height = "match_parent">
    <LinearLayout
        android:orientation = "vertical"
        android:layout_width = "match_parent"
        android:layout_height = "match_parent">
        <android.support.v7.widget.Toolbar
            android:id = "@ + id/toolBar"
            android:layout_width = "match_parent"
            android:layout_height = "wrap_content">
        </android.support.v7.widget.Toolbar >
        <FrameLayout
            android:id = "@ + id/framelayout_content"
            android:layout_width = "match_parent"
            android:layout_height = "match_parent">
        </FrameLayout >
    </LinearLayout >
    <android.support.design.widget.NavigationView
        android:id = "@ + id/navigationView"
        android:layout_width = "wrap_content"
        android:layout_height = "match_parent"
        android:layout_gravity = "left"
        app:headerLayout = "@layout/navigation_headerlayout"
        app:menu = "@menu/drawer">
    </android.support.design.widget.NavigationView >
</android.support.v4.widget.DrawerLayout >
```

在 DrawerLayout 中，第一个子元素必须是主要内容视图，即抽屉没有打开时显示的布局，并且设置它的 layout_width 和 layout_height 属性是 match_parent。第二个子元素是抽屉视图，并且必须设置属性 android：layout_gravity＝"left"或者 android：layout_gravity＝"right"，表示是抽屉在左边还是右边，高度为 match_parent，宽度为固定宽度。

9. 在 res\values 目录里的 strings.xml 中添加字符串定义

添加几个字符串定义，以便在 mainActivity.java 中使用。添加后，strings.xml 内容如下：

```
<resources>
    <string name = "app_name">DrawerLayout1</string>
    <string name = "drawer_open">drawer_open</string>
    <string name = "drawer_close">drawer_close</string>
</resources>
```

10. 编辑 MainActivity.java，并编译运行。

编辑后，MainActivity.java 内容如下：

```
package cuc.drawerlayout1;
import android.os.Bundle;
import android.support.annotation.NonNull;
import android.support.design.widget.NavigationView;
import android.support.v4.app.FragmentTransaction;
```

```
import android.support.v4.widget.DrawerLayout;
import android.support.v7.app.ActionBarDrawerToggle;
import android.support.v7.app.AppCompatActivity;
import android.support.v7.widget.Toolbar;
import android.view.MenuItem;
import android.widget.FrameLayout;
public class MainActivity extends AppCompatActivity {
    private DrawerLayout drawerLayout;
    private Toolbar toolBar;
    private FrameLayout framelayoutContent;
    private NavigationView navigationView;
    private void assignViews() {
        drawerLayout = (DrawerLayout) findViewById(R.id.drawerLayout);
        toolBar = (Toolbar) findViewById(R.id.toolBar);
        framelayoutContent = (FrameLayout) findViewById(R.id.framelayout_content);
       navigationView = (NavigationView) findViewById(R.id.navigationView);
    }
      void initDrawer(){
        setSupportActionBar(toolBar);
        ActionBarDrawerToggle toggle = new ActionBarDrawerToggle ( this, drawerLayout,
toolBar,R.string.drawer_open,R.string.drawer_close);
        toggle.syncState();
        drawerLayout.addDrawerListener(toggle);
        navigationView.setNavigationItemSelectedListener(new NavigationView.OnNavigatio-
nItemSelectedListener() {
            @Override
            public boolean onNavigationItemSelected(@NonNull MenuItem menuItem) {
                FragmentTransaction transaction = getSupportFragmentManager().beginTran-
saction();
                switch (menuItem.getItemId()){
                    case R.id.recentcase:
                        transaction.replace(R.id.framelayout_content, new RecentcaseFr-
agment());
                        toolBar.setTitle("recentcase");
                        break;
                    case R.id.wordspeak:
                        transaction.replace(R.id.framelayout_content, new WordspeakFr-
agment());
                        toolBar.setTitle("wordspeak");
                        break;
                    case R.id.peoplenearby:
                        transaction.replace(R.id.framelayout_content,new PeoplenearbyFr-
agment());
                        toolBar.setTitle("peoplenearby");
                        break;
                }
                transaction.commit();
                menuItem.setCheckable(true);
                drawerLayout.closeDrawers();
```

```
                return true;
            }
        });
    }
    @Override
    protected void onCreate(Bundle savedInstanceState) {
        super.onCreate(savedInstanceState);
        setContentView(R.layout.activity_main1);
        assignViews();
        initDrawer();
    }
}
```

最后编译运行。

11.14　实验：水平翻页

实验目的：

(1) 熟悉 ViewPager 的用法。

(2) 熟悉 FragmentPagerAdapter 的用法。

(3) 熟悉 PagerTabStrip 的用法。

ViewPager＋FragmentPagerAdapter＋PagerTabStrip 实现左右翻页。

1. 新建工程 ViewPager1

具体步骤：Application name 为 ViewPager1，company domain 为 cuc，Project 位置 C:\android\myproject\ViewPager1，单击 Next 按钮，选中 Phone and tablet，minimum sdk 设置为 API19：Android 4.4(minSdkVersion)，单击 Next 按钮，选择 Empty activity，再单击 Next 按钮，勾选 Generate Layout File，勾选 Backwards Compatibility(AppCompat)，单击 Finish 按钮。

2. 在 build.gradle(module：app)中，添加 DataBinding 支持，并同步

在文件 build.gradle(module：app)的 android 语句的花括号内添加：

```
dataBinding.enabled = true
```

由于修改了 build.gradle 文件内容，这个文件窗口的上面就出现提示：

"Gradle files have changed since last project sync. A project sync may be necessary for the IDE to work properly. sync now"，单击 sync now 按钮进行同步。

3. 新建 4 个布局文件

新建布局文件 activity_main1.xml 的具体步骤：在工程结构视图中(单击 Project 按钮可以显示或者隐藏工程结构图，使用工程结构图中的下拉列表框可以选择 Android 以显示工程逻辑结构，或者选择 Project 显示工程实际的目录结构)，右击 layout，依次选择 New | Layout resource file，File name 输入框输入 activity_main1，Root element 输入框输入 layout，Source set 为 main，单击 OK 按钮。layout/activity_main1.xml 内容如下：

```
<?xml version = "1.0" encoding = "utf - 8"?>
< layout xmlns:android = "http://schemas.android.com/apk/res/android">
    < data/>
    < LinearLayout
        android:orientation = "vertical" android:layout_width = "match_parent"
        android:layout_height = "match_parent">
        < android.support.v4.view.ViewPager
            android:id = "@ + id/viewPager"
            android:layout_width = "match_parent"
            android:layout_height = "wrap_content">
            < android.support.v4.view.PagerTabStrip
                android:id = "@ + id/pagerTabStrip"
                android:layout_gravity = "top"
                android:layout_width = "match_parent"
                android:layout_height = "wrap_content">
            </android.support.v4.view.PagerTabStrip >
        </android.support.v4.view.ViewPager >
    </LinearLayout >
</layout >
```

在上面的布局文件中，PagerTabStrip 放在 ViewPager 的内部，作为 ViewPager.DecorView.

在工程结构视图中，右击 layout，选择 New|Layout resource file，File name 输入框输入 layout1，Root element 输入框输入 LinearLayout，Source set 为 main，单击 OK 按钮。layout1.xml 内容如下：

```
<?xml version = "1.0" encoding = "utf - 8"?>
< LinearLayout xmlns:android = "http://schemas.android.com/apk/res/android"
    android:orientation = "vertical" android:layout_width = "match_parent"
    android:layout_height = "match_parent">
    < Button
        android:id = "@ + id/buttonInlayout1"
        android:layout_width = "wrap_content"
        android:layout_height = "wrap_content"
        android:textSize = "30sp"
        android:text = "this is a Button\nplease clickon me\n page 1"/>
</LinearLayout >
```

同样的方法，创建两个布局文件 layout2.xml，layout3.xml。layout2.xml 内容如下：

```
<?xml version = "1.0" encoding = "utf - 8"?>
< LinearLayout xmlns:android = "http://schemas.android.com/apk/res/android"
    android:orientation = "vertical" android:layout_width = "match_parent"
    android:layout_height = "match_parent">
    < TextView
        android:layout_width = "wrap_content"
        android:layout_height = "wrap_content"
        android:textSize = "30sp"
        android:text = "this is a TextView\n page 2"/>
</LinearLayout >
```

Layout3. xml 内容如下：

```
<?xml version = "1.0" encoding = "utf - 8"?>
< LinearLayout xmlns:android = "http://schemas.android.com/apk/res/android"
    android:orientation = "vertical" android:layout_width = "match_parent"
    android:layout_height = "match_parent">
    < TextView
        android:layout_width = "wrap_content"
        android:layout_height = "wrap_content"
        android:textSize = "30sp"
        android:text = "this is a TextView\n page 3"/>
</LinearLayout >
```

4. 根据布局文件 layout1. xml、layout2. xml、layout3. xml 生成 Fragment 类

具体步骤：在 Android 工程结构视图中，右击布局文件 layout1. xml，在弹出菜单中选中 generate Android Code|Fragment，Source path 选择\app\src\main\java，包名采用默认的 cuc. viewpager1，单击 Create File 按钮，就生成了文件 Layout1Fragment. java。可以看到插件生成的 Fragment 有错误，Layout1Fragment extends Fragment implements OnClickListener 中的 OnClickListener 颜色为红色，错误信息为 Cannot resolve symbol 'OnClickListener'，更正错误的方法就是把 OnClickListener 修改成 View. OnClickListener。若把鼠标移到代码上，出现的提示是：

```
? android.view.View.OnClickListener? (multiple choice… ) Alt + Enter
```

可以采用以下方法更正此错误：把鼠标指针移到 OnClickListener 里的任何一个字符后，按 Alt+Enter，单击 Import class 选项，单击 OnClickListener in view(android. view)选项。另外添加文件中 buttonInlayout1 的响应，修改后 Layout1Fragment. java 中的 onClick 方法内容如下：

```
package cuc.viewpager1;
import android.os.Bundle;
import android.support.v4.app.Fragment;
import android.view.LayoutInflater;
import android.view.View;
import android.view.ViewGroup;
import android.widget.Toast;

public class Layout1Fragment extends Fragment implements View.OnClickListener {
    @Override
    public View onCreateView(LayoutInflater inflater, ViewGroup container,
                             Bundle savedInstanceState) {
        return inflater.inflate(R.layout.layout1, null);
    }
    @Override
    public void onViewCreated(View view, Bundle savedInstanceState) {
        super.onViewCreated(view, savedInstanceState);
```

```
        view.findViewById(R.id.buttonInlayout1).setOnClickListener(this);
    }
    @Override
    public void onClick(View view) {
        switch (view.getId()) {
            case R.id.buttonInlayout1:
                Toast.makeText(getActivity(),"buttonInlayout1 has been clicked",Toast.
LENGTH_SHORT).show();
                break;
        }
    }
}
```

生成的文件 Layout2Fragment.java 内容如下：

```
package cuc.viewpager1;
import android.support.v4.app.Fragment;
import android.view.View;
import android.view.LayoutInflater;
import android.view.ViewGroup;
import android.os.Bundle;
public class Layout2Fragment extends Fragment {
    @Override
    public View onCreateView(LayoutInflater inflater, ViewGroup container,
                             Bundle savedInstanceState) {
        return inflater.inflate(R.layout.layout2, null);
    }
    @Override
    public void onViewCreated(View view, Bundle savedInstanceState) {
        super.onViewCreated(view, savedInstanceState);
    }
}
```

生成的文件 Layout3Fragment.java 内容如下：

```
package cuc.viewpager1;
import android.support.v4.app.Fragment;
import android.view.View;
import android.view.LayoutInflater;
import android.view.ViewGroup;
import android.os.Bundle;

public class Layout3Fragment extends Fragment {
    @Override
    public View onCreateView(LayoutInflater inflater, ViewGroup container,
                             Bundle savedInstanceState) {
        return inflater.inflate(R.layout.layout3, null);
    }
```

```
    @Override
    public void onViewCreated(View view, Bundle savedInstanceState) {
        super.onViewCreated(view, savedInstanceState);
    }
}
```

5. 新建一个 FragmentPagerAdapter 类

具体步骤：在工程结构视图中 main 源集 Java 文件夹里，右击 cuc.viewpager1 包，选择 New|Java Class，Name 输入框中输入 MyFragmentPagerAdapter，Superclass 输入框中输入 FragmentPagerAdapter，单击 OK 按钮。生成的文件 MyFragmentPagerAdapter.java 内容如下：

```
package cuc.viewpager1;
import android.support.v4.app.FragmentPagerAdapter;
public class MyFragmentPagerAdapter extends FragmentPagerAdapter {
}
```

光标移动到文件中 MyFragmentPagerAdapter 位置，单击 Android Studio 菜单中的 code|Implement Methods|Select methods to Implement，单击 OK 按钮，就可以实现 getItem 和 getCount 方法。

由于在 ViewPager 中使用了 PagerTabStrip 子控件，为了使 PagerTabStrip 显示页面的标题，PagerAdapter 还必须覆盖 getPageTitle 方法。单击 Android Studio 菜单中的 Code|Override Methods|Select Methods to Override/Implement，选择 getPageTitle 方法，单击 OK 按钮。

再单击 Android Studio 菜单中的 code|Generate Alt+Insert| Constructor，自动生成构造函数。最后对类的内容进行编辑，编辑后 MyFragmentPagerAdapter.java 内容如下：

```
package cuc.viewpager1;
import android.support.v4.app.Fragment;
import android.support.v4.app.FragmentManager;
import android.support.v4.app.FragmentPagerAdapter;
import java.util.ArrayList;
import java.util.List;

public class MyFragmentPagerAdapter extends FragmentPagerAdapter {
    private List<Fragment> fragmentList;
    private List<String> titleList;
    public MyFragmentPagerAdapter(FragmentManager fm) {
        super(fm);
        fragmentList = new ArrayList<>();
        fragmentList.add(new Layout1Fragment());
        fragmentList.add(new Layout2Fragment());
        fragmentList.add(new Layout3Fragment());
        titleList = new ArrayList<>();
        titleList.add("Tom");
        titleList.add("Jenny");
        titleList.add("Herry");
    }
```

```
    @Override
    public Fragment getItem(int position) {
        return fragmentList.get(position);
    }
    @Override
    public int getCount() {
        return fragmentList.size();
    }
    @Override
    public CharSequence getPageTitle(int position) {
        return titleList.get(position);
    }
}
```

6. 编辑 MainActivity.java，然后编译运行

编辑 MainActivity.java，MainActivity.java 内容如下：

```
package cuc.viewpager1;
import android.databinding.DataBindingUtil;
import android.graphics.Color;
import android.support.v4.view.PagerTabStrip;
import android.support.v4.view.ViewPager;
import android.support.v7.app.AppCompatActivity;
import android.os.Bundle;
import cuc.viewpager1.databinding.ActivityMain1Binding;

public class MainActivity extends AppCompatActivity {
    void initPagerTabStrip(){
        ActivityMain1Binding binding = DataBindingUtil.setContentView(this, R.layout.
activity_main1);
        binding.pagerTabStrip.setTabIndicatorColor(Color.GREEN);
        binding.viewPager.setAdapter(new MyFragmentPagerAdapter(getSupportFragmentManager()));
    }
    @Override
    protected void onCreate(Bundle savedInstanceState) {
        super.onCreate(savedInstanceState);
        initPagerTabStrip();
    }
}
```

最后编译运行。

采用 TabLayout 代替 PagerTabStrip 实现选项卡功能。TabLayout 可以使用自定义的 view，功能更为强大。

7. 添加 com.android.support：design 库

因为在布局文件中需要使用 android.support.design.widget.TabLayout，所以需要 design 库的支持。可以通过菜单添加需要的库，库的版本应该和 build.gradle（module：app）文件中 compileSdkVersion 版本一致。具体步骤为：

依次单击 Android Studio 的菜单 File|Project Structure|app|Dependencies，单击右边的“+”，选择 1 Library dependency 选项，在输入栏中输入 design，并且按下回车键进行搜索，搜索到 cardview 库后，如果没有版本信息，就添加版本信息，例如 28.+，单击 OK 按钮完成库的添加，再单击 OK 按钮，关闭 Project Structure 界面。

也可以在 build.gradle(module：app)文件的 dependencies 中直接添加依赖：

```
implementation 'com.android.support:design:28.+'
```

由于 build.gradle 文件内容已被修改，这个文件窗口的上面就出现提示：

“Gradle files have changed since last project sync. A project sync may be necessary for the IDE to work properly. sync now”。单击 sync now 按钮进行同步。

8. 新建布局文件 activity_main2.xml

activity_main2.xml 文件内容如下：

```
<?xml version="1.0" encoding="utf-8"?>
<layout xmlns:android="http://schemas.android.com/apk/res/android"
    xmlns:app="http://schemas.android.com/apk/res-auto">
    <data/>
    <LinearLayout
        android:orientation="vertical" android:layout_width="match_parent"
        android:layout_height="match_parent">
        <android.support.design.widget.TabLayout
            app:tabMode="scrollable"
            android:id="@+id/tabLayout"
            android:layout_width="match_parent"
            android:layout_height="wrap_content"/>
        <android.support.v4.view.ViewPager
            android:id="@+id/viewPager"
            android:layout_width="match_parent"
            android:layout_height="wrap_content">
        </android.support.v4.view.ViewPager>
    </LinearLayout>
</layout>
```

9. 编辑 MainActivity.java

添加方法 initTabLayout，并且在 Oncreate 方法中调用 initTabLayout，最后 MainActivity.java 内容如下：

```
package cuc.viewpager1;
import android.databinding.DataBindingUtil;
import android.graphics.Color;
import android.support.v4.view.PagerTabStrip;
import android.support.v4.view.ViewPager;
import android.support.v7.app.AppCompatActivity;
import android.os.Bundle;
```

```
import cuc.viewpager1.databinding.ActivityMain1Binding;
import cuc.viewpager1.databinding.ActivityMain2Binding;

public class MainActivity extends AppCompatActivity {
    void initPagerTabStrip(){
        ActivityMain1Binding binding = DataBindingUtil.setContentView(this, R.layout.
activity_main1);
        binding.pagerTabStrip.setTabIndicatorColor(Color.GREEN);
        binding.viewPager.setAdapter(new MyFragmentPagerAdapter(getSupportFragmentManager()));
    }
    void initTabLayout(){
        ActivityMain2Binding binding = DataBindingUtil.setContentView(this, R.layout.
activity_main2);
        binding.viewPager.setAdapter(new MyFragmentPagerAdapter(getSupportFragmentManager()));
        binding.tabLayout.setupWithViewPager(binding.viewPager);
        binding.tabLayout.getTabAt(0).setIcon(android.R.drawable.ic_input_get);
        binding.tabLayout.getTabAt(1).setIcon(android.R.drawable.ic_input_add);
        binding.tabLayout.getTabAt(2).setIcon(android.R.drawable.ic_delete);
    }
    @Override
    protected void onCreate(Bundle savedInstanceState) {
        super.onCreate(savedInstanceState);
        initTabLayout();
    }
}
```

10. 编译运行

运行后 Tab 的效果为：Tab 的图标在上，文字在下，若要实现自己定义的图标文字位置，例如图标在左，文字在右，就必须使用 TabLayout.Tab 的 setCustomView(@Nullable View view)或者 setCustomView(@LayoutRes int resId)方法。

使用 setCustomView 设置 Tab 的视图，视图的布局由用户定义。这时 FragmentPagerAdapter 不需要实现 getPageTitle 方法。

11. 新建 Tab 布局文件

新建布局文件 self_tablayout.xml，内容如下：

```
<?xml version="1.0" encoding="utf-8"?>
<LinearLayout xmlns:android="http://schemas.android.com/apk/res/android"
    android:orientation="horizontal" android:layout_width="match_parent"
    android:layout_height="match_parent">
    <ImageView
        android:id="@+id/tab_image"
        android:layout_width="wrap_content"
        android:layout_height="wrap_content" />
    <TextView
        android:textSize="40sp"
        android:id="@+id/tab_text"
        android:layout_width="wrap_content"
        android:layout_height="wrap_content" />
</LinearLayout>
```

12. 编辑 MainActivity.java

编辑 MainActivity，编辑后的 MainActivity.java 内容如下：

```
package cuc.viewpager1;
import android.databinding.DataBindingUtil;
import android.graphics.Color;
import android.support.v4.view.PagerTabStrip;
import android.support.v4.view.ViewPager;
import android.support.v7.app.AppCompatActivity;
import android.os.Bundle;
import android.view.LayoutInflater;
import android.view.View;
import android.widget.ImageView;
import android.widget.TextView;
import cuc.viewpager1.databinding.ActivityMain1Binding;
import cuc.viewpager1.databinding.ActivityMain2Binding;

public class MainActivity extends AppCompatActivity {
    void initPagerTabStrip(){
        ActivityMain1Binding binding = DataBindingUtil.setContentView(this, R.layout.
activity_main1);
        binding.pagerTabStrip.setTabIndicatorColor(Color.GREEN);
        binding.viewPager.setAdapter(new MyFragmentPagerAdapter(getSupportFragmentManager()));
    }
    void initTabLayout(){
        ActivityMain2Binding binding = DataBindingUtil.setContentView(this, R.layout.
activity_main2);
        binding.viewPager.setAdapter(new MyFragmentPagerAdapter(getSupportFragmentManager()));
        binding.tabLayout.setupWithViewPager(binding.viewPager);
        binding.tabLayout.getTabAt(0).setIcon(android.R.drawable.ic_input_get);
        binding.tabLayout.getTabAt(1).setIcon(android.R.drawable.ic_input_add);
        binding.tabLayout.getTabAt(2).setIcon(android.R.drawable.ic_delete);
    }
    private int[] tabicons = {android.R.drawable.ic_input_add,
            android.R.drawable.ic_delete,
            R.mipmap.ic_launcher};
    String[] tabStrings = {"Toma","Jennyb","Rosec"};
    private View getViewsForTab(int i) {
        View view = LayoutInflater.from(this).inflate(R.layout.self_tablayout,null);
        ImageView tabImage = (ImageView) view.findViewById(R.id.tab_image);
        TextView tabText = (TextView) view.findViewById(R.id.tab_text);
        tabImage.setImageResource(tabicons[i]);
        tabText.setText(tabStrings[i]);
        return view;
    }
    void initTabLayout2(){
        ActivityMain2Binding binding = DataBindingUtil.setContentView(this, R.layout.
activity_main2);
        binding.viewPager.setAdapter(new MyFragmentPagerAdapter(getSupportFragmentManager()));
```

```
        binding.tabLayout.setupWithViewPager(binding.viewPager);
        for (int i = 0;i < binding.tabLayout.getTabCount();i++)
            binding.tabLayout.getTabAt(i).setCustomView(getViewsForTab(i));
    }
    @Override
    protected void onCreate(Bundle savedInstanceState) {
        super.onCreate(savedInstanceState);
        initTabLayout2();
    }
}
```

13. 编译运行

可以看出 Tab 采用 self_tablayout.xml 定义的布局：图标在左，文字在右。

11.15 本章主要参考文献

1. https://material.io/design/components/app-bars-top.html#behavior
2. https://developer.android.google.cn/reference/android/widget/ScrollView.html
3. https://developer.android.google.cn/reference/android/support/v4/widget/NestedScrollView.html
4. https://developer.android.google.cn/reference/android/widget/ListView.html
5. https://developer.android.google.cn/reference/android/support/v7/widget/RecyclerView.html
6. https://developer.android.google.cn/reference/android/support/design/widget/CoordinatorLayout.html
7. https://developer.android.google.cn/training/implementing-navigation/nav-drawer.html
8. https://developer.android.google.cn/reference/android/support/v4/widget/DrawerLayout.html
9. https://developer.android.google.cn/reference/android/support/v4/view/ViewPager.html
10. https://developer.android.google.cn/reference/android/support/v4/view/PagerAdapter
11. https://developer.android.google.cn/reference/android/support/v4/app/FragmentPagerAdapter
12. https://developer.android.google.cn/reference/android/support/v4/view/PagerTabStrip.html
13. https://developer.android.google.cn/reference/android/support/design/widget/TabLayout.html
14. https://developer.android.google.cn/reference/android/widget/HorizontalScrollView.html
15. https://developer.android.google.cn/training/implementing-navigation/lateral.html
16. https://material.io/design/layout/spacing-methods.html#spacing

第12章 BroadcastReceiver

CHAPTER 12

当各种系统事件发生时，Android 系统会自动发送广播，例如系统引导、设备开始充电、切换进入或者退出飞行模式。应用也可以发送自定义的广播，例如通知其他应用它们感兴趣的事件(例如新数据已经下载完毕)，也可以接收来自系统和其他应用的广播。应用可以注册接收特定的广播事件，当有广播发送时，系统会把广播发送到那些订阅了接收特定类型广播的应用。广播可以用作正常的流程之外跨应用的消息系统。android. content. BroadcastReceiver 是能够接收和处理广播 Intent 的基类。如果不需要跨应用发送广播，应该使用 android. support. v4. content. LocalBroadcastManager。

定义 BroadcastReceiver，需要重写 onReceive()方法，在这个方法里完成需要处理的工作。onReceive()方法运行在主线程，不能做耗时的操作，否则会导致主线程阻塞，需要处理耗时操作时可以通过启动一个 Service 进行处理。当接收到广播时，会创建 BroadcastReceiver 对象，回调 onReceive()方法，之后 BroadcastReceiver 对象将被销毁。在 Activity 的内部定义 BroadcastReceiver 类，只能使用动态注册的方式。

注册 BroadcastReceiver 有两种方式：一种方式是，在 AndroidManifest. xml 中用 < receiver>标签声明，并在标签内用< intent-filter >标签设置过滤器，这种注册方式 receiver 的生命周期就是 application 级的。另一种方式是，在代码中先创建一个 IntentFilter 对象，然后在合适的地方调用 Context. registerReceiver(BroadcastReceiver，intentFilter)方法注册 BroadcastReceiver 实例，在合适的地方调用 Context. unregisterReceiver(BroadcastReceiver)方法注销动态注册的 BroadcastReceiver。如果在 Activity. onResume()方法中注册了 receiver，那么就需要在 Activity. onPause()方法中注销 receiver，在 paused 状态下，不能接收 Intent，这样可以减少不必要的系统开销，不要在 Activity. onSaveInstanceState()方法中注销 receiver，因为在用户回退到历史栈中的时候，这个方法不会被调用。如果用于动态注册 BroadcastReceiver 的 Context 对象被销毁了，BroadcastReceiver 也注销了。

Android 7.0(API 级别 24)和更高的版本不再发送 ACTION_NEW_PICTURE 和 ACTION_NEW_VIDEO 系统广播。从 Android 8.0(API 级别 26)开始，系统对在清单文件中声明的 BroadcastReceiver 施加了附加限制。如果应用目标是 Android 8.0 或者更高版本时，大多数隐式广播不能在清单文件中声明，只有动态注册的 receiver 才能收到，只有少数隐式广播可以在清单文件中声明，可查阅 https://developer. android. google. cn/guide/components/broadcast-exceptions. html。从 Android 9(API 级别 28)开始，NETWORK_STATE_CHANGED_ACTION 不再接收用户位置信息。

12.1 广播类型

BroadcastReceiver 所对应的广播分两类：普通广播和有序广播。

普通广播：普通广播通过 Context. sendBroadcast(intent)发送，它是完全异步的。如图 12-1 所示，所有的广播接收器以不确定的顺序运行，因此所有的 receivers(接收器)接收 broadcast 的顺序不确定。发送方发出广播后，几乎同时到达多个广播接收者。通过 Context. registerReceiver()注册的接收器都会一次性被处理，因为它们的进程都是活着的，执行的顺序是不固定的，由 Android 的进程调度决定。当动态注册的 BroadcastRecevier 被执行完以后，会按照优先级从高到低依次执行静态注册的 BroadcastReceiver。对于静态注册的接收器，进程不一定是活的，一次性激活很多进程，会急剧恶化系统性能，所以静态注册的 Recevier 按照优先级排列成一个链表，按顺序检查队列里面的 Receiver 是否存活，如果是活着的，则执行该 Recevier 的 OnReceiver 回调，如果进程尚不存在，则首先创建进程，然后执行 BroadcastReceiver 的 OnRecevier 回调。即一个 BroadcastReceiver 执行结束，再去执行 receivers 队列里的下一个 BroadcastReceiver。也就是说，动态注册的接收器(无论多低的优先级)总是比静态注册的接收器(无论多高的优先级)先接收到同一个普通广播。静态注册的接收器中，priority 值大的先接收到广播，相同的 priority 值的接收器，先注册的先收到广播。对于普通广播，BroadcastReceiver 总是不能使用 setResult 方法、getResult 方法及 abort 方法，即某个接收器不能对收到的广播进行一番处理后再传给下一个接收者，并且无法终止广播继续传播。

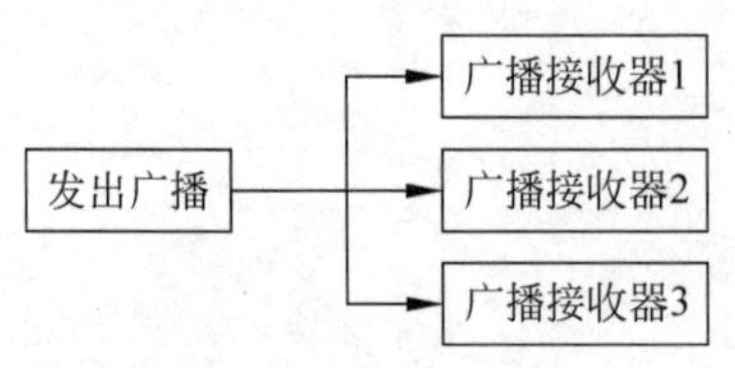

图 12-1 普通广播

有序广播：有序广播通过 Context. sendOrderedBroadcast 来发送。有序是针对广播接收者而言的。广播接收者需要提前设置优先级，优先级数值为－1000～1000，在 AndroidManifest. xml 中通过< intent-filter android：priority＝"xxx">设置。优先级别最高的广播接收器(跟动态注册还是静态注册无关)先接收到某个有序广播，接收完了如果没有丢弃，就继续传给下一个次高优先级别的广播接收器进行处理，依次类推，直到最低优先级的接收器，相同优先级的接收器，动态注册的比静态注册的先接收到广播。同一个 App 中相同优先级的静态接收器，写在清单文件的前面的先接收，如图 12-2 所示。先接收的广播接收器可以对广播进行修改，那么后接收的广播接收器将接收到被修改后的广播。先接收的广播接收者可以对广播进行截断，低优先级的广播接收器就接收不到此广播了。具体地说，BroadcastReceiver 可以使用 setResult 系列方法修改结果，把结果传给下一个 BroadcastReceiver，通过 getResult 系列方法可以取得上一个 BroadcastReceiver 返回的结果，也可以通过 abortBroadcast()函数来让系统丢弃该广播，使得该广播不再传送到别的 BroadcastReceiver。BroadcastReceiver 类中与有序广播相关的方法，如表 12-1 所示。

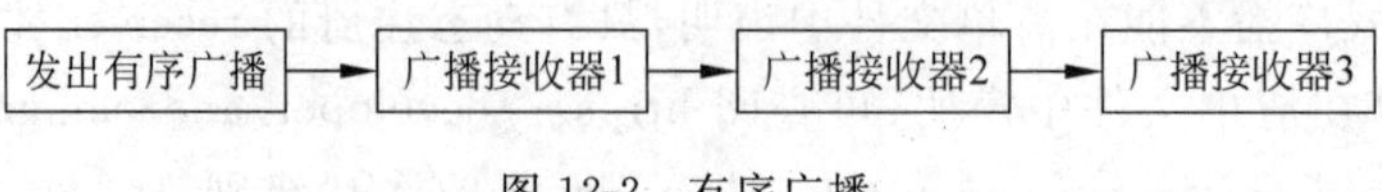

图 12-2 有序广播

表 12-1 BroadcastReceiver 类中与有序广播相关的方法

方　　法	解　　释
final void setResult(int code,String data,Bundle extras)	设置这个广播的所有结果数据
final void setResultCode(int code)	设置这个广播现在的结果编码
final int getResultCode()	获得现在的结果码,这是由前一个接收器设置的
final void setResultData(String data)	设置这个广播的现在的结果数据
finalString getResultData()	获得现在的结果数据,这是由前一个接收器设置的
final void setResultExtras(Bundle extras)	设置这个广播现在的结果 extras
finalBundle getResultExtras(boolean makeMap)	获得现在的结果中的 extra 数据,这是由前一个接收器设置的
final void abortBroadcast()	设置标志,接收器应该放弃现在的广播

需要注意的是,广播机制里的 Intent 和通过 Context. startActivity()启动 Activity 里的 Intent 是完全分开的。BroadcastReceiver 是无法看见或者俘获用来启动 Activity 的 Intent 的;同样,发送的广播 Intent 是不能打开 Activity 的。这两个操作在语法上是完全不同的,使用 Intent 打开一个 Activity 是一个和用户交互的前台操作,而使用 Intent 广播是一个用户平常不会注意到的后台操作。

12.2 LocalBroadcastManager

BroadcastReceiver 可以方便应用程序和系统、应用程序之间、应用程序内的通信。如果需要在同一个应用内的不同组件间发送广播,可以采用以下方式:

(1) 注册接收器时将 exported 属性设置为 false,使得此接收器不能接收其他 App 发出的广播。

(2) 或者在广播的发送端,或者在广播的接收端实施访问权限,用于权限验证。为了在发送广播的时候实施权限,要提供一个非 NULL 的权限变量给 sendBroadcast(Intent, String)或者 sendOrderedBroadcast(Intent, String, BroadcastReceiver, android. os. Handler,int,String,Bundle)。只有获得权限允许的接收者(通过在 AndroidManifest. xm 文件中的< uses-permission >标签)可以接收到广播。为了在接收广播的时候实施权限,需要在调用 registerReceiver(BroadcastReceiver,IntentFilter,String,android. os. Handler)时或者在 AndroidManifest. xml 中的< receiver >标签中,提供一个非 NULL 的权限。只有获得权限允许才可以发送 Intent 给广播接收者(通过在 AndroidManifest. xm 文件中的< uses-permission >标签)。

(3) 从 Android 4. 0. 4(ICE_CREAM_SANDWICH)开始,发送广播时通过 intent. setPackage(packageName)指定包名,此广播将只会发送到指定包中的 App 内与之相匹配的有效广播接收器中。

androidx. localbroadcastmanager. content. LocalBroadcastManager 用于在同一个应用内的不同组件间发送广播,可以代替 Handler 更新 UI,使用方式上与全局广播几乎相同,只是注册/注销广播接收器和发送广播时将 context 变成了 LocalBroadcastManager 的单一实例。

LocalBroadcastManager 注册广播只能通过 Java 代码注册的方式。LocalBroadcastManager 由于没有跨进程的通信，广播数据只在当前 APP 中传播，不会泄露给其他 APP，其他应用无法发广播给当前应用，因此比系统全局广播更加高效，安全性也更高。

12.3 BroadcastReceiver 在清单文件中的语法

```
<receiver android:enabled = ["true" | "false"]
      android:exported = ["true" | "false"]
      android:icon = "drawable resource"
      android:label = "string resource"
      android:name = "string"
      android:permission = "string"
      android:process = "string" >
  ...
</receiver>
```

receiver 元素包含在 application 元素内，receiver 元素可以包含 intent-filter 元素和 meta-data 元素。receiver 元素的属性的含义，如表 12-2 所示。

表 12-2 <receiver>元素的属性的含义

属　　性	解　　释
name	broadcastReceiver 的类名
enabled	广播接收器是否可以被系统实例化，true 表示可以。默认值是 true。<application>元素也有自己的 enabled 属性，它应用到所有的组件，包含广播接收器。只有<application>和<receiver>的 enabled 都设为 true，BroadcastReceiver 才能被系统实例化。如果其中任何一个是 false，BroadcastReceiver 就不能被系统实例化
exported	决定此 BroadcastReceiver 能否接收其他应用发送的广播，true 表示可以。如果设为 false，BroadcastReceiver 仅仅能接收同一应用或者具有相同用户 ID 的应用的组件发送的广播。该属性默认值取决于 receiver 是否包含<intent-filter>，当包含了<intent-filter>时，该属性的默认值为 true，否则为 false。如果没有包含<intent-filter>，说明该接收器仅仅由精确指定类名的显式 Intent 对象唤起，这意味着该接收器仅能用于应用内部使用（其他应用正常情况下不会知道类名），这种情况下，默认值是 false。另一方面，如果广播接收器至少包含一个<intent-filter>，这意味着广播接收器准备接收来自系统或者其他应用的广播，默认值是 true。这个属性不是限制广播接收器暴露给外部应用的唯一方式。也可以通过 permission 来限制可以向该接收器发送消息的外部组件
permission	设置广播发送方的权限，具备相应权限的发送方发送的广播可以被该 receiver 接收。如果这个属性没有设置，<application>元素的 permission 属性将应用到广播接收器。如果<receiver>和<application>都没有设置，<receiver>将不受 permission 的保护
process	设置 receiver 运行时所处的进程的名称。正常情况下，一个应用的所有组件运行在应用创建的默认进程中，和应用的包有相同的名字。<application>元素的 process 属性可以为应用的所有组件设置一个不同于包名的默认值。但是每个组件可以覆盖自己的默认的 process 属性，允许应用跨多个进程。当进程名以冒号（：）开始时表示该进程是应用的私有进程，在必要的时候会创建一个新进程，这时候会把进程名附加到包名后面形成新的进程名。当进程名是以小写字母开头时，则表示这个服务将运行在一个以这个名字命名的全局进程中，但前提是该进程必须具备相应的权限

在 AndroidManifest.xml 里面静态注册的例子：

```
<receiver
    android:name = ".MyReceiver"
    android:enabled = "false"
    android:exported = "true">
    <intent - filter android:priority = "998">
        <action android:name = "android.intent.action.MY_BROADCAST" />
    </intent - filter>
</receiver>
```

<intent-filter>用于指定此广播接收器将用于接收特定的广播类型，<intent-filter>必须包含隐式 Intent 设置的 Action。广播接收器必须通过隐式 Intent 来启动，隐式 Intent 对象必须设置唯一的 action，根据 IntentFilter 匹配隐式 Intent 的要求，IntentFilter 中必须包含隐式 Intent 设置的 action。

12.4 广播接收器的生命周期

一个广播接收器只有在 onReceive(Context，Intent)方法调用的持续期里是有效的。一个正在执行 BroadcastReceiver 的进程(指的是正在执行 onReceive 方法里的代码)被认为是前台进程、除非在系统极度稀缺内存的情况下不会被杀掉的。一旦代码从 onReceive()方法返回，系统就认为 BroadcastReceiver 已经结束并且不再活跃，持有它的进程和在它里面的其他正在运行的应用组件保持同等的重要性。在 onReceive 方法中不能进行任何异步操作，如果进程只持有 BroadcastReceiver(通常是用户不再或最近不和它产生交互)，当从 onReceive()返回时系统会认为这个进程是空的，而且当有其他更重要的进程需要资源时会把它杀掉。这意味着从处理异步操作的方法中返回的时候，BroadcastReceiver 可能不再活跃，系统可能在异步操作完成之前自由地杀掉它的进程。特别地，不能从 BroadcastReceiver 里显示一个对话框或者启动 Service。如果要显示一个对话框，应该使用 NotificationManager API 来代替。如果要启动 Service，可以使用 Context.startService()来发送一个命令给 Service。

12.5 广播从发送到接收的方法调用过程

广播的发送者将广播发送到 ActivityManagerService，ActivityManagerService 接收到这个广播以后，就会在自己的注册中心查看有哪些广播接收器订阅了该广播，然后把这个广播逐一发送到这些广播接收器中，但是 ActivityManagerService 并不等待广播接收器处理这些广播就返回了，因此，广播的发送和处理是异步的。概括来说，广播的发送路径就是从发送者到 ActivityManagerService，再从 ActivityManagerService 到接收者，这中间的两个过程都是通过 Binder 进程间通信机制来完成的。具体的过程如下：

第一阶段：sendBroadcast 把一个广播通过 Binder 进程间通信机制发送给

ActivityManagerService，ActivityManagerService 根据这个广播的 Action 类型找到相应的广播接收器，然后把这个广播放进自己的消息队列中去，就完成第一阶段对这个广播的异步分发了。

```
ContextWrapper 实例的 sendBroadcast(Intent intent)
-->ContextImpl 实例的 sendBroadcast(Intent intent)
-->ActivityManagerProxy.broadcastIntent 方法
-->ActivityManagerService. broadcastIntent 方法
-->ActivityManagerService.broadcastIntentLocked 方法
-->ActivityManagerService.scheduleBroadcastsLocked
-->调用 Handler.sendEmptyMessage 方法
```

第二阶段：ActivityManagerService 在消息循环中处理这个广播，并通过 Binder 进程间通信机制把这个广播分发给注册的广播接收分发器 ReceiverDispatcher，ReceiverDispatcher 把这个广播放进 MainActivity 所在的线程的消息队列中去，ReceiverDispatcher 的内部类 Args 在 MainActivity 所在的线程消息循环中处理这个广播，最终是将这个广播分发给所注册的 BroadcastReceiver 实例的 onReceive 函数进行处理。

```
ActivityManagerService.processNextBroadcast
-->ActivityManagerService.deliverToRegisteredReceiverLocked
-->ActivityManagerService.performReceiveLocked
-->ApplicationThreadProxy.scheduleRegisteredReceiver
-->ApplicaitonThread.scheduleRegisteredReceiver
-->InnerReceiver.performReceive
-->ReceiverDispatcher.performReceive-->Hanlder.post
-->Args.run-->BroadcastReceiver.onReceive:
```

12.6 BroadcastReceiver 实验

实验目的：

(1) 熟悉无序广播的发送方式。

(2) 熟悉有序广播的接收顺序。

(3) 熟悉局部广播的发送方式。

1. 新建工程 Receiver1

具体步骤：运行 Android Studio，单击 Start a new Android Studio project，在 Create New Project 页中，Application name 为 Receiver1，Company domain 为 cuc，Project location 为 C:\android\myproject\Receiver1，单击 Next 按钮，选中 Phone and Tablet，minimum sdk 设置为 API 19：Android 4.4(minSdkVersion)，单击 Next 按钮，选择 Empty Activity，再单击 Next 按钮，勾选 Generate Layout File 选项，勾选 Backwards Compatibility(AppCompat) 选项，单击 Finish 按钮。

2. 新建六个 BroadcastReceiver

在工程视图中 main 源集 Java 文件夹里，右击 cuc. receiver1 包，选择 New|Java Class，Name

输入框中输入 MyReceiver1，Superclass 输入框中输入 android. content. BroadcastReceiver，单击 OK 按钮。编辑 MyBehavior1. java，文件内容如下：

```
package cuc.receiver1;
import android.content.BroadcastReceiver;
import android.content.Context;
import android.content.Intent;
import android.os.Bundle;
import android.util.Log;

public class MyReceiver1 extends BroadcastReceiver {
private static final String TAG = "MyReceiver1";
@Override
public void onReceive(Context context, Intent intent) {
  Log.i(TAG, "onReceive: " + intent.getStringExtra("msg"));
   Log.i(TAG, "onReceive: " + "resiltCode:" + getResultCode()
     + ", resultData:" + getResultData() + ", resultBundle, msg:" + getResultExtras(true).
getString("msg"));
   Bundle bundle = new Bundle();
   bundle.putString("msg", "ResultExtra from MyReceiver1");
   setResult(1, "ResultData from MyReceiver1", bundle);
}
}
```

使用同样的方法可以创建 MyReceiver2、MyReceiver3、MyReceiver4、MyReceiver5 和 MyReceiver6。MyReceiver2 内容如下：

```
package cuc.receiver1;
import android.content.BroadcastReceiver;
import android.content.Context;
import android.content.Intent;
import android.os.Bundle;
import android.util.Log;

public class MyReceiver2 extends BroadcastReceiver {
private static final String TAG = "MyReceiver2";
@Override
public void onReceive(Context context, Intent intent) {
  Log.i(TAG, "onReceive: " + intent.getStringExtra("msg"));
   Log.i(TAG, "onReceive: " + "resiltCode:" + getResultCode()
     + ", resultData:" + getResultData() + ", resultBundle, msg:" + getResultExtras(true).
getString("msg"));
   Bundle bundle = new Bundle();
   bundle.putString("msg","ResultExtra from MyReceiver2");
   setResult(2,"ResultData from MyReceiver2",bundle);
}
}
```

MyReceiver3 内容如下：

```
package cuc.receiver1;
import android.content.BroadcastReceiver;
import android.content.Context;
import android.content.Intent;
import android.os.Bundle;
import android.util.Log;

public class MyReceiver3 extends BroadcastReceiver {
private static final String TAG = "MyReceiver3";
@Override
public void onReceive(Context context, Intent intent) {
  Log.i(TAG, "onReceive: " + intent.getStringExtra("msg"));
   Log.i(TAG, "onReceive: " + "resultCode:" + getResultCode()
      + ", resultData:" + getResultData() + ", resultExtras:msg" + getResultExtras(true).
getString("msg"));
   Bundle bundle = new Bundle();
   bundle.putString("msg","ResultExtra from MyReceiver3");
   setResult(3,"ResultData from MyReceiver3",bundle);
}
}
```

MyReceiver4 内容如下：

```
package cuc.receiver1;
import android.content.BroadcastReceiver;
import android.content.Context;
import android.content.Intent;
import android.os.Bundle;
import android.util.Log;
public class MyReceiver4 extends BroadcastReceiver {
private static final String TAG = "MyReceiver4";
@Override
public void onReceive(Context context, Intent intent) {
  Log.i(TAG, "onReceive: " + intent.getStringExtra("msg"));
   Log.i(TAG, "onReceive: " + "resultCode:" + getResultCode()
      + ", resultData:" + getResultData() + ", resultExtras,msg:" + getResultExtras(true).
getString("msg"));
   Bundle bundle = new Bundle();
   bundle.putString("msg","ResultExtra from MyReceiver4");
   setResult(4,"ResultData from MyReceiver4",bundle);
}
}
```

MyReceiver5 内容如下：

```
package cuc.receiver1;
import android.content.BroadcastReceiver;
import android.content.Context;
```

```
import android.content.Intent;
import android.os.Bundle;
import android.util.Log;
public class MyReceiver5 extends BroadcastReceiver {
private static final String TAG = "MyReceiver5";
@Override
public void onReceive(Context context, Intent intent) {
  Log.i(TAG, "onReceive: " + intent.getStringExtra("msg"));
  Log.i(TAG, "onReceive: " + "resultCode:" + getResultCode()
     + ", resultData:" + getResultData() + ", resultExtras, msg:" + getResultExtras(true).
getString("msg"));
  Bundle bundle = new Bundle();
  bundle.putString("msg","ResultExtra from MyReceiver5");
  setResult(5,"ResultData from MyReceiver5",bundle);
}
}
```

MyReceiver6 内容如下：

```
package cuc.receiver1;
import android.content.BroadcastReceiver;
import android.content.Context;
import android.content.Intent;
import android.util.Log;

public class MyReceiver6 extends BroadcastReceiver {
private static final String TAG = "MyReceiver6";
@Override
public void onReceive(Context context, Intent intent) {
  Log.i(TAG, "onReceive: " + intent.getStringExtra("msg"));
}
}
```

3. 在 AndroidManifest.xml 静态注册 BroadcastReceiver

在前一步创建了 6 个 BroadcastReceiver，其中 4 个在 AndroidManifest.xml 文件中注册，另外两个 MyReceiver5 和 MyReceiver6 在 Java 代码中动态注册。AndroidManifest.xml 内容如下：

```
<?xml version = "1.0" encoding = "utf-8"?>
<manifest xmlns:android = "http://schemas.android.com/apk/res/android"
    package = "cuc.receiver1">
    <application
        android:allowBackup = "true"
        android:icon = "@mipmap/ic_launcher"
        android:label = "@string/app_name"
        android:roundIcon = "@mipmap/ic_launcher_round"
        android:supportsRtl = "true"
        android:theme = "@style/AppTheme">
```

```
        <activity android:name=".MainActivity">
            <intent-filter>
                <action android:name="android.intent.action.MAIN"/>
                <category android:name="android.intent.category.LAUNCHER"/>
            </intent-filter>
        </activity>
        <receiver android:name=".MyReceiver1" android:enabled="true"
            android:exported="true">
            <intent-filter android:priority="500">
                <action android:name="android.intent.action.MY_BROADCAST"/>
            </intent-filter>
        </receiver>
        <receiver android:name=".MyReceiver2" android:enabled="true"
            android:exported="true">
            <intent-filter android:priority="500">
                <action android:name="android.intent.action.MY_BROADCAST"/>
            </intent-filter>
        </receiver>
        <receiver android:name=".MyReceiver3" android:enabled="true"
            android:exported="true">
            <intent-filter android:priority="700">
                <action android:name="android.intent.action.MY_BROADCAST"/>
            </intent-filter>
        </receiver>
        <receiver android:name=".MyReceiver4" android:enabled="true"
            android:exported="true">
            <intent-filter android:priority="800">
                <action android:name="android.intent.action.MY_BROADCAST"/>
            </intent-filter>
        </receiver>
    </application>
</manifest>
```

4. 在 build.gradle(module: app)中，添加 DataBinding 支持，并同步

在文件 build.gradle(module: app)的 android 语句的花括号内添加：

```
dataBinding.enabled = true
```

由于修改了 build.gradle 文件内容，这个文件窗口的上面就出现提示："Gradle files have changed since last project sync. A project sync may be necessary for the IDE to work properly. sync now"，单击 sync now 按钮进行同步。

5. 新建布局文件 activity_main1.xml

具体步骤：在工程结构视图中，右击 layout，依次选择 New|Layout resource file，File name 输入框输入 activity_main1，Root element 输入框输入 layout，Source set 为 main，单击 OK 按钮。layout/activity_main1.xml 内容如下：

```
<?xml version="1.0" encoding="utf-8"?>
```

```
<layout xmlns:android = "http://schemas.android.com/apk/res/android">
    <data/>
    <LinearLayout
        android:orientation = "vertical"
        android:layout_width = "match_parent"
        android:layout_height = "match_parent">
        <Button
            android:layout_width = "match_parent"
            android:layout_height = "wrap_content"
            android:id = "@ + id/sendBroadcastBtn"
            android:text = "sendBroadcast" />
        <Button
            android:layout_width = "match_parent"
            android:layout_height = "wrap_content"
            android:id = "@ + id/sendOrderedBroadcastBtn"
            android:text = "sendOrderedBroadcast" />
        <Button
            android:layout_below = "@id/sendOrderedBroadcastBtn"
            android:layout_width = "match_parent"
            android:layout_height = "wrap_content"
            android:id = "@ + id/sendLocalBroadcastBtn"
            android:text = "sendLocalBroadcast" />
    </LinearLayout>
</layout>
```

6. 编辑 MainActivity.java

前面声明了6个BroadcastReceiver,其中4个(MyReceiver1,MyReceiver2、MyReceiver3和MyReceiver4)已经在AndroidManifest.xml文件中注册,剩下的MyReceiver5和MyReceiver6在Java代码中注册。MyReceiver5通过Context.registerReceiver(@Nullable BroadcastReceiver receiver,IntentFilter filter)注册,MyReceiver6通过LocalBroadcastManager.registerReceiver(BroadcastReceiver receiver,IntentFilter filter)注册。编辑后,MainActivity.java内容如下:

```
package cuc.receiver1;
import android.content.Intent;
import android.content.IntentFilter;
import android.databinding.DataBindingUtil;
import android.os.Bundle;
import android.support.v4.content.LocalBroadcastManager;
import android.support.v7.app.AppCompatActivity;
import android.view.View;
import cuc.receiver1.databinding.ActivityMain1Binding;

public class MainActivity extends AppCompatActivity implements View.OnClickListener{
private static final String TAG  =  "MainActivity";
private static final String ActionName = "android.intent.action.MY_BROADCAST";
private LocalBroadcastManager localBroadcastManager;
private MyReceiver5 myReceiver5;
```

```
private MyReceiver6 myReceiver6;
void registerMyReceiver(){
  myReceiver5 = new MyReceiver5();
  IntentFilter intentFilter5 = new IntentFilter();
  intentFilter5.addAction(ActionName);
  intentFilter5.setPriority(100);
  registerReceiver(myReceiver5,intentFilter5);
  myReceiver6 = new MyReceiver6();
  localBroadcastManager = LocalBroadcastManager.getInstance(this);
  localBroadcastManager.registerReceiver(myReceiver6, intentFilter5);
}

@Override
protected void onCreate(Bundle savedInstanceState) {
  super.onCreate(savedInstanceState);
  ActivityMain1Binding binding = DataBindingUtil.setContentView(this, R.layout.activity_
main1 );
  registerMyReceiver();
  binding.sendBroadcastBtn.setOnClickListener(this);
  binding.sendOrderedBroadcastBtn.setOnClickListener(this);
  binding.sendLocalBroadcastBtn.setOnClickListener(this);
}

@Override
protected void onDestroy() {
  super.onDestroy();
  unregisterReceiver(myReceiver5);
  localBroadcastManager.unregisterReceiver(myReceiver6);
}
@Override
public void onClick(View v) {
  Intent intent = new Intent(ActionName);
  intent.putExtra("msg", "from Broadcast Intent");
  switch (v.getId()) {
   case R.id.sendBroadcastBtn:
     sendBroadcast(intent);
     break;
   case R.id.sendOrderedBroadcastBtn:
     sendOrderedBroadcast(intent, null);
     break;
   case R.id.sendLocalBroadcastBtn:
     localBroadcastManager.sendBroadcast(intent);
     break;
   }
  }
 }
```

7. 编译运行

编译运行,单击 sendBroadcastBtn 按钮,发送普通的广播,Logcat 中 log 输出如下:

```
I/MyReceiver5: onReceive: from Broadcast Intent
    onReceive: resultCode: - 1, resultData:null, resultExtras,msg:null
I/MyReceiver4: onReceive: from Broadcast Intent
    onReceive: resultCode: - 1, resultData:null, resultExtras,msg:null
I/MyReceiver3: onReceive: from Broadcast Intent
    onReceive: resultCode:0, resultData:null, resultExtras:msgnull
I/MyReceiver1: onReceive: from Broadcast Intent
    onReceive: resiltCode:0, resultData:null, resultBundle,msg:null
I/MyReceiver2: onReceive: from Broadcast Intent
    onReceive: resiltCode:0, resultData:null, resultBundle,msg:null
```

从上面结果可以看出，普通广播接收的顺序是 MyReceiver5、MyReceiver4、MyReceiver3、MyReceiver1、MyReceiver2。MyReceiver5(priority=10)是动态注册的，静态注册的先后顺序为 MyReceiver1(priority=500)，MyReceiver2(priority=500)、MyReceiver3(priority=700)和 MyReceiver4(priority=800)。动态注册的接收器(无论多低的优先级)总是比静态注册的接收器先接收到普通广播，静态注册的接收器中，priority 值大的先接收到广播，相同的 priority 值的接收器，先注册的先收到广播。

8. 发送有序广播

在清单文件中，把 MyReceiver3 的 priority 从 700 改为 100，重新编译运行。单击 sendOrderedBroadcastBtn 按钮，发送有序广播，Logcat 中 log 输出如下：

```
I/MyReceiver4: onReceive: from Broadcast Intent
    onReceive: resultCode: - 1, resultData:null, resultExtras,msg:null
I/MyReceiver1: onReceive: from Broadcast Intent
    onReceive: resiltCode: 4, resultData: ResultData from MyReceiver4, resultBundle, msg:
ResultExtra from MyReceiver4
I/MyReceiver2: onReceive: from Broadcast Intent
I/MyReceiver2: onReceive: resiltCode: 1, resultData: ResultData from MyReceiver1,
resultBundle,msg:ResultExtra from MyReceiver1
I/MyReceiver5: onReceive: from Broadcast Intent
    onReceive: resultCode: 2, resultData: ResultData from MyReceiver2, resultExtras, msg:
ResultExtra from MyReceiver2
I/MyReceiver3: onReceive: from Broadcast Intent
    onReceive: resultCode: 5, resultData: ResultData from MyReceiver5, resultExtras:
msgResultExtra from MyReceiver5
```

有序广播接收的顺序是 MyReceiver4、MyReceiver1、MyReceiver2、MyReceiver5、MyReceiver3。MyReceiver5(priority = 100)是动态注册的，静态注册的先后顺序为 MyReceiver1(priority=500)，MyReceiver2(priority=500)、MyReceiver3(priority=100)和 MyReceiver4(priority=800)。优先级高的接收器(跟动态注册还是静态注册无关)先接收到广播，相同优先级的接收器，动态注册的比静态注册的先接收到广播。相同优先级的静态注册的接收器，先注册的先接收到广播。

单击 sendLocalBroadcastBtn 按钮，发送应用内的广播，Logcat 中 log 输出如下：

```
9668 - 9668/cuc.receiver1 I/MyReceiver6: onReceive: from Broadcast Intent
```

可以看到通过 LocalBroadcastManager. sendBroadcast(Intent intent)发送的广播只有 LocalBroadcastManager. registerReceiver(BroadcastReceiver receiver,IntentFilter filter)注册的广播能接收到。

12.7 本章主要参考文献

1. https://developer. android. google. cn/reference/android/content/BroadcastReceiver. html
2. https://developer. android. google. cn/guide/components/broadcasts. html
3. https://developer. android. google. cn/reference/android/support/v4/content/LocalBroad castManager

第 13 章 Handler 与 Service

CHAPTER 13

当应用程序启动时，系统为应用创建一个进程，同时在此进程中开启一个主线程（即 UI 线程），这个主线程专用于运行一个消息队列、负责管理顶级应用程序对象（如 Activity、BroadcastReceiver 等）和它们创建的所有窗口。如果应用需要进行一个耗时的操作，例如联网读取数据，或者读取本地较大的一个文件的时候，这些操作不能放在主线程中。对于这种情况，应用可以创建自己的线程，在子线程中进行耗时的操作。Android 主线程是线程不安全的，也就是说，更新 UI 只能在主线程中更新，在子线程中操作 UI 是危险的。在新创建的线程里调用 post 方法邮寄 Runnable 对象、调用 sendMessage 传递 Message 对象，给定的 Runnable 或 Message 将在 Handler 的 MessageQueue 中被规划，并在合适的时候进行处理，主线程接收子线程发送的 Message，并用此消息配合主线程更新 UI。

13.1 Handler

android. os. Handler 用于发送和处理 Message 和 Runnable 对象，每一个 Handler 实例都关联着一个线程和该线程的 MessageQueue。Handler 的构造函数，如表 13-1 所示。

表 13-1 Handler 的 Public 构造体

构 造 体	解 释
Handler()	默认的构造体，把新创建的 Handler 和创建 Handler 的线程的 Looper 关联起来
Handler(Handler. Callback callback)	把新创建的 Handler 和创建 Handler 的线程的 Looper 关联起来，可以在 callback 里处理消息
Handler(Looper looper)	把新创建的 Handler 与由参数指定的 Looper 关联起来
Handler (Looper looper, Handler. Callback callback)	把新创建的 Handler 与由参数指定的 Looper 关联起来，并提供了 Handler. Callback 用于处理消息

可以看出，Handler 构造函数的参数中如果包含了 android. os. Looper，就关联指定的 Looper，否则关联创建 Handler 的线程的 Looper。Looper 类中包含 MessageQueue 类型的 mQueue 和 Thread 类型的对象 mThread。Handler 类的成员主要有：

```
final Looper mLooper;
 final MessageQueue mQueue;
```

```
final Callback mCallback;
 final boolean mAsynchronous;
 IMessenger mMessenger;
```

Handler 将发送消息和 runnable 对象到 MessageQueue，并且当它们从 MessageQueue 出来时被执行。Handler 类中嵌套的接口 Handler. Callback 定义如下：

```
public interface Callback {
      public boolean handleMessage(Message msg);
 }
```

当实例化一个 Handler 时，实现 Callback 接口，以避免实现 Handler 的子类。表 13-2 列出了 Handler 中的部分方法。

表 13-2　Handler 中的部分 public 方法

方　法	解　释
void dispatchMessage(Message msg)	用于处理消息
final Looper getLooper()	返回跟该 Handler 关联的 Looper
String getMessageName(Message message)	返回指定消息的名字
void handleMessage(Message msg)	Handler 的子类必须实现这个方法以便接收消息
final boolean hasMessages(int what)	检查在消息队列中是否有 code 为'what'的未处理的消息
final boolean hasMessages(int what,Object object)	检查在消息队列中是否有 code 为 what 并且 obj 为 'object'的未处理消息
finalMessage obtainMessage() finalMessage obtainMessage(int what) finalMessage obtainMessage(int what,Object obj) finalMessage obtainMessage(int what,int arg1,int arg2) finalMessage obtainMessage(int what, int arg1, int arg2, Object obj)	返回来自于全局消息池的新的 Message，并使用参数初始化 Message 的对应的成员变量
final boolean post(Runnable r) final boolean postDelayed(Runnable r,long delayMillis) final boolean sendEmptyMessage(int what) final boolean sendMessage(Message msg) final boolean sendEmptyMessageDelayed(int what, long delayMillis) final boolean sendMessageDelayed(Message msg,long delayMillis)	消息执行的时间为系统现在的时间＋delayMillis，若没有指定延时时间，delayMillis 为 0 what 和 r 参数分别用于初始化 Message 的 what 和 callback 成员
final boolean postAtFrontOfQueue(Runnable r) final boolean sendMessageAtFrontOfQueue(Message msg)	将消息插到队列的最前端

续表

方法	解释
final boolean postAtTime(Runnable r,long uptimeMillis) final boolean postAtTime(Runnable r, Object token, long uptimeMillis) final boolean sendEmptyMessageAtTime(int what, long uptimeMillis) boolean sendMessageAtTime(Message msg,long uptimeMillis)	消息执行的时间为 uptimeMillis。若 uptimeMillis 为 0,将消息插到队列的最前端。what、r 和 token 参数分别用于初始化 Message 的 what、callback 和 obj 成员
final void removeCallbacks(Runnable r) final void removeCallbacks(Runnable r,Object token) final void removeMessages(int what) final void removeMessages(int what,Object object)	移除在消息队列中还没处理的满足参数给定条件的消息。如果 object 或者 token 为 null,相当于不用检查 obj,obj 总是满足条件
final void removeCallbacksAndMessages(Object token)	移除在消息队列中还没处理的回调以及所有 obj 为 token 的消息。如果 token 为 null,相当于不用检查 obj,obj 总是满足条件,即移出所有的消息

post、postDelayed、postAtTime、sendEmptyMessage、sendMessage 和 sendMessageDelayed 都是直接或者间接调用 sendMessageAtTime(Message msg,long uptimeMillis)这个方法。下面的源代码可以证实这一点。Handler 的 postAtTime 方法的定义如下:

```
public final boolean postAtTime(Runnable r, Object token, long uptimeMillis)
{
return sendMessageAtTime(getPostMessage(r, token), uptimeMillis);
}
```

上面代码中的 getPostMessage 方法定义如下:

```
private static Message getPostMessage(Runnable r, Object token) {
    Message m = Message.obtain();
    m.obj = token;
    m.callback = r;
    return m;
}
```

从上面 getPostMessage 方法的定义可以看出,postAtTime(Runnable r, Object token, long uptimeMillis)方法中的 Runnable 参数是用来初始化 Message 的 callback 成员的。

Handler 中的 dispatchMessage(Message msg)方法定义如下:

```
public void dispatchMessage(Message msg) {
   if (msg.callback != null) {
      handleCallback(msg);
   } else {
      if (mCallback != null) {
         if (mCallback.handleMessage(msg)) {
```

```
                return;
            }
        }
        handleMessage(msg);
    }
}
```

上面代码中的 handleCallback(Message message)定义如下：

```
private static void handleCallback(Message message) {
    message.callback.run();
}
```

从上面 dispatchMessage(Message msg)方法的定义可以看出，如果 Message 自己定义了 callback，就执行 Message 对象自带的 callback 的 run 方法。否则，如果 Handler 定义了回调函数 mCallback，就执行 mCallback. handleMessage(msg)，该方法的返回结果如果为 true，表示消息已经处理完毕，不需要进一步处理；如果为 false 或者 Handler 没有定义回调，消息还需要由 Handler 子类定义的 handleMessage(Message msg)方法继续处理。

13.1.1 Message

Message 实现了 Parcelable 接口，Message 类的主要成员如下：

```
public int what;
public int arg1;
public int arg2;            // 当需要存储几个整型值时,arg1 和 arg2 是 setData()的低成本的替代
public Object obj;          /* 发送的数据对象。仅当使用 Messenger 跨进程发送 Message 时,才
                               有必要使用 obj。obj 必须是实现了 Parcelable 接口的类。 */
public Messenger replyTo;   // 回复该消息要发送的地址
long when;
Bundle data;                // 在同一进程内使用 Bundle 传送数据
Handler target;             // 发送消息的目标
Runnable callback;          // 消息自定义的回调方法
```

尽管 Message 类的构造体是 public，获取 Message 最好是调用 Message. obtain()方法或者 Handler. obtainMessage()方法，这将从可循环利用的对象池中获取一个 Message 对象。

android. os. Messenger 实现了 Parcelable 接口，持有 Handler 的引用，其他线程可以发消息给这个 Handler。通过它可以实现以 Message 为基础的跨进程通信，可以通过创建指向 Handler 的 Messenger，把该 Messenger 传递到另一个进程。它的实现仅仅是对用于通信的 Binder 的一个简单包装。

13.1.2 Thread

Java 虚拟机允许一个应用有多个线程同时运行。每个线程有一个优先权。具有高优先权的线程优先于具有低优先权的线程被执行。每个线程都可以被标记为守护线程，也可以不标记为守护进程，当在某个线程中运行的代码创建了一个新的 java. lang. Thread 对象

时，新创建的线程的优先权被初始设置为与创建它的线程相同的优先权，当且仅当创建它的线程是守护线程时，新创建的线程才是守护线程。有两种方法创建一个新的执行线程。一种是声明一个 Thread 的子类，这个子类应该覆盖 Thread 类的 run 方法，子类的实例启动，执行子类中定义的 run 方法。示例如下：

```
class PrimeThread extends Thread {
      long minPrime;
      PrimeThread(long minPrime) {
        this.minPrime = minPrime;
      }
      public void run() {      // compute primes larger than minPrime
   }
}
```

下面的代码能够创建线程并且启动它运行：

```
PrimeThread p = new PrimeThread(143);
p.start();
```

另外一种创建线程的方式是声明一个实现了 java.lang.Runnable 接口的类（实现了 run 方法），新建这个类的实例，在创建线程时作为参数传进去，然后启动线程，线程被启动时，执行线程的 Runnable 参数中的 run 方法。Runnable 接口的定义如下：

```
@FunctionalInterface
public interface Runnable {
    public abstract void run();
}
```

如果类的实例希望被线程执行，这个类就应该实现 Runnable 接口。Runnable 接口被设计用来为那些希望在活跃时执行代码的对象提供公共的协议。例如 Thread 类就实现了 Runnable 接口。在大多数情况下，如果仅仅打算覆盖 run()方法，而不覆盖 Thread 的其他方法，就应该使用 Runnable 接口。采用 Runnable 对象作为 Thread 参数创建 Thread 的一个例子如下：

```
class PrimeRun implements Runnable {
      long minPrime;
      PrimeRun(long minPrime) {
        this.minPrime = minPrime;
      }
      public void run() {      // compute primes larger than minPrime
...
      }
   }
```

类 PrimeRun 实现了 Runnable 接口，下面的代码以 PrimeRun 对象作为参数创建线程并且启动它运行：

```
PrimeRun p = new PrimeRun(143);
new Thread(p).start();
```

13.1.3 HandlerThread

android. os. HandlerThread 是 java. lang. Thread 的直接子类，这个类主要用于新建具有 android. os. Looper 的 Thread，在 HandlerThread 的 run 方法中创建 Looper，因此在调用了 Thread 的 start 方法后就创建了属于这个线程的 Looper，然后就可以利用这个 Looper 创建 Handler 类型的对象。

使用 HandlerThread 的步骤主要包括：创建 HandlerThread 对象，调用它的 start 方法，获取它的 Looper，创建 Handler，在必要的时候调用 Looper 的 quit 方法或 quitSafely 方法中止消息的处理。

```
HandlerThread mThread = new HandlerThread("handlerThread");  // 创建 HandlerThread 对象
mThread.start();                                             // 调用 Thread 的 start 方法
Looper mLooper = mThread.getLooper();                        // 获取它的 Looper
mWorkHandler = new Handler(mLooper);                         // 创建 Handler
...
mLooper. quit();                                   // 在必要的时候调用 Looper 的 quit 方法
```

13.2 Service

android. app. Service 是 ContextWrapper 的直接子类，实现了 ComponentCallbacks2 接口。Service 是一个应用程序组件，它表示应用期望在不与用户交互的情况下执行运行时间较长的操作，或者期望提供功能给其他应用使用。Service 能被其他应用程序的组件启动，当用户切换到其他的应用时还能保持后台运行。此外，应用程序组件还能与 Service 绑定，并与 Service 进行交互，甚至能进行进程间通信(IPC)。比如，Service 可以处理网络传输、音乐播放、执行文件 I/O，或者与 content provider 进行交互，所有这些都是后台进行的。

Service 不是运行在单独的进程中，除非使用 process 属性特别指定，它运行在和它所在的应用相同的进程的主线程中，不创建自己的线程，这意味着，应当避免在 Service 里面进行耗时的操作，不能把耗时的操作直接放在 Service 的 onStart 方法中，这样会出现 Application Not Responding。如果 Service 要执行一些很耗 CPU 的工作或者阻塞的操作(比如播放 MP3 或网络操作)，就应该在 Service 中创建一个新的线程来执行这些工作。IntentService 是一个有自己的线程并且可以规划自己要做的工作的独立的标准实现。android. app. IntentService 是 Service 的直接子类，并且本身就是一个抽象类，用于处理异步请求，那么它包含了 Service 的全部特性，与 Service 不同的是，IntentService 在执行 onCreate 操作的时候，内部新建了一个线程，它可以用于在后台执行耗时的异步任务，当任务完成后会自动停止。客户端通过调用 startService(Intent)发出请求，Service 在需要的时候启动，使用一个工作线程顺序处理每个 Intent，在工作运行结束后停止它自己。它的内部通过 HandlerThread 和 Handler 实现异步操作，在 Service 的 onStart(Intent intent，int startId)方法中，把 Intent 封装进 Message 的 obj 成员，并调用 ServiceHandler 的 sendMessage(Message msg)发送消息。为了使用 IntentService，需要继承 IntentService，并且实现两个方法，一个是构造方法，必须传递一个线程名称的字符串，另外一个就是进行异步处理的方法 onHandleIntent(Intent intent)方法。IntentService 将接收 Intent，启动一个工作线程，在适当的时候停止 Service。所有的请求在单一的工作线程中处理，它们可能消耗较长的事件，但不会阻塞应用的主循环，但是一次仅仅处理一个请求。

Service 将状态回传给 Activity，可以通过广播的形式，通常的情况，将 BroadcastReceiver 写成 Activity 的内部类，这个 onReceiver 可以直接调用 Activity 的方法来更新界面。

13.2.1 Service 在 AndroidManifest.xml 中的语法

所有的 Service 都必须在 manifest 文件中通过< service >元素来声明。没有在 manifest 中声明的 Service 将对系统不可见并且不会运行。Service 在清单文件 AndroidManifest.xml 中的语法如下：

```
<service android:enabled = ["true" | "false"]
        android:exported = ["true" | "false"]
        android:icon = "drawable resource"
        android:isolatedProcess = ["true" | "false"]
        android:label = "string resource"
        android:name = "string"
        android:permission = "string"
        android:process = "string" >
  ...
</service>
```

< service >的父元素是< application >，< service >的子元素有：< intent-filter >和< meta-data >。表 13-3 列出了 Service 元素的属性。

表 13-3 < service >元素的属性

属　性	解　释
enabled	设置 Service 能否被系统实例化，默认为 true。< application >元素也有它自身的 enabled 属性，能应用到所有的应用组件，包括< service >。要想这个 Service 被使能(实例化)，< application >和< service >的 enabled 属性都必须为 true(它们的默认值都为 true)。如果有一个为 false，这个< service >就会 disabled，不会被实例化
exported	设置其他应用的组件是否能调用这个 Service 或同它交互。当值为 false 时，只有同一个应用的组件或有相同用户 ID 的应用能启动这个 Service 或绑定它。默认值依赖于< service >是否包含< intent-filter >。< intent-filter >的缺失意味着它只能通过指定它的准确类名来调用它。这就意味着这个服务只能在应用内部被使用(因为其他应用不知道类名)，因此，在这种情况下，默认值是 false。另一方面，至少有一个< intent-filter >意味着这个服务可以在外部被使用，因此，默认值为 true。exported 属性并非是限制 Service 暴露给其他应用的唯一途径，也可以通过 permission 属性来限制跟 Service 交互的外部实体(参见 permisson 属性)

续表

属 性	解 释
icon	代表 Service 的图标。这个属性必须被设置为包含图片的 drawable 资源的引用。如果没有设置,就用 application 的图标(< application >元素的 icon 属性)来代替
isolatedProcess	如果设置为 true,这个服务将运行在专门的进程中,这个进程与系统的剩余部分隔离出来,它自身没有权限,同它唯一的通信方式就是通过 Service API(binding 或 starting)
label	设置 Service 显示给用户的名称。如果这个属性没有设置,将使用< application >元素的 label 属性代替。这个 Service 的 label,无论是通过自身的 label 属性设置的,还是通过< application >元素 label 属性设置的,都是< service > 中的< intent-filter >的默认 label 这个 label 应当设置为 string 资源的引用,以便在用户接口中能同其他 string 资源一样本地化。然而,在开发时为了方便起见,它也能被设置为 raw 字符串
name	Service 子类名称,可以是完整格式的类名(例如"com. example. project. RoomService"),也可以是以点开始的相对类名。如果名称的第一个字符是点(例如". RoomService"),它会被添加到在< manifest >元素中声明的包名后面。name 没有默认值,必须设置
permission	为了启动这个 Service 或绑定到它,一个组件必须要有的权限的名称。如果 Service 没有设置这个属性,由< application >元素的 permission 属性设置的权限就会应用到这个 Service。如果都没有设置,那么这个 Service 就不受 permission 保护。如果 startService(), bindService()或 stopService()的调用者没有获取授权,那么这些方法就不会工作,而且这个 Intent 对象也不会传递到 Service
process	Service 将要运行的进程名称。一般来讲,应用的所有组件都运行在应用创建的默认进程中,它和应用的包名一样。< application >元素的 process 属性能为应用内的所有组件设置默认值。然而,组件能通过它自身的 process 属性覆盖默认值,从而允许扩展应用跨越多个进程。如果 process 属性的值以冒号(:)开始,那么当需要的时候,一个新的、对这个应用私有的进程就被创建,同时这个 Service 就在这个新创建的进程里运行。如果 process 属性的值以小写字符开始,那么这个 Service 将运行在 process 属性值指定的全局进程中,这就允许在不同应用中的组件共享这个进程

在清单文件的< service >标签中,可以对 Service 的全局访问进行限制。同时其他应用需要在它们自己的清单文件中声明对应的< uses-permission >元素,以便能开始、停止或者绑定 Service。

从 GINGERBREAD(android 2.3)开始,当使用 Context. startService(Intent)时,可以在 Intent 上设置 Intent. FLAG_GRANT_READ_URI_PERMISSION 和/或者 Intent. FLAG_GRANT_WRITE_URI_PERMISSION,这允许 Service 临时访问 Intent 里指定的 URI,访问将能持续到 Service 调用 stopSelf(int)方法或者直到 Service 被完全停止。

13.2.2 Service 的两种工作方式

Service 主要有两种工作方式:启动的 Service 和绑定的 Service。

启动的 Service:应用告诉系统它想在后台做某件事情(用户不直接和应用进行交互)。这种使用场景对应于应用组件调用 Context 对象的 startService(Intent)方法启动 Service,一旦被启动,Service 就能在后台一直运行下去,即使启动它的组件已经被销毁了。Service 将继续运行、直到调用 Context 对象的 stopService(Intent)方法或者 Service 操作完成后自己的 stopSelf()被调用为止。通常,started Service 执行单一的操作并且不会向调用者返回结果，比如，它可以通过网络下载或上传文件。

绑定的 Service：应用暴露部分功能给其他应用，这种使用场景对应于应用组件调用 Context 对象的 bindService(Intent, ServiceConnection, int)方法，它允许可以长期存在的连接。Bound Service 提供了一个客户端/服务器接口，允许组件与 Service 进行交互、发送请求、获取结果，甚至可以利用进程间通信(IPC)跨进程执行这些操作。绑定的 Service 的生存期和被绑定的应用组件一致。多个组件可以同时与一个 Service 绑定，只有所有的组件解除绑定后，Service 才会被销毁。和 Service 进行交互，可以把 Service 看成本地 Java 对象、直接进行方法调用，或者使用 AIDL 提供一个完全远端的接口。

一个 Service 可以同时以这两种方式工作：可以是 started(一直运行下去)，同时也能被绑定。两个回调方法都要实现：实现 onStartCommand(Intent, int, int)以允许组件启动 Service、实现 onBind()以允许绑定。使用 startService 方法启动 Service 之后，一定要在某处调用 stopService 停止 Service，不管是否使用 bindService。在调用 bindService 绑定 Service 以后，还应当在某处调用 unbindService(ServiceConnection conn) 解除绑定(尽管 Activity 被 finish 的时候绑定会自动解除，并且 Service 会自动停止)，不管 Service 是否被启动。

如果一个 Service 被启动，也被一个或者多个客户端绑定，启动与绑定的先后顺序无关紧要(与 startService 与 bindService 的调用顺序无关)，只有在该 Service 被停止、且所有的客户端都解除了与该 Service 的绑定后(与 stopService 与 unbindService 的调用顺序无关)，该 Service 才会被销毁。

无论应用程序中 Service 是 started、bound、还是两者都支持，任何应用程序组件都可以使用此 Service(即使是从另一个独立的应用)。不过，也可以在 manifest 文件中设置 < service android: exported= "false">，以便阻止其他应用的访问。无论是上面哪种应用场景，当 Service 组件被创建时，系统实际做的事就是在主线程中调用 onCreate()初始化 Service 组件。实现这些是 Service 的义务，例如创建新的线程做自己的工作。

13.2.3 Service 的生命周期

表 13-4 列出了 Service 的 public 方法。Service 的整个生命周期从调用 onCreate()开始，到 onDestroy()返回时结束，如图 13-1 所示。一个 Service，无论它怎么创建的，都可能允许客户端绑定到它，都可以接收 onBind()和 onUnbind()调用。

表 13-4 Service 类的 public 方法

方　法	解　释
final Application getApplication()	返回拥有这个 Service 的应用
abstract IBinder onBind (Intent intent)	返回到 Service 的通信通道 IBinder。 当组件调用 bindService()绑定到 service 时，系统调用此方法。实现 Service 子类必须重写该方法，以便客户端能够使用它与 Service 通信；但是如果不允许绑定，这个方法应返回 null
void onConfigurationChanged (Configuration newConfig)	当 Service 处于运行中，在设备配置改变时，系统会调用这个方法

续表

方　法	解　释
void onCreate()	Service 被系统在首次创建时,系统调用此方法,执行只运行一次的初始化工作(在调用其他方法如 onStartCommand()或 onBind()之前,Service 必须已经创建)。如果 Service 已经创建,这个方法不会被调用。不管是 startService 还是 bindService,最开始的回调函数肯定是 onCreate(),并且该回调只会执行一次,之后再执行 startService 或者 bindService 都不会再调用 onCreate()方法,除非 Service 被销毁了
void onDestroy()	系统调用此方法通知 Service,它不再被使用正在被销毁。应在此方法中释放 service 占用的资源,比如线程、已注册的接收器等等。这是 Service 收到的最后一个调用。如果一个组件通过调用 startService()启动一个 Service,导致 onStartCommand()被调用,之后 service 会保持运行,直到它通过 stopSelf()停止自己或另外的组件调用 stopService()停止它。如果一个组件调用 bindService()来创建 Service,那么 Service 只是运行在绑定期间。一旦 service 从所有的客户端解除绑定,系统就会调用 OnDestroy()销毁它
void onLowMemory()	当总的系统内存不足时,正在运行的进程应当裁剪存储器使用,这个方法被调用
void onRebind(Intent intent)	在客户端已经被通知断开连接后,新的客户端连接到 Service 时,系统会调用这个方法
void onStart(Intent intent, int startId)	这个方法在 API 级别 5 被废弃。已经被 onStartCommand(Intent, int,int)代替
int onStartCommand(Intent intent, int flags,int startId)	当其他组件(比如 Activity)通过调用 startService(intent)显式启动 Service 时,系统调用这个方法。每次执行 startService()都会调用一次 onStartCommond 方法。如果只需要提供绑定,就不需实现此方法。一旦这个方法执行,Service 就启动并且在后台长期运行。如果实现了此方法,就需要在 Service 完成任务时停止它,通过调用 stopSelf()或 stopService() 第一个参数 intent 是 startService(intent)传递过来的参数;第二个参数 flags 是这次启动请求的附加数据,可取值为 START_FLAG_REDELIVERY 或者 START_FLAG_RETRY。第三个参数 startId,是指这个 Service 是第几次启动
void onTaskRemoved(Intent rootIntent)	如果 Service 正在运行,用户已经移除了来自于 Service 的应用的任务时,系统会调用这个方法
void onTrimMemory(int level)	当操作系统已经决定是进程裁剪不必要的存储器的最佳时候时,系统会调用这个方法
boolean onUnbind(Intent intent)	当所有的客户端都已经从这个 Service 公布的接口断开连接的时候,系统会调用这个方法
final void startForeground(int id,Notification notification)	如果通过 startService(Intent)启动了 Service,使得这个 Service 运行在前台,并且在这个状态时,提供持续的通知显示给用户
final void stopForeground(int flags)	从前台状态移除这个 Service,允许需要更多的存储器的时候,杀死这个 Service
final void stopForeground(boolean removeNotification)	与 stopForeground(int)相同的作用

续表

方　　法	解　　释
final voidstopSelf()	如果这个 Service 以前被启动，就停止这个 Service
final voidstopSelf(int startId)	stopSelfResult(int)的旧版本，它不返回结果
final boolean stopSelfResult(int startId)	如果最近一次被启动的 ID 等于 startId，就停止这个 Service

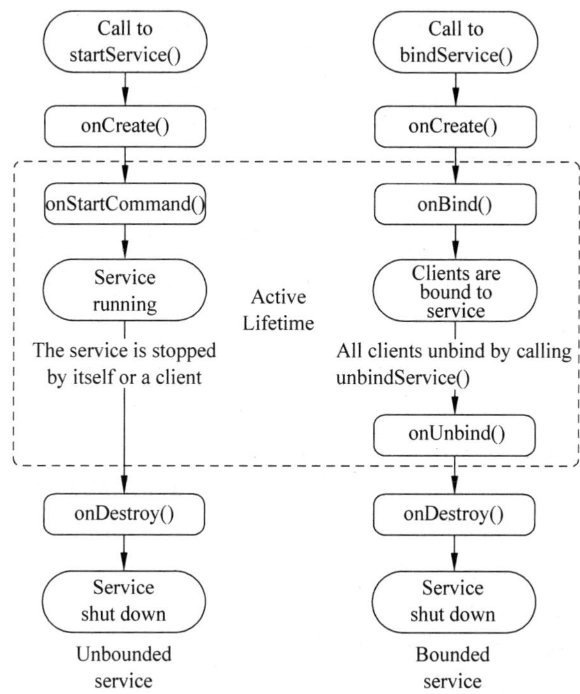

图 13-1　Service 生命周期

1. Started Sevice 的生命周期

如果调用 Context 对象的 startService(Intent)，系统将会重新获得 Service(如果有需要的话，创建它并调用它的 onCreate 方法)，然后调用 onStartCommand(Intent，int，int)方法，该方法的 Intent 参数是客户端调用 startService(Intent)时提供的。Service 将继续运行、直到调用 Context 对象的 stopService(Intent)或者 Service 自身的 stopSelf()被调用为止。服务停止后，系统会将其销毁。Started Sevice 的有效生命周期从调用 onStartCommand()开始，有效生命周期与整个生命周期同时结束(即使在 onStartCommand()返回之后，Service 仍然处于活动状态)。多次调用 Context 对象的 startService(Intent)并没有嵌套，所以不管一个 Service 被启动多少次，虽然每次启动，都能导致 onStartCommand(Intent，int，int)调用，但是一旦 Context 对象的 stopService(Intent)被调用或者 Service 自己的 stopSelf()方法被调用，Service 就会停止。Service 可以使用自己的 stopSelf(int)方法来确保被启动的 Service 处理完毕再停止。

启动的 Service 可以运行在两种附加的操作模式，取决于从 onStartCommand(Intent，int，int)方法的返回值：START_STICKY，START_STICKY 或者 START_REDELIVER_

INTENT。当 Service 从 onStartCommand 返回后，如果 Service 的进程被杀死，则①START_NOT_STICKY，如果没有新的启动 Intent 发送给它，它将离开 started 状态，直到将来客户端显式地调用 Context.startService(Intent)才会创建 Service；②START_STICKY：保留 Service 的状态为开始状态，但不保留客户端传来的 Intent 对象，随后系统会尝试重新创建 Service，由于服务状态为 started 状态，所以创建服务后一定会调用 onStartCommand(Intent, int, int)方法，如果在此期间客户端显式地调用 Context.startService(Intent)，那么 onStartCommand 的参数 Intent 将为 null；③START_REDELIVER_INTENT：系统会自动重启该服务，并把客户端传来的 Intent 对象传入 onStartCommand(Intent,int,int)方法；④START_STICKY_COMPATIBILITY：START_STICKY 的兼容版本，但不保证服务被 kill 后一定能重启。

2. Bound Service 的生命周期

客户端使用 Context 对象的 bindService(Intent, ServiceConnection, int)可以获得一个持续的到 Service 的连接，如果 Service 还没有创建，就会调用 onCreate()创建 Service，但 bindService 并不调用 onStartCommand(Intent, int, int)方法。只要连接没有断开，Service 就保持运行状态，无论客户端是否获得了 Service 的 IBinder 的引用。同一客户端多次调用 bindService(Intent, ServiceConnection, int)并不重复调用 onBind(Intent intent)方法。当所有的客户端都已经从 Service 公布的特定接口断开连接后，系统回调 onUnbind(Intent intent)，参数 Intent 是绑定 Service 时的 Intent，但是 Intent 里的 extras 在这里是看不见的。Bound Service 的有效生命周期从调用 onBind()方法开始，有效生命周期在 onUnbind()返回时结束。当 Service 与所有客户端之间的绑定全部取消时，Android 系统便会销毁 Service(除非还使用 onStartCommand()启动了该服务)。

3. 同时被启动和被绑定的 Service 的生命周期

一个 Service 可以被启动，同时也有连接绑定在它上面。只要它没有被停止，或者有连接绑定在该 Service 上，系统就保持 Service 运行。一旦这两种条件都不具备，Service 的 onDestroy()方法被调用，Service 被销毁。从 onDestroy()方法返回时，所有的清除工作应该完成。

如图 13-2 所示，如果希望新的客户端绑定它的时候调用 onRebind 方法，先前的 onUnbind()方法必须返回 true。Service 类中的 onUnbind 方法默认的实现是返回 false，因此实现 Service 子类时，必须重写 onUnbind(Intent intent)方法并返回 true。如果 Service 的 onUnbind()返回 false，那么重新启动 Activity 来绑定 Service 时，Service 的 onReBind 方法不会执行，但是 ServiceConnection 方法被回调了。

ServiceConnection 定义如下：

```
public interface ServiceConnection {
    void onServiceConnected(ComponentName name, IBinder iBinder);
    void onServiceDisconnected(ComponentName name);
    default void onBindingDied(ComponentName name) {
    }
    default void onNullBinding(ComponentName name) {
    }
}
```

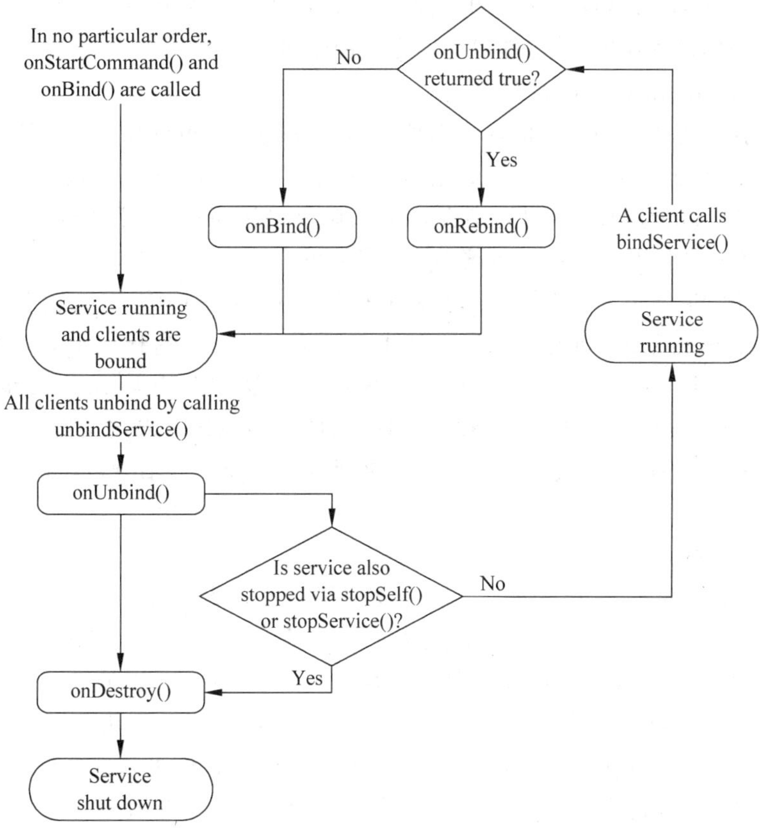

图 13-2 同时被启动和被绑定的 Service 的生命周期

onServiceConnected 方法在连接一个 Service 成功时被调用，以传递 Service 中的 onBind() 方法返回的 IBinder 给客户端。onServiceDisconnected 方法在 Service 崩溃或被杀死导致的连接意外中断时被调用，而如果应用自己解除绑定时则不会调用 onServiceDisconnected 方法。nServiceConnected(ComponentName name，IBinder iBinder)中的参数 iBinder 就是 Servic 的 onBind(Intent)方法返回的 IBinder 对象，用于客户端和 Service 进行通信。

13.2.4 创建绑定的 Service

创建绑定的 Service 时，必须提供客户端用来与 Service 进行交互的编程接口 IBinder。可以通过三种方法定义接口：

第一种方法：扩展 Binder 类

如果 Service 仅供自己的应用使用，并且在与客户端相同的进程中运行(常见情况)，则应该在 Service 中继承 Binder 类并创建它的一个实例，并且重写 Service 的 onBind(Intent intent)方法、返回这个实例。客户端收到 Binder 后，可利用它直接访问 Binder 和 Service 中可用的公共方法。

只有在客户端和 Service 位于同一应用内并且在同一进程内，才能继承 Binder 类创建接口。之所以要求 Service 和客户端必须在同一应用内，是为了便于客户端转换返回的对象和正确调用其 API。Service 和客户端还必须在同一进程内，因为此方法不能进行任何跨进程通信。如果 Service 只是自己应用的后台工作线程,则优先采用这种方法。当 Service 被其他应用或不同的进程使用,就不能以这种方式创建接口。

第二种方法：使用 Messenger

如果想要执行跨越不同应用的客户端请求,客户端和 Service 在不同的进程内,需要进程间通信,但不需要同时处理多个请求,就应该使用 Messenger 来创建接口。在 Service 中可以定义一个 Handler,响应不同类型 Message 对象,Messenger 可以和客户端分享一个 IBinder,允许客户端使用 Message 对象向 Service 发送命令。此外,客户端还可定义自有 Messenger,以便 Service 回传消息。这是执行进程间通信(IPC)的最简单方法,因为 Messenger 会在单一线程中创建包含所有请求 Service 的队列,这样就不必对 Service 进行线程安全设计,而纯粹的 AIDL 接口会同时向 Service 发送多个请求,Service 随后必须应对多线程处理,如果 Service 必须执行多线程处理,则应使用 AIDL 来定义接口。对于大多数应用,不需要 Service 执行多线程处理,可以使用 Messenger 可让 Service 一次处理一个调用。

第三种方法：使用 AIDL

AIDL(Android 接口定义语言)将对象分解成操作系统能够识别的原语,使得这些对象能够跨越进程边界传输。Messenger 的方法实际上是以 AIDL 作为其底层结构。如上所述,Messenger 会在单一线程中创建包含所有客户端请求的队列,以便 Service 一次接收一个请求。不过,如果希望 Service 同时处理多个请求,就可以直接使用 AIDL。只有当不同应用的客户端用 IPC 方式访问 Service,并且想要在 Service 同时处理多个请求(多线程),才有必要使用 AIDL。在此情况下,Service 必须具备多线程处理能力,并采用线程安全式设计。大多数应用都不应该使用 AIDL 来创建绑定的 Service,因为它要求具备多线程处理能力,并导致实现的复杂性增加。

使用 AIDL 时,必须创建一个定义编程接口的.aidl 文件。Android SDK 工具利用该文件生成一个实现接口并处理 IPC 的抽象类,这个抽象类可以在 Service 内对其进行继承。

1. 通过扩展 Binder 类创建绑定的 Service

下面是扩展 Binder 类创建绑定的 Service 的具体步骤：

第一步：在 Service 中,创建一个可满足下列任一要求的 Binder 实例：①Binder 包含客户端可调用的公共方法。②返回当前 Service 实例,Service 包含客户端可调用的公共方法。③返回由 Service 持有的另一个类的实例,这个类包含客户端可调用的公共方法。

第二步：从 onBind()回调方法返回此 Binder 实例。

第三步：客户端中,从 onServiceConnected()回调方法接收 Binder,并使用该 Binder 提供的方法调用绑定的 Service。

例如,下面的 LocalService 类中,LocalBinder 为客户端提供 getService()方法,以获取 LocalService 的当前实例,这样,客户端便可调用 Service 中的公共方法,例如调用 Service 中的方法 getRandomNumber()。

```
import android.app.Service;
import android.content.Intent;
import android.os.Binder;
```

```
import android.os.IBinder;
import java.util.Random;

public class LocalService extends Service {
    private final IBinder mBinder = new LocalBinder();
    private final Random mGenerator = new Random();

    /* 这个 Service 和 client 在在同一应用中,且在相同的进程中,所以不需要处理 IPC */
    public class LocalBinder extends Binder {
        LocalService getService() {
            return LocalService.this;
        }
    }
    @Override
    public IBinder onBind(Intent intent) {
        return mBinder;
    }
    /** method for clients */
    public int getRandomNumber() {
        return mGenerator.nextInt(100);
    }
}
```

下面是 BindingActivity 的代码,BindingActivity 与 LocalService 在同一应用中,且运行在同一进程中,在布局文件 main.xml 中,有一个按钮,onClick 属性设置为 onButtonClick。

```
import android.app.Activity;
import android.content.ComponentName;
import android.content.Context;
import android.content.Intent;
import android.content.ServiceConnection;
import android.os.Bundle;
import android.os.IBinder;
import android.view.View;
import android.widget.Toast;
public class BindingActivity extends Activity {
    LocalService mService;
    boolean mBound = false;
    @Override
    protected void onCreate(Bundle savedInstanceState) {
        super.onCreate(savedInstanceState);
        setContentView(R.layout.main);
    }
    @Override
    protected void onStart() {
        super.onStart();
        Intent intent = new Intent(this, LocalService.class);
        bindService(intent, mConnection, Context.BIND_AUTO_CREATE);
```

```
        }
        @Override
        protected void onStop() {
            super.onStop();
            if (mBound) {
                unbindService(mConnection);
                mBound = false;
            }
        }

        public void onButtonClick(View v) {
            if (mBound) {
                int num = mService.getRandomNumber();
                Toast.makeText(this, "number: " + num, Toast.LENGTH_SHORT).show();
            }
        }

        /** Defines callbacks for service binding, passed to bindService() */
        private ServiceConnection mConnection = new ServiceConnection() {
            @Override
            public void onServiceConnected(ComponentName className,
                                           IBinder iBinder) {
                LocalService.LocalBinder binder = (LocalService.LocalBinder) iBinder;
                mService = binder.getService();
                mBound = true;
            }
            @Override
            public void onServiceDisconnected(ComponentName arg0) {
                mBound = false;
            }
        };
    }
```

在上例中，onStart()方法中将客户端与 Service 绑定，onStop()方法中将客户端与 Service 取消绑定。

2. 使用 Messenger 创建绑定的 Service

如需让 Service 与远端的进程通信，则可使用 Messenger 为 Service 提供接口。客户端并没有直接调用 Service 中的方法，而是传递 Message 对象到 Service 中的 Handler 进行处理。使用 Messenger 创建绑定的 Service 的步骤如下：

第一步，在 Service 中，新建一个 Handler 类型对象 handler，并重写 Handler 的 handleMessage(Message msg)方法。

第二步，使用构造函数 Messenger(Handler target)创建 Messenger 类型对象 messenger，参数为前一步创建的 handler。

第三步，在 Service 的 onBind()方法中，调用 messenger 的 getBinder 方法获取一个 IBinder，并把这个 IBinder 对象返回客户端，即 return messenger.getBinder()。

第四步，客户端调用 bindService(Intent service，ServiceConnection conn，int flags)绑定

Service，在 ServiceConnection 的回调方法 onServiceConnected(ComponentName name, IBinder iBinder)中，调用 Messenger(IBinder iBinder)创建一个 Messenger。因为 iBinder 引用了 Service 中的 Handler，然后使用该 Messenger 对象的 send(Message message)方法发送消息给 Service。Service 中 Handler 接收每个 Message，在 handleMessage()方法中处理。

下面是使用 Messenger 绑定 Service 的一个实例，在 Service 中，首先创建使用当前线程的 Looper 的 Handler，用于处理客户端发来的消息；然后创建 Messenger；最后在 Service 的 onBind(Intent intent)方法中返回 Messenger 的 getBinder()。

```
package cuc.messengerclient;
import android.app.Service;
import android.content.Intent;
import android.os.Handler;
import android.os.IBinder;
import android.os.Message;
import android.os.Messenger;
import android.os.Process;
import android.os.RemoteException;
import android.util.Log;

public class MessengerService extends Service
{
    private static final String TAG = "MessengerService";
    private static final int MSG_SUM = 1;
    Handler handler = new Handler(){
        @Override
        public void handleMessage(Message msg) {
            Message msgToClient = Message.obtain(msg);
            switch (msg.what){
                case MSG_SUM:
                    try {
                        msgToClient.what = MSG_SUM;
                        msgToClient.arg1 = msg.arg1 + msg.arg2;
                        Log.i(TAG, "handleMessage: pid-->" + Process.myPid());
                        msg.replyTo.send(msgToClient);
                    } catch (RemoteException e) {
                        e.printStackTrace();
                    }
                    break;
            }
            super.handleMessage(msg);
        }
    };
    private Messenger messenger = new Messenger(handler);
    @Override
    public IBinder onBind(Intent intent){
        return messenger.getBinder();
    }
}
```

另一应用中的代码调用上面的 Service 的实例如下：客户端要做的就是根据 Service 返回的 iBinder 创建一个 Messenger，然后调用该 Messenger 对象的 send(Message message) 方法发送 Message 给 Service 处理。如果客户端还需要根据 Service 处理消息的结果进行处理，就创建一个 Handler，处理 Service 返回的消息，并用这个 Handler 创建一个 Messenger，把这个 Messenger 作为客户端发送给 Service 的 Message 对象的 replyTo 成员传给 Service。

```
import android.content.ComponentName;
import android.content.Context;
import android.content.Intent;
import android.content.ServiceConnection;
import android.os.Handler;
import android.os.IBinder;
import android.os.Message;
import android.os.Messenger;
import android.os.Process;
import android.os.RemoteException;
import android.support.v7.app.AppCompatActivity;
import android.os.Bundle;
import android.util.AndroidException;
import android.util.Log;
import android.view.View;
import android.widget.Button;
import android.widget.LinearLayout;
import android.widget.TextView;

public class MainActivity extends AppCompatActivity
{
    private static final String TAG = "MainActivity";
    private static final int MSG_SUM = 1;
    private TextView connectTxt;
    private Button addBtn;
    private TextView resultTxt;
    private void assignViews() {
        connectTxt = (TextView) findViewById(R.id.connectTxt);
        addBtn = (Button) findViewById(R.id.addBtn);
        resultTxt = (TextView) findViewById(R.id.resultTxt);
    }
    private boolean isConn;
    Handler handler = new Handler(){
        @Override
        public void handleMessage(Message msg) {
            switch (msg.what){
                case MSG_SUM:

resultTxt.setText(resultTxt.getText() + Integer.toString(msg.arg1));
                    break;
            }
```

```
            super.handleMessage(msg);
        }
    };
    private Messenger messenger = new Messenger(handler);
    private Messenger remoteMessenger;

    private ServiceConnection mConn = new ServiceConnection(){
        @Override
        public void onServiceConnected(ComponentName name, IBinder iBinder){
            remoteMessenger = new Messenger(iBinder);
            isConn = true;
            connectTxt.setText("connected!");
        }

        @Override
        public void onServiceDisconnected(ComponentName name){
            remoteMessenger = null;
            isConn = false;
            connectTxt.setText("disconnected!");
        }
    };

    private int op1 = 1;

    @Override
    protected void onCreate(Bundle savedInstanceState){
        super.onCreate(savedInstanceState);
        setContentView(R.layout.activity_main1);
        assignViews();
        //开始绑定服务
        bindServiceInvoked();
        addBtn.setOnClickListener(new View.OnClickListener(){
            @Override
            public void onClick(View v){
                try{
                  resultTxt.setText(op1 + "+" + op1 + "-->");
                  Message msg = Message.obtain(null, MSG_SUM, op1, op1);
                    op1++;
                    msgFromClient.replyTo = messenger;
                    if (isConn){
                        remoteMessenger.send(msg);
                        Log.i(TAG, "onClick: pid-->" + Process.myPid());
                    }
                 } catch (RemoteException e){
                    e.printStackTrace();
                }
            }
        });
    }
```

```
    private void bindServiceInvoked(){
        Intent intent = new Intent();

intent.setClassName("cuc.messengerclient","cuc.messengerclient.MessengerService");
        bindService(intent, mConn, Context.BIND_AUTO_CREATE);
        Log.i(TAG, "bindService invoked !");
    }
    @Override
    protected void onDestroy(){
        super.onDestroy();
        unbindService(mConn);
    }
}
```

上面的 MainActivity 使用的布局文件 activity_main1 内容如下，单击 addBtn 按钮，就可以把消息发送给 Service。

```
<?xml version = "1.0" encoding = "utf - 8"?>
< LinearLayout xmlns:android = "http://schemas.android.com/apk/res/android"
    android:id = "@ + id/id_ll_container"
    android:orientation = "vertical" android:layout_width = "match_parent"
    android:layout_height = "match_parent">
    < TextView
        android:id = "@ + id/connectTxt"
        android:layout_width = "match_parent"
        android:layout_height = "wrap_content"
        android:text = "Messenger Test!"/>
    < Button android:id = "@ + id/addBtn"
        android:layout_width = "wrap_content"
        android:layout_height = "wrap_content"
        android:text = "add"/>
    < TextView
        android:layout_width = "match_parent"
        android:layout_height = "wrap_content"
        android:id = "@ + id/resultTxt"
        android:text = "result is"/>
</LinearLayout >
```

3. 使用 AIDL 创建绑定的 Service

使用 AIDL 创建绑定 Service 的步骤如下：

第一步，创建.aidl 文件，此文件定义带有方法签名的编程接口。

AIDL 使用 Java 编程语言声明接口，每个.aidl 文件都只能定义单个接口，接口中可以包含一个或多个方法并且只需包含方法签名。①每个方法可带零个或多个参数，返回一个值或 void，方法的参数和返回值可以是任意类型，这些类型包括 Java 编程语言中的所有基本数据类型（如 int、long、char、boolean，等等）、String、CharSequence、List、Map、其他 AIDL 生成的接口，其他未列出的引用类型必须使用 import 语句导入，即使这些类型定义在与接口相同的软件包中。②所有引用类型参数都需要标示数据走向，可以是 in、out 或 inout，表示跨进程通信中数据的流向，其中 in 表示数据只能由客户端流向服务端，out 表示数据只能由服务端流向客户端，而 inout 则表示数据可在服务端与客户端之间双向流通。其中，数据流向是针对在客户端中的那个传入方法的对象而言的。in 表现为服务端将会接收到一个对象的完整数据，但是客户端的那个对象不会因为服务端对传参的修改而发生变动；out 表现为服务端将会接收到那个对象的空对象，但是在服务端对接收到的空对象有任何修改之后客户端将会同步变动；inout 表现为，服务端将会接收到客户端传来对象的完整信息，并且客户端将会同步服务端对该对象的任何变动。Java 中的基本类型和 String，CharSequence 的默认方向为 in，且只能是 in。③. aidl 文件中包括的所有代码注释都将包含在生成的 IBinder 接口中(import 和 package 语句之前的注释除外)。④在 AIDL 中仅仅支持方法、不支持静态的域。

通过 IPC 接口把某个对象从一个进程发送到另一个进程是可以实现的。不过，它必须实现了 Parcelable 接口，并且确保该类的代码对 IPC 通道的另一端可用。

以下是一个. aidl 文件示例，文件名 IRemoteService. aidl。

```
package com.example.android;
interface IRemoteService {
   int getPid();
   void basicTypes(int anInt, long aLong, boolean aBoolean, float aFloat,
        doublc aDouble, String aString);
}
```

AndroidStudio 把. aidl 文件自动保存在项目的 app\src\main\aidl 目录内，单击 Android Studio 的菜单 build|make module'app'，SDK 工具就立即在项目的 app\build\generated\source\aidl\debug 目录下生成 Java 语言编写的 IBinder 接口文件，此接口具有一个名为 Stub 的内部抽象类，该抽象类继承 android. os. Binder 类并实现了从. aidl 文件继承的方法。生成的文件名与. aidl 文件名一致，只是使用了. java 扩展名(例如，IRemoteService. aidl 生成的文件名是 IRemoteService. java)。

第二步，在 Service 类中实现接口，并向客户端公开该接口。

在 Service 类中，创建 Stub 的子类并实现从. aidl 文件继承的方法，最后重写 Service 的 onBind()方法以返回 Stub 类的实例。

以下是一个使用匿名实例实现名为 IRemoteService 的接口(由以上 IRemoteService. aidl 示例定义)的示例：

```
public class RemoteService extends Service {
   @Override
   public void onCreate() {
      super.onCreate();
   }
   private final IRemoteService.Stub stub = new IRemoteService.Stub() {
      public int getPid(){
         return Process.myPid();
      }
```

```
        public void basicTypes(int anInt, long aLong, boolean aBoolean,
            float aFloat, double aDouble, String aString) {
        }
    };
    @Override
    public IBinder onBind(Intent intent) {
        return stub;
    }
}
```

第三步，调用 IPC 方法。客户端必须执行以下步骤，才能调用使用 AIDL 定义的远程接口：

在项目 src\main\aidl 目录中加入. aidl 文件。如果客户端和 Service 在不同的应用内，则必须把 Service 端的. aidl 文件复制到客户端的应用 src\main\aidl 目录内，它生成 android. os. Binder 接口将为客户端提供对 AIDL 方法的访问。

单击 Android Studio 的菜单 build|make module'app'，SDK 工具就立即在项目的 app\build\generated\source\aidl\debug 目录下生成 Java 语言编写的 IBinder 接口文件。

创建 ServiceConnection 接口的实例，并实现接口中的方法。在 onServiceConnected (ComponentName className，IBinder iBinder）实现中，调用 YourAidlInterfaceName. Stub. asInterface(iBinder)，获取 YourAidlInterfaceName 类型的实例。

调用 Context. bindService（Intent，ServiceConnection，int），传入 ServiceConnection 实现。

调用 YourAidlInterfaceName 接口的实例的方法。应该始终捕获 DeadObjectException 异常，它们是在连接中断时引发的；这将是远程方法引发的唯一异常。

如需断开连接，调用 Context. unbindService(ServiceConnection conn)。

```
IRemoteService mIRemoteService;
private ServiceConnection mConnection = new ServiceConnection() {
    public void onServiceConnected(ComponentName className, IBinder iBinder) {
        mIRemoteService = IRemoteService.Stub.asInterface(iBinder);
    }
    // Called when the connection with the service disconnects unexpectedly
    public void onServiceDisconnected(ComponentName className) {
        Log.e(TAG, "Service has unexpectedly disconnected");
        mIRemoteService = null;
    }
```

13.2.5 绑定到 Service

应用组件(客户端)可通过调用 bindService(Intent service，ServiceConnection conn，int flags)绑定到 Service。Android 系统随后调用 Service 的 onBind()方法，该方法返回用于与 Service 交互的 IBinder。从客户端绑定到 Service 的步骤如下。

第一步，新建 ServiceConnection 对象、必须重写它的两个回调方法：

```
public void onServiceConnected(ComponentName className, IBinder iBinder)
public void onServiceDisconnected(ComponentName className)
```

第二步，调用 bindService(Intent intent, ServiceConnection conn, int flags)，传递 ServiceConnection 对象。bindService()的第一个参数是一个 Intent，用于显式命名要绑定的服务。第三个参数 flag 是一个指示绑定选项的标志，通常应该是 BIND_AUTO_CREATE，以便创建尚未激活的服务。其他可能的值为 BIND_DEBUG_UNBIND 和 BIND_NOT_FOREGROUND，或 0(表示无)。绑定是异步的，bindService()并不直接返回 IBinder 给客户端。连接成功后，系统调用 ServiceConnection 的 onServiceConnected(ComponentName name, IBinder iBinder)方法来传递 IBinder 给客户端，以后客户端就可以利用这个 IBinder 与 Service 进行交互了。

第三步，要断开与服务的连接，请调用 unbindService(ServiceConnection conn)。

如果应用在客户端仍绑定到 Service 后销毁客户端，则销毁会导致客户端取消绑定。更好的做法是在客户端与服务交互完成后立即取消绑定客户端，这样可以关闭空闲的 Service。注意，只有 Activity、Service 和 ContentProvider 可以绑定到 Service，不能从 BroadcastReceiver 绑定到 Service。

13.3 PendingIntent 与 TaskStackBuilder

android.app.PendingIntent 实现了 Parcelable 接口，PendingIntent 持有的信息包括 Intent 和当前 App 的 Context。PendingIntent 对象就是系统维护的用于获取原始数据的一个引用。PendingIntent 不是立刻执行某个行为，而是满足某些条件后才执行指定的行为，即其他 App 延时执行 PendingIntent 中描述的 Intent。通过把 PendingIntent 对象传递给其他应用，从而授予它执行指定操作的权利，PendingIntent 仍然拥有原始应用所拥有的权限，就好像其他应用和原始应用一样(相同的身份和权限)。这意味着，即使原始应用进程被杀死，已经授予给其他进程的 PendingIntent 仍然可以通过存在于 PendingIntent 里的 Context 来执行 Intent。PendingIntent 可以供当前 App 或其他 App 调用，而常见的是供其他 App 使用。

可以这样理解：当在 A activity 中启动另一个 B activity，可以选择以下两种情况：

第一种：通过 Intent 配置需要启动的 B activity，然后调用 startActivity()方法立即执行启动操作，跳转过去。

第二种情况是，延时一段时间跳转到 B activity，可以通过 PendingIntent 来实现，PendingIntent 可以包装 Intent，然后通过 AlarmManager 定时 5 分钟之后启动 PendingIntent，实现延时启动。PendingIntent 类的实例可以这样创建：①可以通过 PendingIntent 类的静态方法 getActivity(Context context, int requestCode, Intent intent, int flags)、getActivity(Context, int requestCode, Intent, int)，getActivities(Context, int requestCode, Intent[], int)获取用于启动 Activity 的 PendingIntent 对象。②可以通过 PendingIntent 类的静态方法 getService(Context context, int requestCode, Intent intent, int flags)方法获取用于启动一个 Service 的 PendingIntent 对象。③可以通过 PendingIntent 类

的静态方法 getBroadcast(Context context,int requestCode,Intent intent,int flags)方法获取用于向 BroadcastReceiver 的发送广播的 PendingIntent 对象。返回的 PendingIntent 对象可以传递给其他应用,以便过一段时间后能够执行描述的动作。

因此,创建 PendingIntent 对象时,提供的基本 Intent 应该把组件的名字显式地设置为自己的应用组件。如果创建 PendingIntent 的应用以后获取相同种类的 PendingIntent,它将获取同一标志的 PendingIntent。创建多个仅仅 Intent 的 extra 内容不同的 PendingIntent,获取的 PendingIntent 是相同的。如果在同一时间确实需要多个不同的 PendingIntent 对象有效(同一时间同时显示两个通知),可以设置 Intent 的 filterEquals (Intent other)方法考虑的 Intent 属性,或者提供不同的请求码给 getActivity、getService、getBroadcast 方法。如果在同一时间仅仅需要一个 PendingIntent 对象有效,可以使用标志 FLAG_CANCEL_CURRENT 或者 FLAG_UPDATE_CURRENT 取消或者更改现在的 PendingIntent 关联的 Intent。

TaskStackBuilder 用于构造一个合成的回退栈,主要用于跨任务导航,利用 TaskStackBuilder 的 getPendingIntent(int requestCode, int flags),可将当前任务中的 Activity 和要启动的 Activity 合成为一个新的任务,而用户按返回键的操作就作用在这个新任务中,这相当于实现了跨任务直接跳转。返回键是基于当前任务的局部导航,不能捕获跨任务导航。跨任务导航是通过"最近的 UI"完成的,可以通过单击软件提供的导航栏或者系统栏上的 recent 键,"最近的 UI"在较旧的硬件按钮配置中是通过长按 Home 键来实现的。

TaskStackBuilder 内部包含一个 ArrayList < Intent > mIntents 数组,数组中的每一个 Intent 对象将启动一个 Activity(即一个 Intent 对象对应一个 Activity),mIntents[0]对应 Activity 栈的 root,每一个 Activity 的父 Activity 对应 mIntents 数组中的前一个 Intent。表 13-5 列出了 TaskStackBuilder 的部分分 public 方法。

表 13-5 TaskStackBuilder 的 public 方法(部分)

返回值类型	方法及解释
TaskStackBuilder TaskStackBuilder TaskStackBuilder	addParentStack(ComponentName sourceActivityName) addParentStack(Class <? > sourceActivityClass) addParentStack(Activity sourceActivity) 把 sourceActivity 的父 Activity 链添加到 TaskStackBuilder 对象中。清单文件中 activity 元素的 parentActivityName 属性指定了它的父 Activity
TaskStackBuilder	addNextIntent(Intent nextIntent) 添加一个新的 Intent 到任务栈上,最新添加的 Intent 激活的 Activity 将添加到任务栈的顶部
TaskStackBuilder	addNextIntentWithParentStack(Intent nextIntent) 等价于先 addParentStack(ComponentName sourceActivityName),再调用 addNextIntent(Intent nextIntent),其中 sourceActivityName 由 nextIntent 解析
PendingIntent	getPendingIntent(int requestCode,int flags) 获取一个 PendingIntent,用于启动 TaskStackBuilder 构建的任务

13.4 Notification

android. app. Notification 实现了 Parcelable 接口，代表呈献给用户的通知。通知是可以在应用的常规 UI 外部向用户显示的消息，通知区域位于屏幕的顶部。当调用 NotificationManager 对象的 notify(int id, Notification notification)方法发送通知时，它将先以图标的形式显示在系统的通知区域中。为了查看通知的详细信息，用户必须用手指向下滑动通知区域从而打开抽屉式通知栏。通知区域和抽屉式通知栏均是由系统控制的区域，用户可以随时查看。在平时的使用中，通知主要显示接收到的短消息、及时消息、推送消息等信息，如短信、QQ、微信、广告、版本更新、推荐新闻等，还显示正在运行的程序，例如后台运行的程序，如音乐播放进度、下载进度等。

Notification 对象保存通知相关的数据。使用 Notification. Builder 创建 Notification 对象更容易。由于 Notification. Builder 仅支持 Android 4.1 及之后的版本，为了解决兼容性问题，Google 在 Android Support v4 中加入了 NotificationCompat. Builder 类。对于某些在 Android 4.1 之后才有的特性，即使 NotificationCompat. Builder 支持该方法，在之前的版本中也不能运行。NotificationManager 是通知管理类，它是一个系统服务。调用 NotificationManager 实例的 notify(int id, Notification notification)方法可以向系统发送通知。创建一个简单的 Notification 需要以下五个步骤：

第一步，实例化 NotificationCompat. Builder，为 Notification 设置相关属性和 Action。一个 Notification 对象的必要属性有以下三项：小图标、标题、文本，如果不设置则在运行时会抛出异常。可以调用 setSmallIcon()设置小图标、调用 setContentTitle()设置标题、调用 setContentText()设置文本。小图标用于在状态栏上代表通知，并显示在内容视图的左边(如果没有设置大图标)。用单个手指向下滑动状态栏就可以看到普通视图的通知了。在默认情况下，Notification 仅显示消息图标、标题、消息内容、送达时间这 4 项内容，如图 13-3 所示。当只调用 setSmallIcon()时，在内容视图中，小图标显示在左侧，如图 13-3 所示；当 setSmallIcon()与 setLargeIcon()都调用时，smallIcon 显示在通知的右下角，大图标显示在左侧，如图 13-4 所示。除了以上三项，所有其他属性都是可选的。虽然如此，但还是应该给 Notification 设置一个 action，以便直接跳转到某个 Activity、启动一个 Service 或者发送一个广播。否则，Notification 仅仅只能起到通知的效果，而不能与用户交互。

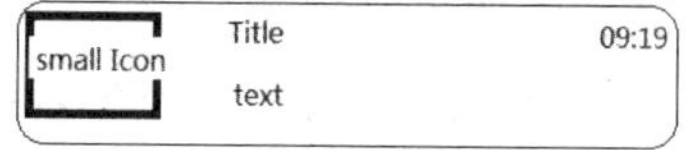

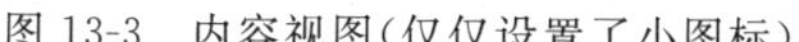
图 13-3 内容视图(仅仅设置了小图标)

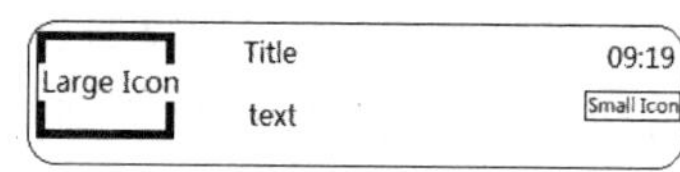

图 13-4 内容视图(同时设置了小图标和大图标)

第二步，将扩展布局应用于通知。

要使通知出现在展开视图中，请先创建一个 NotificationCompat. Builder 对象。大视图的通知在展开前也显示为普通视图。用两个手指按住普通视图向下滑动，将显示大视图。大视图在普通视图的下边增加了详情区域，详情区域根据用途可有多种风格。可以调用

Builder. setStyle(Notification. Style style)设置详情区域的式样。请记住,扩展通知在Android 4.1之前的平台上不可用。如果平台没有提供扩展布局的通知,用户只能看见正常的通知视图。

NotificationCompat. Style 的直接子类有 NotificationCompat. BigPictureStyle, NotificationCompat. BigTextStyle, NotificationCompat. InboxStyle, NotificationCompat. MediaStyle, NotificationCompat. MessagingStyle。BigPictureStyle 的详情区域包含一个256dp 高度的位图。BigTextStyle 的详情区域包含一个大的文字块。InboxStyle 的详情区域包含一个包含多达5个字符串的列表。

例如,以下代码段演示了如何更改前面的创建通知的代码,以便使用扩展布局:

```
NotificationCompat.Builder mBuilder = new NotificationCompat.Builder(this)
   .setSmallIcon(R.drawable.notification_icon)
   .setContentTitle("Event tracker")
   .setContentText("Events received")
NotificationCompat.InboxStyle inboxStyle =
      new NotificationCompat.InboxStyle();
inboxStyle.setBigContentTitle("Event tracker details:");
String[] events = new String[5];
for (int i = 0; i < events.length; i++) {
   inboxStyle.addLine(events[i]);
}
mBuilder.setStyle(inBoxStyle);
...   // Issue the notification here.
```

第三步,调用 NotificationCompat. Builder 对象的 build()方法,它将返回符合规范的 Notification 对象。

第四步,获取 NotificationManager 实例。可通过以下方法获取 NotificationManager 对象:

```
NotificationManager mNotifyManager = (NotificationManager) getSystemService(Context.
NOTIFICATION_SERVICE);
```

第五步:调用 NotificationManager 对象的 notify(int id, Notification notification)方法发送通知,当已经有相同 id 的通知发出并且没有取消时,这个通知将被更新的信息替换。也可以调用 notify(String tag, int id, Notification notification)方法发送通知,当已经有相同 tag 和 id 的通知发出并且没有取消时,这个通知将被更新的信息替换。

当系统接收到通知时,可以通过震动、响铃、呼吸灯等多种方式进行提醒。

在 Android 应用程序中要详细描述一个组件,需要知道该组件所在的应用包名,也就是在 AndroidManifest. xml 文件中 manifest 根结点下的 package="XXX. XXX",还有组件在应用中的完整路径名。可以使用 ComponentName 来封装组件的应用包名和组件的完整路径名。以下代码段演示了一个简单通知,它指定了当用户单击通知时要打开的 Activity,该代码将创建 TaskStackBuilder 对象并使用它来为操作创建 PendingIntent。

```
    TaskStackBuilder stackBuilder = TaskStackBuilder.create(this);
    stackBuilder.addParentStack(ResultActivity.class);
    Intent resultIntent = new Intent(this, ResultActivity.class);
    stackBuilder.addNextIntent(resultIntent);
    PendingIntent resultPendingIntent = stackBuilder.getPendingIntent(0,
        PendingIntent.FLAG_UPDATE_CURRENT);
NotificationCompat.Builder notificationBuilder =
      new NotificationCompat.Builder(this)
      .setSmallIcon(R.drawable.notification_icon)
      .setContentTitle("My notification")
      .setContentText("Hello World!");
    notificationBuilder.setContentIntent(resultPendingIntent);
    NotificationManager notificationManager =
     (NotificationManager) getSystemService(Context.NOTIFICATION_SERVICE);
    int notifyID = 1;
    notificationManager.notify(notifyID, notificationBuilder.build());
```

android.app.Notification.Action类实现了Parcelable接口，封装了一个命名的动作，它将作为通知的一部分显示在通知内容的附近，Action通常被系统显示为按钮。它必须包含一个图标、一个标签和当用户选定动作时要发送的一个PendingIntent，PendingIntent用于启动Activity。一个扩展形式的通知可以显示多达3个Action，按照它们被添加的顺序从左到右显示。Action按钮依赖于扩展的通知，而扩展的通知只有在Android 4.1及更高版本的平台才可以使用，因而Action按钮将不会显示在Android 4.1之前的平台上。为了确保Action按钮的功能总是可以使用，应当首先调用setContentIntent()设置用户单击抽屉式通知区域中的通知文本时启动的Activity的功能，Action按钮的功能在应用的Activity中必须实现，然后调用addAction()实现相同的功能。例如，暂停闹铃或立即答复短信。

Notification.Action的构造函数Notification.Action(int icon, CharSequence title, PendingIntent intent)在API 23中已经被废弃。应该先创建一个Notification.Action.Builder对象，再调用该对象的build方法就能返回一个Notification.Action类型的对象。然后调用Notification.Builder对象的addAction(Notification.Action action)方法给通知添加Action。尽管Notification.Action是可选的，但一个通知至少提供一个Action，应该始终定义一个当用户单击通知时会触发的Action，允许用户从通知直接跳转到应用中的Activity，在Activity中可以执行进一步的操作。可以调用NotificationCompat.Builder对象的setContentIntent方法设置用户单击抽屉式通知区域中的通知文本时启动的Activity，用户单击通知时启动Activity是Notification.Action最常见的使用场景。此外，还可以调用setDeleteIntent(PendingIntent intent)设置用户从通知面板上清除通知时启动的Activity。在Android 4.1及更高版本中，还可以通过Notification.Action按钮启动Activity。

通知的优先级充当一个关于如何显示通知的提示。可以调用NotificationCompat.Builder.setPriority()并传入一个NotificationCompat优先级常量来设置通知的优先级。有五个优先级别：PRIORITY_MIN(−2)，PRIORITY_LOW(−1)、PRIORITY_DEFAULT(0)、PRIORITY_HIGH(1)、PRIORITY_MAX(2)。如果未设置，则优先级默认为PRIORITY_DEFAULT(0)。

为了确保最佳兼容性，请使用 NotificationCompat 及其子类（特别是 NotificationCompat. Builder）创建通知。并非所有通知功能都可用于某特定版本，即使用于设置这些功能的方法位于支持库类 NotificationCompat. Builder 中。例如，依赖于扩展通知的 Action 按钮仅会显示在 Android 4.1 及更高版本的系统中，这是因为扩展通知本身仅在 Android 4.1 及更高版本的系统中可用。无论 Android 系统是哪个版本，都应该为所有用户提供通知的全部功能，为此，必须验证是否可从应用的 Activity 中获得所有功能，这可能需要添加新的 Activity。例如，若要使用 addAction()提供停止和启动媒体播放的控制，应该先在应用的 Activity 中实现此控制，确保所有用户均可通过单击通知启动 Activity 来获得该 Activity 中的功能。为此，请为 Activity 创建 PendingIntent，并调用 setContentIntent()把 PendingIntent 添加到通知里。

13.4.1 管理通知

当需要为同一类型的事件多次发出同一通知时，应当避免创建全新的通知，而是考虑通过更改之前通知的某些值或者添加某些值来更新通知。

1. 更新通知

如果要在发出通知之后更新此通知，请更新或创建 NotificationCompat. Builder 对象，从该对象构建 Notification 对象，并发出与之前所用 Tag 和 ID 相同的 Notification。如果之前的通知仍然可见，则系统会根据 Notification 对象的内容更新该通知。相反，如果之前的通知已被清除，系统则会创建一个新通知。

以下代码段演示了经过更新以反映所发生事件数量的通知，它将通知堆叠并显示摘要：

```
mNotificationManager =
        (NotificationManager) getSystemService(Context.NOTIFICATION_SERVICE);
// Sets an ID for the notification, so it can be updated
int notifyID = 1;
mNotifyBuilder = new NotificationCompat.Builder(this)
    .setContentTitle("New Message")
    .setContentText("You've received new messages.")
    .setSmallIcon(R.drawable.ic_notify_status)
...
   mNotifyBuilder.setContentText("currentText ");
   // Because the ID remains unchanged, the existing notification is updated.
mNotificationManager.notify(notifyID,mNotifyBuilder.build());
```

2. 删除通知

调用 NotificationManager 对象的 cancelAll()方法可以删除它已经发出的所有通知，调用 cancel(int id)或者 cancel(String tag,int id)方法可以删除指定的通知。如果通知是持久的，它将从状态栏中被移除。如果通知是暂时的，它将被隐藏。如果在创建通知时调用了 Notification. Builder 对象的 setAutoCancel(boolean autoCancel)方法、且参数 autoCancel 的取值为 true，当用户在面板上单击通知时，就会自动取消该通知，同时通过 setDeleteIntent(PendingIntent)设置的 PendingIntent 将被广播出去。用户通过手势把通知滑出屏幕从而清除通知。

13.4.2 从通知中启动 Activity 时保留导航

从通知中启动的 Activity 可能是应用的正常工作流程的一部分；也可能仅仅是扩展通知，提供了通知很难显示的信息。两种情况都需要将 PendingIntent 设置为在全新任务中启动。从通知中启动的 Activity 是应用的正常工作流程的一部分，并应该为 PendingIntent 提供返回栈，单击 Back 按钮应该使用户经过应用的正常工作流程、最后回退到主屏幕，而单击 Recent 按钮则应将 Activity 显示为单独的任务；反之，从通知中启动的 Activity 不是应用的正常工作流程的一部分，就没有必要创建回退栈，单击 Back 按钮仍会将用户带到 Home 屏幕。

1. 设置常规 Activity

第一步，在清单文件中定义应用的 Activity 层次结构。

对于 Android 4.1 及更高版本，给正在启动的 Activity 添加 android: parentActivityName 属性。对于 Android 4.0.3 及更低版本，给< activity >元素添加< meta-data >子元素 meta-data 的 name 属性设置为 android. support. PARENT_ACTIVITY，value 属性设置为父 activity 的类名。XML 实例如下：

```
<activity
    android:name = ".MainActivity"
    android:label = "@string/app_name" >
    <intent - filter>
        <action android:name = "android.intent.action.MAIN" />
        <category android:name = "android.intent.category.LAUNCHER" />
    </intent - filter>
</activity>
<activity
    android:name = ".ResultActivity"
    android:parentActivityName = ".MainActivity">
    <meta - data
        android:name = "android.support.PARENT_ACTIVITY"
        android:value = ".MainActivity"/>
</activity>
```

第二步，创建回退栈。

首先调用 TaskStackBuilder 的静态方法 create()创建 TaskStackBuilder 对象，再调用 TaskStackBuilder 对象的 addParentStack 方法添加 Activity。对于在清单文件中所定义层次结构内的每个 Activity，回退栈均包含启动这个 Activity 的 Intent 对象，此方法还会添加一些可在全新任务中启动堆栈的标志。尽管 addParentStack 方法的参数是对被启动 Activity 的引用，但是并不会添加启动该 Activity 的 Intent。然后创建可以从通知中启动 Activity 的 Intent，并调用 TaskStackBuilder 对象的 addNextIntent(Intent nextIntent)，把刚创建的 Intent 对象添加到任务栈上。有些时候，在用户使用 Back 键回退到目标 Activity 时，需要确保此 Activity 显示有意义的数据，这时可以通过调用 TaskStackBuilder 对象的 editIntentAt(int index)向堆栈中的 Intent 对象添加参数。最后调用 TaskStackBuilder 对象的 getPendingIntent(int requestCode,int flags)获得此返回栈的 PendingIntent。

第三步，调用 NotificationCompat. Builder 对象的 setContentIntent（PendingIntent intent）方法，参数是刚从 TaskStackBuilder 获得的 PendingIntent，再调用方法创建通知，最后调用 NotificationManager 对象的 notify 方法发出通知。

以下代码段演示了该流程：

```
TaskStackBuilder stackBuilder = TaskStackBuilder.create(this);
stackBuilder.addParentStack(ResultActivity.class);
Intent resultIntent = new Intent(this, ResultActivity.class);
stackBuilder.addNextIntent(resultIntent);
PendingIntent resultPendingIntent =
        stackBuilder.getPendingIntent(0, PendingIntent.FLAG_UPDATE_CURRENT);
NotificationCompat.Builder builder = new NotificationCompat.Builder(this);
builder.setContentIntent(resultPendingIntent);
NotificationManager mNotificationManager =
    (NotificationManager) getSystemService(Context.NOTIFICATION_SERVICE);
int notifyID = 1;
mNotificationManager.notify(notifyID, builder.build());
```

2. 设置特殊 Activity

特殊 Activity 不需要回退栈，所以没有必要在清单文件中定义其 Activity 层次结构，也不必调用 addParentStack()来构建回退栈。取而代之的是，在清单文件设置 Activity 任务选项，并通过调用 getActivity()创建 PendingIntent。具体步骤如下：

第一步，在清单文件中，这样设置< activity >元素的属性：①设置 android：taskAffinity=""，并与在代码中设置 Intent 的 FLAG_ACTIVITY_NEW_TASK 标志相结合，这可确保此 Activity 不会进入应用的默认任务，不影响任何具有应用默认关联的现有任务；②设置 android：excludeFromRecents="true"，把新任务从 Recents 中排除，这样用户就不会在无意中导航到这个 Activity。

特殊 Activity 在清单文件中的配置实例：

```
<activity android:name = ".ResultActivity" android:launchMode = "singleTask"
    android:taskAffinity = "" android:excludeFromRecents = "true">
</activity>
```

第二步，构建并发出通知。

首先创建启动 Activity 的 Intent 对象，调用 setFlags(int)，传递 FLAG_ACTIVITY_NEW_TASK | FLAG_ACTIVITY_CLEAR_TASK 作为参数，从而将 Activity 设置为在新的空任务中启动；然后调用 PendingIntent 类的静态方法 getActivity(Context context，int requestCode，Intent intent，@Flags int flags)获取一个 PendingIntent。最后调用 NotificationCompat. Builder 对象的 setContentIntent(PendingIntent intent)方法、参数是前一步 getActivity 方法返回的 PendingIntent。

以下代码段演示了该流程：

```
Intent notifyIntent = new Intent(this, ResultActivity.class);
notifyIntent.setFlags(Intent.FLAG_ACTIVITY_NEW_TASK| Intent.FLAG_ACTIVITY_CLEAR_TASK);
```

```
PendingIntent notifyPendingIntent = PendingIntent.getActivity(this,0,
        notifyIntent,PendingIntent.FLAG_UPDATE_CURRENT);
NotificationCompat.Builder builder = new NotificationCompat.Builder(this);
builder.setContentIntent(notifyPendingIntent);
NotificationManager mNotificationManager =
    (NotificationManager) getSystemService(Context.NOTIFICATION_SERVICE);
int notifyID = 1;
mNotificationManager.notify(notifyID, builder.build());
```

13.5 Handler 实验

实验目的：

(1) 熟悉非 UI 线程更新 UI 的方法。

(2) 熟悉 HandlerThread 的使用。

(3) 熟悉 ProgressBar 与 GridLayout 的使用。

1. 新建工程 Handler1

具体步骤：运行 Android Studio，单击 Start a new Android Studio project，在 Create New Project 页中，Application name 为 Handler1，Company domain 为 cuc，Project location 为 C:\android\myproject\Handler1，单击 Next 按钮，选中 Phone and Tablet，minimum sdk 设置为 API 19：Android 4.4(minSdkVersion)，单击 Next 按钮，选择 Empty Activity，再单击 Next 按钮，选中 Generate Layout File 选项，选中 Backwards Compatibility(AppCompat) 选项，单击 Finish 按钮。

2. 添加 DataBinding 支持，并同步

在文件 build.gradle(module：app)的 android 语句的花括号内添加：

```
dataBinding.enabled = true
```

由于修改了 build.gradle 文件内容，这个文件窗口的上面就出现提示："Gradle files have changed since last project sync. A project sync may be necessary for the IDE to work properly. sync now"。单击 sync now 按钮进行同步。

3. 在 res\layout 目录里新建布局文件 activity_main1.xml

具体步骤：在工程结构视图中，右击 layout 文件夹，选择 New | Layout resource file，File name 输入框输入 activity_main1，Root element 输入框输入 layout，Source set 为 main，单击 OK 按钮。layout/activity_main1.xml 内容如下：

```
<?xml version = "1.0" encoding = "utf - 8"?>
< layout xmlns:android = "http://schemas.android.com/apk/res/android">
< data/>
    < GridLayout
        android:orientation = "horizontal"
        android:columnCount = "2"
```

```
        android:layout_width = "match_parent"
        android:layout_height = "match_parent">
    <Button android:id = "@ + id/postRunnableBtn" android:text = "postRunnable"/>
    <Button android:id = "@ + id/removeCallbackBtn" android:text = "removeCallBack"/>
     <Button android:id = "@ + id/postProgrssBtn" android:text = "postProgrss"/>
    <Button android:id = "@ + id/sendProgressBtn" android:text = "sendProgress"/>
        <TextView android:id = "@ + id/progressTxt" android:text = "proress is"/>
        <android.support.v4.widget.Space />
        <ProgressBar android:id = "@ + id/progressBar" android:layout_columnSpan = "2"
            android:layout_gravity = "fill_horizontal"
            style = "?android:attr/progressBarStyleHorizontal"/>
        <Button android:id = "@ + id/sendtoNonUiBtn" android:text = "sendtoNonUi"/>
        <android.support.v4.widget.Space />
        <Button android:id = "@ + id/startSaleThreadBtn" android:text = "startSaleThread" />
        <Button android:id = "@ + id/startSaleRunnableBtn" android:text = "startSaleRunnable"/>
    </GridLayout>
</layout>
```

4. 编辑 MainActivity.java

MainActivity.java 具体代码如下：

```
package cuc.handler1;
import android.databinding.DataBindingUtil;
import android.os.Bundle;
import android.os.Handler;
import android.support.v7.app.AppCompatActivity;
import android.util.Log;
import android.view.View;
import cuc.handler1.databinding.ActivityMain1Binding;

public class MainActivity extends AppCompatActivity implements View.OnClickListener{
 private static final String TAG = "MainActivity";
private ActivityMain1Binding binding;
 private static final int MSG_PRORESS = 1;
 private static final int MSG_LOG = 2;
 private static final String NAME = "name";
 private static final String AGE = "age";
 private static final String CITY = "city";
 private Handler handler1;
 private Handler handler2;
 private Handler handler3;

  @Override
  protected void onCreate(Bundle savedInstanceState) {
   super.onCreate(savedInstanceState);
   setContentView(R.layout.activity_main);
   binding = DataBindingUtil.setContentView(this,R.layout.activity_main1 );
   binding.postRunnableBtn.setOnClickListener(this);
```

```
    binding.removeCallbackBtn.setOnClickListener(this);
     binding.postProgrssBtn.setOnClickListener(this);
     binding.sendProgressBtn.setOnClickListener(this);
     binding.sendtoNonUiBtn.setOnClickListener(this);
     binding.startSaleThreadBtn.setOnClickListener(this);
     binding.startSaleRunnableBtn.setOnClickListener(this);
     handler1 = new Handler();
    }
    Runnable postRunnable = new Runnable() {
     int count = 0;
     @Override
     public void run() {
      Log.i(TAG, "run: " + count++);
      handler1.postDelayed(postRunnable,1000);
     }
    };
    @Override
    public void onClick(View view) {
     switch (view.getId()){
      case R.id.postRunnableBtn:
       Log.i(TAG, "onClick: postBtn");
       handler1.post(postRunnable);
       break;
      case R.id.removeCallbackBtn:
       Log.i(TAG, "onClick: removeCallbackBtn");
       handler1.removeCallbacks(postRunnable);
       break;
     }
    }
   }
```

5. 编译运行

在 MainActivity 类内部，有一个 Handler 类型的成员 handler1，它是通过调用了无参数的构造函数创建的对象，所以它采用了当前线程(UI 主线程)的消息循环 Looper。单击 postBtn 按钮时，调用 handler1.post(postRunnable)发消息，实际上就是指定了 Messsage 的 Runnable 类型的成员变量 callback，也就是说，callback 中包含了处理消息的代码。单击 removeCallbackBtn 按钮，从消息队列移除 postRunnable。

6. 实现 postProgrssBtn 按钮的响应

在 MainActivity 的 onClick 方法的 switch 语句中添加一种 case，处理 postProgrssBtn 按钮的响应。代码如下：

```
case R.id.postProgrssBtn:
   new Thread(){
      int progress = 0;
      @Override
      public void run() {
```

```
                super.run();
                    while(progress < 100){
                        handler1.post(new Runnable() {
                            @Override
                            public void run() {
                                progressTxt.setText("progress -- >" + progress);
                                progressBar.setProgress(progress);
                            }
                        });
                        progress++;
                        try {
                            Thread.sleep(50);
                        } catch (InterruptedException e) {
                            e.printStackTrace();
                        }
                    }
            }
        }.start();
        break;
```

然后编译运行。单击 postProgrssBtn 按钮,新建一个线程,在新建的线程中,使用前一阶段初始化的 handler1 的 post(Runnable r)方法从非 UI 线程往 UI 线程发送消息,post 方法的 Runnable 参数的 run 方法中包含了更新 progressTxt 和 progressBar 的代码。

7. 编辑 MainActivity.java,并编译运行

在类 MainActivity 里的 OnCreate 方法中,添加初始化 handler2 的代码,handler2 也采用了 UI 主线程的 Looper,代码如下:

```
handler2 = new Handler(){
 @Override
 public void handleMessage(Message msg) {
  super.handleMessage(msg);
  switch (msg.what){
   case MSG_PRORESS:
    int progress = msg.arg1;
    binding.progressTxt.setText("progress -- >" + progress);
    binding.progressBar.setProgress(progress);
    break;
  }
 }
};
```

在 MainActivity 的 onClick 方法的 switch 语句中添加一种 case,处理 sendProgressBtn 按钮的响应。代码如下:

```
case R.id.sendProgressBtn:
    new Thread(){
        int progress = 0;
```

```
        @Override
        public void run() {
            super.run();
            while(progress <= 100){
                Message msg = Message.obtain();
                msg.what = MSG_PRORESS;
                msg.arg1 = progress;
                handler2.sendMessage(msg);
                progress++;
                try {
                    Thread.sleep(50);
                } catch (InterruptedException e) {
                    e.printStackTrace();
                }
            }
        }
    }.start();
    break;
```

然后编译运行。单击 sendProgressBtn 按钮，handler2 调用了 sendMessage(Message msg)方法发送了消息，此消息仅仅带有进度数据，消息的类型通过 what 成员来传递，数据可以通过 arg1 和 arg2 传递，也可以调用 Message 的 setData 方法由 Bundle 传递更多数据，Message 没有使用 Runnable 类型对象指定了消息的处理方法，消息的处理方法定义在 handler2 的 handleMessage 方法中，用于更新 progressTxt 和 progressBar。

8. 编辑 MainActivity.java，并编译运行

实验中 handler1 和 handler2 采用了 UI 主线程的 Looper，而 handler3 使用新建 HandlerThread 线程的 Looper，在同一进程里传递消息，并且 Message 使用 Bundle 传递数据，而 Message 本身也没有使用 Runnable 对象指定消息的处理方法，消息数据的处理将在 handler3 的 handleMessage 方法中定义。

首先，在类 MainActivity 里定义以下属性：

```
private static final String NAME = "name";
private static final String AGE = "age";
private static final String CITY = "city";
```

在类 MainActivity 里的 OnCreate 方法中，添加初始化 handler3 的代码，代码如下：

```
HandlerThread handlerThread = new HandlerThread("cuc");
handlerThread.start();
handler3 = new Handler(handlerThread.getLooper()){
    @Override
    public void handleMessage(Message msg) {
        super.handleMessage(msg);
        Bundle bundle = msg.getData();
        Log.i(TAG, "handleMessage: name:" + bundle.getString(NAME) +
                ", age:" + bundle.getInt(AGE) + ", city:" + bundle.getString(CITY) );
```

```
    }
};
```

在 MainActivity 的 onClick 方法的 switch 语句中添加一种 case，处理 sendtoNonUiBtn 按钮的响应。代码如下：

```
case R.id.sendtoNonUiBtn:
    Message msg = handler3.obtainMessage();
    Bundle bundle = new Bundle();
    bundle.putString(NAME,"zhangsan");
    bundle.putInt(AGE,25);
    bundle.putString(CITY,"Beijing");
    msg.setData(bundle);
    msg.sendToTarget();
    break;
```

代码编辑完后，编译运行。单击 sendtoNonUiBtn 按钮，向非 UI 线程发送消息，可以在 Logcat 中看到相应的日志信息。

在这个例子中，不是采用的主线程的 Looper，而是新建了一个 HandlerThread 类型的对象 handlerThread，HandlerThread 对象自己创建了 Looper，handler3 就是采用 HandlerThread 里的 Looper 作为参数创建的 Handler。另外，Message 中没有 Runnable 对象，即没有指定处理消息的方法，而只有消息数据（姓名、年龄和城市），怎么处理消息定义在了新建的 Handler 的 handleMessage 方法中。

9. 新建一个 Thread 类 SaleThread

具体步骤：在工程视图中 main 源集 java 文件夹里，右击 cuc.handler1 包，选择 New|Java Class，在 Name 输入框中输入 SaleThread，在 Superclass 输入框中输入 java.lang.Thread，单击 OK 按钮。编辑 SaleThread.java，内容如下：

```
package cuc.handler1;
import android.util.Log;
public class SaleThread extends Thread {
    private static final String TAG  =  "SaleThread";
    private int ticketTotal = 2;
    private String name;
    public SaleThread(String name) {
        super(name);
        this.name = name;
    }
    @Override
    public void run() {
        super.run();
        while (ticketTotal >= 0){
            Log.i(TAG, "run: " + name + ": ticket left:" + ticketTotal -- );
            try {
                Thread.sleep(100);
```

```
            } catch (InterruptedException e) {
                e.printStackTrace();
            }
        }
    }
}
```

10. 编辑 MainActivity，并编译运行

首先，在 MainActivity 的 onClick 方法的 switch 语句中添加一种 case，处理 sendtoNonUiBtn 按钮的响应。代码如下：

```
case R.id.startSaleThreadBtn:
    new SaleThread("window A").start();
    new SaleThread("window B").start();
    break;
```

代码编辑完后，编译运行。单击 startSaleThreadBtn 按钮，可以在 Logcat 中看到相应的日志信息。

```
30942 - 31001 I/SaleThread: run: window A: ticket left:2
30942 - 31002 I/SaleThread: run: window B: ticket left:2
30942 - 31001 I/SaleThread: run: window A: ticket left:1
30942 - 31002 I/SaleThread: run: window B: ticket left:1
30942 - 31001 I/SaleThread: run: window A: ticket left:0
30942 - 31002 I/SaleThread: run: window B: ticket left:0
```

这个例子中，创建 Thread 的方法是：声明一个 Thread 的子类，这个子类覆盖 Thread 类的 run 方法。子类的实例启动，执行子类中定义的 run 方法。

11. 新建一个类 SaleRunnable，实现 Runnable 接口

具体步骤：在工程视图中 main 源集 Java 文件夹里，右击 cuc.handler1 包，依次选择 New|Java Class，在 Name 输入框中输入 SaleRunnable，在 Interface(s)输入框中输入 java.lang.Runnable，单击 OK 按钮。编辑 SaleRunnable.java，内容如下：

```
package cuc.handler1;
import android.util.Log;
public class SaleRunnable implements Runnable {
    private static final String TAG = "SaleRunnable";
    private int ticketTotal = 2;
    private String name;

    public SaleRunnable(String name) {
        this.name = name;
    }
    @Override
    public void run() {
```

```
        while (ticketTotal >= 0){
            Log.i(TAG, "run: " + name + ": ticket left:" + ticketTotal -- );
            try {
                Thread.sleep(100);
            } catch (InterruptedException e) {
                e.printStackTrace();
            }
        }
    }
}
```

12. 编辑 MainActivity.java，并编译运行

首先，在 MainActivity 的 onClick 方法的 switch 语句中添加一种 case，处理 sendtoNonUiBtn 按钮的响应。代码如下：

```
case R.id.startSaleRunnableBtn:
    SaleRunnable saleRunnable = new SaleRunnable("window A");
    new Thread(saleRunnable).start();
    new Thread(saleRunnable).start();
    break;
```

编辑后，MainActivity.java 内容如下，它包含了这个实验所有的代码：

```
package cuc.handler1;
import android.os.Handler;
import android.os.HandlerThread;
import android.os.Message;
import android.support.v7.app.AppCompatActivity;
import android.os.Bundle;
import android.util.Log;
import android.view.View;
import android.widget.Button;
import android.widget.ProgressBar;
import android.widget.TextView;

public class MainActivity extends AppCompatActivity implements View.OnClickListener {
    private static final String TAG = "MainActivity";
    private static final int MSG_PRORESS = 1;
    private static final int MSG_LOG = 2;
    private static final String NAME = "name";
    private static final String AGE = "age";
    private static final String CITY = "city";
    private Handler handler1;
    private Handler handler2;
    private Handler handler3;

    private Button postBtn;
    private Button removeCallbackBtn;
```

```
    private Button postProgrssBtn;
    private Button sendProgressBtn;
    private TextView progressTxt;
    private ProgressBar progressBar;
    private Button sendtoNonUiBtn;
    private Button startSaleThreadBtn;
    private Button startSaleRunnableBtn;

    private void assignViews() {
        postBtn = (Button) findViewById(R.id.postBtn);
        removeCallbackBtn = (Button) findViewById(R.id.removeCallbackBtn);
        postProgrssBtn = (Button) findViewById(R.id.postProgrssBtn);
        sendProgressBtn = (Button) findViewById(R.id.sendProgressBtn);
        progressTxt = (TextView) findViewById(R.id.progressTxt);
        progressBar = (ProgressBar) findViewById(R.id.progressBar);
        sendtoNonUiBtn = (Button) findViewById(R.id.sendtoNonUiBtn);
        startSaleThreadBtn = (Button) findViewById(R.id.startSaleThreadBtn);
    startSaleRunnableBtn = (Button) findViewById(R.id.startSaleRunnableBtn);
    }
    void setListener(){
        postBtn.setOnClickListener(this);
        removeCallbackBtn.setOnClickListener(this);
        postProgrssBtn.setOnClickListener(this);
        sendProgressBtn.setOnClickListener(this);
        sendtoNonUiBtn.setOnClickListener(this);
        startSaleThreadBtn.setOnClickListener(this);
        startSaleRunnableBtn.setOnClickListener(this);
    }
    @Override
    protected void onCreate(Bundle savedInstanceState) {
        super.onCreate(savedInstanceState);
        setContentView(R.layout.gridlayout);
        assignViews();
        setListener();
        handler1 = new Handler();
        handler2 = new Handler(){
            @Override
            public void handleMessage(Message msg) {
                super.handleMessage(msg);
                switch (msg.what){
                    case MSG_PRORESS:
                        int progress = msg.arg1;
                        progressTxt.setText("progress -->" + progress);
                        progressBar.setProgress(progress);
                        break;
                }
            }
        };
        HandlerThread handlerThread = new HandlerThread("cuc");
```

```
        handlerThread.start();
        handler3 = new Handler(handlerThread.getLooper()){
            @Override
            public void handleMessage(Message msg) {
                super.handleMessage(msg);
                Bundle bundle = msg.getData();
                Log.i(TAG, "handleMessage: name:" + bundle.getString(NAME) +
                        ", age:" + bundle.getInt(AGE) + ", city:" + bundle.getString(CITY) );

            }
        };
    }
    Runnable postRunnable = new Runnable() {
        int count = 0;
        @Override
        public void run(){
            Log.i(TAG, "run: " + count++);
            handler1.postDelayed(postRunnable,1000);
        }
    };
    @Override
    public void onClick(View view) {
        switch (view.getId()){
            case R.id.postBtn:
                Log.i(TAG, "onClick: postBtn");
                handler1.post(postRunnable);
                break;
            case R.id.removeCallbackBtn:
                Log.i(TAG, "onClick: removeCallbackBtn");
                handler1.removeCallbacks(postRunnable);
                break;
            case R.id.postProgrssBtn:
                new Thread(){
                    int progress = 0;
                    @Override
                    public void run() {
                        super.run();
                        while(progress < 100){
                            handler1.post(new Runnable() {
                                @Override
                                public void run() {
                                    progressTxt.setText("progress -->" + progress);
                                    progressBar.setProgress(progress);
                                }
                            });
                            progress++;
                            try {
                                Thread.sleep(50);
                            } catch (InterruptedException e) {
```

```
                        e.printStackTrace();
                    }
                }
            }
        }.start();
        break;
    case R.id.sendProgressBtn:
        new Thread(){
            int progress = 0;
            @Override
            public void run() {
                super.run();
                while(progress <= 100){
                    Message msg = Message.obtain();
                    msg.what = MSG_PRORESS;
                    msg.arg1 = progress;
                    handler2.sendMessage(msg);
                    progress++;
                    try {
                        Thread.sleep(50);
                    } catch (InterruptedException e) {
                        e.printStackTrace();
                    }
                }
            }
        }.start();
        break;
    case R.id.sendtoNonUiBtn:
        Message msg = handler3.obtainMessage();
        Bundle bundle = new Bundle();
        bundle.putString(NAME,"zhangsan");
        bundle.putInt(AGE,25);
        bundle.putString(CITY,"Beijing");
        msg.setData(bundle);
        msg.sendToTarget();
        break;
    case R.id.startSaleThreadBtn:
        new SaleThread("window A").start();
        new SaleThread("window B").start();
        break;
    case R.id.startSaleRunnableBtn:
        SaleRunnable saleRunnable = new SaleRunnable("window A");
        new Thread(saleRunnable).start();
        new Thread(saleRunnable).start();
        break;
    }
  }
}
```

代码编辑完后，编译运行。单击 startSaleRunnableBtn 按钮，可以在 Logcat 中看到相应的日志信息。

```
12776 - 21327 I/SaleRunnable: run: window A: ticket left:2
12776 - 21328 I/SaleRunnable: run: window A: ticket left:1
12776 - 21327 I/SaleRunnable: run: window A: ticket left:0
```

在这个例子中，创建 Thread 的方法是：声明一个实现了 Runnable 接口的类(实现了 run 方法)，新建这个类的实例，在创建线程时作为参数传进去，然后启动线程，线程被启动时，执行线程的 Runnable 参数中的 run 方法。

13.6 Notification 实验

实验目的：熟悉 GridLayout 和 TaskStackBuilder 的用法。

1. 新建工程 Notification1

具体步骤：运行 Android Studio，单击 Start a new Android Studio project，在 Create New Project 页中，Application name 为 Notification1，Company domain 为 cuc，Project location 为 C:\android\myproject\ Notification1，单击 Next 按钮，选中 Phone and Tablet，minimum sdk 设置为 API 19：Android 4.4(minSdkVersion)，单击 Next 按钮，选择 Empty Activity，再单击 Next 按钮，选中 Generate Layout File 选项，选中 Backwards Compatibility (AppCompat)选项，单击 Finish 按钮。

2. 添加 DataBinding 支持，并同步

在文件 build.gradle(module：app)的 android 语句的花括号内添加：

```
dataBinding.enabled = true
```

由于修改了 build.gradle 文件内容，这个文件窗口的上面就出现提示："Gradle files have changed since last project sync. A project sync may be necessary for the IDE to work properly. sync now"。单击 sync now 按钮进行同步。

3. 新建两个布局文件

新建布局文件 activity_main1.xml 的具体步骤：在工程结构视图中，右击 layout，选择 New|Layout resource file，在 File name 输入框输入 activity_main1，在 Root element 输入框输入 layout，Source set 为 main，单击 OK 按钮。layout/activity_main1.xml 内容如下：

```
<?xml version = "1.0" encoding = "utf - 8"?>
< layout >
    < data/>
    < GridLayout xmlns:android = "http://schemas.android.com/apk/res/android"
        android:layout_width = "match_parent" android:layout_height = "match_parent"
        android:orientation = "horizontal" android:columnCount = "2">
        < Button android:id = "@ + id/noticeBtn" android:text = "notice" />
        < Button android:id = "@ + id/noticeTagBtn" android:text = "noticeTag" />
```

```
        <Button
            android:id = "@ + id/noticeActionTagBtn" android:text = "noticeActionTag"/>
        <Button android:id = "@ + id/send3Btn" android:text = "send 3" />
        <Button android:id = "@ + id/noticeNoClearBtn" android:text = "noticeNoClear"/>
        <Button android:id = "@ + id/noticeOngoingBtn" android:text = "noticeOngoing" />
        <Button
            android:id = "@ + id/noticeAutoCancelBtn" android:text = "noticeAutoCancel" />
    <Button android:id = "@ + id/noticeBigTextBtn" android:text = "noticeBigText"/>
        <Button android:id = "@ + id/noticeBigPictureBtn" android:text = "noticeBigPicture"/>
      <Button android:id = "@ + id/noticeInboxBtn" android:text = "noticeInbox"/>
        <Button android:layout_marginTop = "10dp"
            android:id = "@ + id/removeBtn" android:text = "remove" />
        <Button android:layout_marginTop = "10dp"
            android:id = "@ + id/removeTagBtn" android:text = "removeTag" />
        <Button android:id = "@ + id/removeAllBtn" android:text = "removeAll" />
    </GridLayout>
</layout>
```

activity_main1. xml 中，GridLayout 的 orientation 设置为 horizontal，水平方向被首先处理，该布局中的按钮，用于发送通知或者清除通知。

新建布局文件 hello. xml，用作 Activity_a、Activity_b 和 Activity_c 的布局文件。hello. xml 文件内容如下：

```
<?xml version = "1.0" encoding = "utf - 8"?>
<LinearLayout xmlns:android = "http://schemas.android.com/apk/res/android"
    android:orientation = "vertical" android:layout_width = "match_parent"
    android:layout_height = "match_parent">
    <TextView
        android:layout_width = "match_parent" android:layout_height = "wrap_content"
        android:textSize = "40sp" android:id = "@ + id/helloTxt"/>
</LinearLayout>
```

4. 新建 3 个 Activity

在工程视图中 main 源集 Java 文件夹里，右击 cuc. notification1 包，选择 New | Java Class，在 Name 输入框中输入 Activity_a，在 Superclass 输入框中输入 android. support. v7. app. AppCompatActivity，单击 OK 按钮。编辑 Activity_a. java，文件内容如下：

```
package cuc.notification1;
import android.support.v7.app.AppCompatActivity;
import android.os.Bundle;
import android.widget.TextView;
public class Activity_a extends AppCompatActivity {
    @Override
    protected void onCreate(Bundle savedInstanceState) {
        super.onCreate(savedInstanceState);
        setContentView(R.layout.hello);
        ((TextView)findViewById(R.id.helloTxt)).setText("Activity_a");
    }
}
```

Activity_b. java 文件内容如下：

```
package cuc.notification1;
import android.support.v7.app.AppCompatActivity;
import android.os.Bundle;
import android.widget.TextView;
public class Activity_b extends AppCompatActivity {
    @Override
    protected void onCreate(Bundle savedInstanceState) {
        super.onCreate(savedInstanceState);
        setContentView(R.layout.hello);
        ((TextView)findViewById(R.id.helloTxt)).setText("Activity_b");
    }
}
```

Activity_c. java 文件内容如下：

```
package cuc.notification1;
import android.support.v7.app.AppCompatActivity;
import android.os.Bundle;
import android.widget.TextView;
public class Activity_c extends AppCompatActivity {
    @Override
    protected void onCreate(Bundle savedInstanceState) {
        super.onCreate(savedInstanceState);
        setContentView(R.layout.hello);
        ((TextView)findViewById(R.id.helloTxt)).setText("Activity_c");
    }
}
```

5. 在清单文件中注册 Activity

由于 Activity_a、Activity_b 和 Activity_c 都是通过 New|Java Class 创建的，系统没有自动在清单文件中注册，需要手动注册。如果通过 New|Activity 创建，系统就自动在清单文件中注册。AndroidManifest. xml 文件内容如下：

```
<?xml version="1.0" encoding="utf-8"?>
<manifest xmlns:android="http://schemas.android.com/apk/res/android"
    package="cuc.notification1">
    <application
        android:allowBackup="true"
        android:icon="@mipmap/ic_launcher"
        android:label="@string/app_name"
        android:roundIcon="@mipmap/ic_launcher_round"
        android:supportsRtl="true"
        android:theme="@style/AppTheme">
        <activity android:name=".MainActivity">
            <intent-filter>
                <action android:name="android.intent.action.MAIN" />
```

```
            <category android:name = "android.intent.category.LAUNCHER"/>
        </intent-filter>
    </activity>
    <activity android:name = ".Activity_a" />
    <activity android:name = ".Activity_b" />
    <activity android:name = ".Activity_c" />
</application>
```

6. 编辑 MainActivity.java

编辑后，MainActivity.java 文件内容如下：

```
package cuc.notification1;
import android.app.Notification;
import android.app.NotificationManager;
import android.app.PendingIntent;
import android.app.TaskStackBuilder;
import android.content.Intent;
import android.databinding.DataBindingUtil;
import android.graphics.BitmapFactory;
import android.os.Bundle;
import android.support.v4.app.NotificationCompat;
import android.support.v7.app.AppCompatActivity;
import android.view.View;
import cuc.notification1.databinding.ActivityMain1Binding;

public class MainActivity extends AppCompatActivity implements View.OnClickListener{
    private NotificationManager manager;
    private static final int NOTIFY_ID = 1;
    private static final String NOTIFY_TAG = "NOTIFY_TAG";

    @Override
    protected void onCreate(Bundle savedInstanceState) {
        super.onCreate(savedInstanceState);
        ActivityMain1Binding binding =
         DataBindingUtil.setContentView(this,R.layout.activity_main1 );
      manager = (NotificationManager)getSystemService(NOTIFICATION_SERVICE);
        binding.noticeActionTagBtn.setOnClickListener(this);
        binding.noticeAutoCancelBtn.setOnClickListener(this);
        binding.noticeBigPictureBtn.setOnClickListener(this);
        binding.noticeBigTextBtn.setOnClickListener(this);
        binding.noticeBtn.setOnClickListener(this);
        binding.noticeInboxBtn.setOnClickListener(this);
        binding.noticeNoClearBtn.setOnClickListener(this);
        binding.noticeOngoingBtn.setOnClickListener(this);
        binding.noticeTagBtn.setOnClickListener(this);
        binding.notice3Btn.setOnClickListener(this);
        binding.removeBtn.setOnClickListener(this);
        binding.removeTagBtn.setOnClickListener(this);
```

```
        binding.removeAllBtn.setOnClickListener(this);
    }
    @Override
    public void onClick(View view) {
        NotificationCompat.Builder builder = new NotificationCompat.Builder(this,"myFirstNotice");
        PendingIntent pendingIntent = PendingIntent.getActivity(this,0,
                new Intent(this,MainActivity.class),0);

builder.setSmallIcon(R.mipmap.ic_launcher_round).setContentTitle("Title").setContentText
("Text").setContentIntent(pendingIntent);
        switch (view.getId()) {
            case R.id.noticeBtn:
                manager.notify(NOTIFY_ID,builder.build());
                break;
            case R.id.noticeTagBtn:
                manager.notify(NOTIFY_TAG,NOTIFY_ID,builder.build());
                break;
            case R.id.noticeActionTagBtn:
                TaskStackBuilder stackBuilder1 = TaskStackBuilder.create(this);
            stackBuilder1.addNextIntent(new Intent(this,MainActivity.class));
            stackBuilder1.addNextIntent(new Intent(this,Activity_a.class));
            stackBuilder1.addNextIntent(new Intent(this,Activity_b.class));
            stackBuilder1.addNextIntent(new Intent(this,Activity_c.class));
                PendingIntent pendingIntent1 = stackBuilder1.getPendingIntent(0,PendingIntent.
FLAG_UPDATE_CURRENT);

                TaskStackBuilder stackBuilder2 = TaskStackBuilder.create(this);
            stackBuilder2.addNextIntent(new Intent(this,MainActivity.class));
             stackBuilder2.addNextIntent(new Intent(this,Activity_a.class));
                PendingIntent pendingIntent2 =  stackBuilder2.getPendingIntent(0,PendingIntent.
FLAG_UPDATE_CURRENT);
                builder.addAction(new NotificationCompat.Action
                        .Builder(android.R.drawable.ic_input_add,"c-b-a-main",
pendingIntent1).build());
                builder.addAction(new NotificationCompat.Action
                        .Builder(android.R.drawable.ic_input_delete,"a-main",
pendingIntent2).build());
                manager.notify(NOTIFY_TAG,NOTIFY_ID,builder.build());
                break;
            case R.id.notice3Btn:
                for(int i = 1;i <= 3;i++){

builder.setContentTitle("Title" + i).setContentText("text" + i);
                    manager.notify(i,builder.build());
                }
                break;
            case R.id.noticeNoClearBtn:
                Notification notification = builder.build();
                notification.flags |= Notification.FLAG_NO_CLEAR;
                manager.notify(NOTIFY_ID,notification);
```

```
                break;
            case R.id.noticeOngoingBtn:
                builder.setOngoing(true);
                manager.notify(NOTIFY_ID,builder.build());
                break;
            case R.id.noticeAutoCancelBtn:
                builder.setAutoCancel(true);
                manager.notify(NOTIFY_ID,builder.build());
                break;
            case R.id.noticeBigTextBtn:
                NotificationCompat.BigTextStyle bigTextStyle = new NotificationCompat.BigTextStyle();

    bigTextStyle.setBigContentTitle("Title: bigTextStyle").setSummaryText("Summary:
    bigTextStyle").bigText("big Text\n big Text\n ");
                builder.setStyle(bigTextStyle);
                manager.notify(NOTIFY_ID,builder.build());
                break;
            case R.id.noticeBigPictureBtn:
                NotificationCompat.BigPictureStyle bigPictureStyle = new NotificationCompat.
    BigPictureStyle();

    bigPictureStyle.setBigContentTitle("Title: BigPictureStyle").setSummaryText("Summary:
    BigPictureStyle").bigPicture(BitmapFactory.decodeResource(getResources(),R.mipmap.ic_
    launcher_round));
                builder.setStyle(bigPictureStyle);
                manager.notify(NOTIFY_ID,builder.build());
                break;
            case R.id.noticeInboxBtn:
                NotificationCompat.InboxStyle inboxStyle = new NotificationCompat.InboxStyle();

    inboxStyle.setBigContentTitle("Title:InboxStyle").setSummaryText("summary:InboxStyle")
                        .addLine("task 1").addLine("task 2").addLine("task 3");
                builder.setStyle(inboxStyle);
                manager.notify(NOTIFY_ID,builder.build());
                break;
            case R.id.removeBtn:
                manager.cancel(NOTIFY_ID);
                break;
            case R.id.removeTagBtn:
                manager.cancel(NOTIFY_TAG,NOTIFY_ID);
                break;
            case R.id.removeAllBtn:
                manager.cancelAll();
                break;
        }
    }
}
```

7. 编译运行

通过实验可以看出，①使用notify(String tag,int id,Notification notification)发出的通知可以通过cancel(String tag,int id)和cancelAll()取消通知，但不能通过cancel(int id)取消该通知。使用notify(int id,Notification notification)发出的通知可以通过cancel(int id)和cancelAll()取消通知，但不能通过cancel(String tag,int id)取消该通知。②若Notification设置FLAG_NO_CLEAR，将不能通过手势把它滑出屏幕。③若Notification设置FLAG_ONGOING_EVENT，将此通知放到通知栏的"正在运行"分组中，通知就会像QQ一样一直在状态栏中显示，不能通过手指把它滑出屏幕。④TaskStackBuilder对象addNextIntent(Intent nextIntent)把一个新的Intent添加到TaskStackBuilder的ArrayList<Intent>mIntents的最后面，最近添加的Intent将在任务栈的顶部激活Activity。

8. 编辑清单文件

在清单文件中，给Activity_c、Activity_b、Activity_a添加parentActivityName属性，添加后这3个Activity的声明如下：

```
<activity android:name = ".Activity_a" android:parentActivityName = ".MainActivity"/>
<activity android:name = ".Activity_b" android:parentActivityName = ".Activity_a"/>
<activity android:name = ".Activity_c" android:parentActivityName = ".Activity_b"/>
```

9. 编辑MainActivity.java，并编译运行

在MainActivity.java文件的onClick(View view)方法R.id.noticeActionTagBtn分支中，pendingIntent1采用以下内容初始化：

```
TaskStackBuilder stackBuilder1 = TaskStackBuilder.create(this);
    stackBuilder1.addParentStack(Activity_c.class);
    stackBuilder1.addNextIntent(new Intent(this,Activity_c.class));
    PendingIntent pendingIntent1 = stackBuilder1.getPendingIntent(0, PendingIntent.FLAG_
UPDATE_CURRENT);
```

代码stackBuilder1.addParentStack(Activity_c.class)把Activity_c.class的父链添加到了TaskStackBuilder的ArrayList<Intent>mIntents中，具体过程为：首先在mIntents的最后位置(这时stackBuilder1的mIntents中还没有Intent对象，所以index=0)添加一个Intent，用于启动Activity_c的父Activity_b；还是在原来的位置(index=0)插入一个Intent，用于启动Activity_b的父Activity_a；还是在原来的位置(index=0)插入一个Intent，用于启动Activity_a的父MainActivity；MainActivity没有指定父，到此为止。

最后编译运行，可以看出两种初始化pendingIntent1的效果完全一样。

13.7 Service开始和绑定实验

本实验分八个阶段，各阶段实验目的如下：

第一阶段：理解启动Service生命周期。

第二阶段：理解绑定Service生命周期。

第三阶段：理解 Service 既被启动又被绑定的情况。

第四阶段：前台 Service。

第五阶段：返回本地 Service 实例创建绑定的 Service。

第六阶段：启动 IntentService。

第七阶段：使用 Messenger 创建绑定的 Service。

第八阶段：使用 AIDL 创建绑定的 Service。

第一阶段：理解启动 Service 生命周期

1. 新建工程 Service1

具体步骤：运行 Android Studio，单击 Start a new Android Studio project，在 Create New Project 页中，Application name 为 Service1，Company domain 为 cuc，Project location 为 C:\android\myproject\Service1，单击 Next 按钮，选中 Phone and Tablet，minimum sdk 设置为 API 19：Android 4.4(minSdkVersion)，单击 Next 按钮，选择 Empty Activity，再单击 Next 按钮，选中 Generate Layout File 选项，选中 Backwards Compatibility(AppCompat) 选项，单击 Finish 按钮。

2. 新建两个 Service 类

在工程视图中 main 源集 java 文件夹里，右击 cuc.service1 包，选择 New | Service | Service，在 Class Name 输入框中输入 MyLocalService，Exported 选项和 Enabled 选项保持默认状态（选中），单击 Finish 按钮。系统新建名为 MyLocalService 的类，并完成了在 AndroidManifest.xml 中的注册。编辑 MyLocalService.java，内容如下：

```
package cuc.service1;
import android.app.Service;
import android.content.Intent;
import android.os.Binder;
import android.os.IBinder;
import android.os.Parcel;
import android.os.RemoteException;
import android.util.Log;

public class MyLocalService extends Service {
    private static final String TAG = "MyLocalService";
    public MyLocalService() {
    }
    @Override
    public int onStartCommand(Intent intent, int flags, int startId) {
        Log.i(TAG, "onStartCommand:school:" + intent.getStringExtra("school") + ", flags:" +
flags + ", startId:" + startId + ",threadID:" + Thread.currentThread().getId());
        return START_NOT_STICKY;
    }
    @Override
    public void onCreate() {
        super.onCreate();
        Log.i(TAG, "onCreate: threadID:" + Thread.currentThread().getId());
    }
```

```
    @Override
    public void onDestroy() {
        Log.i(TAG, "onDestroy: ");
        super.onDestroy();
    }
    @Override
    public boolean onUnbind(Intent intent) {
        Log.i(TAG, "onUnbind: ");
        return true;
    }
    @Override
    public void onRebind(Intent intent) {
        Log.i(TAG, "onRebind: threadID:" + Thread.currentThread().getId());
        super.onRebind(intent);
    }
    @Override
    public IBinder onBind(Intent intent) {
        Log.i(TAG, "onBind: ");
        return new MyBinder();
    }
    class MyBinder extends Binder {
        @Override
        protected boolean onTransact(int code, Parcel data, Parcel reply, int flags) throws
RemoteException {
            Log.i(TAG, "onTransact: from parcel data, " + data.readString() + ", " + data.
readInt() + ", " + data.readFloat() );
            reply.writeString("dianxin");
            reply.writeInt(101);
            reply.writeFloat(432.1f);
            return super.onTransact(code, data, reply, flags);
        }
    }
}
```

MyLocalService 类重写了 Service 类的各种回调方法，以便理解启动的 Service 的生命周期、绑定的 Service 的生命周期，以及一个 Service 同时被启动和绑定的情况。MyLocalService 的内部类 MyBinder 重写了 onTransact 方法，Binder 包含客户端可调用的公共方法。

同样的方法创建 MyLocalService2。在 MyLocalService2 中，Binder 仅仅包含一个客户端可调用的公共方法，用于获取它所在的 Service 实例，Service 包含客户端可调用的公共方法 getRandomNumber。MyLocalService2.java 文件内容如下：

```
package cuc.service1;
import android.app.Service;
import android.content.Intent;
import android.os.Binder;
import android.os.IBinder;
```

```
import java.util.Random;

public class MyLocalService2 extends Service {
    public MyLocalService2() {
    }
    private final Random mGenerator = new Random();
    /* 这个 Service 和 client 在在同一应用中,且在相同的进程中,所以不需要处理 IPC */
    public class MyBinder extends Binder {
        MyLocalService2 getService() {
/* 返回 MyLocalService2 实例,以便客户端调用它的 public 方法 */
            return MyLocalService2.this;
        }
    }
    @Override
    public IBinder onBind(Intent intent) {
        return new MyBinder ();
    }
    public int getRandomNumber() {
        return mGenerator.nextInt(100);
    }
}
```

3. 新建一个 IntentService 类 MyIntentService

具体步骤：在工程视图中 main 源集 java 文件夹里，右击 cuc.service1 包，选择 New|Service|Service(IntentService)，Class Name 输入框中输入 MyIntentService，Exported 选项和 Enabled 选项保持默认状态（选中），取消 Include helper start methods 选项前面的选中，单击 Finish 按钮。系统新建名为 MyIntentService 的类，并完成了在 AndroidManifest.xml 中的注册。编辑 MyIntentService.java，内容如下：

```
package cuc.service1;
import android.app.IntentService;
import android.content.Intent;
import android.content.Context;
import android.support.v4.content.LocalBroadcastManager;
import android.util.EventLogTags;
import android.util.Log;
public class MyIntentService extends IntentService {
    private static final String TAG = "MyIntentService";
    private LocalBroadcastManager broadcastManager;
    public static final String MY_ACTION_SERVICE = "MY_ACTION_SERVICE";
    public static final String MY_ACTION_THREAD = "MY_ACTION_THREAD";
    public static final String STATUS = "status";
    public static final String COUNT = "COUNT";
    public static final String PROGRESS = "PROGRESS";
    private void sendServiceStatus(String status){
        Intent intent = new Intent(MY_ACTION_SERVICE);
        intent.putExtra(STATUS, status);
```

```
        broadcastManager.sendBroadcast(intent);
    }
    private void sendThreadStatus(String status, int progress){
        Intent intent = new Intent(MY_ACTION_THREAD);
        intent.putExtra(STATUS, status);
        intent.putExtra(PROGRESS, progress);
        broadcastManager.sendBroadcast(intent);
    }
    public MyIntentService() {
        super("MyIntentService");
    }
    @Override
    public void onCreate() {
        super.onCreate();
        broadcastManager = LocalBroadcastManager.getInstance(this);
        sendServiceStatus("onCreate");
        Log.i(TAG, "onCreate: ");
    }
    @Override
    public void onDestroy() {
        super.onDestroy();
        sendServiceStatus("Service:onDestroy");
        Log.i(TAG, "onDestroy: ");
    }
    @Override
    protected void onHandleIntent(Intent intent) {
        sendServiceStatus("Service:onHandleIntent");
        if (intent != null) {
            int progress = 0;
            sendThreadStatus("Thread start:", progress);
            while(progress < 100){
                progress += 20;
                try {
                    Thread.sleep(1000);
                } catch (InterruptedException e) {
                    e.printStackTrace();
                }
                sendThreadStatus("Thread running:", progress);
                Log.i(TAG, "onHandleIntent: -->" + progress);
            }
            sendThreadStatus("Thread Over", progress);
        }
    }
}
```

MyIntentService 继承自 IntentService，在后台执行耗时的异步任务，当任务完成后会自动停止。客户端通过调用 startService(Intent)发出请求，Service 在需要的时候启动，使用一个工作线程顺序处理每个 Intent，在工作运行结束后停止它自己。MyIntentService 包含了 LocalBroadcastManager 实例，用于发送局部广播，更新 UI。

4. 新建一个 Service 类 MessengerService

在工程视图中 main 源集 Java 文件夹里，右击 cuc. service1 包，选择 New | Service | Service，Class Name 输入框中输入 MessengerService，Exported 选项和 Enabled 选项保持默认状态（选中），单击 Finish 按钮。系统新建名为 MessengerService 的类，并完成了在 AndroidManifest. xml 中的注册。编辑 MessengerService. java，内容如下：

```
package cuc.service1;
import android.app.Service;
import android.content.Intent;
import android.os.Handler;
import android.os.IBinder;
import android.os.Message;
import android.os.Messenger;
import android.os.Process;
import android.os.RemoteException;
import android.util.Log;

public class MessengerService extends Service {
    private static final String TAG = "MessengerService";
    public MessengerService() {
    }
    private static final int MSG_SUM = 1;
    Handler handler = new Handler(){
        @Override
        public void handleMessage(Message msg) {
            Message msgToClient = Message.obtain(msg);
            switch (msg.what){
                case MSG_SUM:
                    try {
                        msgToClient.what = MSG_SUM;
                        msgToClient.arg1 = msg.arg1 + msg.arg2;
                        Log.i(TAG, "handleMessage: pid-->" + Process.myPid());
                        msg.replyTo.send(msgToClient);
                    } catch (RemoteException e) {
                        e.printStackTrace();
                    }
                    break;
            }
            super.handleMessage(msg);
        }
    };
    private Messenger messenger = new Messenger(handler);
    @Override
    public IBinder onBind(Intent intent){
        return messenger.getBinder();
    }
}
```

MessengerService 用于客户端和 Service 不在同一进程的情况，客户端在另一个应用

cuc.client1 中，两个进程间通过 Messenger 进行通信，Service 顺序处理客户端的请求。

5. 添加 DataBinding 支持，并同步

在文件 build.gradle(module: app)的 android 语句的花括号内添加：

```
dataBinding.enabled = true
```

由于修改了 build.gradle 文件内容，这个文件窗口的上面就出现提示：Gradle files have changed since last project sync. A project sync may be necessary for the IDE to work properly. sync now。单击 sync now 按钮进行同步。

6. 新建布局文件 activity_main1.xml

具体步骤：在工程结构视图中，右击 layout，依次选择 New|Layout resource file，File name 在输入框输入 activity_main1，在 Root element 输入框输入 layout，Source set 为 main，单击 OK 按钮。layout/activity_main1.xml 内容如下：

```
<?xml version = "1.0" encoding = "utf - 8"?>
< layout xmlns:android = "http://schemas.android.com/apk/res/android">
    < data/>
    < GridLayout
        android:orientation = "horizontal" android:columnCount = "2"
        android:layout_width = "match_parent" android:layout_height = "match_parent">
      < Button android:id = "@ + id/startServiceBtn" android:text = "startService"/>
        < Button android:id = "@ + id/stopServiceBtn" android:text = "stopService"/>
        < Button android:id = "@ + id/bindServiceBtn" android:text = "bindService"/>
    < Button android:id = "@ + id/unbindServiceBtn" android:text = "unbindService"/>
      < Button android:id = "@ + id/transactBtn" android:text = "transact" />
        < Space />
      < Button android:id = "@ + id/bindService2Btn" android:text = "bindService2"/>
  < Button android:id = "@ + id/unbindService2Btn" android:text = "unbindService2"/>
        < Button android:id = "@ + id/service2Btn" android:text = "Random" />
        < TextView android:id = "@ + id/service2Txt" android:text = "int:" />
        < Button android:id = "@ + id/startIntentServiceBtn" android:text = "startIntentService"/>
        < Button android:id = "@ + id/stopIntentServiceBtn" android:text = "stopIntentService"/>
        < TextView android:id = "@ + id/serviceTxt" />
        < TextView android:id = "@ + id/threadTxt" />
        < ProgressBar android:id = "@ + id/progressBar" style = "@android:style/Widget.
ProgressBar.Horizontal"
            android:layout_columnSpan = "2" android:layout_gravity = "fill_horizontal" />
    </GridLayout >
</layout >
```

activity_main1.xml 中的 startServiceBtn、stopServiceBtn、bindServiceBtn、unbindServiceBtn、transactBtn 分别用于启动 MyLocalService、停止 MyLocalService、绑定 MyLocalService、解除绑定 MyLocalService、与 MyLocalService 通信。

bindService2Btn、unbindService2Btn 分别绑定、解除绑定 MyLocalService2。service2Btn 用于调用 MyLocalService2 的 public 方法 getRandomNumber 获取随机数，service2Txt 用于显示这个随机数。

startIntentServiceBtn 用于启动 MyIntentService 执行长时间的任务，stopIntentServiceBtn 用于停止 MyIntentService。progressBar 用于显示执行的进度，serviceTxt 和 threadTxt 分别用于显示 Service 的状态和线程的状态。

7. 编辑 MainActivity.java

编辑后，MainActivity.java 内容如下：

```
package cuc.service1;
import android.content.BroadcastReceiver;
import android.content.ComponentName;
import android.content.Context;
import android.content.Intent;
import android.content.IntentFilter;
import android.content.ServiceConnection;
import android.databinding.DataBindingUtil;
import android.os.Binder;
import android.os.Bundle;
import android.os.IBinder;
import android.os.Parcel;
import android.os.RemoteException;
import android.support.v4.content.LocalBroadcastManager;
import android.support.v7.app.AppCompatActivity;
import android.util.Log;
import android.view.View;
import cuc.service1.databinding.ActivityMain1Binding;

public class MainActivity extends AppCompatActivity implements View.OnClickListener {
    private ActivityMain1Binding binding;
    private static final String TAG = "MainActivity";
    private LocalBroadcastManager broadcastManager;
    private Binder binder;
    private boolean isConnMyLocalService;
    private ServiceConnection connMyLocalService = new ServiceConnection() {
        @Override
        public void onServiceConnected(ComponentName name, IBinder iBinder) {
            MainActivity.this.binder = (Binder) iBinder;
            isConnMyLocalService = true;
            Log.i(TAG, "onServiceConnected: ");
        }
        @Override
        public void onServiceDisconnected(ComponentName name) {
            isConnMyLocalService = false;
            Log.i(TAG, "onServiceDisconnected: ");
        }
    };
    private MyLocalService2 myLocalService2;
    private boolean isConnMyLocalService2;
    private ServiceConnection connMyLocalService2 = new ServiceConnection() {
        @Override
```

```
        public void onServiceConnected(ComponentName name, IBinder iBinder) {
            myLocalService2 = ((MyLocalService2. MyBinder)iBinder).getService();
            isConnMyLocalService2 = true;
            Log.i(TAG, "onServiceConnected: ");
        }
        @Override
        public void onServiceDisconnected(ComponentName name) {
            isConnMyLocalService2 = false;
            Log.i(TAG, "onServiceDisconnected: ");
        }
    };

    private MyReceiver myReceiver;
    void myRegisterReceiver(){
        broadcastManager = LocalBroadcastManager.getInstance(this);
        myReceiver = new MyReceiver();
        IntentFilter intentFilter = new IntentFilter();
        intentFilter.addAction(MyIntentService.MY_ACTION_SERVICE);
        intentFilter.addAction(MyIntentService.MY_ACTION_THREAD);
        broadcastManager.registerReceiver(myReceiver, intentFilter);
    }
    class MyReceiver extends BroadcastReceiver{
        @Override
        public void onReceive(Context context, Intent intent) {
            switch (intent.getAction()){
                case MyIntentService.MY_ACTION_SERVICE:

binding.serviceTxt.setText(intent.getStringExtra(MyIntentService.STATUS));
                    break;
                case MyIntentService.MY_ACTION_THREAD:
                    int progess = intent.getIntExtra(MyIntentService.PROGRESS,0);
                    binding.progressBar.setProgress(progess);

binding.threadTxt.setText(intent.getStringExtra(MyIntentService.STATUS) + progess);
                    break;
            }
        }
    }
    @Override
    protected void onCreate(Bundle savedInstanceState) {
        super.onCreate(savedInstanceState);
        binding = DataBindingUtil.setContentView(this,R.layout.activity_main1 );
        myRegisterReceiver();
        binding.startServiceBtn.setOnClickListener(this);
        binding.stopServiceBtn.setOnClickListener(this);
        binding.bindServiceBtn.setOnClickListener(this);
        binding.transactBtn.setOnClickListener(this);
        binding.unbindServiceBtn.setOnClickListener(this);
        binding.bindService2Btn.setOnClickListener(this);
```

```
        binding.service2Btn.setOnClickListener(this);
        binding.unbindService2Btn.setOnClickListener(this);
        binding.startIntentServiceBtn.setOnClickListener(this);
        binding.stopIntentServiceBtn.setOnClickListener(this);
    }

    @Override
    protected void onDestroy() {
        super.onDestroy();
        broadcastManager.unregisterReceiver(myReceiver);
    }
    @Override
    public void onClick(View view) {
        Intent intent;
        switch (view.getId()){
            case R.id.startServiceBtn:
                Log.i(TAG, "onClick startServiceBtn: thread id:" + Thread.currentThread().getId());
                intent = new Intent(MainActivity.this,MyLocalService.class);
                intent.putExtra("school","cuc");
                startService(intent);
                break;
            case R.id.stopServiceBtn:
                Log.i(TAG, "onClick stopServiceBtn: thread id:" + Thread.currentThread().getId());
                intent = new Intent(MainActivity.this,MyLocalService.class);
                stopService(intent);
                break;
            case R.id.bindServiceBtn:
                Log.i(TAG, "onClick bindServiceBtn: thread id:" + Thread.currentThread().getId());
                intent = new Intent(MainActivity.this,MyLocalService.class);
                bindService(intent,connMyLocalService,BIND_AUTO_CREATE);
                break;
            case R.id.unbindServiceBtn:
                Log.i(TAG, "onClick unbindServiceBtn: thread id:" + Thread.currentThread().getId());
                unbindService(connMyLocalService);
                break;
            case R.id.transactBtn:
                Parcel data = Parcel.obtain();
                data.writeString("cuc");
                data.writeInt(100);
                data.writeFloat(123.4f);
                Parcel reply = Parcel.obtain();
                try {
                    if (isConnMyLocalService)binder.transact(0,data,reply,0);
                    Log.i(TAG, "after transact in onClick: from parcel reply, "
                            + reply.readString() + ", " + reply.readInt() + ", " + reply.readFloat() );
                } catch (RemoteException e) {
                    e.printStackTrace();
                }
                break;
```

```
            case R.id.bindService2Btn:
                Log.i(TAG, "onClick bindServiceBtn: thread id:" + Thread.currentThread().getId());
                intent = new Intent(MainActivity.this,MyLocalService2.class);
                bindService(intent,connMyLocalService2,BIND_AUTO_CREATE);
                break;
            case R.id.unbindService2Btn:
                Log.i(TAG, "onClick unbindServiceBtn: thread id:" + Thread.currentThread().getId());
                unbindService(connMyLocalService2);
                break;
            case R.id.service2Btn:

    binding.service2Txt.setText("int:" + myLocalService2.getRandomNumber());
                break;
            case R.id.startIntentServiceBtn:
                Log.i(TAG, "onClick: startIntentServiceBtn");
                intent = new Intent(MainActivity.this,MyIntentService.class);
                intent.putExtra("school","cuc");
                startService(intent);
                break;
            case R.id.stopIntentServiceBtn:
                Log.i(TAG, "onClick: stopIntentServiceBtn");
                intent = new Intent(MainActivity.this,MyIntentService.class);
                stopService(intent);
                break;
        }
    }
}
```

8. 编译运行

为了便于观察应用的log输出，先新建一个过滤器，过滤多余的信息。如果Logcat窗口没显示，就单击Android Studio集成环境最下边的“6：Logcat”，弹出的界面如图13-5所示。Logcat界面的左上角显示的是正在调试的Android设备，如果连接的有多个设备，可以单击它右边的向下箭头选择要显示的设备。由于应用中的日志输出采用的是Log.i，所以通过下拉箭头选择日志的级别为info。为了显示MainActivity类和各个Service类里的Log.i输出，增加一个过滤器，步骤如下：先单击Logcat界面右上角的向下箭头，选择Edit Filter configuration选项，弹出Create New Logcat Filter界面，在Log tag输入框中输入MainActivity|^(My)，其中“|”在正则表达式中是“或者”的意思，即tag名是MainActivity或者以My开头，所以Log tag后面的正则表达式Regex要选中，以便支持正则表达式。在Package name输入框中输入cuc.service1，由于在正则表达式中的点可以匹配除换行符\n之外的任何单字符，所以可以匹配cuc.service1。Package Name后面的Regex选项可以选中，也可以不选中。不选中表示精确的包名为cuc.service1。在Filter Name中输入名字，例如cuc.service1，单击OK按钮，完成过滤器的配置。

单击Logcat界面中的Logcat Header配置按钮（在Logcat界面的左侧），弹出Configure Logcat Header界面，如图13-6所示，可以不选Show date and time选项和Show package name选项，这样Logcat Header信息更简洁。

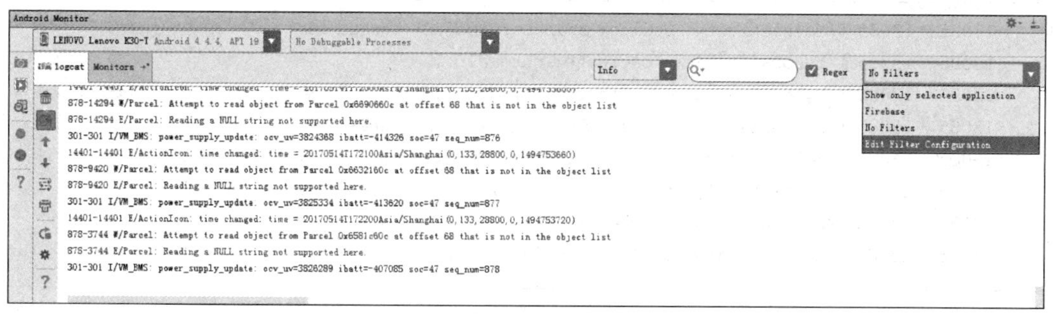

图 13-5　Logcat 界面

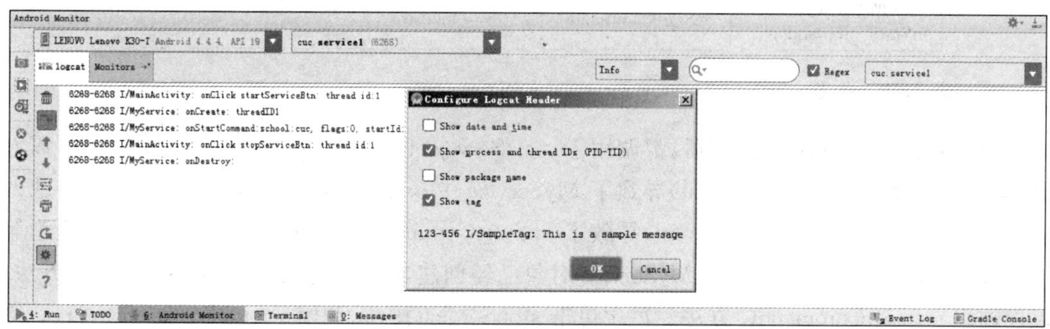

图 13-6　Logcat 中的 Configure Logcat Header 界面

上面主要介绍 Logcat 中过滤器的配置和 Logcat Header 的配置，通过下面步骤，观察 Logcat 输出，以便了解 Service 的回调函数和生命周期。

(1) 依次单击界面中的 startService 按钮和 stopService 按钮，对应的 Service 中的回调方法，如图 13-7 所示，Logcat 中 log 输出如下：

```
13085 - 13085 I/MainActivity: onClick startServiceBtn: thread id:1
13085 - 13085 I/MyLocalService: onCreate: threadID:1
13085 - 13085 I/MyLocalService: onStartCommand:school:cuc, flags:0, startId:1,threadID:1
13085 - 13085 I/MainActivity: onClick stopServiceBtn: thread id:1
13085 - 13085 I/MyLocalService: onDestroy
```

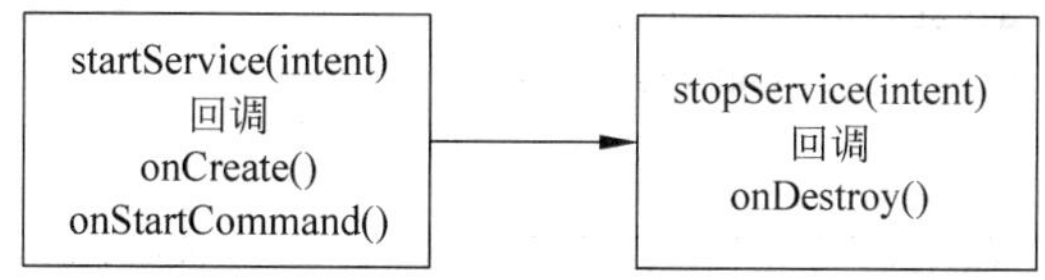

图 13-7　单击 startService 和 stopService 按钮后的方法调用

从上面 log 输出可以看到，单击 startService 按钮后，调用了 startService(intent)，从而导致了 MyService 中 onCreate()、onStartCommand()方法的调用。单击 stopService 按钮，调用了 stopService(intent)，从而导致了 MyService 中 onDestroy()方法的调用。

(2) 依次单击界面中的 startService、startService、stopService 和 stopService 按钮，对应的 Service 中的回调方法，如图 13-8 所示，Logcat 输出 log 输出如下：

```
13085 - 13085 I/MainActivity: onClick startServiceBtn: thread id:1
13085 - 13085 I/MyLocalService: onCreate: threadID:1
13085 - 13085 I/MyLocalService: onStartCommand:school:cuc, flags:0, startId:1,threadID:1
13085 - 13085 I/MainActivity: onClick startServiceBtn: thread id:1
13085 - 13085 I/MyLocalService: onStartCommand:school:cuc, flags:0, startId:2,threadID:1
13085 - 13085 I/MainActivity: onClick stopServiceBtn: thread id:1
13085 - 13085 I/MyLocalService: onDestroy:
13085 - 13085 I/MainActivity: onClick stopServiceBtn: thread id:1
```

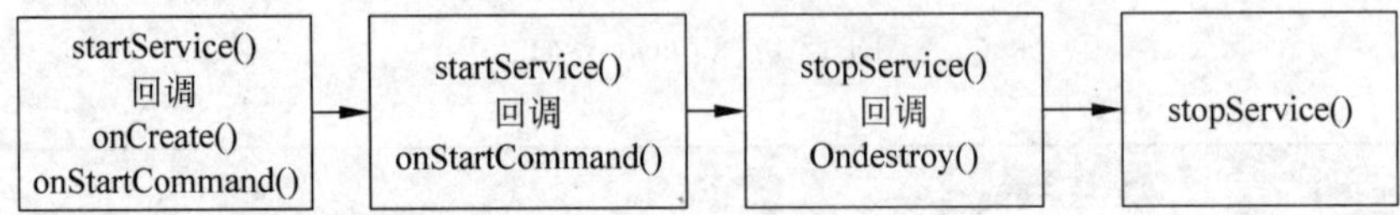

图 13-8 单击 startService、startService、stopService 和 stopService 按钮后的方法调用

每次单击 startService 按钮后，都调用了 startService(intent)。从(3)的 Logcat 输出可以看出，第一次调用 startService(intent)导致了 MyService 中 onCreate()、onStartCommand()方法的调用，第二次调用 startService(intent)导致了 onStartCommand()方法的再次调用，这是因为第二次调用 startService(intent)前，MyService 对象已经创建，因而不会重复调用 onCreate()方法、仅仅调用 onStartCommand()方法。每次单击 stopService 按钮，调用了 stopService(intent)。第一次调用 stopService(intent)导致了 MyService 中 onDestroy()方法的调用，第二次调用 stopService(intent)不会调用 MyService 中 onDestroy()方法，因为 MyService 对象已经被销毁。

第二阶段：理解绑定 Service 生命周期

(3) 依次单击界面中的 bindService、transact 和 stopService 按钮，对应的 Service 中的回调方法，如图 13-9 所示，Logcat 输出 log 输出如下：

```
13085 - 13085 I/MainActivity: onClick bindServiceBtn: thread id:1
13085 - 13085 I/MyLocalService: onCreate: threadID:1
13085 - 13085 I/MyLocalService: onBind:
13085 - 13085 I/MainActivity: onServiceConnected:
13085 - 13085 I/MyLocalService: onTransact: from parcel data, cuc, 100,123.4
13085 - 13085 I/MainActivity: after transact in onClick: from parcel reply, dianxin, 101, 432.1
13085 - 13085 I/MainActivity: onClick unbindServiceBtn: thread id:1
13085 - 13085 I/MyLocalService: onUnbind:
13085 - 13085 I/MyLocalService: onDestroy:
```

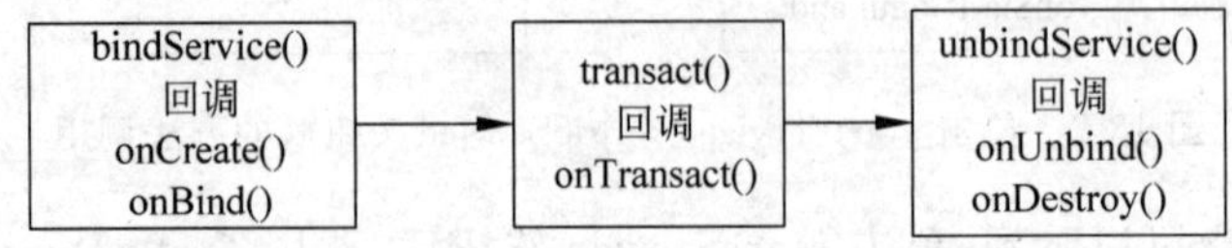

图 13-9 单击 bindService、transact 和 stopService 按钮的方法调用

从上面的 log 输出可以看到，单击 bindService 按钮后，调用了 bindService(intent, conn,BIND_AUTO_CREATE)，从而导致了 MyService 中 onCreate()、onBind()方法的调

用。单击 transact 按钮，调用了 binder.transact(0,data,reply,0)，从而导致了 MyService 中 onTransact 方法的调用。单击 unbindService 按钮后，调用了 unbindService(conn)的调用，从而导致了 MyService 中 onUnbind 和 onDestroy 方法的调用。

(4) 依次单击界面中的 bindService、bindService、unbindService 和 unbindService 按钮，对应的 Service 中的回调方法，如图 13-10 所示，Logcat 输出 log 输出如下：

```
13085 - 13085 I/MainActivity: onClick bindServiceBtn: thread id:1
13085 - 13085 I/MyLocalService: onCreate: threadID:1
13085 - 13085 I/MyLocalService: onBind:
13085 - 13085 I/MainActivity: onServiceConnected:
13085 - 13085 I/MainActivity: onClick bindServiceBtn: thread id:1
13085 - 13085 I/MainActivity: onClick unbindServiceBtn: thread id:1
13085 - 13085 I/MyLocalService: onUnbind:
13085 - 13085 I/MyLocalService: onDestroy:
13085 - 13085 I/MainActivity: onClick unbindServiceBtn: thread id:1
```

从上面的 log 输出可以看到，同一客户端多次调用 bindService(Intent,ServiceConnection,int)并不重复调用 onBind(Intent intent)方法。第二次单击 unbindService，手机上提示"很抱歉，Service1 已停止运行"，这时整个应用已经停止运行。

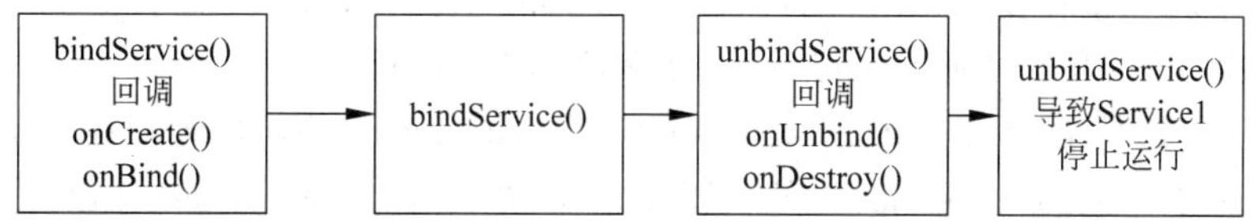

图 13-10 依次 bindService、bindService、unbindService 和 unbindService 按钮的方法调用

第三阶段：理解 Service 既被启动又被绑定的情况

(5) 依次单击图 13-10 中的 startService、bindService、unbindService 和 stopService 按钮，对应的 Service 中的回调方法，如图 13-11 所示，Logcat 输出 log 输出如下：

```
21473 - 21473 I/MainActivity: onClick startServiceBtn: thread id:1
21473 - 21473 I/MyLocalService: onCreate: threadID:1
21473 - 21473 I/MyLocalService: onStartCommand:school:cuc, flags:0, startId:1,threadID:1
21473 - 21473 I/MainActivity: onClick bindServiceBtn: thread id:1
21473 - 21473 I/MyLocalService: onBind:
21473 - 21473 I/MainActivity: onServiceConnected:
21473 - 21473 I/MainActivity: onClick unbindServiceBtn: thread id:1
21473 - 21473 I/MyLocalService: onUnbind:
21473 - 21473 I/MainActivity: onClick stopServiceBtn: thread id:1
21473 - 21473 I/MyLocalService: onDestroy:
```

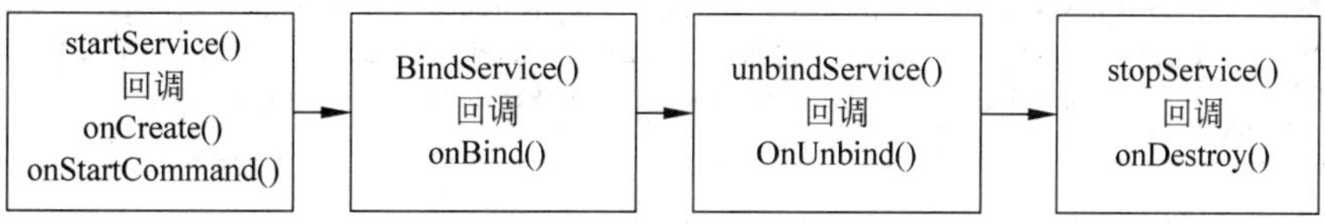

图 13-11 单击 startService、bindService、unbindService 和 stopService 按钮的方法调用

(6) 依次单击界面中的 startService、bindService、stopService 和 unbindService 按钮，对应的 Service 中的回调方法，如图 13-12 所示，Logcat 输出 log 输出如下：

```
21473 - 21473 I/MainActivity: onClick startServiceBtn: thread id:1
21473 - 21473 I/MyLocalService: onCreate: threadID:1
21473 - 21473 I/MyLocalService: onStartCommand:school:cuc, flags:0, startId:1,threadID:1
21473 - 21473 I/MainActivity: onClick bindServiceBtn: thread id:1
21473 - 21473 I/MyLocalService: onBind:
21473 - 21473 I/MainActivity: onServiceConnected:
21473 - 21473 I/MainActivity: onClick stopServiceBtn: thread id:1
21473 - 21473 I/MainActivity: onClick unbindServiceBtn: thread id:1
21473 - 21473 I/MyLocalService: onUnbind:
21473 - 21473 I/MyLocalService: onDestroy:
```

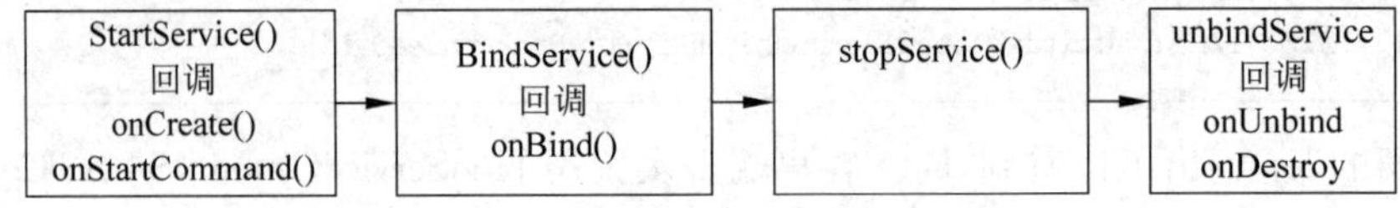

图 13-12 单击 startService、bindService、stopService 和 unbindService 按钮的方法调用

从(5)(6)的 log 输出可以看到，同一个 Service 实例可以被启动和绑定，仅仅调用 onCreate 一次，仅当既解除绑定又停止后，才调用 onDestroy。

(7) 依次单击界面中的 startService、bindService、unbindService、bindService、unbindService 和 stopService 按钮，对应的 Service 中的回调方法，如图 13-13 所示，Logcat 输出 log 输出如下：

```
21473 - 21473 I/MainActivity: onClick startServiceBtn: thread id:1
21473 - 21473 I/MyLocalService: onCreate: threadID:1
21473 - 21473 I/MyLocalService: onStartCommand:school:cuc, flags:0, startId:1,threadID:1
21473 - 21473 I/MainActivity: onClick bindServiceBtn: thread id:1
21473 - 21473 I/MyLocalService: onBind:
21473 - 21473 I/MainActivity: onServiceConnected:
21473 - 21473 I/MainActivity: onClick unbindServiceBtn: thread id:1
21473 - 21473 I/MyLocalService: onUnbind:
21473 - 21473 I/MainActivity: onClick bindServiceBtn: thread id:1
21473 - 21473 I/MainActivity: onServiceConnected:
21473 - 21473 I/MyLocalService: onRebind: threadID:1
21473 - 21473 I/MainActivity: onClick unbindServiceBtn: thread id:1
21473 - 21473 I/MyLocalService: onUnbind:
21473 - 21473 I/MainActivity: onClick stopServiceBtn: thread id:1
21473 - 21473 I/MyLocalService: onDestroy:
```

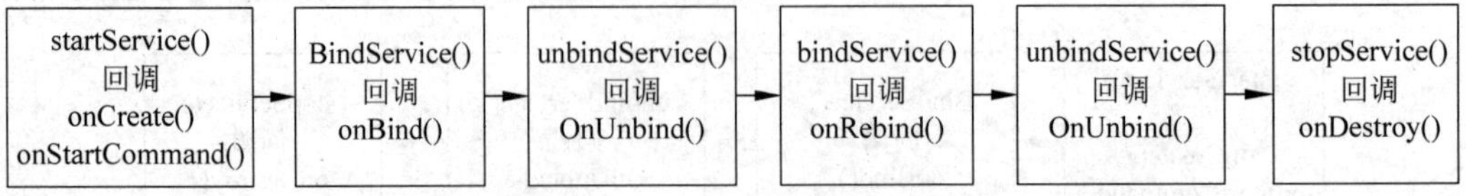

图 13-13 单击 startService、bindService、unbindService、bindService、unbindService 和 stopService 按钮的方法调用

从上面的 log 输出可以看到，当所有的客户端都已经从 Service 公布的特定接口断开连接后，系统回调用 onUnbind(Intent intent)，如果此时 Service 实例还存在(这个实例还没有被停止)，并且希望新的客户端绑定它的时候调用 onRebind 方法，必须重写 onUnbind (Intent intent)方法并返回 true。

(8) 把 MyLocalService 类的 onUnbind 方法的返回值改为 false，修改后 onUnbind 的代码如下：

```
@Override
public boolean onUnbind(Intent intent) {
    Log.i(TAG, "onUnbind: ");
    return false;
}
```

重新编译运行，依次单击界面中的 startService、bindService、unbindService、bindService、unbindService 和 stopService 按钮，对应的 Service 中的回调方法，如图 13-14 所示，Logcat 输出 log 输出如下：

```
14836 - 14836 I/MainActivity: onClick startServiceBtn: thread id:1
14836 - 14836 I/MyLocalService: onCreate:
14836 - 14836 I/MyLocalService: onStartCommand:school:cuc, flags:0, startId:1,threadID:1
14836 - 14836 I/MainActivity: onClick bindServiceBtn: thread id:1
14836 - 14836 I/MyLocalService: onBind:
14836 - 14836 I/MainActivity: onServiceConnected:
14836 - 14836 I/MainActivity: onClick unbindServiceBtn: thread id:1
14836 - 14836 I/MyLocalService: onUnbind:
14836 - 14836 I/MainActivity: onClick bindServiceBtn: thread id:1
14836 - 14836 I/MainActivity: onServiceConnected:
14836 - 14836 I/MainActivity: onClick unbindServiceBtn: thread id:1
14836 - 14836 I/MainActivity: onClick stopServiceBtn: thread id:1
14836 - 14836 I/MyLocalService: onDestroy:
```

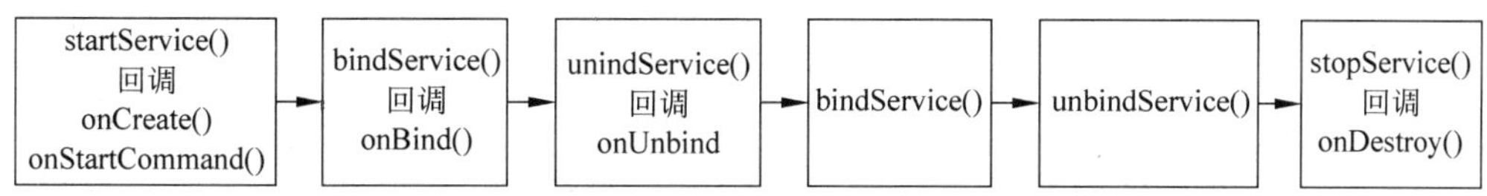

图 13-14 单击 startService、bindService、unbindService、bindService、unbindService 和 stopService 按钮的方法调用

(1)(2)是启动的 Service 的方法调用情况，(3)(4)是绑定的 Service 的方法调用情况，(5)(6)(7)(8)是一个 Service 同时被启动和被绑定的方法调用情况。

第四阶段：前台 Service

9. 创建前台 Service

Service 几乎都是在后台运行的，Service 的系统优先级还是比较低的，当系统出现内存不足情况时，就有可能会回收掉正在后台运行的 Service。如果希望通过 Context # startService(Intent)启动的 Service 可以一直保持运行状态，不会由于系统内存不足的原因导致被回收，就可以考虑调用 startForeground(int id,Notification notification)使 Service 运行在前台。仅仅调用 startForeground(int id, Notification notification)并不能启动 Service，必须先调用 Context#startService(Intent)启动 Service，再调用 startForeground(int id,Notification notification)使 Service 运行在前台。前台 Service 和普通 Service 最大的区别就在于，它会一直有一个正在运行的图标在系统的状态栏显示，下拉状态栏后可以看到更加详细的信息，非常类似于通知的效果。

编辑 MyLocalService 的 onCreate()方法，其他方法不变，编辑后 MyLocalService.java 文件内容如下：

```
package cuc.service1;
import android.app.Notification;
import android.app.PendingIntent;
import android.app.Service;
import android.content.Intent;
import android.os.Binder;
import android.os.IBinder;
import android.os.Parcel;
import android.os.RemoteException;
import android.support.v4.app.NotificationCompat;
import android.util.Log;

public class MyLocalService extends Service {
    private static final String TAG = "MyLocalService";
    public MyLocalService() {
    }
    @Override
    public int onStartCommand(Intent intent, int flags, int startId) {
        Log.i(TAG, "onStartCommand:school:" + intent.getStringExtra("school") + ", flags:" +
flags + ", startId:" + startId + ",threadID:" + Thread.currentThread().getId());
        return START_NOT_STICKY;
    }
    @Override
    public void onCreate() {
        int requestCode = 0;
        super.onCreate();
        Log.i(TAG, "onCreate: ");
        Intent intent = new Intent(this,MainActivity.class);
        PendingIntent pendingIntent = PendingIntent.getActivity(this,requestCode,intent,0);
        NotificationCompat.Builder builder = new NotificationCompat.Builder(this);
        builder.setContentIntent(pendingIntent)
                .setSmallIcon(R.mipmap.ic_launcher_round)
                .setContentTitle("title:MyService")
                .setContentText("text:Myservice")
                .setWhen(System.currentTimeMillis());
        Notification notification = builder.build();
        startForeground(1,notification);
    }
```

```
    @Override
    public void onDestroy() {
        Log.i(TAG, "onDestroy: ");
        super.onDestroy();
    }
    @Override
    public boolean onUnbind(Intent intent) {
        Log.i(TAG, "onUnbind: ");
        return true;
    }
    @Override
    public void onRebind(Intent intent) {
        Log.i(TAG, "onRebind: threadID:" + Thread.currentThread().getId());
        super.onRebind(intent);
    }
    @Override
    public IBinder onBind(Intent intent) {
        Log.i(TAG, "onBind: ");
        return new MyBinder();
    }
    class MyBinder extends Binder {
        @Override
        protected boolean onTransact(int code, Parcel data, Parcel reply, int flags) throws
RemoteException {
            Log.i(TAG, "onTransact: from parcel data, " + data.readString() + ", " + data.
readInt() + ", " + data.readFloat() );
            reply.writeString("dianxin");
            reply.writeInt(101);
            reply.writeFloat(432.1f);
            return super.onTransact(code, data, reply, flags);
        }
    }
}
```

这里只是修改了 MyLocalService 中 onCreate()方法的代码。可以看到，首先创建了一个 Notification 对象，并设置了单击通知后就打开 MainActivity。然后调用 startForeground()方法就可以让 MyLocalService 变成一个前台 Service，并会将通知的图片显示出来。

现在重新运行一下程序，并单击 Start Service 或 Bind Service 按钮，MyService 就会以前台 Service 的模式启动了，并且在系统状态栏会弹出一个通知图标，下拉状态栏后可以看到通知的详细内容，单击通知后就打开了 MainActivity。

第五阶段：返回本地 Service 实例创建绑定的 Service

在 MyLocalService2 中，Binder 仅仅包含一个客户端可调用的公共方法，用于获取它所在的 Service 实例，Service 包含客户端可调用的公共方法 getRandomNumber。

单击 bindService2Btn 按钮绑定 MyLocalService2，onBind(Intent intent)方法返回了 MyLocalService2.LocalBinder 实例，调用 LocalBinder 实例的 getService()方法返回该实例所在

的 MyLocalService2 对象。单击 service2Btn 按钮调用 MyLocalService2 实例的方法 getRandomNumber()获取随机数，service2Txt 显示这个随机数。最后单击 unbindService2Btn 按钮解除绑定 MyLocalService2。

第六阶段：IntentService 被启动

startIntentServiceBtn 按钮、stopIntentServiceBtn 按钮分别用于启动和停止 MyIntentService，progressBar 用于显示执行的进度，serviceTxt 和 threadTxt 分别用于显示 Service 的状态和线程的状态。

单击 startIntentServiceBtn 按钮，MyIntentService 开始运行，线程计数完成后线程结束，随后 MyIntentService 结束。

```
6376 - 6376 I/MainActivity: onClick: startIntentServiceBtn
6376 - 6376 I/MyIntentService: onCreate:ThreadID: 1
6376 - 6457 I/MyIntentService: onHandleIntent: ThreadID:351
6376 - 6457 I/MyIntentService: onHandleIntent: -->25
6376 - 6457 I/MyIntentService: onHandleIntent: -->50
6376 - 6457 I/MyIntentService: onHandleIntent: -->75
6376 - 6457 I/MyIntentService: onHandleIntent: -->100
6376 - 6457 I/MyIntentService: onHandleIntent: Thread Over
6376 - 6376 I/MyIntentService: onDestroy:
```

单击 startIntentServiceBtn，MyIntentService 开始运行，在线程计数完成前，单击 stopIntentServiceBtn 按钮调用 stopService(intent)，服务先结束，随后线程计数完成后线程结束。

```
6376 - 6376 I/MainActivity: onClick: startIntentServiceBtn
6376 - 6376 I/MyIntentService: onCreate:ThreadID: 1
6376 - 6544 I/MyIntentService: onHandleIntent: ThreadID:352
6376 - 6544 I/MyIntentService: onHandleIntent: -->25
6376 - 6376 I/MainActivity: onClick: stopIntentServiceBtn
6376 - 6376 I/MyIntentService: onDestroy:
6376 - 6544 I/MyIntentService: onHandleIntent: -->50
6376 - 6544 I/MyIntentService: onHandleIntent: -->75
6376 - 6544 I/MyIntentService: onHandleIntent: -->100
6376 - 6544 I/MyIntentService: onHandleIntent: Thread Over
```

连续单击两次 startIntentServiceBtn 按钮，MyIntentService 开始运行，第一次线程计数完成后，进行第二次线程计数，两次完成后，线程和服务都结束。

```
6376 - 6376 I/MainActivity: onClick: startIntentServiceBtn
6376 - 6376 I/MyIntentService: onCreate:ThreadID: 1
6376 - 6656 I/MyIntentService: onHandleIntent: ThreadID:353
6376 - 6656 I/MyIntentService: onHandleIntent: -->25
6376 - 6656 I/MyIntentService: onHandleIntent: -->50
6376 - 6376 I/MainActivity: onClick: startIntentServiceBtn
6376 - 6656 I/MyIntentService: onHandleIntent: -->75
```

```
6376 - 6656 I/MyIntentService: onHandleIntent: -->100
6376 - 6656 I/MyIntentService: onHandleIntent: Thread Over
6376 - 6656 I/MyIntentService: onHandleIntent: ThreadID:353
6376 - 6656 I/MyIntentService: onHandleIntent: -->25
6376 - 6656 I/MyIntentService: onHandleIntent: -->50
6376 - 6656 I/MyIntentService: onHandleIntent: -->75
6376 - 6656 I/MyIntentService: onHandleIntent: -->100
6376 - 6656 I/MyIntentService: onHandleIntent: Thread Over
6376 - 6376 I/MyIntentService: onDestroy:
```

单击两次 startIntentServiceBtn 按钮，MyIntentService 开始运行，在第一次线程计数完成前，单击 stopIntentServiceBtn，MyIntentService 结束，第一次线程计数结束后不进行第二次计数。

```
6376 - 6376 I/MainActivity: onClick: startIntentServiceBtn
6376 - 6376 I/MyIntentService: onCreate:ThreadID: 1
6376 - 6763 I/MyIntentService: onHandleIntent: ThreadID:355
6376 - 6763 I/MyIntentService: onHandleIntent: -->25
6376 - 6376 I/MainActivity: onClick: startIntentServiceBtn
6376 - 6763 I/MyIntentService: onHandleIntent: -->50
6376 - 6376 I/MainActivity: onClick: stopIntentServiceBtn
6376 - 6376 I/MyIntentService: onDestroy:
6376 - 6763 I/MyIntentService: onHandleIntent: -->75
6376 - 6763 I/MyIntentService: onHandleIntent: -->100
6376 - 6763 I/MyIntentService: onHandleIntent: Thread Over
```

10. 新建一个 AIDL 接口文件

具体步骤：在工程结构视图中，右击 App，选择 New | AIDL | AIDL File，Interface Name 采用默认值 IMyAidlInterface.aidl，Target Source Set 采用默认值 main，单击 Finish 按钮。IMyAidlInterface.aidl 文件内容如下：

```
// IMyAidlInterface.aidl
package cuc.service1;
// Declare any non-default types here with import statements
interface IMyAidlInterface {
            int getPid();
            int getRandomNumber();
}
```

11. Make Module 'app' 生成 AIDL 类

具体步骤：单击工程 Service1 窗口的菜单 Build | Make Module 'app'，生成的 cuc.service1.IMyAidlInterface 位于\app\build\generated\source\aidl\debug 文件夹下，\app\build\generated\source\aidl\debug\cuc\service1\IMyAidlInterface.java 内容如下：

```
/*
 * This file is auto-generated. DO NOT MODIFY.
```

```
 * Original file: C:\\android\\myproject\\Service1\\app\\src\\main\\aidl\\cuc\\service1\\IMyAidlInterface.aidl
  */
package cuc.service1;
// Declare any non-default types here with import statements

public interface IMyAidlInterface extends android.os.IInterface
{
/** Local-side IPC implementation stub class. */
public static abstract class Stub extends android.os.Binder implements cuc.service1.IMyAidlInterface
{
private static final java.lang.String DESCRIPTOR = "cuc.service1.IMyAidlInterface";
/** Construct the stub at attach it to the interface. */
public Stub()
{
this.attachInterface(this, DESCRIPTOR);
}
/**
  * Cast an IBinder object into an cuc.service1.IMyAidlInterface interface,
  * generating a proxy if needed.
  */
public static cuc.service1.IMyAidlInterface asInterface(android.os.IBinder obj)
{
if ((obj==null)) {
return null;
}
android.os.IInterface iin = obj.queryLocalInterface(DESCRIPTOR);
if (((iin!=null)&&(iin instanceof cuc.service1.IMyAidlInterface))) {
return ((cuc.service1.IMyAidlInterface)iin);
}
return new cuc.service1.IMyAidlInterface.Stub.Proxy(obj);
}
@Override public android.os.IBinder asBinder()
{
return this;
}
@Override public boolean onTransact(int code, android.os.Parcel data, android.os.Parcel reply, int flags) throws android.os.RemoteException
{
switch (code)
{
case INTERFACE_TRANSACTION:
{
reply.writeString(DESCRIPTOR);
return true;
}
case TRANSACTION_getPid:
{
data.enforceInterface(DESCRIPTOR);
```

```
int _result = this.getPid();
reply.writeNoException();
reply.writeInt(_result);
return true;
}
case TRANSACTION_getRandomNumber:
{
data.enforceInterface(DESCRIPTOR);
int _result = this.getRandomNumber();
reply.writeNoException();
reply.writeInt(_result);
return true;
}
}
return super.onTransact(code, data, reply, flags);
}
private static class Proxy implements cuc.service1.IMyAidlInterface
{
private android.os.IBinder mRemote;
Proxy(android.os.IBinder remote)
{
mRemote = remote;
}
@Override public android.os.IBinder asBinder()
{
return mRemote;
}
public java.lang.String getInterfaceDescriptor()
{
return DESCRIPTOR;
}
@Override public int getPid() throws android.os.RemoteException
{
android.os.Parcel _data = android.os.Parcel.obtain();
android.os.Parcel _reply = android.os.Parcel.obtain();
int _result;
try {
_data.writeInterfaceToken(DESCRIPTOR);
mRemote.transact(Stub.TRANSACTION_getPid, _data, _reply, 0);
_reply.readException();
_result = _reply.readInt();
}
finally {
_reply.recycle();
_data.recycle();
}
return _result;
}
@Override public int getRandomNumber() throws android.os.RemoteException
```

```
{
android.os.Parcel _data = android.os.Parcel.obtain();
android.os.Parcel _reply = android.os.Parcel.obtain();
int _result;
try {
_data.writeInterfaceToken(DESCRIPTOR);
mRemote.transact(Stub.TRANSACTION_getRandomNumber, _data, _reply, 0);
_reply.readException();
_result = _reply.readInt();
}
finally {
_reply.recycle();
_data.recycle();
}
return _result;
}
}
static final int TRANSACTION_getPid = (android.os.IBinder.FIRST_CALL_TRANSACTION + 0);
static final int TRANSACTION_getRandomNumber = (android.os.IBinder.FIRST_CALL_TRANSACTION + 1);
}
public int getPid() throws android.os.RemoteException;
public int getRandomNumber() throws android.os.RemoteException;
}
```

12. 新建一个 Service 类 MyAidlService

具体步骤：在工程视图中 main 源集 Java 文件夹里，右击 cuc.service1 包，依次选择 New|Service|Service，在 Class Name 输入框中输入 MyAidlService，Exported 选项和 Enabled 选项保持默认状态(选中)，单击 Finish 按钮。系统新建名为 MyAidlService 的类，并完成了在 AndroidManifest.xml 中的注册。编辑 MyAidlService.java，内容如下：

```
package cuc.service1;
import android.app.Service;
import android.content.Intent;
import android.os.IBinder;
import android.os.Process;
import android.os.RemoteException;
import java.util.Random;

public class MyAidlService extends Service {
    public MyAidlService() {
    }
    private final Random random = new Random();
IMyAidlInterface.Stub stub = new IMyAidlInterface.Stub() {
    @Override
    public int getPid() throws RemoteException {
        return Process.myPid();
    }
    @Override
```

```
        public int getRandomNumber() throws RemoteException {
            return random.nextInt(50);
        }
    };
        @Override
        public IBinder onBind(Intent intent) {
            return stub;
        }
    }
```

13. 第十四步：编译安装到手机

把Service1应用安装到手机后，MyAidlService就被注册到Andriod系统中，Clinet1应用就可以访问MyAidlService。

14. 新建工程Client1

具体步骤：运行Android Studio，单击Start a new Android Studio project，在Create New Project页中，Application name为Client1，Company domain为cuc，Project location为C:\android\myproject\ Client1，单击Next按钮，选中Phone and Tablet，minimum sdk设置为API 19：Android 4.4(minSdkVersion)，单击Next按钮，选中Empty Activity，再单击Next按钮，选中Generate Layout File选项，选中Backwards Compatibility(AppCompat)选项，单击Finish按钮。

15. 拷贝工程Service1的aidl文件夹到Client1工程中

具体步骤：把工程Service1的工程结构视图从Android视图切换到Project视图，选中app\src\main\aidl，右击，在弹出的界面中单击Copy。

再把工程Client1的工程结构视图从Android视图切换到Project视图，选中app\src\main，右击，在弹出的界面中单击Paste，弹出Copy界面，单击界面中的OK按钮，就完成了aidl文件夹的拷贝。

最后把工程结构视图从Project视图切换回Android视图。

16. 在Client1工程中make module'app'

具体步骤：单击工程Client1窗口的菜单Build|Make Module'app'，生成IMyAidlInterface.java。

17. 添加DataBinding支持，并同步

在文件build.gradle(module：app)的android语句的花括号内添加：

```
dataBinding.enabled = true
```

由于修改了build.gradle文件内容，这个文件窗口的上面就出现提示："Gradle files have changed since last project sync. A project sync may be necessary for the IDE to work properly. sync now"。单击sync now按钮进行同步。

18. 在工程Client1新建布局文件activity_main1.xml

具体步骤：在工程Client1工程结构视图中，右击layout，选择New|Layout resource file，在File name输入框输入activity_main1，在Root element输入框输入layout，Source set为main，单击OK按钮。layout/activity_main1.xml内容如下：

```
<?xml version="1.0" encoding="utf-8"?>
<layout xmlns:android="http://schemas.android.com/apk/res/android">
    <data/>
    <LinearLayout android:orientation="vertical" android:layout_width="match_parent"
        android:layout_height="match_parent">
        <Button
            android:layout_width="match_parent"
            android:layout_height="wrap_content"
            android:id="@+id/bindMyAidlServiceBtn"
            android:text="bindMyAidlService" />
        <Button
            android:layout_width="match_parent"
            android:layout_height="wrap_content"
            android:id="@+id/unbindMyAidlServiceBtn"
            android:text="unbindMyAidlService" />
        <Button
            android:layout_width="match_parent"
            android:layout_height="wrap_content"
            android:id="@+id/getPidBtn"
            android:text="getPid" />
        <Button
            android:layout_width="match_parent"
            android:layout_height="wrap_content"
            android:id="@+id/getRandomBtn"
            android:text="getRandom" />
        <Button
            android:layout_width="match_parent"
            android:layout_height="wrap_content"
            android:id="@+id/bindMessengerServiceBtn"
            android:text="bindMessengerService" />
        <Button
            android:layout_width="match_parent"
            android:layout_height="wrap_content"
            android:id="@+id/unbindMessengerServiceBtn"
            android:text="unbindMessengerService" />
        <Button android:id="@+id/addBtn"
            android:layout_width="wrap_content"
            android:layout_height="wrap_content"
            android:text="add"/>
        <TextView
            android:layout_width="match_parent"
            android:layout_height="wrap_content"
            android:id="@+id/resultTxt" />
    </LinearLayout>
</layout>
```

19. 编辑 MainActivity.java

MainActivity.java 内容如下：

```
package cuc.client1;
import android.content.ComponentName;
import android.content.Intent;
import android.content.ServiceConnection;
import android.databinding.DataBindingUtil;
import android.os.Bundle;
import android.os.Handler;
import android.os.IBinder;
import android.os.Message;
import android.os.Messenger;
import android.os.RemoteException;
import android.support.v7.app.AppCompatActivity;
import android.util.Log;
import android.view.View;
import cuc.client1.databinding.ActivityMain1Binding;
import cuc.service1.IMyAidlInterface;

public class MainActivity extends AppCompatActivity implements View.OnClickListener{
    private static final String TAG = "MainActivity";
    private ActivityMain1Binding binding;
    @Override
    protected void onCreate(Bundle savedInstanceState) {
        super.onCreate(savedInstanceState);
      binding = DataBindingUtil.setContentView(this,R.layout.activity_main1 );
        binding.bindMyAidlServiceBtn.setOnClickListener(this);
        binding.bindMessengerServiceBtn.setOnClickListener(this);
        binding.unbindMyAidlServiceBtn.setOnClickListener(this);
        binding.unbindMessengerServiceBtn.setOnClickListener(this);
        binding.getRandomBtn.setOnClickListener(this);
        binding.getPidBtn.setOnClickListener(this);
        binding.addBtn.setOnClickListener(this);
    }

    private IMyAidlInterface iMyAidlInterface;
    private boolean isConnectingAIDL;
    private ServiceConnection connMyAidlService = new ServiceConnection() {
        @Override
        public void onServiceConnected(ComponentName componentName, IBinder iBinder) {
            iMyAidlInterface = IMyAidlInterface.Stub.asInterface(iBinder);
            isConnectingAIDL = true;
            Log.i(TAG, "onServiceConnected: MyAidlService");
        }

        @Override
        public void onServiceDisconnected(ComponentName componentName) {
            isConnectingAIDL = false;
            Log.i(TAG, "onServiceDisconnected: MyAidlService");
```

```
        }
    };
    private Messenger remoteMessenger;
    private boolean isConnectingMessenger;
    private static final int MSG_SUM = 1;
    private int operand1 = 1;
    private Messenger messenger = new Messenger(new Handler(){
        @Override
        public void handleMessage(Message msg) {
            super.handleMessage(msg);
            if (msg.what == MSG_SUM){
                binding.resultTxt.setText(binding.resultTxt.getText() + Integer.toString(msg.arg1));
            }
        }
    });
    private ServiceConnection connMessengerService = new ServiceConnection() {
        @Override
        public void onServiceConnected(ComponentName componentName, IBinder iBinder) {
            remoteMessenger = new Messenger(iBinder);
            isConnectingMessenger = true;
        }

        @Override
        public void onServiceDisconnected(ComponentName componentName) {
            isConnectingMessenger = false;
        }
    };
    @Override
    public void onClick(View view) {
        Intent intent;
        switch (view.getId()){
            case R.id.bindMyAidlServiceBtn:
                intent = new Intent();

intent.setClassName("cuc.service1","cuc.service1.MyAidlService");
                bindService(intent,connMyAidlService,BIND_AUTO_CREATE);
                break;
            case R.id.unbindMyAidlServiceBtn:
                unbindService(connMyAidlService);
                break;
            case R.id.getRandomBtn:
                if (isConnectingAIDL)
                    try {
                        binding.resultTxt.setText("random: " + iMyAidlInterface.getRandomNumber());
                    } catch (RemoteException e) {
                        e.printStackTrace();
                    }
```

```
                break;
            case R.id.getPidBtn:
                if (isConnectingAIDL)
                    try {
                        binding.resultTxt.setText("pid: " + iMyAidlInterface.getPid());
                    } catch (RemoteException e) {
                        e.printStackTrace();
                    }
                break;
            case R.id.bindMessengerServiceBtn:
                intent = new Intent();

intent.setClassName("cuc.service1","cuc.service1.MessengerService");
                bindService(intent,connMessengerService,BIND_AUTO_CREATE);
                break;
            case R.id.unbindMessengerServiceBtn:
                unbindService(connMessengerService);
                break;
            case R.id.addBtn:
                binding.resultTxt.setText(operand1 + " + " + operand1 + " = ");
                if(isConnectingMessenger){
                    Message message = Message.obtain(null,MSG_SUM,operand1,operand1);
                    operand1++;
                    message.replyTo = messenger;
                    try {
                        remoteMessenger.send(message);
                    } catch (RemoteException e) {
                        e.printStackTrace();
                    }
                }
        }
    }
}
```

20. 编译运行

第七阶段：使用 Messenger 创建绑定的 Service

单击界面中的 bindMessengerServiceBtn 按钮绑定 MessengerService，单击 addBtn 按钮把被加数和加数作为 Message 的参数，通过 Messenger 传给 MessengerService，MessengerService 处理后再通过另一 Messenger 回传给客户端，更新 resultTxt。最后单击 unbindMessengerServiceBtn 按钮解除与 MessengerService 的绑定。

第八阶段：使用 AIDL 创建绑定的 Service

单击界面中的 bindMyAidlServiceBtn 按钮绑定 MyAidlService，单击 getRandomBtn 按钮调用 IMyAidlInterface 接口中的 getRandomNumber()，单击 getPidBtn 按钮调用 IMyAidlInterface 接口中的 getPid()，就好像在本地一样。resultTxt 用于显示方法的返回值。最后单击 unbindMyAidlServiceBtn 按钮解除与 MyAidlService 的绑定。

13.8 本章主要参考文献

1. https://developer.android.google.cn/reference/android/os/Handler.html
2. https://developer.android.google.cn/reference/android/os/Message.html
3. https://developer.android.google.cn/reference/android/app/Service.html
4. https://developer.android.google.cn/guide/topics/manifest/service-element.html
5. https://developer.android.google.cn/guide/components/services#Lifecycle
6. https://developer.android.google.cn/guide/components/bound-services.html
7. https://developer.android.google.cn/guide/components/aidl.html
8. https://developer.android.google.cn/guide/topics/ui/notifiers/notifications.html
9. https://developer.android.google.cn/reference/android/app/PendingIntent.html
10. https://developer.android.google.cn/reference/android/support/v4/app/TaskStackBuilder.html

第 14 章 数据存储

CHAPTER 14

Android 提供了多种方法用于保存永久性应用数据：①Shared Preferences 以键值对的形式存储私有原始数据；②Internal Storage 用于在设备的内部存储器中存储私有数据；③External Storage 用于在共享的外部存储中存储公共数据；④SQLite Databases 用于在私有数据库中存储结构化数据；⑤Network Connection 使用网络服务器存储数据；⑥ContentProvider 是一个读写应用数据的组件，它能将数据（甚至是私有数据）公开给其他应用。

14.1 SharedPreferences

接口 android.content.SharedPreferences 提供了保存和读取原始数据类型的键值对的通用框架，原始数据类型包括布尔值、浮点值、整型值、长整型和字符串。一般使用 SharedPreference 来存储应用程序的配置信息，一般存储在应用程序的私有目录下（data/data/包名/shared_prefs/），文件权限是私有的，也就是说只能供写入者读取，不支持跨进程使用。这个类可能会让 APP 运行变慢，经常变化的属性或者可以容忍丢失的属性应该使用别的措施存储数据。android.content.Context 类中的 getSharedPreferences(String name, @PreferencesMode int mode) 是个抽象的方法。android.app.ContextImpl 提供了这个方法的具体实现。使用 SharedPreferences 的步骤如下：

第一步，调用 Context 的 getSharedPreferences(String name, int mode) 方法或者调用 Activity 的 getPreferences(int mode) 方法获取 SharedPreferences 对象。Activity 的 getPreferences(int mode) 方法的定义如下：

```
public SharedPreferences getPreferences(@Context.PreferencesMode int mode) {
        return getSharedPreferences(getLocalClassName(), mode);
}
```

getPreferences 直接调用 getSharedPreferences(String name, int mode)，传递 Activity 的类名作为 SharedPreferences 文件的名字。如果需要多个按名称识别的 SharedPreferences 文件，请调用 getSharedPreferences(String name, int mode) 方法，参数 name 指定了 SharedPreferences 文件的名称。

第二步，保存数据或者读取数据到 SharedPreferences 文件。

为了把数据保存到第一步获取的 SharedPreferences 对象中，步骤如下：① 调用

SharedPreferences 对象的 edit() 方法获取 SharedPreferences. Editor 对象；② 使用 SharedPreferences. Editor 对象的 putBoolean() 和 putString() 等方法添加值；③ 使用 commit()提交新值。

为了读取 SharedPreferences 对象中的值，请使用 SharedPreferences 接口中 getBoolean()和 getString()等方法。

以下是保存静音按键模式首选项的示例：

```
public class Calc extends Activity {
  public static final String PREFS_NAME = "MyPrefsFile";
  @Override
  protected void onCreate(Bundle state){
    super.onCreate(state);
    // Restore preferences
    SharedPreferences preferences = getSharedPreferences(PREFS_NAME, 0);
    boolean silent = preferences.getBoolean("silentMode", false);
    …
  }
  @Override
  protected void onStop(){
   super.onStop();
   SharedPreferences preferences = getSharedPreferences(PREFS_NAME, 0);
   SharedPreferences.Editor editor = preferences.edit();
   editor.putBoolean("silentMode", mSilentMode);
   editor.commit();
  }
}
```

14.2 使用内部存储

可以直接在设备的内部存储器中保存文件。默认情况下，保存到内部存储的文件是应用的私有文件，其他应用和用户都不能访问这些文件。当用户卸载应用时，这些文件也会被移除。可以通过以下步骤创建私有文件并写入到内部存储。

(1) 调用 Context 对象的 openFileOutput(String name, int mode)，将返回一个 FileOutputStream，参数分别为文件名称和操作模式。从 API 级别 17 开始，mode 取值只能是 MODE_PRIVATE 和 MODE_APPEND，常量 MODE_WORLD_READABLE 和 MODE_WORLD_WRITEABLE 已被弃用。MODE_PRIVATE 将会创建文件(或替换具有相同名称的文件)，MODE_APPEND 用于在已经存在的文件最后添加新内容。如果应用需要与其他应用共享私有文件，唯一安全的方式是发送文件内容的 URI，并且授权访问这个 URI 的临时访问许可，把 FileProvider 与 FLAG_GRANT_READ_URI_PERMISSION 配合使用。如果应用之间仅仅需要共享少量的文本或者数字，可以使用 Intent 携带少量数据。

(2) 调用 FileOutputStream 的 write 方法把数据写入文件。

(3) 调用 FileOutputStream 的 close 方法关闭流。

例如：

```
String FILENAME = "hello_file";
String string = "hello world!";
FileOutputStream fos = null;
try {
    fos = openFileOutput(FILENAME, Context.MODE_PRIVATE);
    fos.write(string.getBytes());
    fos.close();
} catch (Exception e) {
    e.printStackTrace();
}
```

从内部存储读取文件的步骤如下：①调用 Context 对象的 openFileInput(String name)并向其传递要读取的文件名称，这将返回一个 FileInputStream；②调用 FileInputStream 对象的 read()读取文件字节；③调用 FileInputStream 对象的 close()关闭流。

FileOutputStream 用于把原始字节(例如图像数据)写入 File，FileInputStr 用于从文件读取原始字节。如果要读写入字符流，可以使用 FileReader 和 FileWriter。FileWriter 类的继承关系，如图 14-1 所示。

java.lang.Object			
	java.io.Writer		
		java.io.OutputStreamWriter	
			java.io.FileWriter

图 14-1　FileWriter 类的继承关系

FileWriter 是为了方便写入字符文件的类，这个类的构造函数使用了默认的字符编码和默认的字节缓冲区大小。如果需要自己指定这些值，使用 FileOutputStream 构造 OutputStreamWriter。OutputStreamWriter 是字符流转换为字节流的桥，字符使用指定的字符集编码成字节。OutputStreamWriter 的构造函数，如表 14-1 所示。

表 14-1　OutputStreamWriter 的构造函数

构造函数	解释
OutputStreamWriter(OutputStream out, String charsetName)	使用给定的字符集的名字创建 OutputStreamWriter
OutputStreamWriter(OutputStream out)	使用默认的字符集创建 OutputStreamWriter
OutputStreamWriter(OutputStream out, Charset cs)	使用给定的字符集的名字创建 OutputStreamWriter
OutputStreamWriter(OutputStream out, CharsetEncoder enc)	使用给定的字符编码器创建 OutputStreamWriter

内部存储中应用的主要私有目录，如表 14-2 所示，其中包名用点隔开，不是斜线，调用的方法都是 Context 实例的方法。

表 14-2　内部存储中应用的主要私有目录位置

目　　录	存取方法
/data/data/应用包名/files	使用 getFilesDir()获取该目录,openFileInput()和 openFileOutput()函数在该目录下操作文件,fileList()列出该目录下的所有文件,deleteFile(String name)用来删除该目录下的文件
/data/data/应用包名/cache	使用 getCacheDir()获取该目录
/data/data/应用包名/lib	存储该应用的.so 静态库文件
/data/data/应用包名/databases	使用 getDatabasePath(String name)获取该目录下由 name 参数指定的数据库
/data/data/应用包名/shared_prefs	使用 getSharedPreferences(String name, int mode)或者 getPreferences(int mode)方法获取该文件夹下 SharedPreferences
/data/data/应用包名/	使用 getDir(String name,int mode)在该目录下创建子目录,参数中的 name 就是子目录名字

14.3　使用外部存储

机身内置存储是指手机自身携带的存储空间,不能被移除,出厂时就已经有了。在 Android 4.4 版本以前,机身内置存储就是内部存储,外置 SD 卡就是外置存储。在 Android 4.4 及更高版本中,机身内置存储包含了内部存储和外部存储,外置 SD 卡也是外部存储。

Android 的设备支持共享"外部存储",它可能是可移除的存储介质(例如 SD 卡)或不可移除的机身内置存储。如果用户移走了存储介质或者在计算机上加载了外部存储,则外部存储可能变为不可用状态,并且保存在外部存储中的文件没有实施任何安全性,所有应用都能读取和写入外部存储里的文件,并且用户可以删除这些文件。在计算机上启用 USB 大容量存储以传输文件后,可由用户修改这些文件。

要读取或写入外部存储上的文件,应用必须获取 READ_EXTERNAL_STORAGE 或 WRITE_EXTERNAL_STORAGE 系统权限。如果同时需要读取和写入文件,则只需请求 WRITE_EXTERNAL_STORAGE 权限,因为此权限也隐含了读取权限要求。从 Android 4.4(API 19)开始,当读取或写入外部存储中应用私有目录中的文件,不再需要 READ_EXTERNAL_STORAGE 或 WRITE_EXTERNAL_STORAGE 权限。因此,可以通过添加 maxSdkVersion 属性来声明,只需在较低版本的 Android 中请求该权限。

```
<manifest ...>
    <uses-permission android:name = "android.permission.WRITE_EXTERNAL_STORAGE"
            android:maxSdkVersion = "18" />
    ...
</manifest>
```

14.3.1 保存应用私有文件到外部存储

如果正在处理的文件不适合其他应用使用，则应该调用 Context 对象的 getExternalFilesDir (String type)或者 getExternalFilesDirs(String type)来访问外部存储上的私有存储目录，type 参数指定子目录的类型(例如 DIRECTORY_MOVIES)，见表 14-3。如果不需要特定的媒体目录，传递 null 参数，使用应用私有目录的根目录。

Context 对象的 getExternalFilesDir(String type)方法将获取主要的外部存储器。不过，从 Android 4.4 开始，可通过调用 getExternalFilesDirs(String type)来同时获取两个位置，该方法将会返回包含各个位置条目的 File 数组，数组中的第一个条目被视为主要的外部存储器；在 Android 4.3 和更低版本中，使用支持库中的静态方法 ContextCompat.getExternalFilesDirs()，此方法也会返回一个 File 数组，但其中始终仅包含一个条目。尽管 MediaStore 和 ContentProvider 不能访问 Context 对象的 getExternalFilesDir ( ) 和 getExternalFilesDirs()所提供的外部存储里的私有目录，但其他具有 READ_EXTERNAL_STORAGE 权限的应用仍可访问外部存储上的所有文件。如果需要完全限制私有文件的访问权限，则应该将文件写入到内部存储。

表 14-3 外部存储中应用的主要私有目录位置

方法	方法获取的目录
getExternalCacheDir()	/storage/emulated/0/Android/data/应用包名/cache
getExternalCacheDirs()	/storage/emulated/0/Android/data/应用包名/cache/storage/sdcard1/Android/data/应用包名/cache
getExternalFilesDir (Environment. DIRECTORY_PICTURES)	/storage/emulated/0/Android/data/应用包名/files/Pictures
getExternalFilesDirs (Environment. DIRECTORY_PICTURES)	/storage/emulated/0/Android/data/应用包名/files/Pictures/storage/sdcard1/Android/data/应用包名/files/Pictures
getExternalFilesDir(null)	/storage/emulated/0/Android/data/应用包名/files
getExternalFilesDirs(null)	/storage/emulated/0/Android/data/应用包名/files/storage/sdcard1/Android/data/应用包名/files
getExternalMediaDirs()	/storage/emulated/0/Android/media/应用包名/storage/sdcard1/Android/media/应用包名

14.3.2 保存可与其他应用共享的文件

用户通过应用获取的新文件可以保存到设备上的“公共”位置，以便其他应用能够访问这些文件，并且用户也能轻松地复制这些文件。共享的公共目录包括 Music/、Pictures/和 Ringtones/等。如果希望在媒体扫描程序中隐藏某个文件夹里的所有媒体文件，只需要这个文件夹内包含名为.nomedia 的空文件(文件名中有点前缀)，这将阻止媒体扫描程序读取这个文件夹里的媒体文件、通过 MediaStore 和 ContentProvider 提供给其他应用。

要获取公共目录里的 File，请调用 Environment 类的静态方法 getExternalStoragePublicDirectory(String type)，参数是需要的目录类型，例如 DIRECTORY_MUSIC、

DIRECTORY_PODCASTS、DIRECTORY_RINGTONES、DIRECTORY_ALARMS、DIRECTORY_NOTIFICATIONS、DIRECTORY_PICTURES、DIRECTORY_MOVIES、DI-RECTORY_DOWNLOADS、DIRECTORY_DCIM或者DIRECTORY_DOCUMENTS。例如type为DIRECTORY_PICTURES，对应的路径为/storage/emulated/0/Pictures。通过将文件保存到相应的媒体类型目录，系统的媒体扫描程序可以在系统中正确地归类这些文件（例如铃声在系统设置中显示为铃声而不是音乐）。调用Environment类的静态方法getExternalStorageDirectory()获取的是外部公共存储的根目录/storage/emulated/0。

例如，以下方法在公共图片目录中创建了一个用于新相册的目录：

```
public File getAlbumStorageDir(String albumName) {
  File file = new File(Environment.getExternalStoragePublicDirectory(
       Environment.DIRECTORY_PICTURES), albumName);
  if (!file.mkdirs()) {
    Log.e(LOG_TAG, "Directory not created");
  }
  return file;
}
```

14.3.3 使用作用域目录访问

在Android 7.0（API 24）或更高版本中，可以使用作用域目录访问外部存储上的特定目录，它简化了应用访问标准外部存储目录（例如Pictures目录）的方式，并提供简单的权限UI解释应用正在请求访问的目录。作用域目录访问和Android 6.0访问内部储存空间一样，需要应用程序主动向用户请求读写权限。不同的是，作用域目录访问不再要求应用声明android.permission.WRITE_EXTERNAL_STORAGE权限，也限制了应用程序访问内部储存空间行为，只能在请求的作用域内进行读写操作（包括文件、子文件夹）。使用作用域目录访问的步骤如下：

（1）获取StorageManager的实例。

（2）调用StorageManager的实例的getStorageVolume(File file)方法获取包含指定文件的StorageVolume，或者调用getStorageVolumes()方法获取所有可得到的StorageVolume的列表，包含可移除的媒体卷，或者调用getPrimaryStorageVolume()方法获取主要的StorageVolume。

（3）调用这个StorageVolume实例的createAccessIntent()方法创建一个Intent对象。调用createAccessIntent()方法时，次要卷（例如外部SD卡）使用null作为参数可以访问整个次要卷，若主要卷采用null做参数将返回null。传递无效的文件夹的名字作为参数将返回null。

（4）调用Activity实例的startActivityForResult(Intent intent,int requestCode)方法，第一个参数就是前一步创建的Intent，系统请求用户授予访问标准文件夹或者整个卷的权限，请求的结果将通过onActivityResult(int requestCode,int resultCode,Intent data)处理。如果用户授予了访问权限，onActivityResult的参数resultCode为RESULT_OK，Intent中将包含URI；如果用户没有授予访问权限，onActivityResult的参数resultCode为

RESULT_CANCELED,Intent 数据为 null。获取了访问某个文件夹的权限也同时获取了读取子文件夹的权限。

StorageVolume 是 Android 7.0(API 24)新添加进来的,它代表对于特定用户的存储卷的逻辑视图。一个 Android 设备总是有唯一的主要的 StorageVolume,但可以有多个额外的 StorageVolume,例如 SD 卡或者 USB 存储设备。对于同一个物理的存储卷,不同的用户有不同的逻辑视图。StorageVolume 是没有必要加载的,应用可以使用 getState()验证它的状态。为了读写 StorageVolume,应用需要获取授权:使用 createAccessIntent(String),访问标准的文件夹(例如 DIRECTORY_PICTURES)。为了访问任意目录,可以使用存储器访问框架(Storage Acess Framework)的 API,例如 Intent. ACTION_OPEN_DOCUMENT 和 Intent. ACTION_OPEN_DOCUMENT_TREE。

下面的代码是访问主要的共享存储设备上的 Pictures 的示例。

```
StorageManager sm = (StorageManager)getSystemService(Context.STORAGE_SERVICE);
StorageVolume volume = sm.getPrimaryStorageVolume();
Intent intent = volume.createAccessIntent(Environment.DIRECTORY_PICTURES);
startActivityForResult(intent, request_code);
```

14.3.4 访问可移动介质上的目录

为了使用作用域目录访问可移除媒介上的文件夹,可以先注册一个 BroadcastReceiver 监听 MEDIA_MOUNTED 事件,例如:

```
<receiver
    android:name = ".MediaMountedReceiver"
    android:enabled = "true"
    android:exported = "true" >
    <intent - filter >
        <action android:name = "android.intent.action.MEDIA_MOUNTED" />
        <data android:scheme = "file" />
    </intent - filter >
</receiver >
```

当用户加载了可移除媒介(例如 SD 卡)后,系统将发出 MEDIA_MOUNTED 广播,广播 Intent 的数据中提供了一个 StorageVolume 对象,可以用来访问可移除媒介上的文件夹。例如下面的代码访问可移除媒介上的 Pictures 文件夹。

```
// BroadcastReceiver has already cached the MEDIA_MOUNTED
// notification Intent in mediaMountedIntent
StorageVolume volume = (StorageVolume)
    mediaMountedIntent.getParcelableExtra(StorageVolume.EXTRA_STORAGE_VOLUME);
volume.createAccessIntent(Environment.DIRECTORY_PICTURES);
startActivityForResult(intent, request_code);
```

14.4 SQLite 数据库

SQLite 是自给自足的、无服务器的、零配置的、事务型的 SQL 数据库引擎。SQLite 数据库引擎实现了主要的 SQL92 标准,支持 SQL92(SQL2)标准的大多数查询语言。引擎本身只有一个文件,但是并不作为一个独立的进程运行,而是动态或者静态的链接到其他应用程序中。SQLite 不需要安装或配置。SQLite 本身是 ANSI-C 语言开发的,开源也跨平台,可在 UNIX(Linux,Mac OS-X,Android,iOS)和 Windows(Windows 32,Windows CE,Windows RT)中运行。它生成的数据库文件是一个普通的磁盘文件,可以放置在任何目录下。

SQLite 的 sqlite3 命令被用来创建新的 SQLite 数据库,创建数据库不需要任何特殊的权限。在 Linux shell 环境下、创建数据库的基本语法如下:

```
sqlite3 testDB.db
```

上面的命令将在当前目录下创建一个文件 testDB.db,该文件将被 SQLite 引擎用作数据库。sqlite3 命令在成功创建数据库文件之后,将提供一个 sqlite>提示符。一旦数据库被创建,就可以使用 SQLite 的.databases 命令来检查它是否在数据库列表中。

可以使用以下步骤进入 Android 模拟器的 sqlite3 命令行环境:首先启动 Android 模拟器。按下 Win+R 组合键,输入 cmd.exe,按回车键进入命令行模式,输入 adb shell 进入 Linux 环境,并输入命令 cd /sdcard 进入可读写的文件夹/sdcard。输入 sqlite3 test.db,这时就进入到 SQLite 命令行环境,提示如下:

```
C:\Users\Administrator>adb shell
root@generic_x86:/ # cd sdcard
root@generic_x86:/sdcard # sqlite3 test.db
SQLite version 3.7.11 2012-03-20 11:35:50
Enter ".help" for instructions
Enter SQL statements terminated with a ";"
sqlite>
```

它显示了如下信息:版本号,每一条 SQL 语句必须用分号结尾,还有命令行帮助是.help。

在命令行环境下输入.help 按回车键,显示所有可使用的命令以及这些命令的帮助,所有的命令开头都是一个点。

在 Android 中,某个应用程序创建的数据库,位于 Android 设备/data/data/package_name/databases 文件夹中,只有应用本身可以访问,其他应用程序是不能访问的。有两点需要注意,访问手机的 data/data 这个目录是需要 root 权限的,adb shell 进入手机后,不是每个手机都能找到 sqlite3 这个命令的。

14.4.1 SQLite存储类型

SQLite3支持NULL、INTEGER(有符号整数,根据值的大小可以是1,2,3,4,6或者8字节)、REAL(8字节的IEEE浮点数字)、TEXT(字符串文本)和BLOB(二进制对象)存储类型,它支持的存储类型虽然只有五种,但实际上sqlite3也接受varchar(n)、char(n)、decimal(p,s)等数据类型,只不过在运算或保存时会转成对应的五种存储类型。

SQLite和其他数据库最大的不同就是对数据类型的支持,创建一个表时,可以在CREATE TABLE语句中指定字段的数据类型,但是可以把任何数据类型放入任何字段中,无论这个字段声明的数据类型是什么。当某个值插入数据库时,SQLite将检查它的类型,如果该类型与关联的字段不匹配,则SQLite会尝试将该值转换成该字段的类型。如果不能转换,则该值直接存储。例如:可以在Integer字段中存放字符串,或者在布尔型字段中存放浮点数,或者在字符型字段中存放日期型值。但有一种情况例外:定义为INTEGER PRIMARY KEY的字段只能存储64位整数,当向这种字段中保存除整数以外的数据时,将会产生错误。SQLite称这为"弱类型"。

14.4.2 SQLite运算符

该运算符主要用于SQLite语句的WHERE子句中执行操作,指定SQLite语句中的条件,并在语句中连接多个条件。运算符包括算术运算符(+,-,*,/,%),SQLite位运算符(&,|,~,≪,≫),比较运算符、逻辑运算符。位运算符则将操作数转化为二进制,并按位进行相应运算,并相应地返回1或0,运算完成后再重新转换为数字。例如,假设变量A=60(二进制111100),变量B=13(二进制1101),(A & B)将得到12(二进制01100),(A|B)将得到61(二进制0011 1101),A≪2将得到240(二进制11110000),A≫2将得到15(二进制0000 1111)。SQLite比较运算符如表14-4所示,SQLite逻辑运算符如表14-5所示。

表14-4 SQLite比较运算符

运 算 符	描 述	实例(假设变量A=10、B=20)
==	检查两个操作数的值是否相等,如果相等则条件为真	(a == b)不为真
=	检查两个操作数的值是否相等,如果相等则条件为真	(a =b)不为真
!=	检查两个操作数的值是否相等,如果不相等则条件为真	(a != b)为真
<>	检查两个操作数的值是否相等,如果不相等则条件为真	(a <> b)为真
>	如果左操作数的值大于右操作数的值,则条件为真	(a > b)不为真
<	如果左操作数的值小于右操作数的值,则条件为真	(a < b)为真
>=	如果左操作数的值大于等于右操作数的值,则条件为真	(a >= b)不为真
<=	如果左操作数的值小于等于右操作数的值,则条件为真	(a <= b)为真
!<	如果左操作数的值不小于右操作数的值,则条件为真	(a !< b)为假
!>	如果左操作数的值不大于右操作数的值,则条件为真	(a !> b)为真

表 14-5 SQLite 逻辑运算符

运 算 符	描 述
AND	逻辑与：两个条件都为真，结果为真
BETWEEN	选择在给定最小值和最大值范围内的一系列值
WHERE EXISTS(subquery)	如果子查询返回至少一行，就认为条件满足。主要使用在 SELECT、UPDATE 或者 DELETE 语句中。在 SQLite 中使用 EXISTS 条件是非常没有效率的
IN	expression [NOT] IN (value_list|subquery) 表达式是否匹配值列表中的任何值，返回 true 或 false。要查询非列表中的值匹配，请使用 NOT IN 运算符。其中 value_list 是固定值列表或子查询返回的一列的结果集。例如 SELECT ID, AGE, NAME, ADDRESS FROM student WHERE ID IN (1,3,5); SELECT ID, AGE, NAME, ADDRESS FROM student WHERE ID NOT IN (SELECT emp_id FROM department);
LIKE	用于把某个值与使用通配符的相似值进行比较，有两个通配符与 LIKE 运算符一起使用：百分号(%)和下画线(_)。百分号表示零个、一个或多个数字或字符。下画线代表一个单一的数字或字符。这些符号可以被组合使用。例如 WHERE SALARY LIKE '200%'
GLOB	用于把某个值与使用通配符相似值进行比较。GLOB 与 LIKE 不同之处在于，GLOB 运算符遵循 UNIX 的语法，它是大小写敏感的，使用以下通配符：星号(*)和问号(?)。星号表示零个或多个数字或字符。问号表示单个数字或字符。 SELECT FROM table_name WHERE column GLOB '*XXXX*'
NOT	not 用于把逻辑操作符的含义取反，例如：NOT EXISTS, NOT BETWEEN, NOT IN, 等
OR	用于组合 where 子句中的多个条件，例如 WHERE AGE >= 25 OR SALARY >= 65000
IS	IS 运算符与 = 相似
IS NOT	IS NOT 运算符与!=相似
||	用于将运算符两侧的两个不同的字符串连接创建为一个新的字符串

14.4.3 SQLite 语句语法

所有的 SQLite 语句可以以任何关键字开始，如 SELECT、INSERT、UPDATE、DELETE、ALTER、DROP 等，所有的语句以分号(;)结束。

1. CREATE TABLE 创建表语句

SQLite 的 CREATE TABLE 语句用于在任何给定的数据库创建一个新表。CREATE TABLE 后跟着表的唯一的名称或标识(也可以选择指定 database_name)、列的名字及每一列的数据类型。CREATE TABLE 语句的完整语法如图 14-2 所示。其中 column-def 语法如图 14-3 所示，column-constraint 语法如图 14-4 所示，table-constraint 语法如图 14-5 所示。

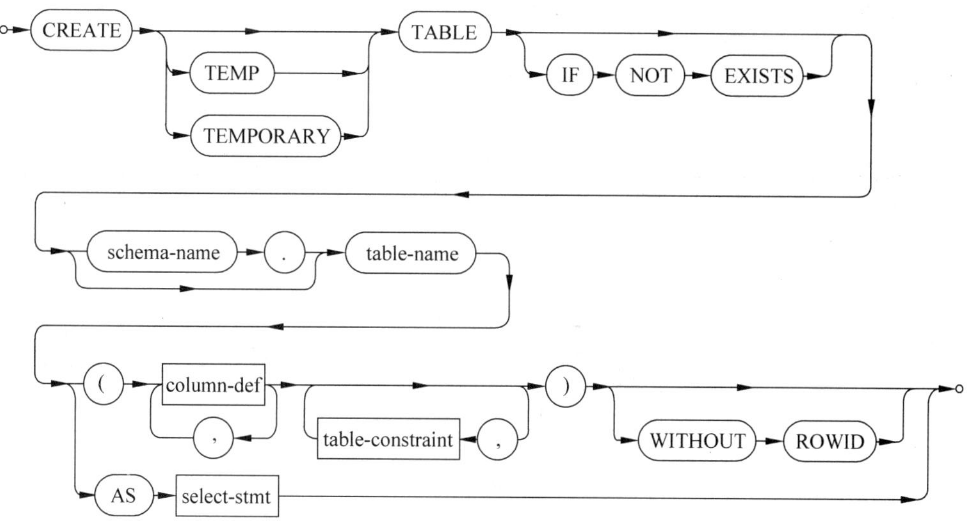

图 14-2 CREATE TABLE 语句语法

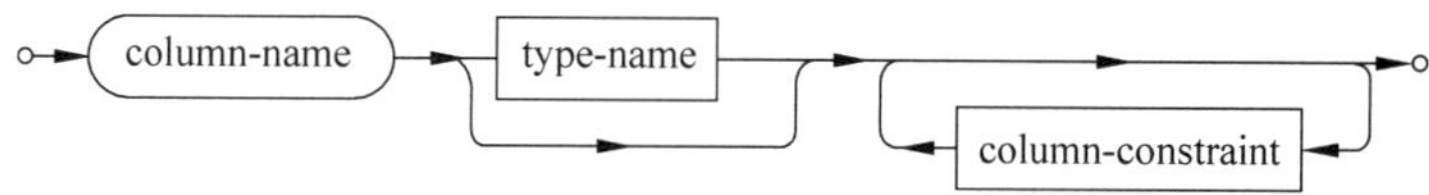

图 14-3 column-def 语法

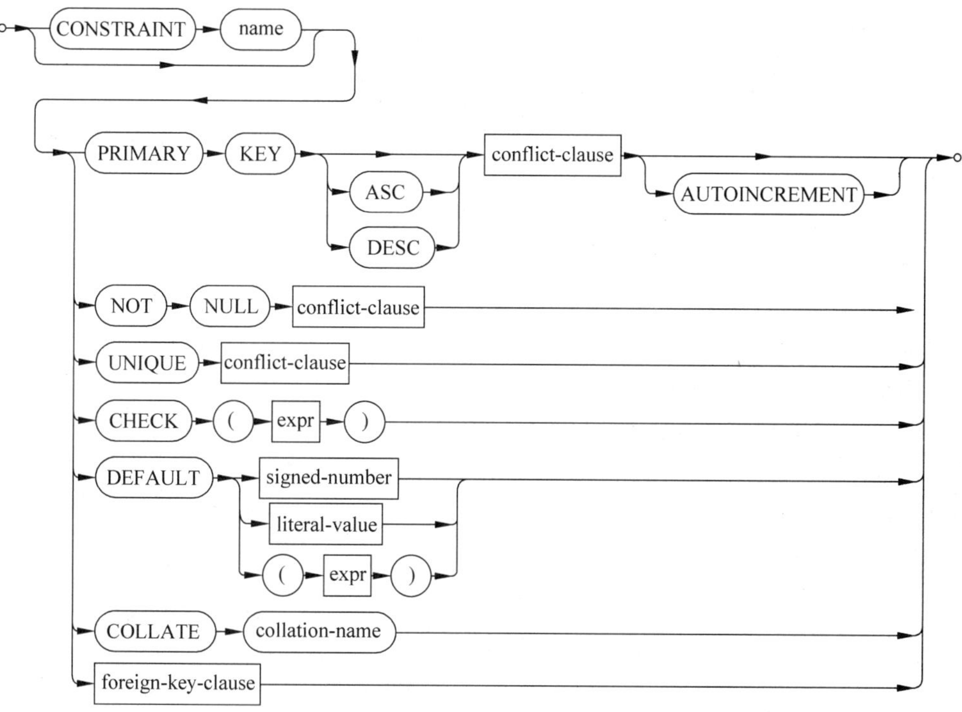

图 14-4 column-constraint 语法

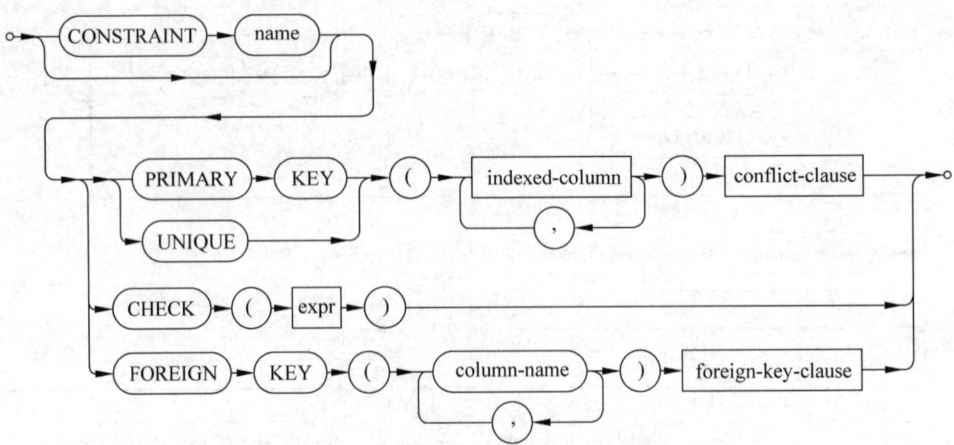

图 14-5　table-constraint 语法

创建表的最基本形式如下：

```
CREATE TABLE table_name(column1 datatype PRIMARY KEY,
  column2 datatype, columnN datatype);
```

例如：

```
sqlite> CREATE TABLE DEPARTMENT (ID INT PRIMARY KEY NOT NULL,
  DEPT CHAR(50) NOT NULL, EMP_ID INT NOT NULL);
```

. tables 命令(没有参数)用于列出当前数据库中的所有表。

". tables"命令后也可以跟一个 pattern 参数，这样命令就只列出数据库中表名和该参数匹配的表。

". schema"命令(没有参数)，显示最初用于创建数据库的 CREATE TABLE 和 CREATE INDEX 的 SQL 语句。

". schema"命令可以包含一个 pattern 参数，这时只会显示满足条件的表和所有索引的 SQL 语句。

2. DROP TABLE 删除表语句

SQLite 的 DROP TABLE 语句用来删除某个数据库中的表及其所有相关数据、索引、触发器、约束和该表的权限规范。若没有指定数据库，删除的是当前数据库中的表。使用此命令时要特别慎重，因为一旦一个表被删除，表中所有信息也将永远丢失。DROP TABLE 语法如图 14-6 所示。

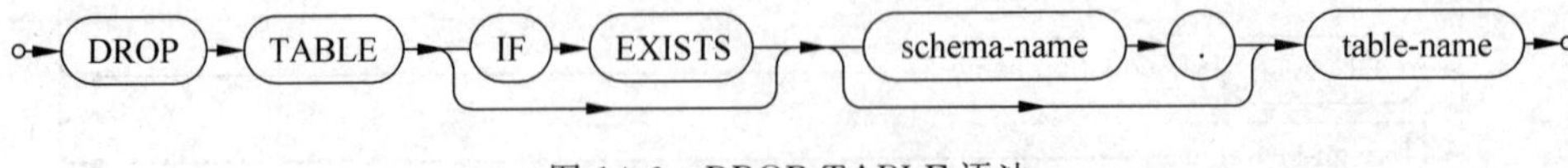

图 14-6　DROP TABLE 语法

3. INSERT 语句

INSERT INTO 语句用于向数据库的某个表中添加新的数据行。INSERT INTO 语句的语法如图 14-7 所示。

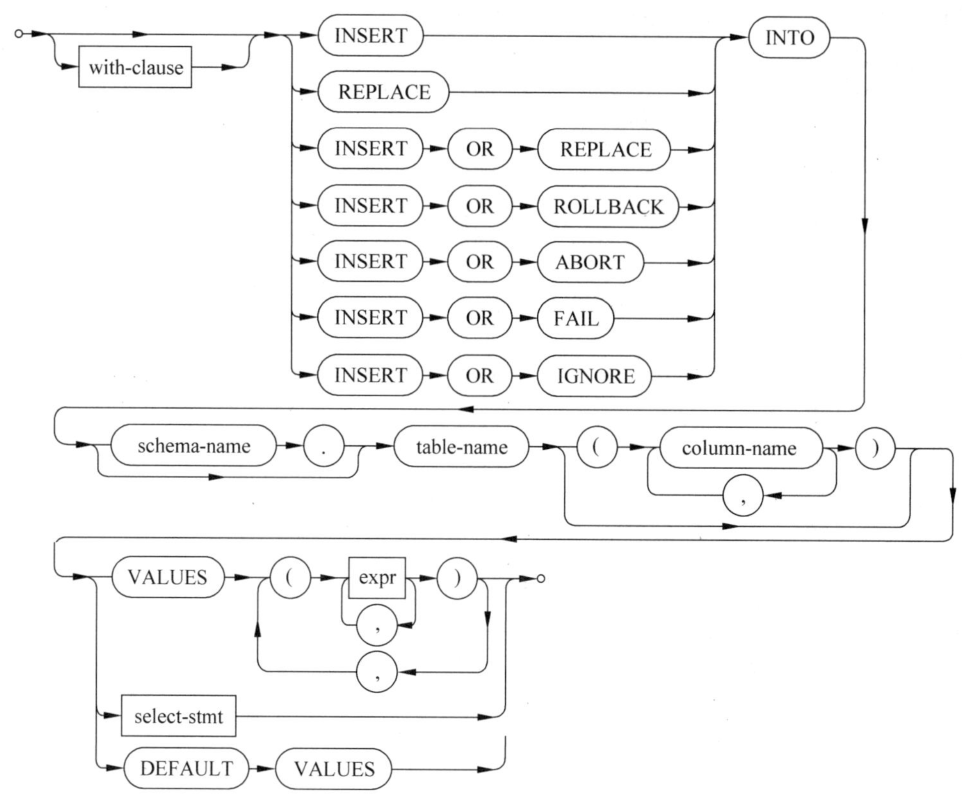

图 14-7 INSERT INTO 语句的语法

最常用的形式如下，其中 column1，column2…，columnN 是要插入数据的表中的列的名称。

```
INSERT INTO TABLE_NAME (column1, column2, column3,...columnN)
VALUES (value1, value2, value3,...valueN);
```

如果要为表中的所有列添加值，可以不指定列名称，但要确保值的顺序与列在表中的顺序一致。

```
INSERT INTO TABLE_NAME VALUES (value1,value2,value3,...valueN);
```

例如：

```
sqlite> CREATE TABLE COMPANY ( ID INT PRIMARY KEY NOT NULL,NAME TEXT NOT NULL,
   AGE INTNOT NULL,ADDRESS CHAR(50),SALARY REAL);
INSERT INTO COMPANY (ID,NAME,AGE,ADDRESS,SALARY) VALUES (1, 'Paul', 32, 'CA', 2000 );
INSERT INTO COMPANY VALUES (7, 'James', 24, 'Houston', 10000.00 );
```

4. 查询语句 SELECT

SELECT 语句用于从 SQLite 数据库中获取数据，以结果表的形式返回数据。这些结果表也被称为结果集。SELECT 语句的完整语法如图 14-8 所示，其中 result-column 语法如图 14-9 所示，其中各字段描述如表 14-6 所示。

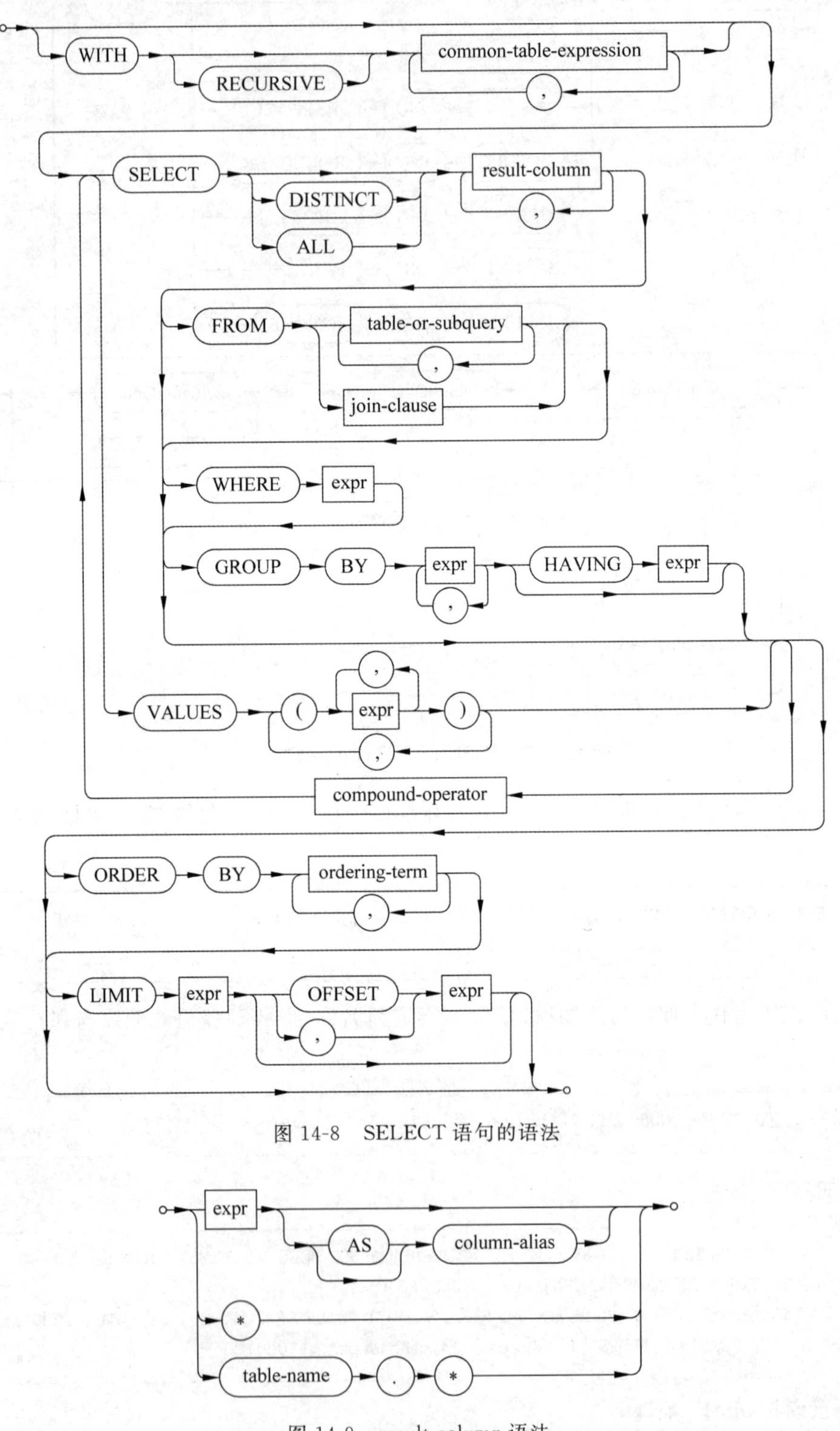

图 14-8 SELECT 语句的语法

图 14-9 result-column 语法

表 14-6　Sqlite select 语句各字段的描述

关　键　字	描　　述
ALL	可选的。如果指定，返回所有匹配的行
DISTINCT	可选的。如果指定，从结果集中删除重复的行
result-column	希望获得的列名或者计算，如果希望获得所有的列，使用*
table-or-subquery	FROM 子句指定一个表或者子查询，指定从哪里获得记录
WHERE conditions	可选的。符合条件的记录将被选择
GROUP BY expressions	SELECT expression1,... expression_n, aggregate_function (expression) FROM tables GROUP BY expression_m; aggregate_function 包含 count, min, max, or avg 等聚合函数
HAVING condition	WHERE 子句，是针对行的过滤。要对分组结果进行过滤，必须使用 HAVING 子句。HAVING 和 GROUPBY 字句组合，先根据 GROUP BY 字句把结果分组，再返回满足 HAVING 子句条件的组。例如 sqlite > SELECT * FROM COMPANY GROUP BY name HAVING count(name) < 2；返回 name 列为同一个值的行数小于 2 的记录
ORDER BY expression	可选的。用于把结果集排序
LIMIT number_rows OFFSET offset_value	可选的。用于控制结果集的最大数量。结果集最多返回 number_rows 指定的行数。第一行由 offset_value 决定

select 语句的最基本的用法如下：

```
SELECT column1, column2, columnN FROM table_name;
```

在这里，column1，column2，...是期望获取的表的字段，如果想获取所有的字段，result-column 用星号代替，形式为：

```
SELECT * FROM table_name;
```

带有 WHERE 子句的 SELECT 语句的基本语法如下：

```
SELECT column1, column2, columnN FROM table_name WHERE [condition];
```

在 WHERE 子句使用比较或逻辑运算符指定条件，比如>、<、=、LIKE、NOT 等。举例：

```
sqlite>.header on
sqlite>.mode column
sqlite> SELECT ID, NAME, SALARY FROM COMPANY;
sqlite> SELECT * FROM COMPANY;
sqlite> SELECT * FROM COMPANY WHERE AGE >= 25 AND SALARY >= 5000;
说明：AGE 大于等于 25 且工资大于等于 5000 的所有记录
sqlite> SELECT * FROM COMPANY WHERE AGE >= 25 OR SALARY >= 5000;
说明：AGE 大于等于 25 或工资大于等于 5000 的所有记录
sqlite> SELECT * FROM COMPANY WHERE AGE IS NOT NULL;
说明：AGE 不为 NULL 的所有记录

sqlite> SELECT * FROM COMPANY WHERE NAME LIKE 'Ki%';
```

```
sqlite> SELECT * FROM COMPANY WHERE AGE IN ( 25, 27 );
说明：AGE 的值为 25 或 27 的所有记录

sqlite> SELECT * FROM COMPANY WHERE AGE NOT IN ( 25, 27 );
说明：AGE 的值既不是 25 也不是 27 的所有记录

sqlite> SELECT * FROM COMPANY WHERE AGE BETWEEN 25 AND 27;
AGE 的值在 25 与 27 之间的所有记录
sqlite> SELECT employee_id, last_name, first_name FROM employees
WHERE employee_id < 50 ORDER BY last_name ASC, employee_id DESC;
对结果集排序,last_name 升序, employee_id 降序
select * from COMPANY limit 5 offset 3;
或者
select * from Account limit 3,5;
说明：获取 5 条记录,跳过前面 3 条记录
SELECT * FROM COMPANY group by NAME having count( * )> 2;
说明：按 NAME 分组,返回分组中记录数大于 2 的分组
```

5. 多表查询

left join(左连接)返回包括左表中的所有记录和右表中连接字段相等的记录。right join(右连接)返回包括右表中的所有记录和左表中连接字段相等的记录。left join 中的左表的数据全部显示(即使在右表中没有匹配的记录),左表相当于主表,右表就相当于从表;right join 中右表的数据全部显示(即使在左表中没有匹配的记录),右表相当于主表,左表就相当于从表。从表数据则只显示关联部分匹配的数据,无匹配的数据用 null 补全。

full join 在两张表进行连接查询时,返回左表和右表中所有的行(包括没有匹配的行),查询结果是 left join 和 right join 的并集。

inner join(等值连接)只返回两个表中连接字段相等的行,查询结果是 left join 和 right join 的交集。

实例：假设有 Persons 和 Orders 两张表,如表 14-7 所示。

表 14-7 多表查询实例中的两张表

Person 表

Id_P	Name	City
1	John	Landon
2	Gorge	Shanghai
3	Rose	Beijing

Orders 表

Id_O	OrderNo	Id_P
1	001	3
2	002	3
3	003	1
4	004	1
5	005	65

查询语句如下,其中左连接的查询结果,如表 14-8 所示。右连接的查询结果,如表 14-9 所示。全连接的查询结果,如表 14-10 所示。内连接的查询结果,如表 14-11 所示。

```
sqlite> SELECT Persons.Name, Orders.OrderNo FROM Persons LEFT JOIN Orders
ON Persons.Id_P = Orders.Id_P ORDER BY Persons.Name;
```

```
sqlite> SELECT Persons.Name, Orders.OrderNo FROM Persons RIGHT JOIN Orders
ON Persons.Id_P = Orders.Id_P ORDER BY Persons.Name;
sqlite> SELECT Persons.Name, Orders.OrderNo FROM Persons FULL JOIN Orders
ON Persons.Id_P = Orders.Id_P ORDER BY Persons.Name;
sqlite> SELECT Persons.Name, Orders.OrderNo FROM Persons INNER JOIN Orders ON Persons.Id_P =
Orders.Id_P ORDER BY Persons.Name;
```

表 14-8 左连接的查询结果

Name	OrderNO
John	003
John	004
Rose	002
Gorge	NULL

表 14-9 右连接的查询结果

Name	OrderNO
John	003
John	004
Rose	001
Rose	002
NULL	005

表 14-10 全连接的查询结果

Name	OrderNO
John	003
John	004
John	001
Rose	002
Gorge	NULL
NULL	005

表 14-11 内连接的查询结果

Name	OrderNO
John	003
John	004
Rose	001
Rose	002

6. 修改语句 UPDATE

SQLite 的 UPDATE 用于修改表中已有的记录。如果 UPDATE 带有 WHERE 子句，只对满足 WHERE 子句的记录进行更新，否则所有的记录都会被更新。update 语句语法如图 14-10 所示。

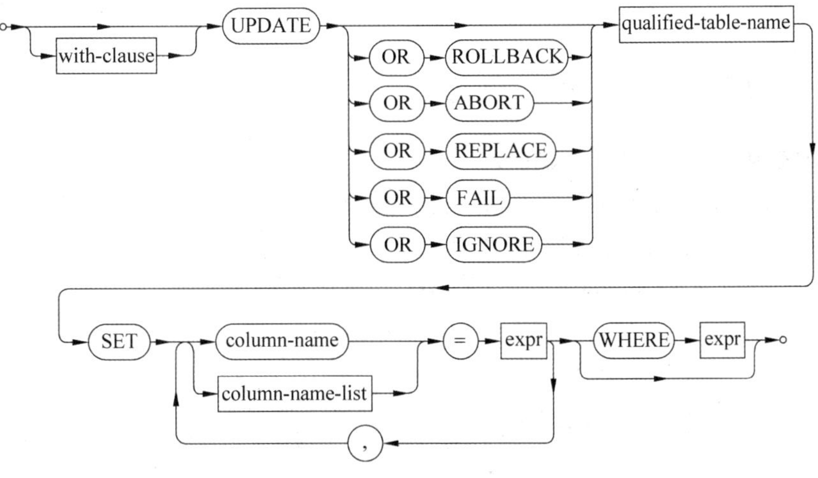

图 14-10 UPDATE 语句语法

带有 WHERE 子句的 UPDATE 的基本形式如下：

```
UPDATE table_name SET column1 = value1, column2 = value2..., columnN = valueN
WHERE [condition];
```

下面是一个实例，它会更新 ID 为 6 的客户地址：

```
sqlite> UPDATE COMPANY SET ADDRESS = 'Texas' WHERE ID = 6;
```

如果想修改 COMPANY 表中 ADDRESS 和 SALARY 列的所有值，则不需要使用 WHERE 子句，UPDATE 语句如下：

```
sqlite> UPDATE COMPANY SET ADDRESS = 'Texas', SALARY = 2000;
```

7. DELETE 语句

DELETE 语句用于删除表中已有的记录。可以使用带有 WHERE 子句的 DELETE 语句来删除满足指定条件的记录，否则所有的记录都会被删除。DELETE 语句语法如图 14-11 所示。

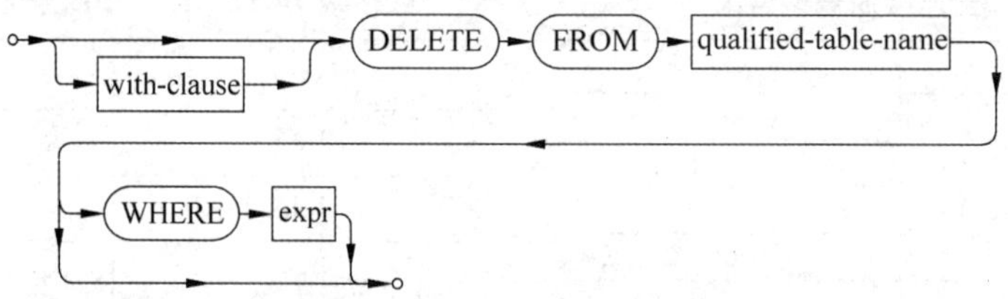

图 14-11　DELETE 语句的语法

下面是一个实例，它会删除 COMPANY 表中 ID 为 7 的客户：

```
sqlite> DELETE FROM COMPANY WHERE ID = 7;
```

如果想要从 COMPANY 表中删除所有记录，则不需要使用 WHERE 子句，DELETE 语句如下

```
sqlite> DELETE FROM COMPANY;
```

8. 更改表格语句 ALTER TABLE

ALTER TABLE 语句允许用户给表格重命名、在已经存在的表中新增加一列等。ALTER TABLE 语句的语法如图 14-12 所示。

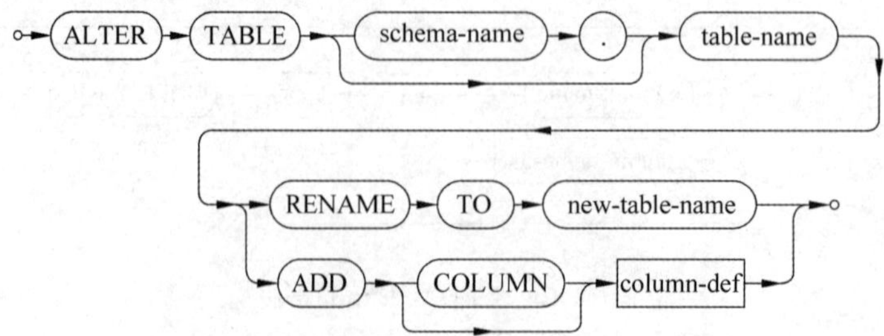

图 14-12　ALTER TABLE 语句的语法

14.5 Android 系统中的 SQLiteDatabase

SQLiteDatabase 提供了管理 SQLite 数据库的方法：例如创建数据库、删除数据库、执行 SQL 命令等。android. database. sqlite. SQLiteDatabase 类提供了创建或者删除数据库以及执行 SQL 命令的方法。可以调用 SQLiteDatabase 的静态方法 openDatabase 或者 openOrCreateDatabase 打开或者创建数据库，还可以通过 Context 对象调用 openOrCreateDatabase (String dbName,int mode,CursorFactory factory)方法，在私有数据库目录中创建或打开一个名为 name 的数据库，参数 mode 的默认值为 MODE_PRIVATE，可选值有 MODE_ENABLE_WRITE_AHEAD_LOGGING 和 MODE_NO_LOCALIZED_COLLATORS。参数 factory 当 query 函数被调用时，会使用该工厂类返回一个 Cursor，可为 null，使用默认的工厂类。

android. database. sqlite. SQLiteOpenHelper 是一个用于数据库创建和版本管理的帮助类。要创建 SQLiteOpenHelper 的子类，通常必须实现构造函数、onCreate(SQLiteDatabase)方法和 onUpgrade(SQLiteDatabase,int,int)这三个方法，其中 onCreate 和 onUpgrade 是抽象方法，还可以选择重写 onOpen(SQLiteDatabase)。这个类使得 ContentProvider 实现延迟打开和升级数据库到至第一次使用时变得容易，避免长时间的数据库升级阻碍应用的启动。SQLiteOpenHelper 的构造函数，有以下两种形式：

```
SQLiteOpenHelper(Context context, String name, SQLiteDatabase.CursorFactory factory, int
version)
SQLiteOpenHelper(Context context, String name, SQLiteDatabase.CursorFactory factory, int
version, DatabaseErrorHandler errorHandler)
```

参数 context 是用于打开或者创建数据库的 Context。参数 name 是数据库文件的名字，null 代表内存中的数据库。参数 factory：工厂类，当 SQLiteDatabase 实例的 query 函数被调用时，会使用该工厂类返回一个 Cursor。参数 factory 可为 null，使用默认的 Cursor。参数 version 是数据库的版本。参数 errorHandler 当 SQLite 报告数据库崩溃时使用。

SQLiteOpenHelper 的 public 方法，如表 14-12 所示。

表 14-12　SQLiteOpenHelper 的方法(部分)

方　　法	解　　释
void close()	关闭打开的数据库对象
SQLiteDatabase getReadableDatabase()	打开数据库。调用 getReadableDatabase 方法返回的并不总是只读数据库对象，一般来说该方法和 getWriteableDatabase 方法的返回的对象相同，只有在数据库仅开放只读权限或磁盘已满时才会返回一个只读的数据库对象 在调用 getReadableDatabase 或 getWritableDatabase 时，SQLiteOpenHelper 会自动判断指定的数据库是否存在，若不存在就调用 onCreate 创建，onCreate 只在数据库第一次创建时才执行，开发人员无须再自己判断数据库是否存在

续表

方　法	解　释
SQLiteDatabase getWritableDatabase()	打开一个用于读写的数据库。该方法第一次被调用时,数据库将被打开,一旦打开成功后,数据库将被缓存,每次需要写数据库时、就可以调用该方法。记住当不再需要数据库时,一定要调用 close()关闭数据库
abstract void onCreate(SQLiteDatabase db)	该方法在第一次创建数据库的时候被调用,而只有当调用 SQLiteOpenHelper 对象的 getReadableDataBase()或者 getWritableDataBase()时,才会调用该方法。一般在 onCreate 里面做一些创建表或表的初始化植入的操作。通过该方法创建的数据库存放的目录是固定的,其路径为/data/data/packageName/databases/
abstract void onUpgrade(SQLiteDatabase db, int oldVersion, int newVersion)	当打开数据库时传入的版本比当前的版本更新、需要升级时被调用,可以更新数据库的表结构,如增加表、删除表、给表重命名、增加表中的列字段等操作。在软件升级前,最好对原有数据进行备份,在新表建好后把数据导入新表中
void onDowngrade (SQLiteDatabase db, int oldVersion, int newVersion)	当打开数据库时传入的版本比当前的版本更旧、需要降级时被调用
void onOpen(SQLiteDatabase db)	当数据库已经打开后被调用

SQLiteOpenHelper 类的基本用法是:当需要创建或打开一个数据库并获得数据库对象时,首先使用指定的数据库名作为参数创建一个 SQLiteOpenHelper 对象,然后调用该对象的 getWritableDatabase 或 getReadableDatabase 方法获 SQLiteDatabase 对象。调用 getReadableDatabase 方法返回的并不总是只读数据库对象,一般来说该方法和 getWriteableDatabase 方法的返回情况相同,只有在数据库仅开放只读权限或磁盘已满时才会返回一个只读的数据库对象。可以采用以下方法打开数据库:

```
//第一种
SQLiteDatabase db = mySqliteOpenHelper.getReadableDatabase();
//第二种
SQLiteDatabase db1 = this.openOrCreateDatabase("info.db", MODE_PRIVATE,null);
//第三种,获取 info.db 的路径
File file = new File(getApplication().getDatabasePath("info.db").getPath());
SQLiteDatabase db2 = SQLiteDatabase.openOrCreateDatabase(file,null);
```

表 14-13 列出了 SQLiteDatabase 类的部分方法。

表 14-13　SQLiteDatabase 类的方法(部分)

方　法	解　释
void execSQL(String sql)	执行单条 SQL 语句,不需要结尾的分号,不支持用分号隔开的多条 SQL 语句,不能执行任何需要返回数据的 SQL 语句,如 SELECT。例如,尽可能使用 insert(String, String, ContentValues), update(String, ContentValues,String,String[])方法
void execSQL(String sql,Object[]bindArgs)	执行单条 SQL 语句,不能执行 SELECT/INSERT/UPDATE/DELETE 语句或任何需要返回数据的 SQL 语句,适合的语句有 ALTER TABLE、CREATE or DROP table/trigger/view/index/virtual table、REINDEX、RELEASE、SAVEPOINT、PRAGMA

续表

方　法	解　释
Cursor rawQuery(String sql, String[] selectionArgs)	用于执行 SELECT 语句,第一个参数 sql 是 SQL SELECT 语句,但不需要结尾的分号。在 WHERE 子句中,可能含有占位符(?),第二个参数 selectionArgs 为 SQL 语句中占位符参数的值,参数值在数组中的顺序要和占位符的位置对应
long insert(String table, String nullColumnHack, ContentValues values)	如果成功则返回 INSERT 的行的 ROW ID,反之返回 −1。nullColumnHack 用于指定空值字段的名称。当第三个参数 values 为 NULL 或者元素个数为 0 时,必然要添加一条除了主键之外其他字段为 NULL 值的记录,INSERT 是会失败的(底层数据库不允许插入一个空行)。为了防止这种情况,就会将 nullColumnHack 的值设为 NULL,然后再向数据库中插入。若不添加 nullColumnHack 则 SQL 语句最终的结果将会类似 insert into tableName()values(),显然这不满足标准 SQL 的语法。若添加上 nullColumnHack 则 SQL 语句将会变成 insert into tableName (nullColumnHack) values (null),这是可以的。对于字段名,建议使用主键之外的字段,如果使用了主键字段,该主键字段值也不会为 NULL。如果第三个参数 values 不为 NULL 并且元素的个数大于 0,可以把第二个参数设置为 NULL。ContentValues 其实就是一个 Map,Key 值是字段名称,Value 值是字段的值。通过 ContentValues 的 put 方法就可以把数据放到 ContentValues 对象中,然后插入到表中去
long insertOrThrow(String table, String nullColumnHack, ContentValues values)	和 insert()功能基本相同,只是当 exception 发生时,它会 throw SQLException
int delete(String table, String whereClause, String[] whereArgs)	参数 table 是要查询的表的名字。参数 whereClause 是删除记录时的 WHERE 语句,注意,不需要带"WHERE",当有 whereClause 时,返回受影响的行数;当 whereClause 传入 1 时,删除所有的行、并且返回删除行的数量;当 whereClause 传入 null 时,将全部删除,返回 0。参数 whereArgs 是参数 whereClause 中占位符的对应的值
int update(String table, ContentValues values, String whereClause, String[] whereArgs)	参数的含义与 INSERT/DELETE 方法中的参数相同
Cursor query(boolean distinct, String table, String[] columns, String selection, String [ ] selectionArgs, String groupBy, String having, String orderBy, String limit)	distinct:取值为 true 将移除结果中重复的行,取值为 false,将选择重复的行。 参数 table 是要查询的表的名字。 Columns:要返回的列的名字的列表。若这个参数为 NULL,则返回所有列,不鼓励从数据库读出不使用的数据。 selection:SQL WHERE 子句,但不包含 WHERE 本身。若这个参数为 NULL,则返回表的所有的行 selectionArgs:在前一个参数 selection 中可能包含?占位符,这些占位符将被这个参数中的对应的值(根据在 selection 中的出现顺序对应)所替换 group by:SQL 的 group by 字句,但不包含 group by 本身,用于把满足 where 条件记录分组。若这个参数为 null,所有的记录都不被分组

续表

方　法	解　释
Cursor query(boolean distinct, String table, String[] columns, String selection, String[] selectionArgs, String groupBy, String having, String orderBy, String limit)	having: SQL 的 HAVE 子句,但不包含 HAVE 本身。当 groupBy 被使用时,HAVING 用于声明哪些组应该被包含在 cursor 中。若这个参数为 null,将使得 cursor 中包含所有的组 orderBy: SQL 的 order by 字句,但不包含 order by 本身,用于声明怎样对行进行排序。若这个参数为 NULL,将使用默认的排序方式,也可能不排序 limit: 限制返回的记录的数量。若这个参数为 NULL,表示没有限制
query(String table, String[] columns, String selection, String[] selectionArgs, String groupBy, String having, String orderBy, String limit)	参数含义同上

对于熟悉 SQL 语法的程序员而言,可以直接使用 execSQL()和 rawQuery()方法执行 SQL 语句就能完成数据的添加、删除、更新和查询操作。SQLiteDatabase 还专门提供了对应于添加、删除、更新、查询的操作方法: insert()、delete()、update()和 query()。

SQLiteDatabase 的方法举例:

```
String CREATE_TABLE = "create table 表名(列名,列名,……)";
mSQLiteDatabase.execSQL (CREATE_TABLE);
mSQLiteDatabase.execSQL ("drop table 表名");
```

SQLiteDatabase 的 rawQuery()用于查询,rawQuery()方法的第一个参数为 SELECT 语句;第二个参数为 SELECT 语句中占位符参数的值,如果 select 语句没有使用占位符,该参数可以设置为 NULL。不带占位符参数的 select 语句使用例子如下:

```
SQLiteDatabase db = ....;
Cursor cursor = db.rawQuery("select * from person ", null);
while (cursor.moveToNext()) {
int personid = cursor.getInt(0);         //获取第一列的值,第一列的索引从 0 开始
String name = cursor.getString(1);       //获取第二列的值
int age = cursor.getInt(2);              //获取第三列的值
}
cursor.close();
db.close();
```

带有占位符的查询语句例子如下:

```
Cursor cursor = db.rawQuery("select * from person where name like ? and age = ?", new String
[]{"%zhangsanfeng%", "4"});
```

Cursor 接口提供了对数据库查询返回的结果集的随机读写。Cursor 是结果集游标,使用 moveToNext()方法可以将游标从当前行移动到下一行,如果已经移过了结果集的最后

一行,返回结果为 false,否则为 true。另外 Cursor 还有常用的 moveToPrevious()方法(用于将游标从当前行移动到上一行,如果已经移过了结果集的第一行,返回值为 false,否则为 true)、moveToFirst()方法(用于将游标移动到结果集的第一行,如果结果集为空,返回值为 false,否则为 true)和 moveToLast()方法(用于将游标移动到结果集的最后一行,如果结果集为空,返回值为 false,否则为 true)。

Insert()方法用于添加数据,各个字段的数据使用 ContentValues 进行存放。android.content.ContentValues 实现了 Parcelable 接口,类似于 MAP,它提供了存取数据对应的 put(String key,Xxx value)和 getAsXxx(String key)方法,key 为字段名称,value 为字段值,Xxx 指的是各种常用的数据类型,如:String、Integer 等。执行 Insert()方法不管第三个参数是否包含数据,必然会添加一条记录,如果第三个参数为空,会添加一条除主键之外其他字段值为 Null 的记录。

```
SQLiteDatabase db = databaseHelper.getWritableDatabase();
ContentValues values = new ContentValues();
values.put("name", "zhangsan");
values.put("age", 4);
long rowid = db.insert("person", null, values);
```

14.6 Room 持久库

Room 持久库提供了一个 SQLite 抽象层,以便在发挥 SQLite 能力的同时,数据库访问更加稳健流畅。该库帮助应用程序在设备上创建数据缓存。这个缓存是应用程序唯一的真实的数据源,允许用户查看应用程序中关键信息的一致副本,而不管用户是否有 Internet。处理复杂的结构化数据的应用程序可以极大地受益于本地数据的持久化。最常见的用例是缓存相关的数据碎片。这样,当设备无法访问网络时,用户仍然可以在离线时浏览该内容。在设备返回联机之后,任何用户发起的内容更改都会同步到服务器。建议使用 Room 而不是直接使用 SQLite API,因为 raw SQL 查询没有编译时验证,当数据映射改变时,需要手动更新受影响的 SQL 查询,这个过程非常耗时。

Room 有 3 个主要的组件:entity class、DAO interface 和 Database 抽象类。

14.6.1 entity class

Entity 类代表了数据库里的一张表。Room 给每个 entity 创建一张数据库表。默认情况下,Room 使用类名作为数据库的表名,如果希望表有一个不同的名称,设置@Entity 注解的 tableName 属性(取值为字符串)。注意:SQLite 中的表名是大小写敏感的。默认情况下,Room 为定义在 entity 中的每个字段创建一个列,如果 entity 的一些字段不想持久化,可以使用@Ignore 注解忽略它们。Room 默认使用字段名称作为列名称。如果希望一个列有不同的名称,为字段增加@ColumnInfo 注解。为了持久化一个字段,Room 必须有它的入口,可以设置字段为 public,或者提供 setter 和 getter 方法。如果使用 setter 和 getter 方法,记住在 Room 中遵守 Java Beans 的惯例。Entity 必须提供一个无参数的构造函数，以便 Room 使用，其他的构造函数前应该加上@Ignore，Room 忽略添加了@Ignore 注解的构造函数，其他 Java 类仍然可以使用添加了@Ignore 注解的构造函数。

每个 entity 必须定义一个主键，除非它的父类定义了已经定义了一个主键。如果 entity 和父类都定义了主键，子类的定义覆盖父类的定义。即使 entity 只有一个字段，仍然需要使用@PrimaryKey 注解这个字段。并且，如果希望 Room 给 entities 动态分配 ID，可以设置@PrimaryKey 注解的 autoGenerate 属性（取值为布尔量）。如果需要定义复合主键，可以使用@Entity 注解的 primaryKeys 的属性（取值为字符串数组）。如果 PrimaryKey 定义在一个 Embedded 域上，则从 Embedded 域继承的所有的列作为复合主键。

为了提高查询的效率，可能想为特定的字段建立索引。添加索引可以加快 select 查询，但使 insert 或者 update 变慢。有 2 种方式定义索引，一种是给 ColumnInfo 注解添加 index 属性（取值为布尔量）索引单个域，另一种是给 Entity 注解添加 indices 属性（取值为 Index 数组）定义索引。

示例 1：在属性前添加@PrimaryKey 标识主键。

```
@Entity(tableName = "users")
class User {
    @PrimaryKey (autoGenerate = true)
    public int id;

    @ColumnInfo(name = "first_name")
    public String firstName;

    @ColumnInfo(name = "last_name", index = true)
    public String lastName;

    @Ignore
    Bitmap picture;
}
```

实例 2：在@Entity 注解中标识主键。

```
@Entity(primaryKeys = {"firstName", "lastName"})
class User {
    public String firstName;
    public String lastName;
    @Ignore
    Bitmap picture;
}
```

实例 3：在@Entity 注解里添加 indices 属性。

```
@Entity(indices = {@Index(value = {"first_name","last_name"},unique = true),@Index
("address")})
class User {
   @PrimaryKey(autoGenerate = true)
    public int id;
```

```
    @ColumnInfo(name = "first_name")
    public String firstName;

    @ColumnInfo(name = "last_name")
    public String lastName;
    public String address;
    @Ignore
    Bitmap picture;
}
```

SQLite 是一个关系数据库，可以指定对象之间的关系。Room 严禁实体对象相互引用，Room 仍然允许在实体之间定义 ForeignKey 约束。主表(也称父表)设置主键(primary key)用于与其他表相关联，并且作为在主表中的唯一性标识。从表(也称字表)设置外键与主表进行关联，以主表的主键(primary key)值作为外键(ForeignKey)。B 表引用 A 表的字段作为外键，那么 A 表是主表，B 表是从表。比如现有某学校三个数据表：学生表(学号，姓名)，课程(课程编号，课程名)，选课(学号，课程号，成绩)。选课表中的"学号"，"课程号"必须是另外两个表中存在的数据，才有意义；而且一旦另外两表中的某一学生或课程被删除，选课表中的相应学号或课程号必须自动删除。

例如，如果有另一个名为 Book 的实体，可以使用@ForeignKey 注释来定义与 User 实体的关系，如以下代码片段所示：

```
@Entity(foreignKeys = @ForeignKey(entity = User.class,
                                  parentColumns = "id",
                                  childColumns = "user_id"))
class Book {
    @PrimaryKey
    public int bookId;
    public String title;
    @ColumnInfo(name = "user_id")
    public int userId;
}
```

ForeignKey 非常强大，可以指定引用实体更新时的操作。例如，如果通过在@ForeignKey 注释中包含 onDelete=CASCADE 删除了相应的 User 实例，则可以告诉 SQLite 删除用户的所有 book。

有时候，希望在数据库逻辑中表达一个实体或普通的 Java 对象(POJO)，即使对象包含多个字段。在这些情况下，可以使用@Embedded 注释来表示要在表中分解为子字段的对象。然后，像其他单独的列一样查询嵌入的字段。

例如，User 类可以包含一个 Address 类型的字段，它表示一个名为 street，city，state 和 postCode 的字段的组合。

```
class Address {
    public String street;
    public String state;
```

```
    public String city;
    @ColumnInfo(name = "post_code")
    public int postCode;
}

@Entity
class User {
    @PrimaryKey
    public int id;
    public String firstName;
    @Embedded
    public Address address;
}
```

User 对象的表包含以下名称的列：id，firstName，street，state，city 和 post_code。

14.6.2 DAO interface

DAO(data access object)接口定义用于访问数据库的方法。DAO 是 Room 中的主要组件，被注解为@Database 的类必须包含一个没有参数的抽象方法并且返回注解为@Dao 的类，编译时 Room 生成这个类的实现代码。当测试应用的时候，DAO 允许轻松地模拟数据库访问。给类加上@Dao 注解用于生成一个 DAO。在添加了@Dao 注解的类中，方法前可以加上@Insert、@Update、@Delete 和@Query 注解。使用@Insert、@Update 和@Delete注解的方法，所有参数只能是@Entity 注解的类或者集合、数组。使用@Query 注解的方法，参数类型没有限制。

当创建了一个方法并且添加了@Insert 注解，Room 生成一个实现，将所有的参数在一次事务中插入数据库。如果@Insert 方法只接收到一个参数，它可以返回一个 long，这是插入项的新 rowId。如果参数是数组或集合，它应该返回 long[]或 List<Long>。

@Update 注解定义一个更新一系列 entity 的方法。它根据每个 entity 的主键作为更新的依据。可以让这个方法返回一个 int 类型的值，表示更新影响的行数，虽然通常并没有这个必要。

@Delete 注解定义一个删除一系列 entity 的方法，它使用主键找到要删除的 entity。

```
@Dao
public interface MyDao {
    @Insert(onConflict = OnConflictStrategy.REPLACE)
    public void insertUsers(User… users);
    @Insert
    public void insertBothUsers(User user1, User user2);
    @Insert
    public void insertUsersAndFriends(User user, List<User> friends);
    @Update
    public void updateUsers(User… users);
    @Delete
    public void deleteUsers(User… users);
```

```
    @Delete
    void deleteUser(User user);
    @Delete
     public void deleteAll(User user1, User user2);
     @Delete
     public void deleteWithFriends(User user, List<User> friends);
}
```

@Query 是 DAO 类中使用的主要注解。它允许对数据库执行查询操作。@Query 支持 3 类 SQL 语句：SELECT、UPDATE 和 DELETE。每个@Query 方法都在编译时被验证，如果只有一些字段名称匹配，则会发出警告；如果没有字段名称匹配，则会给出错误。执行查询时，经常希望应用程序的 UI 在数据更改时自动更新。要实现这一点，请在查询方法描述中使用类型为 LiveData 的返回值，当数据库更新时，Room 会生成所有必需的代码来更新 LiveData。Room 还可以从定义的查询中返回 RxJava2 Publisher 和 Flowable 对象，要使用这个功能，在 Gradle dependencies 中添加 android. arch. persistence. room:rxjava2。可以从查询中返回一个 Cursor 对象。非常不鼓励使用 Cursor API，仅当已经具有期望 Cursor 的代码并且不能轻易重构时，才返回 Cursor 对象。

```
@Dao
public interface MyDao {
  @Query("SELECT * FROM user WHERE age BETWEEN :minAge AND :maxAge")
  public User[] loadAllUsersBetweenAges(int minAge, int maxAge);

  @Query("SELECT * FROM user WHERE first_name LIKE :search "
          + "OR last_name LIKE :search")
  public List<User> findUserWithName(String search);

  @Query("SELECT * FROM user WHERE first_name LIKE : first "
          + "OR last_name LIKE : last")
   public List<User> findUsersByNameAndLastName(String first, String last);


   @Query("SELECT first_name, last_name FROM user WHERE region IN (:regions)")
  public List<NameTuple> loadUsersFromRegions(List<String> regions);

  @Query("SELECT first_name, last_name FROM user WHERE region IN (:regions)")
  public LiveData<List<User>> loadUsersFromRegionsSync(List<String> regions);

  @Query("SELECT * from user where id in (:ids)")
  public LiveData<List<User>> loadUserByIds(int[] ids);

  @Query("SELECT * from user where id = :id LIMIT 1")
  public Flowable<User> loadUserById(int id);

  @Query("SELECT * FROM user WHERE age > :minAge LIMIT 5")
  public Cursor loadRawUsersOlderThan(int minAge);
```

```
    @Query("delete from user where id = :id")
    public void deleteUserByid(int id);

    @Query("delete from user")
    public void deleteAllusers();
}
```

大多数查询,只需要返回 entity 的部分字段。比如,图形界面也许只需显示 User 的姓名,而不是用户的详细信息,只获取 UI 需要的字段可以节省可观的资源,查询也更快。只要结果的字段和返回的对象匹配,Room 允许返回任何的 Java 对象。比如,可以创建如下的 POJO 获取 User 的姓名:

```
public class NameTuple {
    @ColumnInfo(name = "first_name")
    public String firstName;
    @ColumnInfo(name = "last_name")
    public String lastName;
}

@Dao
public interface MyDao {
    @Query("SELECT first_name, last_name FROM user")
    public List<NameTuple> loadFullName();
}
```

某些查询可能需要根据多个表查询出结果。Room 允许书写任何查询,表连接(join)也是可以的。而且如果响应是一个可观察的数据类型,比如 Flowable 或者 LiveData,Room 将观察查询中涉及到的所有表。下面的代码演示了如何执行一个表连接查询来查出借阅图书的 user 与被借出图书之间的信息。

```
@Dao
public interface MyDao {
    @Query("SELECT * FROM book "
            + "INNER JOIN loan ON loan.book_id = book.id "
            + "INNER JOIN user ON user.id = loan.user_id "
            + "WHERE user.name LIKE :userName")
    public List<Book> findBooksBorrowedByNameSync(String userName);
}
```

14.6.3 Database 抽象类

Database 抽象类用于创建一个数据库容器。这个类必须满足以下条件:①它必须是一个抽象的继承自 RoomDatabase 的类;②它必须包含无参数、返回值为以@DAO 注解的类的抽象方法;③它必须使用@Database 注解,且注解中包含一个或者多个 entity。

App 使用 Room 数据库获取 DAO。然后 App 使用 DAO 获取数据库中的 entity,或者把 entity 的变化保存到数据库。最后 App 使用 entity 获取或者设置与数据库表的列对应的值。Room 组件之间的关系如图 14-13 所示。

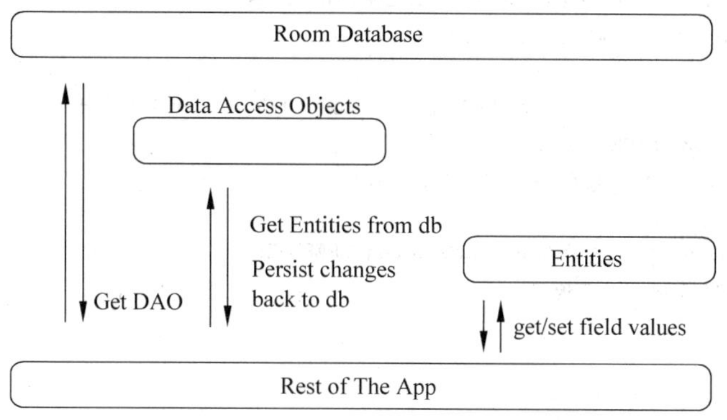

图 14-13 Room 结构图

Room 不允许通过主线程上访问数据库,除非在构建器上调用 allowMainThreadQueries(),因为它可能会长时间锁定用户界面。异步查询(返回 LiveData 或 RxJava Flowable 的查询)将免除此规则,因为它们在需要时异步地在后台线程上运行查询。

下面的示例中,AppDatabase 数据库包含一个 entity 和一个 DAO。User. java 定义了一个 entity,文件内容如下:

```
@Entity (tableName = "users")

public class User {
    @PrimaryKey
    public int uid;

    @ColumnInfo(name = "first_name")
    public String firstName;

    @ColumnInfo(name = "last_name")
    public String lastName;

    @Ignore
    Bitmap picture;

    // Getters and setters are ignored for brevity,
    // but they're required for Room to work.
}
```

UserDao. java 定义了数据库的增删改查方法,用于从数据库获取数据或者更新数据库里的数据,文件内容如下:

```
@Dao
public interface UserDao {
```

```
    @Query("SELECT * FROM user")
    List<User> getAll();

    @Query("SELECT * FROM user WHERE uid IN (:userIds)")
    List<User> loadAllByIds(int[] userIds);

    @Query("SELECT * FROM user WHERE first_name LIKE :first AND " +
        "last_name LIKE :last LIMIT 1")
    User findByName(String first, String last);

    @Insert(onConflict = OnConflictStrategy.REPLACE)
    void insertAll(User... users);

    @Delete
    void delete(User user);
}
```

AppDatabase.java 定义了一个 Database，@Database 注解定义了数据库包含的所有 entity，类中定义了无参数抽象方法，用于获取 DAO，文件内容如下：

```
@Database(entities = {User.class}, version = 1)
public abstract class AppDatabase extends RoomDatabase {
  public abstract UserDao userDao();
}
```

调用 Room.databaseBuilder()或者 Room.inMemoryDatabaseBuilder()获取数据库实例。

```
public class MyApplication extends Application {
   private static Context context;
   private static final String DATABASE_NAME = "MyDatabase";
   private static AppDatabase appDatabase;
   @Override public void onCreate() {
       super.onCreate();
       context = getApplicationContext();
       new Thread(new Runnable() {
           @Override public void run() {
               appDatabase = Room.databaseBuilder(getApplicationContext(), AppDatabase.
class,
                   DATABASE_NAME).build(); } }).start();
   }
   public static AppDatabase getDB() {
      return appDatabase;
   }
}
```

需要注意的是不能在主线程中初始化，必须新开启一个线程进行初始化，否则会报错，或者无法创建数据库。如果应用在单一的进程中运行，实例化 AppDatabase 对象时应该遵守单例设计模式。创建 RoomDatabase 实例既耗时又耗内存，而且在一个进程中很少需要访问多个实例。如果应用在多个进程中运行，在数据库构建时包含 enableMultiInstanceInvalidation()，当在每个进程中有一个 AppDatabase 实例时，可以使一个进程中的共享数据文件无效，这种无效自动传播到其他进程的 AppDatabase 实例。

14.6.4 类型转换

有时候希望 Room 把自定义数据类型存储在单个列中，可以定义一个转换类，包含成对的两个方法：一个方法把自定义类型转换为 Room 可以保留的已知类型，另一个方法则完成相反的转换，把数据库存储的类型转换为原始的自定义类型，在每个方法前都要添加 @TypeConverter 注解。然后给 Database 类添加@TypeConverters 注解，说明 Database 要用到的所有转换类。使用这些转换器，查询时就可以像使用原始类型一样使用自定义类型。以下代码片段演示了类型转换的使用。

```
public class Converters {
    @TypeConverter
    public static Date fromTimestamp(Long value) {
        return value == null ? null : new Date(value);
    }

    @TypeConverter
    public static Long dateToTimestamp(Date date) {
        return date == null ? null : date.getTime();
    }
}

@Database(entities = {User.java}, version = 1)
@TypeConverters({Converter.class})
public abstract class AppDatabase extends RoomDatabase {
    public abstract UserDao userDao();
}

@Entity
public class User {
    ...
    private Date birthday;
}

@Dao
public interface UserDao {
    ...
    @Query("SELECT * FROM user WHERE birthday BETWEEN :from AND :to")
    List<User> findUsersBornBetweenDates(Date from, Date to);
}
```

注意：Entities 能够有一个空的构造函数(如果 dao 类能够访问每个持久化的字段)或者一个参数带有匹配 entity 中的字段的类型和名称的构造函数，例如一个只接收其中一些字段的构造函数。

14.7 SharedPreference 实验

1. 新建工程 SharedPreference1

具体步骤：运行 Android Studio，单击 Start a new Android Studio project，在 Create New Project 页中，Application name 为 SharedPreference1，Company domain 为 cuc，Project location 为 C:\android\myproject\ SharedPreference1，单击 Next 按钮，选中 Phone and Tablet，minimum sdk 设置为 API 19：Android 4.4(minSdkVersion)，单击 Next 按钮，选择 Empty Activity，再单击 Next 按钮，选中 Generate Layout File 选项，选中 Backwards Compatibility(AppCompat)选项，单击 Finish 按钮。

2. 添加 DataBinding 支持，并同步

在文件 build.gradle(module：app)的 Android 语句的花括号内添加：

```
dataBinding.enabled = true
```

由于修改了 build.gradle 文件内容，这个文件窗口的上面就出现提示："Gradle files have changed since last project sync. A project sync may be necessary for the IDE to work properly. sync now"。单击 sync now 按钮进行同步。

3. 在 res\layout 目录新建布局文件 activity_main1.xml

具体步骤：在工程结构视图中，右击 layout，选择 New | Layout resource file，在 File name 输入框输入 gridlayout，在 Root element 输入框输入 layout，Source set 为 main，单击 OK 按钮。layout/activity_main1.xml 内容如下：

```
<?xml version = "1.0" encoding = "utf - 8"?>
< layout xmlns:android = "http://schemas.android.com/apk/res/android">
< data/>
< LinearLayout
    android:orientation = "vertical"
    android:layout_width = "match_parent"
    android:layout_height = "match_parent">
    < EditText
        android:layout_width = "match_parent"
        android:layout_height = "wrap_content"
        android:id = "@ + id/nameEdt"
        android:hint = "name"
        android:inputType = "text" />
    < RadioGroup
        android:checkedButton = "@id/beijingRadioBtn"
        android:id = "@ + id/cityRadioGroup"
        android:layout_width = "match_parent"
        android:layout_height = "wrap_content"
        android:orientation = "horizontal">
```

```
        <RadioButton
            android:layout_width="wrap_content"
            android:layout_height="wrap_content"
            android:text="Beijing"
            android:id="@+id/beijingRadioBtn" />
        <RadioButton
            android:layout_width="wrap_content"
            android:layout_height="wrap_content"
            android:text="shanghai"
            android:id="@+id/shanghaiRadioBtn" />
        <RadioButton
            android:layout_width="wrap_content"
            android:layout_height="wrap_content"
            android:text="guangzhou"
            android:id="@+id/guangzhouRadioBtn" />
    </RadioGroup>
    <CheckBox
        android:layout_width="match_parent"
        android:layout_height="wrap_content"
        android:text="swim"
        android:id="@+id/swimCheckBox" />
    <CheckBox
        android:layout_width="match_parent"
        android:layout_height="wrap_content"
        android:id="@+id/singCheckBox"
        android:text="sing" />
    <Button
        android:layout_width="match_parent"
        android:layout_height="wrap_content"
        android:id="@+id/saveBtn"
        android:text="save" />
    <Button
        android:layout_width="match_parent"
        android:layout_height="wrap_content"
        android:id="@+id/readBtn"
        android:text="read" />
</LinearLayout>
</layout>
```

4. 编辑 MainActivity.java

MainActivity.java 内容如下：

```
package cuc.sharedpreference1;
import android.content.SharedPreferences;
import android.databinding.DataBindingUtil;
import android.os.Bundle;
import android.support.v7.app.AppCompatActivity;
import android.view.View;
```

```
import android.widget.CompoundButton;
import android.widget.RadioButton;
import android.widget.RadioGroup;
import android.widget.Toast;
import cuc.sharedpreference1.databinding.ActivityMain1Binding;

public class MainActivity extends AppCompatActivity implements View.OnClickListener,
CompoundButton.OnCheckedChangeListener, RadioGroup.OnCheckedChangeListener {
private ActivityMain1Binding binding;
 private static final String NAME = "name";
 private static final String SWIM = "swim";
 private static final String SING = "SING";
 private static final String BEIJING = "Beijing";
 private static final String SHANGHAI = "Shanghai";
 private static final String GUANGZHOU = "Guangzhou";

 @Override
 protected void onCreate(Bundle savedInstanceState) {
  super.onCreate(savedInstanceState);
  binding = DataBindingUtil.setContentView(this,R.layout.activity_main1 );
  binding.saveBtn.setOnClickListener(this);
  binding.readBtn.setOnClickListener(this);
  binding.singCheckBox.setOnCheckedChangeListener(this);
  binding.swimCheckBox.setOnCheckedChangeListener(this);
  binding.cityRadioGroup.setOnCheckedChangeListener(this);
 }
 void saveSharedPreference(){
  SharedPreferences preference = getPreferences(MODE_PRIVATE);
  SharedPreferences.Editor editor = preference.edit();
  editor.putString(NAME,binding.nameEdt.getText().toString());
  editor.putBoolean(SWIM,binding.swimCheckBox.isChecked());
  editor.putBoolean(SING,binding.singCheckBox.isChecked());
  editor.putBoolean(BEIJING,binding.beijingRadioBtn.isChecked());
  editor.putBoolean(SHANGHAI,binding.shanghaiRadioBtn.isChecked());
  editor.putBoolean(GUANGZHOU,binding.guangzhouRadioBtn.isChecked());
  editor.commit();
 }
 void readSharedPreference(){
  SharedPreferences preference = getPreferences(MODE_PRIVATE);
  binding.nameEdt.setText(preference.getString(NAME,null));
  binding.swimCheckBox.setChecked(preference.getBoolean(SWIM,false));
  binding.singCheckBox.setChecked(preference.getBoolean(SING,false));
  binding.beijingRadioBtn.setChecked(preference.getBoolean(BEIJING,false));
  binding.shanghaiRadioBtn.setChecked(preference.getBoolean(SHANGHAI,false));
  binding.guangzhouRadioBtn.setChecked(preference.getBoolean(GUANGZHOU,false));
 }

 @Override
 public void onClick(View view) {
```

```
        switch (view.getId()){
         case R.id.saveBtn:
          saveSharedPreference();
          break;
         case R.id.readBtn:
          readSharedPreference();
          break;
        }
      }

      @Override
      public void onCheckedChanged(CompoundButton buttonView, boolean isChecked) {
    Toast.makeText (MainActivity.this, buttonView.getText ( ) + " checkbox" + (isChecked?
    "checked":"unckecked"), Toast.LENGTH_SHORT).show();
      }

      @Override
      public void onCheckedChanged(RadioGroup group, int checkedId) {
       Toast.makeText(MainActivity.this, ((RadioButton)findViewById(checkedId)).getText(),
    Toast.LENGTH_SHORT).show();
      }
    }
```

5. 编译运行

运行后，单击界面中的 Save 按钮，保存 RadioButton 和 CheckBox 的状态以及 EditText 的文字到 SharedPreference 中。再改变 RadioButton 和 CheckBox 的状态以及 EditText 的文字，单击 Read 按钮，就可以从 SharedPreference 中读出 RadioButton 和 CheckBox 的状态以及 EditText 的文字，并显示在界面上。SharedPreference 文件 userInfo.xml 保存在 /data/data/cuc.sharedpreference1/shared_prefs/文件夹中。

6. 给 RadioGroup 加上边框，再编译运行

这一步把 RadioGroup 的边框画出来，以便可以明显地看出 Beijing、Shanghai 和 Guangzhou 这几个 RadioButton 属于同一个 RadioGroup。在 res 的 drawable 中定义一个 shape，设置该 shape 的 stroke 用于描边，RadioGroup 设置 android:background 属性值为定义的 shape。

具体步骤：在工程结构视图中，右击 res，选择 New|Android Resource File，在 File name 输入框输入 radio_group_shape，Resource type 从下拉列表中选择 Drawable，在 Root element 输入框输入 shape，Source set 为 main，单击 OK 按钮。drawable/radio_group_shape.xml 内容如下：

```
<?xml version = "1.0" encoding = "utf-8"?>
<shape xmlns:android = "http://schemas.android.com/apk/res/android">
    <stroke android:dashWidth = "10dp" android:dashGap = "0dp" android:width = "1.5dp"
        android:color = "@android:color/holo_red_light"/>
</shape>
```

再在 activity_main1. xml 文件中，给 RadioGroup 加上如下属性

```
android:background = "@drawable/radio_group_shape"
```

再次编译运行，发现 RadioGroup 多了边框，边框颜色为红色。

14.8 SQLite 实验

1. 新建工程 SQLite1

具体步骤：运行 Android Studio，单击 Start a new Android Studio project，在 Create New Project 页中，Application name 为 SQLite1，Company domain 为 cuc，Project location 为 C:\android\myproject\SQLite1，单击 Next 按钮，选中 Phone and Tablet，minimum sdk 设置为 API 19：Android 4.4(minSdkVersion)，单击 Next 按钮，选择 Empty Activity，再单击 Next 按钮，选中 Generate Layout File 选项，选中 Backwards Compatibility(AppCompat) 选项，单击 Finish 按钮。

2. 添加 DataBinding 支持，并同步

在文件 build. gradle(module：app)的 Android 语句的花括号内添加：

```
dataBinding.enabled = true
```

由于修改了 build. gradle 文件内容，这个文件窗口的上面就出现提示："Gradle files have changed since last project sync. A project sync may be necessary for the IDE to work properly. sync now"。单击 sync now 按钮进行同步。

3. 在 res\layout 目录新建布局文件 activity_main1. xml

具体步骤：在工程结构视图中，右击 layout，选择 New | Layout resource file，在 File name 输入框输入 activity_main1，在 Root element 输入框输入 layout，Source set 为 main，单击 OK 按钮。layout/activity_main1. xml 内容如下：

```
<?xml version = "1.0" encoding = "utf - 8"?>
<layout xmlns:android = "http://schemas.android.com/apk/res/android">
    <data/>
    <TableLayout
        android:layout_width = "match_parent" android:layout_height = "match_parent">
        <TableRow>
            <TextView android:text = "id"/>
            <EditText android:id = "@ + id/idEdt"/>
        </TableRow>
        <TableRow>
            <TextView android:text = "name"/>
            <EditText android:id = "@ + id/nameEdt"/>
        </TableRow>
        <TableRow>
            <TextView android:text = "city"/>
            <EditText android:id = "@ + id/cityEdt"/>
```

```
            </TableRow>
            <TableRow>
                <Button android:id = "@ + id/insertBtn" android:text = "insert"/>
            </TableRow>
            <TableRow>
                <Button android:id = "@ + id/updateByIdBtn" android:text = "updateById"/>
                <Button android:id = "@ + id/deleteByIdBtn" android:text = "deleteById"/>
            </TableRow>
            <Button android:id = "@ + id/queryAllBtn" android:text = "queryAll"/>
            <Button android:id = "@ + id/queryByIdBtn" android:text = "queryById"/>
        </TableLayout>
</layout>
```

4. 在 cuc. sqlite1 包里新建一个 Java 类 MySQLiteOpenHelper

具体步骤：在工程结构视图中 main 源集 Java 文件夹里，右击 cuc. sqlite1 包，依次选择 New|Java Class，Name 输入框中输入 MySQLiteOpenHelper，在 Superclass 输入框中输入 android. database. sqlite. SQLiteOpenHelper，单击 OK 按钮。

在 MySQLiteOpenHelper 的编辑窗口中，把光标移到类名所在的行，按 Alt+Enter 组合键，单击 Implement methods，选择 onCreate 和 onUpgrade 两个方法，单击 OK 按钮。在 MySQLiteOpenHelper 的编辑窗口中，把光标移到类 MySQLiteOpenHelper 的里面，单击 Android Studio 的菜单 Code|Generate|Constructor，在弹出的窗口中，选中要实现的构造函数，单击 OK 按钮。编辑 MySQLiteOpenHelper. java，内容如下：

```
package cuc.sqlite1;
import android.content.Context;
import android.database.sqlite.SQLiteDatabase;
import android.database.sqlite.SQLiteOpenHelper;
import android.util.Log;

public class MySQLiteOpenHelper extends SQLiteOpenHelper {
    private final static int VERSION = 1;
    private static final String TAG  =  "MySQLiteOpenHelper";

    public MySQLiteOpenHelper(Context context, String name, SQLiteDatabase.CursorFactory
factory, int version) {
        super(context, name, factory, version);
    }
    public MySQLiteOpenHelper(Context context, String name, SQLiteDatabase.CursorFactory
factory) {
        this(context, name, factory,VERSION);
    }
    public MySQLiteOpenHelper(Context context, String name) {
        this(context, name, null);
    }
    @Override
    public void onCreate(SQLiteDatabase db) {
```

```
        Log.i(TAG, "onCreate: ");
        db.execSQL("create table student (name varchar(20),id int,city varchar(20))");
    }
    @Override
    public void onUpgrade(SQLiteDatabase sqLiteDatabase, int i, int i1) {
        Log.i(TAG, "onUpgrade: ");
    }
}
```

5. 编辑 MainActivity.java

编辑后，MainActivity.java 如下：

```
package cuc.sqlite1;
import android.content.ContentValues;
import android.database.Cursor;
import android.database.sqlite.SQLiteDatabase;
import android.databinding.DataBindingUtil;
import android.os.Bundle;
import android.support.v7.app.AppCompatActivity;
import android.util.Log;
import android.view.View;
import android.widget.Toast;
import cuc.sqlite1.databinding.ActivityMain1Binding;

public class MainActivity extends AppCompatActivity implements View.OnClickListener {
private static final String TAG = "MainActivity";
private static final String TABLE = "student";
private static final String NAME = "name";
private static final String ID = "id";
private static final String CITY = "city";
private ActivityMain1Binding binding;

@Override
protected void onCreate(Bundle savedInstanceState) {
  super.onCreate(savedInstanceState);
  binding = DataBindingUtil.setContentView(this, R.layout.activity_main1);
}

@Override
public void onClick(View view) {
  Cursor cursor;
  MySQLiteOpenHelper helper = new MySQLiteOpenHelper(this,"myfirst.db");
  SQLiteDatabase db = helper.getWritableDatabase();
  switch (view.getId()){
   case R.id.deleteByIdBtn:
    cursor = db.query(TABLE,new String[]{ID},"id == ?",
      new String[]{binding.idEdt.getText().toString()},null,null,null);
     if(cursor.getCount() == 0){
```

```
        Toast.makeText(this, "No records to deleteById!", Toast.LENGTH_SHORT).show();
        break;
      }
      db.delete(TABLE,"id == ?",new String[]{binding.idEdt.getText().toString()});
      Toast.makeText(this, "deleteById finished!", Toast.LENGTH_SHORT).show();
      break;
    case R.id.insertBtn:
      ContentValues values = new ContentValues();
      values.put(NAME,binding.nameEdt.getText().toString());
      values.put(ID,Integer.parseInt(binding.idEdt.getText().toString()));
      values.put(CITY,binding.cityEdt.getText().toString());
      db.insert(TABLE,null,values);
      Toast.makeText(this, "insert finished!", Toast.LENGTH_SHORT).show();
      break;
    case R.id.queryAllBtn:
      cursor = db.rawQuery("select * from student",null);
      if(cursor.getCount() == 0){
        Log.i(TAG, "queryAll: No records!");
        break;
      }
      while(cursor.moveToNext()){
        String name = cursor.getString(cursor.getColumnIndex("name"));
        int id = cursor.getInt(cursor.getColumnIndex("id"));
        String city = cursor.getString(cursor.getColumnIndex("city"));
        Log.i(TAG, "queryAll: id:" + id + ", name:" + name + ", city:" + city);
      }
      break;
    case R.id.queryByIdBtn:
      cursor = db.query(TABLE,new String[]{NAME,ID,CITY},"id == ?",
          new String[]{binding.idEdt.getText().toString()},null,null,null);
      if(cursor.getCount() == 0){
        Log.i(TAG, "queryById: No records!");
        break;
      }
      while(cursor.moveToNext()){
        String name = cursor.getString(cursor.getColumnIndex("name"));
        int id = cursor.getInt(cursor.getColumnIndex("id"));
        String city = cursor.getString(cursor.getColumnIndex("city"));
        Log.i(TAG, "queryById: id:" + id + ", name:" + name + ", city:" + city);
      }
      break;
    case R.id.updateByIdBtn:
      cursor = db.query(TABLE,new String[]{ID},"id == ?",
          new String[]{binding.idEdt.getText().toString()},null,null,null);
      if(cursor.getCount() == 0){
        Toast.makeText(this, "No records to updateById!", Toast.LENGTH_SHORT).show();
        break;
      }
      ContentValues values1 = new ContentValues();
```

```
        values1.put(NAME,binding.nameEdt.getText().toString());
        values1.put(ID,Integer.parseInt(binding.idEdt.getText().toString()));
        values1.put(CITY,binding.cityEdt.getText().toString());
        db.update(TABLE,values1,"id == ?",new String[]{binding.idEdt.getText().toString()});
        Toast.makeText(this, "updateById finished!", Toast.LENGTH_SHORT).show();
        break;
    }
  }
}
```

最后编译运行。

14.9 LiveData 与 Room 实验

实验目的：

（1）熟悉 ViewModel 和 LiveData 的使用。

（2）熟悉给 RecyclerView 添加 Item 点击事件的方法。

（3）熟悉 Room 框架的使用。

1. 新建工程 Room1

具体步骤：运行 Android Studio，单击 Start a new Android Studio project，在 Create New Project 页中，Application name 为 Room1，Company domain 为 cuc，Project location 为 C:\android\myproject\Room1，单击 Next 按钮，选中 Phone and Tablet，minimum sdk 设置为 API 19：Android 4.4（minSdkVersion），单击 Next 按钮，选择 Empty Activity，再单击 Next 按钮，选中 Generate Layout File 选项，选中 Backwards Compatibility（AppCompat）选项，单击 Finish 按钮。

2. 添加 DataBinding 支持，并同步

在工程的根目录里的 gradle.properties 文件末尾，添加如下内容：

```
android.databinding.enableV2 = true
```

在文件 build.gradle（module：app）的 Android 语句的花括号内添加：

```
dataBinding.enabled = true
```

由于修改了 build.gradle 文件内容，这个文件窗口的上面就出现提示："Gradle files have changed since last project sync. A project sync may be necessary for the IDE to work properly. sync now"。单击 sync now 按钮进行同步。

3. 添加 Java8 支持，以便支持 Lamba 表达式

单击 Android Studio 的菜单 File | Project Structure | app | properties，从 Source Compatibility 后的下拉列表框中选择 1.8，从 Target Compatibility 后的下拉列表框中选择 1.8，单击 OK 按钮。

4. 添加依赖

在文件 build. gradle(module: app)的 dependencies 语句的花括号内添加:

```
implementation 'android.arch.persistence.room:runtime:1.1.1'
annotationProcessor 'android.arch.persistence.room:compiler:1.1.1'
implementation 'android.arch.lifecycle:extensions:1.1.1'
annotationProcessor 'android.arch.lifecycle:compiler:1.1.1'
implementation 'com.android.support:recyclerview-v7:28.0.0'
implementation 'com.android.support:cardview-v7:28.0.0'
```

5. 在 cuc. room1 包里新建 db、viewmodel 和 ui 包,在 db 内新建 entity、dao 和 converter 包

具体步骤:在工程结构视图中 main 源集 Java 文件夹里,右击 cuc. room1 包,选择 New|Package,输入包名 db、viewmodel、ui、db. entity、db. converter 或者 db. dao,单击 OK 按钮,就能创建相应的包。

6. 在 cuc. room1. db. entity 包里新建 ProductEntity 类

ProductEntity. java 文件内容如下:

```
package cuc.room1.db.entity;

import android.arch.persistence.room.Entity;
import android.arch.persistence.room.PrimaryKey;
@Entity(tableName = "products")
public class ProductEntity {
    @PrimaryKey
    private int id;
    private String name;
    private String description;
    private int price;

    public ProductEntity() {
    }

    public ProductEntity(int id, String name, String description, int price) {
        this.id = id;
        this.name = name;
        this.description = description;
        this.price = price;
    }

    public int getId() {
        return id;
    }

    public void setId(int id) {
        this.id = id;
    }
```

```
    public String getName() {
        return name;
    }

    public void setName(String name) {
        this.name = name;
    }

    public String getDescription() {
        return description;
    }

    public void setDescription(String description) {
        this.description = description;
    }

    public int getPrice() {
        return price;
    }

    public void setPrice(int price) {
        this.price = price;
    }
}
```

7. 在 cuc.room1.db.entity 包里新建 CommentEntity 类

CommentEntity.java 文件内容如下：

```
package cuc.room1.db.entity;
import android.arch.persistence.room.Entity;
import android.arch.persistence.room.ForeignKey;
import android.arch.persistence.room.Ignore;
import android.arch.persistence.room.Index;
import android.arch.persistence.room.PrimaryKey;
import java.util.Date;

@Entity(tableName = "comments",
        foreignKeys = {
                @ForeignKey(entity = ProductEntity.class,
                        parentColumns = "id",
                        childColumns = "productId",
                        onDelete = ForeignKey.CASCADE)},
        indices = {@Index(value = "productId")
        })
public class CommentEntity {
    @PrimaryKey(autoGenerate = true)
    private int id;
    private int productId;
```

```
    private String text;
    private Date postedAt;

    public CommentEntity() {
    }
@Ignore
    public CommentEntity(int id, int productId, String text, Date postedAt) {
        this.id = id;
        this.productId = productId;
        this.text = text;
        this.postedAt = postedAt;
    }

    public int getId() {
        return id;
    }

    public void setId(int id) {
        this.id = id;
    }

    public int getProductId() {
        return productId;
    }

    public void setProductId(int productId) {
        this.productId = productId;
    }

    public String getText() {
        return text;
    }

    public void setText(String text) {
        this.text = text;
    }

    public Date getPostedAt() {
        return postedAt;
    }

    public void setPostedAt(Date postedAt) {
        this.postedAt = postedAt;
    }
}
```

8. 在 cuc.room1.db.converter 包里，新建 DateConverter 类

CommentEntity 类中的成员变量 postedAt 是 java.util.Date，转换为 Long 存储到数据库，从数据库读出后再进行逆转换。DateConverter.java 文件内容如下：

```
package cuc.room1.db.converter;
import android.arch.persistence.room.TypeConverter;
import java.util.Date;

public class DateConverter {
    @TypeConverter
    public static Date toDate(Long timestamp) {
        return timestamp == null ? null : new Date(timestamp);
    }
    @TypeConverter
    public static Long toTimestamp(Date date) {
        return date == null ? null : date.getTime();
    }
}
```

9. 在 cuc.room1.db.dao 包里新建 ProductDao 类

ProductDao.java 文件内容如下：

```
package cuc.room1.db.dao;
import android.arch.lifecycle.LiveData;
import android.arch.persistence.room.Dao;
import android.arch.persistence.room.Insert;
import android.arch.persistence.room.OnConflictStrategy;
import android.arch.persistence.room.Query;
import java.util.List;
import cuc.room1.db.entity.ProductEntity;

@Dao
public interface ProductDao {
    @Query("SELECT * FROM products")
    LiveData<List<ProductEntity>> loadAllProducts();

    @Insert(onConflict = OnConflictStrategy.REPLACE)
    void insertAll(List<ProductEntity> products);

    @Query("select * from products where id = :productId")
    LiveData<ProductEntity> loadProduct(int productId);

    @Query("select * from products where id = :productId")
    ProductEntity loadProductSync(int productId);
}
```

10. 在 cuc.room1.db.dao 包里新建 CommentDao 类

CommentDao.java 文件内容如下：

```
package cuc.room1.db.dao;
import android.arch.lifecycle.LiveData;
import android.arch.persistence.room.Dao;
```

```
import android.arch.persistence.room.Insert;
import android.arch.persistence.room.OnConflictStrategy;
import android.arch.persistence.room.Query;
import java.util.List;
import cuc.room1.db.entity.CommentEntity;

@Dao
public interface CommentDao {
    @Query("SELECT * FROM comments where productId = :productId")
    LiveData<List<CommentEntity>> loadComments(int productId);

    @Query("SELECT * FROM comments where productId = :productId")
    List<CommentEntity> loadCommentsSync(int productId);

    @Insert(onConflict = OnConflictStrategy.REPLACE)
    void insertAll(List<CommentEntity> comments);
}
```

11. 在 cuc.room1.db 包里新建 DataGenerator 类

DataGenerator.java 用于生成一些产品及评论的数据，用于测试。文件内容如下：

```
package cuc.room1.db;
import java.util.ArrayList;
import java.util.Date;
import java.util.List;
import java.util.Random;
import java.util.concurrent.TimeUnit;
import cuc.room1.db.entity.CommentEntity;
import cuc.room1.db.entity.ProductEntity;

public class DataGenerator {
    private static final String[] FIRST = new String[]{
            "Expensive", "Cheap","Big","Small"};
    private static final String[] SECOND = new String[]{
            "apple", "pear","orange","watermelon", "banana", "pineapple"};
    private static final String[] DESCRIPTION = new String[]{
            "is on the table", "is on the desk",
            "is in the bag", "is delicious", "is beautiful", "is unpalatable"};
    private static final String[] COMMENTS = new String[]{
            "Comment 1", "Comment 2", "Comment 3", "Comment 4", "Comment 5", "Comment 6"};

    public static List<ProductEntity> generateProducts() {
        List<ProductEntity> products = new ArrayList<>(FIRST.length * SECOND.length);
        Random rnd = new Random();
        for (int i = 0; i < FIRST.length; i++) {
            for (int j = 0; j < SECOND.length; j++) {
                ProductEntity product = new ProductEntity();
                product.setName(FIRST[i] + " " + SECOND[j]);
```

```
                product.setDescription(product.getName() + " " + DESCRIPTION[j]);
                product.setPrice(rnd.nextInt(100));
                product.setId(FIRST.length * i + j + 1);
                products.add(product);
            }
        }
        return products;
    }

    public static List<CommentEntity> generateCommentsForProducts(
            final List<ProductEntity> products) {
        List<CommentEntity> comments = new ArrayList<>();
        Random rnd = new Random();

        for (ProductEntity product : products) {
            int commentsNumber = rnd.nextInt(5) + 1;
            for (int i = 0; i < commentsNumber; i++) {
                CommentEntity comment = new CommentEntity();
                comment.setProductId(product.getId());
                comment.setText(COMMENTS[i] + " for " + product.getName());
                comment.setPostedAt(new Date(System.currentTimeMillis()
                        - TimeUnit.DAYS.toMillis(commentsNumber - i) + TimeUnit.HOURS.
toMillis(i)));
                comments.add(comment);
            }
        }

        return comments;
    }
}
```

12. 在 cuc.room1 包里新建 AppExecutors 类

AppExecutors.java 用于创建几个 Executor，不同的任务使用不同的 Executor 执行。文件内容如下：

```
package cuc.room1;
import android.os.Handler;
import android.os.Looper;
import android.support.annotation.NonNull;
import java.util.concurrent.Executor;
import java.util.concurrent.Executors;

public class AppExecutors {
    private final Executor diskIO;
    private final Executor networkIO;
    private final Executor mainThread;
    public AppExecutors() {
        diskIO = Executors.newSingleThreadExecutor();
```

```
            networkIO = Executors.newFixedThreadPool(3);
            mainThread = new MainThreadExecutor();
        }

        public Executor getDiskIO() {
            return diskIO;
        }
        public Executor getNetworkIO() {
            return networkIO;
        }
        public Executor getMainThread() {
            return mainThread;
        }

        private class MainThreadExecutor implements Executor{
            private Handler handler = new Handler(Looper.getMainLooper());
            @Override
            public void execute(@NonNull Runnable command) {
                handler.post(command);
            }
        }
    }
```

13. 在 build.gradle(module: app)中添加 schema 输出位置,并同步

在文件 build.gradle(module: app)的 defaultConfig 语句的花括号内添加:

```
javaCompileOptions {
    annotationProcessorOptions {
        arguments = ["room.schemaLocation":
                     "$projectDir/schemas".toString()]
    }
}
```

现在 build.gradle(module: app)文件内容如下:

```
apply plugin: 'com.android.application'

android {
    dataBinding.enabled = true
    compileSdkVersion 28
    defaultConfig {
        javaCompileOptions {
            annotationProcessorOptions {
                arguments = ["room.schemaLocation":
                             "$projectDir/schemas".toString()]
            }
        }
        applicationId "cuc.room1"
```

```
        minSdkVersion 19
        targetSdkVersion 28
        versionCode 1
        versionName "1.0"
        testInstrumentationRunner "android.support.test.runner.AndroidJUnitRunner"
    }
    buildTypes {
        release {
            minifyEnabled false
            proguardFiles getDefaultProguardFile('proguard-android.txt'), 'proguard-rules.pro'
        }
    }
    compileOptions {
        sourceCompatibility JavaVersion.VERSION_1_8
        targetCompatibility JavaVersion.VERSION_1_8
    }
}
dependencies {
    implementation fileTree(include: ['*.jar'], dir: 'libs')
    implementation 'com.android.support:appcompat-v7:28.0.0'
    implementation 'com.android.support.constraint:constraint-layout:1.1.3'
    implementation 'android.arch.persistence.room:runtime:1.1.1'
    annotationProcessor 'android.arch.persistence.room:compiler:1.1.1'
    implementation 'android.arch.lifecycle:extensions:1.1.1'
    annotationProcessor 'android.arch.lifecycle:compiler:1.1.1'
    implementation 'com.android.support:recyclerview-v7:28.0.0'
    implementation 'com.android.support:cardview-v7:28.0.0'
    testImplementation 'junit:junit:4.12'
    androidTestImplementation 'com.android.support.test:runner:1.0.2'
    androidTestImplementation 'com.android.support.test.espresso:espresso-core:3.0.2'
}
```

由于修改了 build.gradle 文件内容，这个文件窗口的上面就出现提示："Gradle files have changed since last project sync. A project sync may be necessary for the IDE to work properly. sync now"。单击 sync now 按钮进行同步。

14. 在 cuc.room1.db 包里新建 AppDatabase 类

AppDatabase.java 文件内容如下：

```
package cuc.room1.db;
import android.arch.lifecycle.LiveData;
import android.arch.lifecycle.MutableLiveData;
import android.arch.persistence.db.SupportSQLiteDatabase;
import android.arch.persistence.room.Database;
import android.arch.persistence.room.Room;
import android.arch.persistence.room.RoomDatabase;
import android.arch.persistence.room.TypeConverters;
import android.content.Context;
```

```
import android.support.annotation.NonNull;
import android.support.annotation.VisibleForTesting;
import java.util.List;
import cuc.room1.AppExecutors;
import cuc.room1.db.converter.DateConverter;
import cuc.room1.db.dao.CommentDao;
import cuc.room1.db.dao.ProductDao;
import cuc.room1.db.entity.CommentEntity;
import cuc.room1.db.entity.ProductEntity;

@Database(entities = {ProductEntity.class, CommentEntity.class}, version = 1)
@TypeConverters(DateConverter.class)
public abstract class AppDatabase extends RoomDatabase {
    private static AppDatabase sInstance;

    @VisibleForTesting
    public static final String DATABASE_NAME = "basic-sample-db";
    public abstract ProductDao productDao();
    public abstract CommentDao commentDao();

    private final MutableLiveData<Boolean> mIsDatabaseCreated = new MutableLiveData<>();
    private void updateDatabaseCreated(final Context context) {
        if (context.getDatabasePath(DATABASE_NAME).exists()) {
            setDatabaseCreated();
        }
    }

    private void setDatabaseCreated(){
        mIsDatabaseCreated.postValue(true);
    }
    private static void insertData(final AppDatabase database, final List<ProductEntity>
products,
                                   final List<CommentEntity> comments) {
        database.runInTransaction(() -> {
            database.productDao().insertAll(products);
            database.commentDao().insertAll(comments);
        });
    }

    private static void addDelay() {
        try {
            Thread.sleep(4000);
        } catch (InterruptedException ignored) {
        }
    }

    public LiveData<Boolean> getDatabaseCreated() {
        return mIsDatabaseCreated;
    }
```

```
    private static AppDatabase buildDatabase(final Context appContext,
                                             final AppExecutors executors) {
        return Room.databaseBuilder(appContext, AppDatabase.class, DATABASE_NAME)
                .addCallback(new Callback() {
                    @Override
                    public void onCreate(@NonNull SupportSQLiteDatabase db) {
                        super.onCreate(db);
                        executors.getDiskIO().execute(() -> {
                      // Add a delay to simulate a long-running operation
                            addDelay();
                      // Generate the data for pre-population
                            AppDatabase database = AppDatabase.getInstance(appContext, executors);
                            List<ProductEntity> products = DataGenerator.generateProducts();
                            List<CommentEntity> comments =
                            DataGenerator.generateCommentsForProducts(products);

                            insertData(database, products, comments);
        /* notify that the database was created and it's ready to be used */
                            database.setDatabaseCreated();
                        });
                    }
                })
                .build();
    }
     public static AppDatabase getInstance(final Context context, final AppExecutors
executors) {
        if (sInstance == null) {
            synchronized (AppDatabase.class) {
                if (sInstance == null) {
                    sInstance = buildDatabase(context.getApplicationContext(), executors);
                    sInstance.updateDatabaseCreated(context.getApplicationContext());
                }
            }
        }
        return sInstance;
    }
}
```

15. 在 cuc.room1 包里新建 DataRepository 类

DataRepository.java 用于把各种数据源合并，例如本地数据和网络数据，这个例子中只有本地数据。文件内容如下：

```
package cuc.room1;
import android.arch.lifecycle.LiveData;
import android.arch.lifecycle.MediatorLiveData;
import java.util.List;
import cuc.room1.db.AppDatabase;
import cuc.room1.db.entity.CommentEntity;
```

```
import cuc.room1.db.entity.ProductEntity;

public class DataRepository {
    private static DataRepository sInstance;
    private final AppDatabase appDatabase;
    private MediatorLiveData<List<ProductEntity>> mObservableProducts;

    private DataRepository(final AppDatabase database) {
        appDatabase = database;
        mObservableProducts = new MediatorLiveData<>();

    mObservableProducts.addSource(appDatabase.productDao().loadAllProducts(),
            productEntities -> {
                if (appDatabase.getDatabaseCreated().getValue() != null) {
                    mObservableProducts.postValue(productEntities);
                }
            });
    }

    public static DataRepository getInstance(final AppDatabase database) {
        if (sInstance == null) {
            synchronized (DataRepository.class) {
                if (sInstance == null) {
                    sInstance = new DataRepository(database);
                }
            }
        }
        return sInstance;
    }

    /** Get the list of products from the database and get notified when the data changes. */
    public LiveData<List<ProductEntity>> getProducts() {
        return mObservableProducts;
    }

    public LiveData<ProductEntity> loadProduct(final int productId) {
        return appDatabase.productDao().loadProduct(productId);
    }

    public LiveData<List<CommentEntity>> loadComments(final int productId) {
        return appDatabase.commentDao().loadComments(productId);
    }
}
```

16. 定制 Application，并在清单文件中注册

在 cuc.room1 包里新建 BasicApp 类，BasicApp.java 文件内容如下：

```
package cuc.room1;
import android.app.Application;
import cuc.room1.db.AppDatabase;
```

```
public class BasicApp extends Application {
    private AppExecutors appExecutors;

    @Override
    public void onCreate() {
        super.onCreate();
        appExecutors = new AppExecutors();
    }
    public AppDatabase getDatabase() {
        return AppDatabase.getInstance(this, appExecutors);
    }
    public DataRepository getRepository() {
        return DataRepository.getInstance(getDatabase());
    }
}
```

在清单文件中，给<application>元素添加如下 name 属性：

```
android:name = ".BasicApp"
```

如果清单文件中没有定义<application>元素的名字，系统就使用默认的 Application。

17. 在 cuc.room1.ui 包里，新建 CommentClickCallback 接口

CommentClickCallback.java 文件内容如下：

```
package cuc.room1.ui;
import cuc.room1.db.entity.CommentEntity;
public interface CommentClickCallback {
    void onClick(CommentEntity commentEntity);
}
```

18. 在 cuc.room1.ui 包里，新建 ProductClickCallback 接口

ProductClickCallback.java 文件内容如下：

```
package cuc.room1.ui;
import cuc.room1.db.entity.ProductEntity;
public interface ProductClickCallback {
    void onClick(ProductEntity productEntity);
}
```

19. 在 cuc.room1.ui 包里，新建 dataBinding 的适配器 BindingAdapters

在 product_and_comments.xml 和 products_list.xml 中，visibleGone 属性是自定义的属性取值为数据绑定表达式。另外 TextView 的 text 属性值是数据绑定表达式，表达式类型 java.util.Date，需要转换为 setText 方法的参数类型。BindingAdapters.java 文件内容如下：

```
package cuc.room1.ui;
import android.databinding.BindingAdapter;
import android.databinding.BindingConversion;
```

```
import android.view.View;
import java.text.SimpleDateFormat;
import java.util.Date;

public class BindingAdapters {
    @BindingAdapter("visibleGone")
    public static void showHide(View view, boolean show) {
        view.setVisibility(show ? View.VISIBLE : View.GONE);
    }
    @BindingConversion
    public static String convertDateToString(Date date){
        SimpleDateFormat sdf = new SimpleDateFormat("yyyy-MM-dd HH:mm:ss");
        return sdf.format(date);
    }
}
```

20. 新建布局文件 comment_item.xml

comment_item.xml 内容如下：

```
<?xml version="1.0" encoding="utf-8"?>
<layout xmlns:android="http://schemas.android.com/apk/res/android"
    xmlns:app="http://schemas.android.com/apk/res-auto">
    <data>
        <variable
            name="callback"
            type="cuc.room1.ui.CommentClickCallback"/>
        <variable
            name="commentEntity"
            type="cuc.room1.db.entity.CommentEntity"/>
    </data>
    <android.support.v7.widget.CardView
    android:onClick="@{()->callback.onClick(commentEntity)}"
        android:id="@+id/cardView"
        android:layout_width="match_parent"
        android:layout_height="wrap_content"
        app:contentPadding="5dp"
        app:cardCornerRadius="20dp"
        app:cardElevation="3dp"
        app:cardUseCompatPadding="true">
        <LinearLayout
            android:orientation="vertical"
            android:layout_width="match_parent"
            android:layout_height="wrap_content">
            <TextView
                android:text="@{commentEntity.text}"
                android:layout_width="match_parent"
                android:layout_height="wrap_content" />
            <TextView
```

```
                android:text = "@{commentEntity.postedAt}"
                android:layout_width = "match_parent"
                android:layout_height = "wrap_content" />
        </LinearLayout>
    </android.support.v7.widget.CardView>
</layout>
```

21. 新建布局文件 product_item.xml

在 values/strings.xml 文件中 resource 元素内添加如下字符串定义：

```
<string name = "price">Price: $ %d</string>
```

新建布局文件 product_item.xml，内容如下：

```
<?xml version = "1.0" encoding = "utf - 8"?>
<layout xmlns:android = "http://schemas.android.com/apk/res/android"
    xmlns:app = "http://schemas.android.com/apk/res - auto">
    <data>
        <variable
            name = "callback"
            type = "cuc.room1.ui.ProductClickCallback"/>
        <variable
            name = "productEntity"
            type = "cuc.room1.db.entity.ProductEntity"/>
    </data>
    <android.support.v7.widget.CardView
        android:onClick = "@{() -> callback.onClick(productEntity)}"
        android:id = "@ + id/cardView"
        android:layout_width = "match_parent"
        android:layout_height = "wrap_content"
        app:contentPadding = "5dp"
        app:cardCornerRadius = "20dp"
        app:cardElevation = "3dp"
        app:cardUseCompatPadding = "true">
        <LinearLayout
            android:orientation = "vertical"
            android:layout_width = "match_parent"
            android:layout_height = "wrap_content">
            <TextView
                android:text = "@{productEntity.name}"
                android:layout_width = "match_parent"
                android:layout_height = "wrap_content" />
            <TextView
                android:text = "@{@string/price(productEntity.price)}"
                android:layout_width = "match_parent"
                android:layout_height = "wrap_content" />
            <TextView
                android:text = "@{productEntity.description}"
                android:layout_width = "match_parent"
```

```
                android:layout_height = "wrap_content" />
        </LinearLayout>
    </android.support.v7.widget.CardView>

</layout>
```

22. 在 cuc.room1.viewmodel 包里新建 ProductAndCommentsViewModel

ProductAndCommentsViewModel 用于为 ProductAndCommentsFragment 提供数据。ProductAndCommentsViewModel.java 文件内容如下：

```
package cuc.room1.viewmodel;
import android.app.Application;
import android.arch.lifecycle.AndroidViewModel;
import android.arch.lifecycle.LiveData;
import android.arch.lifecycle.ViewModel;
import android.arch.lifecycle.ViewModelProvider;
import android.databinding.ObservableField;
import android.support.annotation.NonNull;
import java.util.List;
import cuc.room1.BasicApp;
import cuc.room1.DataRepository;
import cuc.room1.db.entity.CommentEntity;
import cuc.room1.db.entity.ProductEntity;

public class ProductAndCommentsViewModel extends AndroidViewModel {
    private final int mProductId;
    private final LiveData<ProductEntity> mObservableProduct;
    private final LiveData<List<CommentEntity>> mObservableComments;
    public ObservableField<ProductEntity> product = new ObservableField<>();

    public ProductAndCommentsViewModel(@NonNull Application application, DataRepository
repository,
                        final int productId) {
        super(application);
        mProductId = productId;
        mObservableComments = repository.loadComments(mProductId);
        mObservableProduct = repository.loadProduct(mProductId);
    }

    /*** Expose the LiveData Comments query so the UI can observe it. */
    public LiveData<List<CommentEntity>> getComments() {
        return mObservableComments;
    }

    public LiveData<ProductEntity> getObservableProduct() {
        return mObservableProduct;
    }
```

```
    public void setProduct(ProductEntity product) {
        this.product.set(product);
    }

    /**
     * A creator is used to inject the product ID into the ViewModel.
  This creator is to showcase how to inject dependencies into ViewModels. It's not actually
necessary in this case, as the product ID can be passed in a public method. */
    public static class Factory extends ViewModelProvider.NewInstanceFactory {
        @NonNull
        private final Application mApplication;
        private final int mProductId;
        private final DataRepository mRepository;

        public Factory(@NonNull Application application, int productId) {
            mApplication = application;
            mProductId = productId;
            mRepository = ((BasicApp) application).getRepository();
        }

        @Override
        public <T extends ViewModel> T create(Class<T> modelClass) {
            //noinspection unchecked
            return (T) new ProductAndCommentsViewModel(mApplication, mRepository, mProductId);
        }
    }
}
```

23. 在 cuc.room1.ui 里新建 CommentAdapter

CommentAdapter 用于给 product_and_comments.xml 文件中的 RecyclerView 提供适配器。

CommentAdapter.java 内容如下：

```
package cuc.room1.ui;
import android.databinding.DataBindingUtil;
import android.support.annotation.Nullable;
import android.support.v7.util.DiffUtil;
import android.support.v7.widget.RecyclerView;
import android.view.LayoutInflater;
import android.view.ViewGroup;
import java.util.List;
import java.util.Objects;
import cuc.room1.R;
import cuc.room1.databinding.CommentItemBindingImpl;
import cuc.room1.db.entity.CommentEntity;

public class CommentAdapter extends RecyclerView.Adapter<CommentAdapter.CommentViewHolder> {
    private List<? extends CommentEntity> mCommentList;
```

```java
    @Nullable
    private final CommentClickCallback mCommentClickCallback;

    public CommentAdapter(@Nullable CommentClickCallback commentClickCallback) {
        mCommentClickCallback = commentClickCallback;
    }

    public void setCommentList(final List<? extends CommentEntity> comments) {
        if (mCommentList == null) {
            mCommentList = comments;
            notifyItemRangeInserted(0, comments.size());
        } else {
            DiffUtil.DiffResult diffResult = DiffUtil.calculateDiff(new DiffUtil.Callback() {
                @Override
                public int getOldListSize() {
                    return mCommentList.size();
                }
                @Override
                public int getNewListSize() {
                    return comments.size();
                }
                @Override
                public boolean areItemsTheSame(int oldItemPosition, int newItemPosition) {
                    CommentEntity old = mCommentList.get(oldItemPosition);
                    CommentEntity comment = comments.get(newItemPosition);
                    return old.getId() == comment.getId();
                }

                @Override
                public boolean areContentsTheSame(int oldItemPosition, int newItemPosition) {
                    CommentEntity old = mCommentList.get(oldItemPosition);
                    CommentEntity comment = comments.get(newItemPosition);
                    return old.getId() == comment.getId()
                            && old.getPostedAt() == comment.getPostedAt()
                            && old.getProductId() == comment.getProductId()
                            && Objects.equals(old.getText(), comment.getText());
                }
            });
            mCommentList = comments;
            diffResult.dispatchUpdatesTo(this);
        }
    }

    @Override
    public CommentViewHolder onCreateViewHolder(ViewGroup parent, int viewType) {
        CommentItemBindingImpl binding = DataBindingUtil
                .inflate(LayoutInflater.from(parent.getContext()), R.layout.comment_item,
                        parent, false);
        binding.setCallback(mCommentClickCallback);
```

```
            return new CommentViewHolder(binding);
        }

        @Override
        public void onBindViewHolder(CommentViewHolder holder, int position) {
            holder.binding.setCommentEntity(mCommentList.get(position));
            holder.binding.executePendingBindings();
        }
        @Override
        public int getItemCount() {
            return mCommentList == null ? 0 : mCommentList.size();
        }

        static class CommentViewHolder extends RecyclerView.ViewHolder {
            final CommentItemBindingImpl binding;

            CommentViewHolder(CommentItemBindingImpl binding) {
                super(binding.getRoot());
                this.binding = binding;
            }
        }
    }
```

24. 新建 product_and_comments.xml

product_and_comments.xml，用于显示产品和所有该产品的评论，文件内容如下：

```
<?xml version = "1.0" encoding = "utf - 8"?>
<layout xmlns:android = "http://schemas.android.com/apk/res/android"
    xmlns:app = "http://schemas.android.com/apk/res - auto">
    <data>
        <variable
            name = "isLoading"
            type = "boolean"/>
        <variable
            name = "viewModel"
            type = "cuc.room1.viewmodel.ProductAndCommentsViewModel"/>
    </data>
    <LinearLayout
        android:orientation = "vertical"
        android:layout_width = "match_parent"
        android:layout_height = "match_parent">
        <include
            layout = "@layout/product_item"
            app:productEntity = "@{viewModel.product}"/>
        <TextView
            app:visibleGone = "@{isLoading}"
            android:layout_width = "match_parent"
            android:layout_height = "wrap_content" />
```

```
        <android.support.v7.widget.RecyclerView
            android:id="@+id/commentsRecyclerView"
            android:layout_width="match_parent"
            android:layout_height="match_parent"
            app:visibleGone="@{!isLoading}">
        </android.support.v7.widget.RecyclerView>
    </LinearLayout>
</layout>
```

25. 在 cuc.room1.ui 里新建 ProductAndCommentsFragment

ProductAndCommentsFragment 的视图由 product_and_comments.xml 提供。ProductAndCommentsFragment.java 内容如下：

```
package cuc.room1.ui;
import android.arch.lifecycle.Observer;
import android.arch.lifecycle.ViewModelProviders;
import android.databinding.DataBindingUtil;
import android.os.Bundle;
import android.support.annotation.Nullable;
import android.support.v4.app.Fragment;
import android.support.v7.widget.LinearLayoutManager;
import android.view.LayoutInflater;
import android.view.View;
import android.view.ViewGroup;
import java.util.List;
import cuc.room1.R;
import cuc.room1.databinding.ProductAndCommentsBinding;
import cuc.room1.db.entity.CommentEntity;
import cuc.room1.db.entity.ProductEntity;
import cuc.room1.viewmodel.ProductAndCommentsViewModel;


public class ProductAndCommentsFragment extends Fragment {
    private static final String KEY_PRODUCT_ID = "product_id";
    private ProductAndCommentsBinding mBinding;
    private CommentAdapter mCommentAdapter;

    @Nullable
    @Override
    public View onCreateView(LayoutInflater inflater, @Nullable ViewGroup container,
                             @Nullable Bundle savedInstanceState) {

        mBinding = DataBindingUtil.inflate(inflater, R.layout.product_and_comments, container, false);
        mCommentAdapter = new CommentAdapter(mCommentClickCallback);
        mBinding.commentsRecyclerView.setLayoutManager(new LinearLayoutManager(getActivity()));
        mBinding.commentsRecyclerView.setAdapter(mCommentAdapter);
        return mBinding.getRoot();
    }
```

```
    private final CommentClickCallback mCommentClickCallback = new CommentClickCallback() {
        @Override
        public void onClick(CommentEntity commentEntity) {
            // no-op
        }
    };

    @Override
    public void onActivityCreated(@Nullable Bundle savedInstanceState) {
        super.onActivityCreated(savedInstanceState);

        ProductAndCommentsViewModel.Factory factory = new ProductAndCommentsViewModel.Factory(
                getActivity().getApplication(), getArguments().getInt(KEY_PRODUCT_ID));
        final ProductAndCommentsViewModel model = ViewModelProviders.of(this, factory)
                .get(ProductAndCommentsViewModel.class);
        mBinding.setViewModel(model);
        subscribeToModel(model);
    }

    private void subscribeToModel(final ProductAndCommentsViewModel model) {

        // Observe product data
        model.getObservableProduct().observe(this, new Observer<ProductEntity>() {
            @Override
            public void onChanged(@Nullable ProductEntity productEntity) {
                model.setProduct(productEntity);
            }
        });

        // Observe comments
        model.getComments().observe(this, new Observer<List<CommentEntity>>() {
            @Override
            public void onChanged(@Nullable List<CommentEntity> commentEntities) {
                if (commentEntities != null) {
                    mBinding.setIsLoading(false);
                    mCommentAdapter.setCommentList(commentEntities);
                } else {
                    mBinding.setIsLoading(true);
                }
            }
        });
    }

    /** Creates product fragment for specific product ID */
    public static ProductAndCommentsFragment forProduct(int productId) {
        ProductAndCommentsFragment fragment = new ProductAndCommentsFragment();
        Bundle args = new Bundle();
        args.putInt(KEY_PRODUCT_ID, productId);
        fragment.setArguments(args);
        return fragment;
    }
}
```

26. 新建布局文件 products_list. xml

products_list. xml 用于显示所有的产品，文件内容如下：

```
<?xml version = "1.0" encoding = "utf - 8"?>
<layout xmlns:android = "http://schemas.android.com/apk/res/android"
    xmlns:app = "http://schemas.android.com/apk/res - auto">
    <data>
        <variable
            name = "isLoading"
            type = "boolean"/>
    </data>
    <LinearLayout
        android:orientation = "vertical"
        android:layout_width = "match_parent"
        android:layout_height = "match_parent">
        <TextView
            android:id = "@ + id/isLoadingTxt"
            android:layout_width = "match_parent"
            android:layout_height = "wrap_content"
            android:text = "Loading Products..."
            app:visibleGone = "@{isLoading}"/>
        <android.support.v7.widget.RecyclerView
            android:id = "@ + id/recyclerView"
            android:layout_width = "match_parent"
            android:layout_height = "match_parent"
            app:visibleGone = "@{!isLoading}">
        </android.support.v7.widget.RecyclerView>
    </LinearLayout>
</layout>
```

27. 在 cuc. room1. ui 里新建 ProductsAdapter

ProductsAdapter 用于 products_list. xml 文件中的 RecyclerView 提供适配器。ProductsAdapter. java 内容如下：

```
package cuc.room1.ui;
import android.databinding.DataBindingUtil;
import android.support.annotation.Nullable;
import android.support.v7.util.DiffUtil;
import android.support.v7.widget.RecyclerView;
import android.view.LayoutInflater;
import android.view.ViewGroup;
import java.util.List;
import java.util.Objects;
import cuc.room1.R;
import cuc.room1.databinding.ProductItemBindingImpl;
import cuc.room1.db.entity.ProductEntity;

public class ProductsAdapter extends RecyclerView.Adapter<ProductsAdapter.ProductViewHolder> {
```

```
List<? extends ProductEntity> mProductList;
@Nullable
private final ProductClickCallback mProductClickCallback;

public ProductsAdapter(@Nullable ProductClickCallback clickCallback) {
    mProductClickCallback = clickCallback;
    setHasStableIds(true);
}

public void setProductList(final List<? extends ProductEntity> productList) {
    if (mProductList == null) {
        mProductList = productList;
        notifyItemRangeInserted(0, productList.size());
    } else {
        DiffUtil.DiffResult result = DiffUtil.calculateDiff(new DiffUtil.Callback() {
            @Override
            public int getOldListSize() {
                return mProductList.size();
            }

            @Override
            public int getNewListSize() {
                return productList.size();
            }

            @Override
            public boolean areItemsTheSame(int oldItemPosition, int newItemPosition) {
                return mProductList.get(oldItemPosition).getId() ==
                        productList.get(newItemPosition).getId();
            }

            @Override
            public boolean areContentsTheSame(int oldItemPosition, int newItemPosition) {
                ProductEntity newProduct = productList.get(newItemPosition);
                ProductEntity oldProduct = mProductList.get(oldItemPosition);
                return newProduct.getId() == oldProduct.getId()
                        && Objects.equals(newProduct.getDescription(), oldProduct.getDescription())
                        && Objects.equals(newProduct.getName(), oldProduct.getName())
                        && newProduct.getPrice() == oldProduct.getPrice();
            }
        });
        mProductList = productList;
        result.dispatchUpdatesTo(this);
    }
}

@Override
public ProductViewHolder onCreateViewHolder(ViewGroup parent, int viewType) {
    ProductItemBindingImpl binding = DataBindingUtil
```

```
                .inflate(LayoutInflater.from(parent.getContext()), R.layout.product_item,
                        parent, false);
        binding.setCallback(mProductClickCallback);
        return new ProductViewHolder(binding);
    }

    @Override
    public void onBindViewHolder(ProductViewHolder holder, int position) {
        holder.binding.setProductEntity(mProductList.get(position));
        holder.binding.executePendingBindings();
    }

    @Override
    public int getItemCount() {
        return mProductList == null ? 0 : mProductList.size();
    }

    @Override
    public long getItemId(int position) {
        return mProductList.get(position).getId();
    }

    static class ProductViewHolder extends RecyclerView.ViewHolder {
        final ProductItemBindingImpl binding;
        public ProductViewHolder(ProductItemBindingImpl binding) {
            super(binding.getRoot());
            this.binding = binding;
        }
    }
}
```

28. 新建 activity_main1.xml

```
<?xml version="1.0" encoding="utf-8"?>
<FrameLayout xmlns:android="http://schemas.android.com/apk/res/android"
    android:id="@+id/fragmentContainer"
    android:layout_width="match_parent" android:layout_height="match_parent">
</FrameLayout>
```

29. 从 cuc.room1 包移动 MainActivity.java 到 cuc.room1.ui 包

在工程结构视图里，右击 MainActivity.java，选择 Refactor|Move，在 To package 右边的下拉列表框中，选择 cuc.room1.ui，单击 Refactor 按钮。

编辑 MainActivity.java，内容如下：

```
package cuc.room1.ui;
import android.os.Bundle;
import android.support.v7.app.AppCompatActivity;
import cuc.room1.R;
```

```
import cuc.room1.db.entity.ProductEntity;

public class MainActivity extends AppCompatActivity {

    @Override
    protected void onCreate(Bundle savedInstanceState) {
        super.onCreate(savedInstanceState);
        setContentView(R.layout.activity_main1);
    }
    /** Shows the product detail fragment */
    public void show(ProductEntity productEntity) {
            ProductAndCommentsFragment fragment = ProductAndCommentsFragment.forProduct
(productEntity.getId());
        getSupportFragmentManager()
                .beginTransaction()
                .addToBackStack("product")
                .replace(R.id.fragmentContainer,
                        fragment, null).commit();
    }
}
```

30. 在 cuc.room1.viewmodel 包里新建 ProductsListViewModel 类

ProductsListViewModel 为 ProductsListFragment 提供数据。

ProductsListViewModel.java 内容如下：

```
package cuc.room1.viewmodel;
import android.app.Application;
import android.arch.lifecycle.AndroidViewModel;
import android.arch.lifecycle.LiveData;
import android.arch.lifecycle.MediatorLiveData;
import java.util.List;
import cuc.room1.BasicApp;
import cuc.room1.DataRepository;
import cuc.room1.db.entity.ProductEntity;

public class ProductsListViewModel extends AndroidViewModel {
    private final DataRepository mRepository;
    private final MediatorLiveData<List<ProductEntity>> mObservableProducts;

    public ProductsListViewModel(Application application) {
        super(application);
        mObservableProducts = new MediatorLiveData<>();
        /* set by default null, until we get data from the database. */
        mObservableProducts.setValue(null);
        mRepository = ((BasicApp) application).getRepository();
        LiveData<List<ProductEntity>> products = mRepository.getProducts();
    /* observe the changes of the products from database and forward them */
```

```
            mObservableProducts.addSource(products, mObservableProducts::setValue);
        }

        /** Expose the LiveData Products query so the UI can observe it. */
        public LiveData<List<ProductEntity>> getProducts() {
            return mObservableProducts;
        }
    }
```

31. 在 cuc.room1.ui 包里新建 ProductsListFragment 类

products_list.xml 为 ProductsListFragment 提供视图。

ProductsListFragment.java 内容如下：

```
package cuc.room1.ui;
import android.arch.lifecycle.Lifecycle;
import android.arch.lifecycle.LiveData;
import android.arch.lifecycle.Observer;
import android.arch.lifecycle.ViewModelProviders;
import android.databinding.DataBindingUtil;
import android.os.Bundle;
import android.support.annotation.Nullable;
import android.support.v4.app.Fragment;
import android.support.v7.widget.DividerItemDecoration;
import android.support.v7.widget.LinearLayoutManager;
import android.view.LayoutInflater;
import android.view.View;
import android.view.ViewGroup;
import java.util.List;
import cuc.room1.R;
import cuc.room1.databinding.ProductsListBindingImpl;
import cuc.room1.db.entity.ProductEntity;
import cuc.room1.viewmodel.ProductsListViewModel;

public class ProductsListFragment extends Fragment {
    public static final String TAG = "ProductsListViewModel";
    private ProductsAdapter mProductsAdapter;
    private ProductsListBindingImpl mBinding;

    @Nullable
    @Override
    public View onCreateView(LayoutInflater inflater, @Nullable ViewGroup container,
                             @Nullable Bundle savedInstanceState) {
        mBinding = DataBindingUtil.inflate(inflater, R.layout.products_list, container,
false);
        mProductsAdapter = new ProductsAdapter(mProductClickCallback);
        mBinding.recyclerView.setLayoutManager(new LinearLayoutManager(getActivity()));
        mBinding.recyclerView.setAdapter(mProductsAdapter);
        mBinding.recyclerView.addItemDecoration(new DividerItemDecoration(getActivity(),
DividerItemDecoration.VERTICAL));
```

```
        return mBinding.getRoot();
    }

    @Override
    public void onActivityCreated(@Nullable Bundle savedInstanceState) {
        super.onActivityCreated(savedInstanceState);
        final ProductsListViewModel viewModel =
                ViewModelProviders.of(this).get(ProductsListViewModel.class);
        subscribeUi(viewModel.getProducts());
    }

    private void subscribeUi(LiveData<List<ProductEntity>> liveData) {
        // Update the list when the data changes
        liveData.observe(this, new Observer<List<ProductEntity>>() {
            @Override
            public void onChanged(@Nullable List<ProductEntity> myProducts) {
                if (myProducts != null) {
                    mBinding.setIsLoading(false);
                    mProductsAdapter.setProductList(myProducts);
                } else {
                    mBinding.setIsLoading(true);
                }
                // espresso does not know how to wait for data binding's loop so we execute changes
                // sync.
                mBinding.executePendingBindings();
            }
        });
    }

    private final ProductClickCallback mProductClickCallback = new ProductClickCallback
() {
        @Override
        public void onClick(ProductEntity productEntity) {
            if (getLifecycle().getCurrentState().isAtLeast(Lifecycle.State.STARTED)) {
                ((MainActivity) getActivity()).show(productEntity);
            }
        }
    };
}
```

32. 编辑 MainActivity.java

在 onCreate()方法里添加如下代码：

```
if (savedInstanceState == null) {
    ProductsListFragment fragment = new ProductsListFragment();
    getSupportFragmentManager().beginTransaction()
            .add(R.id.fragmentContainer, fragment, ProductsListFragment.TAG).commit();
}
```

MainActivity.java 内容如下：

```
package cuc.room1.ui;
import android.os.Bundle;
import android.support.v7.app.AppCompatActivity;
import cuc.room1.R;
import cuc.room1.db.entity.ProductEntity;

public class MainActivity extends AppCompatActivity {

    @Override
    protected void onCreate(Bundle savedInstanceState) {
        super.onCreate(savedInstanceState);
        setContentView(R.layout.activity_main1);
        if (savedInstanceState == null) {
            ProductsListFragment fragment = new ProductsListFragment();
            getSupportFragmentManager().beginTransaction()
                    .add(R.id.fragmentContainer, fragment, ProductsListFragment.TAG).commit();
        }
    }
    /** Shows the product detail fragment */
    public void show(ProductEntity productEntity) {
            ProductAndCommentsFragment fragment = ProductAndCommentsFragment.forProduct
(productEntity.getId());
        getSupportFragmentManager()
                .beginTransaction()
                .addToBackStack("product")
                .replace(R.id.fragmentContainer,
                        fragment, null).commit();
    }
}
```

33. 编译运行

14.10 本章主要参考文献

1. https://developer.android.google.cn/guide/topics/data/data-storage.html#pref
2. https://developer.android.google.cn/reference/android/content/SharedPreferences.html
3. https://developer.android.google.cn/reference/android/content/Context#MODE_WORLD_READABLE
4. https://developer.android.google.cn/reference/android/os/storage/StorageVolume
5. https://developer.android.google.cn/training/articles/scoped-directory-access
6. https://developer.android.google.cn/guide/topics/providers/document-provider
7. https://developer.android.google.cn/reference/android/widget/RadioGroup.html
8. http://www.sqlite.org/datatype3.html
9. https://www.sqlite.org/lang_createtable.html
10. https://www.sqlite.org/lang_select.html
11. https://www.sqlite.org/syntax/result-column.html
12. https://www.sqlite.org/syntax/table-or-subquery.html

13. https://developer.android.google.cn/reference/android/database/sqlite/SQLiteDatabase
14. http://www.w3school.com.cn/sql/sql_join_inner.asp
15. https://developer.android.google.cn/reference/android/database/sqlite/SQLiteDatabase
16. https://developer.android.google.cn/reference/android/database/sqlite/SQLiteOpenHelper
17. https://developer.android.google.cn/topic/libraries/architecture/room
18. https://developer.android.google.cn/training/data-storage/room/index.html
19. https://github.com/googlesamples/android-architecture-components

第 15 章 CHAPTER 15 ContentProvider

抽象的 android. content. ContentProvider 类用于管理结构化数据集的访问。ContentProvider 封装了数据,以一个或多个表(与关系型数据库中的表类似)的形式将数据呈现给外部应用,行表示某种数据类型的实例,行中的每个列表示为实例的一项数据。ContentProvider 在外部表现为一组表,与关系型数据库类似,但这并不是对 ContentProvider 内部实现的要求。其他应用使用 ContentResolver 对象来访问 ContentProvider 中的数据。ContentProvider 与 ContentResolver 共同提供一致的标准数据接口,该接口还可处理跨进程通信并保护数据访问的安全性。当 ContentResolver 对象发出请求时,系统解析给定 URI 的 authority,把请求传递给登记了这个 authority 的 ContentProvider,ContentProvider 能够解释 URI 的其余部分,执行请求的操作并返回结果。android. content. UriMatcher 可以用来帮助分析 URI。应用使用 ContentResolver 对象访问数据,此对象的方法调用 ContentProvider 对象中的相同名字的方法。要访问 ContentProvider,应用通常需要在其清单文件中请求特定权限。

15.1 设计数据的原始存储方式

ContentProvider 以三种方式提供数据:文件数据、“结构化”数据和网络数据。

文件数据是指数据存储在文件中,如照片、音频或视频。Android 提供了各种面向文件的 API,将文件存储在应用的私有空间内,ContentProvider 可以提供文件句柄以响应其他应用发出的文件请求。

“结构化”数据是指存储在数据库、数组或类似结构中的数据,以兼容于行列表的形式存储数据。行代表实体,如人员或库存项目;列代表实体的某项数据,如人员的姓名或商品的价格。此类数据通常存储在 SQLite 数据库中,但也可以使用任何类型的持久存储。数据表应始终设置一个主键,ContentProvider 将其作为与每行对应的唯一数字值加以维护。可以使用此值将该行链接到其他表中的相关行(将其用作“外键”)。尽管此列可以使用任何名称,但使用 BaseColumns._ID 是最佳选择,因为将 ContentProvider 查询的结果与 ListView 绑定的条件是,检索到的其中一个列的名称必须是_ID。ListView 使用的 CursorAdapter 要求 Cursor 中的其中一个列必须是_ID。应用从支持 ListView 的 Cursor 中选取对应行,获取该行的_ID 值,将其追加到内容 URI 结尾,然后向 ContentProvider 发送访问请求。然后,ContentProvider 便可对用户选取的特定行执行查询或修改。如果想要提供位图图像或其他非常庞大的文件导向型数据，数据应该存储在一个文件中，然后间接提供这些数据，而不是直接把这些数据存储在表中。

使用二进制大型对象(BLOB)数据类型存储大小或结构会发生变化的数据。例如，可以使用 BLOB 列来存储协议缓冲区或 JSON 结构。也可以使用 BLOB 来实现独立于 schema 协议的表。在这类表中，需要以 BLOB 形式定义一个主键列、一个 MIME 类型列以及一个或多个通用列。这些 BLOB 列中数据的含义通过 MIME 类型列中的值指示。这样做，就可以在同一个表中存储不同类型的行。举例来说，联系人 ContentProvider 的"数据"表 ContactsContract. Data 便是一个独立于 Schema 协议的表。

可以使用 java. net 和 android. net 中的类访问基于网络的数据，也可以将基于网络的数据与本地数据存储(如数据库)同步，然后以表或文件的形式提供数据。

15.2 设计 Content URI

Content URI 包括整个 ContentProvider 的符号名称(授权)、一个指向表或者文件的名称(路径)、可选 ID 指向表中的单个行。内容 URI 的路径部分来选择要访问的表，ContentProvider 通常会为其公开的每个表显示一条路径。内容 URI 的形式为 content://authority/path/id。content 是 URI 的架构。authority 是一个字符串，用于标识整个 ContentProvider，整个 ContentProvider 的所有的内容 URI 都从 authority 开始。每个 ContentProvider 的授权必须保证唯一性。为了避免与其他 ContentProvider 发生冲突，建议使用互联网网域所有权(反向)作为 ContentProvider 授权的基础，由于此建议也适用于 Android 软件包名称，因此可以将 ContentProvider 授权定义为包含该 ContentProvider 的软件包名称的扩展名。例如，如果 Android 软件包名称为 com. example. < appname >，则授权为 com. example. < appname >. < providername >。path 包含一个或者多个段，由向前的斜线(/)分隔，用来标识 ContentProvider 数据的子集。大多数 ContentProvider 使用 path 标识不同的表。路径并不限定于单个段，也无须为每一级路径都创建一个表。开发者通常在授权后追加指向单个表的路径来创建内容 URI。例如，如果有两个表：table1 和 table2，内容 URI 可以表示为 com. example. < appname >. < providername >/table1 和 com. example. < appname >. < providername >/table2。id 是一个单一的数字，用于标识前面路径指定的数据子集中的一行。大多数 ContentProvider 能够识别带有 ID 的 URI，并给予特殊处理。按照惯例，ContentProvider 通过接受末尾具有 ID 值的内容 URI 来提供对表中单个行的访问，将 ID 值与表的 _ID 列进行匹配，并对匹配的行执行请求的访问。ContentProvider 的每一个数据访问方法都将内容 URI 作为参数；当调用客户端方法来访问 ContentProvider 中的表时，该表的内容 URI 将是其参数之一。ContentResolver 对象会分析出 URI 的授权，将该授权与已知 ContentProvider 的授权进行比较，然后，ContentResolver 可以将查询参数分派给指定的 ContentProvider。

Uri 和 Uri. Builder 类提供了一些根据字符串构建 URI 对象的方法。ContentUris 提供了一些可以将 ID 值轻松追加到 URI 末尾的方法。

例如，Android 平台的内置的 ContentProvider 用户字典为例，"字词"表的完整 URI 是：content://user_dictionary/words。其中 content 表示架构，user_dictionary 是 Content-

Provider 的授权，words 字符串是表的路径。许多 ContentProvider 都允许将 ID 值追加到表的 URI 末尾来访问表中的单个行。例如，要从用户字典中检索_ID 为 4 的行，则可使用此内容 URI：

```
Uri uri = ContentUris.withAppendedId(UserDictionary.Words.CONTENT_URI,4);
```

UriMatcher 对象的 addURI(String authority, String path, int code) 可以把内容 URI 的模式映射到整型值。在 switch 语句中使用这些整型值，为匹配特定模式的一个或多个内容 URI 选择所需操作。内容 URI 模式使用通配符匹配内容 URI，* 匹配包含任意长度的任何有效字符的字符串，# 匹配包含任意长度的数字字符的字符串。假设一个具有授权 com.example.app.provider 的提供程序能提供以下指向表的内容 URI：

content://com.example.app.provider/table1：一个名为 table1 的表

content://com.example.app.provider/table2/dataset1：一个名为 dataset1 的表

content://com.example.app.provider/table2/dataset2：一个名为 dataset2 的表

content://com.example.app.provider/table3：一个名为 table3 的表

ContentProvider 也能识别追加了行 ID 的内容 URI，例如，content://com.example.app.provider/table3/1 与 table3 中 ID 为 1 的行对应。

可以使用以下内容 URI 模式：

content://com.example.app.provider/*：匹配 ContentProvider 中的任何内容 URI。

content://com.example.app.provider/table2/*：匹配 table2 中的表 dataset1 和表 dataset2 的内容 URI，但不匹配 table1 或 table3 的内容 URI。

content://com.example.app.provider/table3/#：匹配 table3 中单个行的内容 URI，如 content://com.example.app.provider/table3/6 与 table3 中 ID 为 6 的行对应。

一个实例：

```
public class ExampleProvider extends ContentProvider {...
    // Creates a UriMatcher object.
    private static final UriMatcher sUriMatcher = new UriMatcher(UriMatcher.NO_MATCH);

    static {
/* 设置访问 table3 对应整数值 1 */
        sUriMatcher.addURI("com.example.app.provider", "table3", 1);
/* 设置访问 table3 的单个行对应整数值 2 */
        sUriMatcher.addURI("com.example.app.provider", "table3/#", 2);
    }
    // Implements ContentProvider.query()
    public Cursor query(Uri uri,String[] projection,String selection,
        String[] selectionArgs,String sortOrder) {...
        switch (sUriMatcher.match(uri)) {
            case 1: // 如果输入 URI 是访问整个 table3
                if (TextUtils.isEmpty(sortOrder)) sortOrder = "_ID ASC";
                break;
            // If the incoming URI was for a single row
            case 2: // 如果输入 URI 是访问 table3 中的单一行
/* 访问单个行,URI 的路径的最后一段对应_ID,把这个值添加到 WHERE 子句 */
```

```
            selection = selection + "_ID = " + uri.getLastPathSegment();
            break;
        default: // If URI 不能被识别,添加一些错误处理
...
    }
    // 实际的查询功能
}
```

15.3 实现 ContentProvider 类

创建 ContentProvider 主要包含三个步骤：

(1) 设计数据的原始存储方式。

(2) 定义 ContentProvider 的授权字符串、其内容 URI 以及列名称。此外，还要定义想要访问数据的应用必须具备的权限。通常在一个单独的协定类中将所有这些值定义为常量；以后可以将此类公开给其他开发者。协定类是一种 public final 类，其中包含对 URI、列名称、MIME 类型以及其他与提供程序有关的元数据的常量定义。该类可确保即使 URI、列名称等数据的实际值发生变化，也可以正确访问提供程序，从而在提供程序与其他应用之间建立协定。协定类对开发者也有帮助，因为其常量通常采用助记名称，因此可以降低开发者为列名称或 URI 使用错误值的可能性。

(3) 定义 ContentProvider 类及其所需方法的具体实现。抽象类 ContentProvider 定义了六个抽象方法：

```
public abstract Cursor query(Uri uri, String[ ] projection, String selection, String[ ]
selectionArgs, String sortOrder);
public abstract Uri insert(Uri uri, ContentValues values) ;
public abstract int update(Uri uri, ContentValues values, String selection, String[ ]
selectionArgs);
public abstract int delete(Uri uri, String selection, String[ ] selectionArgs);
public abstract String getType(Uri uri);
public abstract boolean onCreate();
public String[ ]getStreamTypes(Uri uri, String mimeTypeFilter);
```

具体子类必须实现这些方法。所有这些方法(onCreate 除外)都由访问 ContentProvider 的客户端应用调用，这些方法的签名与同名的 ContentResolver 方法相同。

query 方法用于检索数据，将数据作为 Cursor 对象返回，参数包括：要查询的表、要返回的行和列，以及结果的排序方式。如果失败，就会抛出 Exception。如果使用 SQLite 数据库作为数据存储，就只需要返回由 SQLiteDatabase 类的 query()方法返回的 Cursor。如果查询结果没有匹配行，应该返回一个 Cursor 实例，其 getCount()方法返回 0。只有当查询过程中出现内部错误时，才应该返回 null。为了获得更好的查询性能，调用者应该遵循以下原则：①提供显式的 projection 参数，阻止从存储器读出无用的数据；②使用问号参数标记符例如'phone＝?'，代替 selection 参数中的显式的值，以便仅仅那些值不同的查询用作 cache 用途时被识别为相同的。

insert 方法，向内容 Uri 指定的表中插入一行，如果 ContentValues 中未包含列名称，可能在 ContentProvider 中或数据库架构中提供其默认值，返回新插入行的内容 URI，可以调用 ContentUris 类的静态方法 withAppendedId(Uri contentUri, long id)向表的内容 URI 追加新行的_ID(或其他主键)值作为新插入行的 URI。

update 方法用于更新 ContentProvider 中的现有行，返回已更新的行数。

delete 方法用于删除一行或者多行，返回已删除的行数。delete()方法并不需要从数据存储中实际物理地删除行。如果 android. content. AbstractThreadedSyncAdapter 与 ContentProvider 一起使用，应该考虑为已删除的行添加删除标志，而不是将行整个移除。AbstractThreadedSyncAdapter 可以检查是否存在已删除的行，并将它们从服务器中移除，然后再将它们从 ContentProvider 中删除。

getType 用于返回给定 Uri 的 MIME 类型，形式为 type/subtype。Uri 参数可以是模式，而不是特定 Uri，这时应该返回与匹配该模式的内容 Uri 关联的数据类型。对于文本、HTML 或 JPEG 等常见数据类型，getType()应该为该数据返回标准 MIME 类型。例如 text/html 具有 text 类型和 html 子类型，如果 provider 返回此类型，这表示面向此 URI 的查询将返回带有 HTML 标记的文本。对于指向一个或多个表数据行的内容 URI，getType()应该以 Android 供应商特有(vendor-specific)的 MIME 类型：vnd. android. cursor. dir 或者 vnd. android. cursor. item。vnd. android. cursor. dir 适用于 URI 模式为多个行，vnd. android. cursor. item 适用于 URI 模式为单个行。MIME 子类型的形式为 vnd. < name >. < type >，对于特定的 ContentProvider，name 值应具有全局唯一性，type 值应在对应的 URI 模式中具有唯一性。建议选择公司的名称或 Android 应用软件包名称的某个部分作为< name >，选择与 URI 关联的表的标识字符串作为< type >。例如，如果 ContentProvider 的授权是 com. example. app. provider，并且它公开了一个名为 table1 的表，则 table1 中多个行的 MIME 类型是：

```
vnd.android.cursor.dir/vnd.com.example.provider.table1
```

对于 table1 的单个行，MIME 类型是：

```
vnd.android.cursor.item/vnd.com.example.provider.table1
```

当 ContentProvider 提供文件时，实现 getStreamTypes()方法，返回给定 Uri 的数据流的 MIME 类型。通过 mimeTypeFilter 参数过滤出需要的 MIME 类型，返回过滤后的给定 URI 位置的数据流的类型。例如，假设 ContentProvider 以. jpg、. png 和. gif 格式提供照片，如果 mimeTypeFilter 为 image/ *，则返回数组：{"image/jpeg","image/png","image/gif"}。如果 mimeTypeFilter 为 * \/jpeg，则返回：{"image/jpeg"}。如果 ContentProvider 未提供过滤器字符串中请求的任何 MIME 类型，则 getStreamTypes()返回 null。

当 ContentResolver 对象尝试访问 ContentProvider 时，系统才会创建 ContentProvider。Android 系统会在创建 ContentProvider 后立即调用 onCreate 方法初始化 ContentProvider，在此方法中只能执行快速的初始化任务，将数据库创建和数据加载推迟到 ContentProvider 实际收到数据请求时进行。如果在 onCreate()中执行长时间的任务，使得 ContentProvider

的启动速度变慢。例如，如果使用 SQLite 数据库，可以在 ContentProvider 的 onCreate()中创建 SQLiteOpenHelper 对象，然后在首次打开数据库时创建 SQL 表。为简化这一过程，在首次调用 getWritableDatabase()时，它会自动调用 SQLiteOpenHelper.onCreate()方法。

在实现这些方法时应考虑以下事项：所有这些方法必须实现，但实现代码可以不执行任何其他操作，仅仅返回要求的数据类型，例如，可能想防止其他应用向某些表插入数据，要实现此目的，可以忽略 insert()调用并返回 0。所有这些方法(onCreate 除外)都可由多个线程同时调用，因此它们必须是线程安全的。避免在 onCreate()中执行长时间操作，将初始化任务推迟到实际需要时进行。

以下代码段展示了 ContentProvider.onCreate()与 SQLiteOpenHelper.onCreate()之间的交互。

```
public class ExampleProvider extends ContentProvider{
    private MainDatabaseHelper mOpenHelper;
    private static final String DBNAME = "mydb";
    private SQLiteDatabase db;
    public boolean onCreate() {
        mOpenHelper = new MainDatabaseHelper(getContext(), DBNAME, null,1);
        return true;
    }
    public Cursor insert(Uri uri, ContentValues values) {
/* Gets a writeable database. This will trigger its creation if it doesn't already exist. */
        db = mOpenHelper.getWritableDatabase();
        // Insert code here to determine which table to open, and so forth
    }

    private static final String SQL_CREATE_MAIN = "CREATE TABLE " +
    "main " + "(" + " _ID INTEGER PRIMARY KEY," + "WORD TEXT,"
    "FREQUENCY INTEGER," + "LOCALE TEXT )";
    protected static final class MainDatabaseHelper extends SQLiteOpenHelper {
    /* Do not do database creation and upgrade here. */
    MainDatabaseHelper(Context context) {
        super(context, DBNAME, null, 1);
    }
    public void onCreate(SQLiteDatabase db) {
        db.execSQL(SQL_CREATE_MAIN);              // Creates the main table
    }
}
```

15.4 在清单文件中注册 ContentProvider

与 Activity 和 Service 组件类似，ContentProvider 也必须在清单文件中注册。在清单文件中注册 ContentProvider 的语法如下：

```
<provider android:authorities = "list"
      android:enabled = ["true" | "false"]
      android:exported = ["true" | "false"]
```

```
        android:grantUriPermissions = ["true" | "false"]
        android:icon = "drawable resource"
        android:initOrder = "integer"
        android:label = "string resource"
        android:multiprocess = ["true" | "false"]
        android:name = "string"
        android:permission = "string"
        android:process = "string"
        android:readPermission = "string"
        android:syncable = ["true" | "false"]
        android:writePermission = "string" >
    . . .
</provider>
```

< provider >包含在< application >中。< provider >还可以包含 meta-data、grant-uri-permission 和 path-permission 子元素。path-permission 元素的语法如下：

```
<path - permission android:path = "string"
          android:pathPrefix = "string"
          android:pathPattern = "string"
          android:permission = "string"
          android:readPermission = "string"
          android:writePermission = "string" />
```

< grant-uri-permission >元素的语法如下：

```
<grant - uri - permission android:path = "string"
          android:pathPattern = "string"
          android:pathPrefix = "string" />
```

provider 元素的 name 属性设置为 ContentProvider 类名，authorities 属性设置 ContentProvider 的内容 URI 中的 authority，grantUriPermssions 属性设置整个 ContentProvider 的临时访问权限，默认值为 false。permission 属性设置整个 ContentProvider 的读取/写入权限，readPermission 和 writePermission 属性分别设置整个 ContentProvider 的读取权限和写入权限，enabled 属性设置是否允许系统实例化该 ContentProvider，exported 属性设置是否允许其他应用使用 ContentProvider，initOrder 属性设置该 ContentProvider 的启动顺序，当多个 ContentProvider 之间有依赖时，为每一个 ContentProvider 设置 initOrder 属性，以便它们按照依赖性所需要的顺序创建，属性值是简单的整数，取值大的先初始化。multiProcess 属性设置是否允许系统在每个客户端进程中创建 ContentProvider 的实例，默认值是 false。如果 ContentProvider 实例可以运行在多个线程中，设置为 true。process 属性设置进程的名称，ContentProvider 将在这个进程中运行。syncable 属性设置 ContentProvider 的数据是否与服务器上的数据同步，如果同步设置为 true。

15.4.1 实现 ContentProvider 的权限

默认情况下,保存到外部存储的数据文件是公用并可全局读取的数据文件,无法使用 ContentProvider 来限制对外部存储内文件的访问,因为其他应用可以使用其他 API 调用来对它们执行读取和写入操作。

默认情况下,存储在设备内部存储上的数据文件是应用和 ContentProvider 的私有数据文件;应用创建的 SQLiteDatabase 数据库是应用和 ContentProvider 的私有数据库。如果将内部文件或数据库用作 ContentProvider 的数据仓库,并向其授予 world-readable 或者 world-writeable 访问权限,则在清单文件中为 ContentProvider 设置的权限不会保护这些内部文件或者数据库。内部存储中文件和数据库的默认访问权限是私有的,如果用作 ContentProvider 的数据仓库,就不应改更改此权限。在默认情况下,ContentProvider 并没有设置权限。即使底层数据为私有数据,仍然可以通过清单文件中的一个或多个 permission 元素,设置适用于整个 ContentProvider、某些表、甚至某些记录的权限。

从适用于整个 ContentProvider 的权限开始,然后逐渐细化,更细化的权限优先于作用域较大的权限:①provider 元素的 permission 属性能同时控制对整个 ContentProvider 的读写权限,而 provider 元素的 readPermission 和 writePermission 属性分别控制整个 ContentProvider 的读权限和写权限,它们优先于 permission 属性指定的权限。②provider 元素的 path-permission 子元素设置的读写、读或写权限,优先于 ContentProvider 级别的权限,分开的读权限或者写权限优先于读写权限。

要使权限对 ContentProvider 具有唯一性,权限应该使用 Java 风格作用域。例如,将读取权限命名为 com. example. app. provider. permission. READ_PROVIDER。

15.4.2 临时权限

如果< provider >元素的 grantUriPermissions 属性设置为 true,则系统会向整个 ContentProvider 授予临时权限,该权限将覆盖 ContentProvider 级别或路径级别权限设置的任何其他权限;如果 grantUriPermissions 属性设置为 false,则必须向< provider >元素添加< grant-uri-permission >子元素。每个 grant-uri-permission 元素设置授予一个或多个内容 URI 临时权限。

如果 URI 关联了临时权限时,可以调用 Context. revokeUriPermission()移除对某个 URI 权限。要向应用授予临时访问权限,Intent 必须包含 FLAG_GRANT_READ_URI_PERMISSION 和/或 FLAG_GRANT_WRITE_URI_PERMISSION 标志,它们通过 setFlags()方法进行设置。临时访问权限可以减少应用需要在其清单文件中请求的权限数量。

假设需要实现电子邮件 ContentProvider 的权限,如果希望外部图像查看器应用能够显示邮件 ContentProvider 中的照片附件,为了在不请求权限的情况下为图像查看器提供必要的访问权限,可以为照片的内容 URI 设置临时权限。用户想要显示照片时,向图像查看器发送一个 Intent,Intent 包含照片的内容 URI 以及权限标志。图像查看器可随后查询邮件 ContentProvider 中的照片附件,即使查看器不具备对邮件 ContentProvider 的正常读取权限,也不受影响。

15.5 FileProvider

androidx. core. content. FileProvider 和 android. support. v4. content. FileProvider 是 android. content. ContentProvider 的直接子类，通过创建 content://Uri 来替代 file:///Uri。Content URI 允许使用临时访问权限进行读写访问。当创建一个包含 content URI 的 Intent 时，必须调用 Intent. setFlags()添加权限访问客户端应用。如果 Intent 是发送到 Activity，只要接收 Intent 的 Activity 是活跃的，它就一致拥有这些权限。如果 Intent 是发送到 Service，只要接收 Intent 的 Service 还在运行，它就一致拥有这些权限。

对于面向 Android 7.0 及更新版本的应用，Android 框架执行的 StrictMode API 政策禁止在应用外部公开 filel://URI，否则出现 FileUriExposedException 异常。如果使用 file:///Uri，必须控制该文件的文件系统权限，任何应用都能获得提供的权限。使用 content://Uri 增加了系统的安全权限。FileProvider 的使用，通常包含以下四个步骤：

第一步，在清单文件中注册 FileProvider。ContentProvider 的子类 FileProvider，也同样需要使用元素 provider 在 Manifest 文件中添加注册信息，并按照要求设置相关属性值。

```
<manifest ...>
    <application>
        ...
        <provider
            android:name = "android.support.v4.content.FileProvider"
            android:authorities = "com.mydomain.fileprovider"
            android:exported = "false"
            android:grantUriPermissions = "true">
                <meta - data
                    android:name = "android.support.FILE_PROVIDER_PATHS"
                    android:resource = "@xml/file_paths" />
     </provider>
    </application>
</manifest>
```

既然 FileProvider 默认的功能包含文件的 content://Uri 的生成，就没有必要在代码中定义子类。为了添加一个 FileProvider 组件，在清单文件中添加一个< provider >元素，它的 name 属性应该设置为 ContentProvider 的完全限定类名：android. support. v4. content. FileProvider 或者 androidx. core. content. FileProvider。根据应用控制的域设置 authorities 属性，例如应用控制的域为 mydomain. com，就设置 android:authorities="com. mydomain. fileprovider"。FileProvider 不需要公开，所以设置 android:exported="false"。为了授予文件的临时访问权限，设置 android:grantUriPermissions="true"。meta-data 配置的是可以访问的文件的路径配置信息，它的 name 属性必须设置为 android. support. FILE_PROVIDER_PATHS，resource 属性设置为配置路径信息的配置文件的资源描述符。路径配置文件通常存储在 res\xml 子文件夹下，格式如下：

```
<paths xmlns:android = "http://schemas.android.com/apk/res/android">
      <files - path name = "my_images" path = "images/"/>
      <cache - path name = "name" path = "path" />
<external - path name = "name" path = "path" />
<external - files - path name = "name" path = "path" />
<external - cache - path name = "name" path = "path" />
<external - media - path name = "name" path = "path" />
</paths>
```

<paths>元素必须包含一个或者多个子元素，这些子元素用于指定共享文件的目录路径，共享文件的目录路径由元素名称和它的 path 属性值确定，name 属性设置该目录对应的 URI 的一段，元素标签可以是 files-path、cache-path、external-path、external-files-path、external-cache-path 或者 external-media-path。元素<files-path>的目录就是 Context. getFilesDir()的返回值，即应用的内部存储区域的 files 目录。<cache-path>的目录就是 Context. getCacheDir()的返回值，即应用的内部存储区域的 cache 目录。<external-path>的目录就是 Environment. getExternalStorageDirectory()的返回值，即外部存储区域的根目录。<external-files-path>的目录就是 Context＃getExternalFilesDir(String)和 Context. getExternalFilesDir(null)的返回值，即应用的外部存储区域的根目录。<external-cache-path>的目录就是 Context. getExternalCacheDir()的返回值，即应用的外部 cache 目录。<external-media-path>的目录就是 Context. getExternalMediaDirs()返回的数组中的第一个，即应用的外部 media 目录，这个目录仅仅在 API21＋的设备上可得到。这些元素的 path 属性设置了共享的子目录的实际名字，注意，不能通过文件名共享单个文件，也不能通过通配符设置文件的子集。name 属性设置了 path 子目录在 Content URI 的对应的一段路径，这是为了安全。

第二步，创建路径配置文件。在 res/xml 文件夹下，创建清单文件里指定的配置文件，例如 file_paths. xml，文件内容的一个示例如下：

```
<paths xmlns:android = "http://schemas.android.com/apk/res/android">
      <files - path name = "my_images" path = "images/"/>
      <files - path name = "my_docs" path = "docs/"/>
</paths>
```

上面的配置文件中，paths 元素包含两个子元素。第一个元素设置应用的内部存储区域的 files 目录的 images 子目录为共享目录，第二个元素设置应用的内部存储区域的 files 目录的 docs 子目录为共享目录，对应的 content://Uri 中的路径分别为 my_images 和 my_docs。

第三步，生成 Content URI，授予 Content URI 访问权限。使用 FileProvider 类提供的公有静态方法 getUriForFile(@NonNull Context context，@NonNull String authority，@NonNull File file)生成 Content URI。生成 Content URI 对象后，需要对其授权访问权限。授权方式有两种：

第一种方式，使用 Context 对象的 grantUriPermission(package，Uri，mode_flags)方法向其他应用授权访问 URI 对象，三个参数分别表示授权访问 URI 对象的其他应用包名，授权访问的 Uri 和授权类型。其中，授权类型为 Intent 类提供的常量：

```
FLAG_GRANT_READ_URI_PERMISSION、FLAG_GRANT_WRITE_URI_PERMISSION
```

或者二者同时授权。这种形式的授权方式，权限有效期截至发生设备重启或者手动调用 revokeUriPermission()方法撤销授权时。

第二种方式，配合 Intent 使用。Intent 对象先调用 setData(Uri data) 方法添加 Content URI，再调用 setFlags(@Flags int flags)或者 addFlags(@Flags int flags)方法设置读写权限，权限的取值同上。这种形式的授权方式，权限有效期截至接收 Activity 所处的任务栈销毁，并且一旦授权给某一个组件后，该应用的其他组件拥有相同的访问权限。

采用 Intent 授予权限的实例如下：

```
File imagePath = new File(Context.getFilesDir(), "images");
File newFile = new File(imagePath, "default_image.jpg");
Uri contentUri = FileProvider.getUriForFile(getContext(), "com.mydomain.fileprovider",
newFile);
Intent intent = new Intent();
intent.setData(contentUri);
intent.setFlags(Intent.FLAG_GRANT_READ_URI_PERMISSION | Intent.FLAG_GRANT_WRITE_URI_
PERMISSION);
setResult(Activity.RESULT_OK, intent);
...
```

上面代码中的 imagePath 就是配置文件中的第一个共享路径，getUriForFile 方法的第二个参数就是清单文件中 FileProvider 的 authorities 属性的值，getUriForFile 方法返回的 content URI 为 content://com.mydomain.fileprovider/my_images/default_image.jpg.

第四步，提供 Content URI 给其他应用。服务器端给其他应用提供文件 content URI 有很多方式。一种方式为，其他应用(客户端应用)调用 startActivityForResult(Intent intent, int requestCode)启动应用中的一个 Activity，请求访问文件。Intent 包含 Intent.ACTION_PICK 或者 Intent.ACTION_GET_CONTENT 动作，并且设置客户端能处理的 MIME 类型。作为响应，应用可以直接通过 setResult(int resultCode, Intent intent)返回 content URI，也可以提供一个用户接口，允许用户选择文件，一旦用户选择文件后，应用马上通过 setResult(int resultCode, Intent intent)返回 content URI，其中 Intent 参数包含 content URI。服务端通过 setResult(int resultCode, Intent intent)设置的 Intent 在客户端应用的 onActivityResult(int, int, android.content.Intent))中返回。一旦客户端应用获得了文件的 content URI，就能调用 ContentResolver 对象的 openFileDescriptor(@NonNull Uri uri, @NonNull String mode)方法返回文件的 ParcelFileDescriptor，从 ParcelFileDescriptor 可以获取普通的 FileDescriptor，使用该 FileDescriptor 可以直接读写文件，也可以用它创建 FileOutputStream 或者 FileInputStream，以便使用文件流更加强大的功能。文件在这个过程中是安全的，因为客户端应用收到的 content URI 不包含文件路径，客户端的应用无法发现和获取服务端应用的其他文件。客户端只能访问获取的文件，而这个访问只能由服务端应用授权。由于这个权限是是临时的，因此一旦客户端的任务栈结束后，这个文件不能再被服务端应用之外的应用访问。实例如下：

```
import android.content.Intent;
import android.net.Uri;
import android.os.Bundle;
import android.os.ParcelFileDescriptor;
import android.support.annotation.Nullable;
import android.support.v7.app.AppCompatActivity;
import java.io.FileDescriptor;
import java.io.FileNotFoundException;
import java.io.FileOutputStream;

public class MainActivity extends AppCompatActivity {
    private static final int MY_REQUEST_CODE = 0;
    @Override
    protected void onCreate(Bundle savedInstanceState) {
        super.onCreate(savedInstanceState);
        setContentView(R.layout.main1);
        Intent mRequestFileIntent = new Intent(Intent.ACTION_PICK);
        mRequestFileIntent.setType("image/jpg");
        startActivityForResult(mRequestFileIntent, MY_REQUEST_CODE);
    }
    @Override
    protected void onActivityResult(int requestCode, int resultCode, @Nullable Intent
returnIntent) {
        super.onActivityResult(requestCode, resultCode, returnIntent);
        if ((resultCode == RESULT_OK)&&(requestCode == MY_REQUEST_CODE) ){
            Uri returnUri = returnIntent.getData();
            try {
              ParcelFileDescriptor pfd = getContentResolver().openFileDescriptor(returnUri, "rw");
                FileDescriptor fd = pfd.getFileDescriptor();
                FileOutputStream fos = new FileOutputStream(fd);
...
            } catch (FileNotFoundException e) {
                e.printStackTrace();
            }
        }
    }
}
```

如果需要一次传递多个 URI 对象，可以多次调用 ClipData 对象的 addItem(Item item) 方法、从而添加多个 URI，再调用 Intent 对象的 setClipData(@Nullable ClipData clip) 方法，并且 Intent 的 setFlags 或者 addFlags 方法设置的临时权限适用于所有 Content URIs。

15.6 ContentProvider 实验

1. 新建工程 ContentProvider1

具体步骤：运行 Android Studio，单击 Start a new Android Studio project，在 Create

New Project 页中，Application name 为 ContentProvider1，Company domain 为 cuc，Project location 为 C:\android\myproject\ContentProvider1，单击 Next 按钮，选中 Phone and Tablet，minimum sdk 设置为 API 19：Android 4.4(minSdkVersion)，单击 Next 按钮，选择 Empty Activity，再单击 Next 按钮，选中 Generate Layout File 选项，选中 Backwards Compatibility(AppCompat)选项，单击 Finish 按钮。

2. 新建一个 Java 类 Student

具体步骤：在工程结构视图中 main 源集 Java 文件夹里，右击 cuc.contentprovider1 包，依次选择 New|Java Class，Name 输入框中输入 Student，依次单击 OK 按钮。先在 Student 内部添加三个成员变量：name、id 和 city，光标移到 Student 类内部，依次单击 Android Studio 的菜单 Code|Generate|Getter and Setter，选中所有的成员变量，依次单击 OK 按钮，以便生成 Getter 和 Setter 方法。依次单击 Android Studio 的菜单 Code|Generate|toString，选中所有的成员变量，单击 OK 按钮，以便生成 toString 方法。依次单击 Android Studio 的菜单 Code|Generate|Construct，选中需要初始化的成员变量，单击 OK 按钮，以便生成构造方法。

```
package cuc.contentprovider1;
public class Student {
    private String name;
    private int id;
    private String city;
    public Student(String name, int id, String city) {
        this.name = name;
        this.id = id;
        this.city = city;
    }
    public String getName() {
        return name;
    }
    public void setName(String name) {
        this.name = name;
    }
    public int getId() {
        return id;
    }
    public void setId(int id) {
        this.id = id;
    }
    public String getCity() {
        return city;
    }
    public void setCity(String city) {
        this.city = city;
    }
    @Override
    public String toString() {
        return "Student{" +
```

```
            "name = '" + name + '\'' +
            ", id = " + id +
            ", city = '" + city + '\'' +
            '}';
    }
}
```

3. 添加 DataBinding 支持，并同步

在文件 build. gradle(module: app)的 android 语句的花括号内添加：

```
dataBinding.enabled = true
```

由于修改了 build. gradle 文件内容，这个文件窗口的上面就出现提示："Gradle files have changed since last project sync. A project sync may be necessary for the IDE to work properly. sync now"。单击 sync now 按钮进行同步。

4. 新建布局文件 activity_main1. xml

具体步骤：：在工程结构视图中，右击 layout，依次选择 New |Layout resource file，在 File name 输入框输入 activity_main1，在 Root element 输入框输入 layout，Source set 为 main，单击 OK 按钮。layout/activity_main1. xml 内容如下：

```
<?xml version = "1.0" encoding = "utf - 8"?>
<layout xmlns:android = "http://schemas.android.com/apk/res/android">
    <data>
        <variable
            name = "student"
            type = "cuc.contentprovider1.Student"/>
    </data>
    <TableLayout
        android:stretchColumns = "0,1"
        android:layout_width = "match_parent"
        android:layout_height = "match_parent">
        <TableRow>
            <TextView android:text = "please input name:"/>
            <EditText android:id = "@ + id/nameEdt" android:inputType = "text"/>
        </TableRow>
        <TableRow>
           <TextView android:text = "please input id:"/>
            <EditText android:id = "@ + id/idEdt" android:inputType = "numberDecimal"/>
        </TableRow>
        <TableRow>
            <TextView android:text = "please input city:"/>
            <EditText android:id = "@ + id/cityEdt" android:inputType = "text"/>
        </TableRow>
        <TableRow>
            <Button android:textAllCaps = "false" android:text = "updateById" android:id =
"@ + id/updateByIdBtn"/>
            <Button android:textAllCaps = "false" android:text = "deleteById" android:id =
"@ + id/deleteByIdBtn"/>
```

```
        </TableRow>
        <Button android:text="insert" android:id="@+id/insertBtn"/>
        <Button android:textAllCaps="false" android:text="queryAll" android:id="@+
id/queryAllBtn" />
        <Button android:textAllCaps="false" android:id="@+id/queryByIdBtn" android:
text="queryById"/>
    </TableLayout>
</layout>
```

该文件采用表格布局(TableLayout 作为根元素),每个 TableRow 的元素占一行,最后两个 Button: insertBtn 和 queryBtn 都是 TableLayout 的直接子元素,因而都单独占一行。

5. 在 cuc.contentprovider1 包里新建一个 Java 类

具体步骤:在工程结构视图中 main 源集 Java 文件夹里,右击 cuc.contentprovider1 包,选择 New|Java Class,在 Name 输入框中输入 MyContentProviderMeta,单击 OK 按钮。

类 MyContentProviderMeta 主要定义了一些常量:Contentprovider 相关的 authority,数据库的名字、表的名字和 Uri、表中的字段名等信息。MyContentProviderMeta.java 内容如下:

```
package cuc.contentprovider1;
import android.net.Uri;
public class MyContentProviderMeta {
    public static final String DATABASE_NAME = "myfirst.db";
    public static final String AUTHORITY =
            "cuc.contentprovider1.MyContentProvider";
    public static final class TableSudentMeta{
        public static final String TABLE_NAME = "student";
        public static final Uri CONTENT_URI = Uri.parse("content://"
                + AUTHORITY + "/" + TABLE_NAME);
        public static final String CONTENT_TYPE_DIR =
                "vnd.android.cursor.dir/" + TABLE_NAME;
        public static final String CONTENT_TYPE_ITEM =
                "vnd.android.cursor.item/" + TABLE_NAME;
        public static final String NAME = "name";
        public static final String ID = "id";
        public static final String CITY = "city";
        public static final String ORDERBY = "id desc";
    }
}
```

6. 新建一个 Java 类 MySQLiteOpenHelper

具体步骤:在工程结构视图中 main 源集 Java 文件夹里,右击 cuc.contentprovider1 包,选择 New|Java Class,Name 输入框中输入 MySQLiteOpenHelper,Superclass 输入框中输入 ndroid.database.sqlite.SQLiteOpenHelper,单击 OK 按钮。

在 MySQLiteOpenHelper 的编辑窗口中,把光标移到类名所在的行,按 Alt+Enter 两个键,单击 Implement methods,选择 onCreate 和 onUpgrade 两个方法,单击 OK 按钮。在

MySQLiteOpenHelper 的编辑窗口中，把光标移到类 MySQLiteOpenHelper 的里面，单击 Android Studio 的菜单 Code|Generate|Constructor，在弹出的窗口中，选中要实现的构造函数，单击 OK 按钮。编辑 MySQLiteOpenHelper.java，内容如下：

```
package cuc.contentprovider1;
import android.content.Context;
import android.database.sqlite.SQLiteDatabase;
import android.database.sqlite.SQLiteOpenHelper;
import android.util.Log;

public class MySQLiteOpenHelper extends SQLiteOpenHelper {
    public MySQLiteOpenHelper(Context context, String name, SQLiteDatabase.CursorFactory
factory, int version) {
        super(context, name, factory, version);
    }
    public MySQLiteOpenHelper(Context context, String name, SQLiteDatabase.CursorFactory
factory){
        this(context, name, factory, 1);
    }
    public MySQLiteOpenHelper(Context context, String name){
        this(context, name, null);
    }
    private static final String TAG = "MySQLiteOpenHelper";
    @Override
    public void onCreate(SQLiteDatabase db) {
        Log.i(TAG, "onCreate: ");
        db.execSQL("create table student (name varchar(20)," +
                "id int,city varchar(20))");
    }
    @Override
    public void onUpgrade(SQLiteDatabase sqLiteDatabase, int i, int i1) {
        Log.i(TAG, "onUpgrade: ");
    }
}
```

7. 新建 ContentProvider

这里有两种方法：一种是新建普通 Java 类，一种是新建 ContentProvider。新建 Java 类的步骤如下：在工程结构视图中 main 源集 java 文件夹里，右击 cuc.contentprovider1 包，选择 New|Java Class，Name 输入框中输入 MyContentProvider，Superclass 输入框中输入 android.content.ContentProvider，单击 OK 按钮。这种方法还需要手动在 AndroidManifest.xml 文件中的< application >添加如下代码，完成 ContentProvider 的注册。

```
<provider
    android:authorities="cuc.contentprovider1.MyContentProvider"
    android:name=".MyContentProvider"/>
```

其中 android: authorities 设置为这个 ContentProvider 的完全限定的类名。

新建 ContentProvider 的步骤如下：在工程结构视图中 main 源集 Java 文件夹里，右击 cuc.contentprovider1 包，依次选择 New | Other | Content Provider，Class Name 输入框中输入 MyContentProvider，URI Authorities 输入框中输入 cuc.contentprovider1.MyContentProvider，然后单击 Finish 按钮。系统自动在 AndroidManifest.xml 中添加如下的注册代码，就不用手动在 AndroidManifest.xml 中注册该 ContentProvider 了。自动生成的注册 provider 的代码如下：

```
<provider
    android:name = ".MyContentProvider"
    android:authorities = "cuc.contentprovider1.MyContentProvider"
    android:enabled = "true"
    android:exported = "true"></provider>
```

MyContentProvider.java 内容如下：

```
package cuc.contentprovider1;
import android.content.ContentProvider;
import android.content.ContentUris;
import android.content.ContentValues;
import android.content.UriMatcher;
import android.database.Cursor;
import android.database.SQLException;
import android.database.sqlite.SQLiteDatabase;
import android.database.sqlite.SQLiteQueryBuilder;
import android.net.Uri;
import android.text.TextUtils;
import android.util.Log;
import java.util.HashMap;

public class MyContentProvider extends ContentProvider {
   private static final String TAG  =  "MyContentProvider";
   private MySQLiteOpenHelper helper;
   private static final UriMatcher uriMatcher = new UriMatcher(UriMatcher.NO_MATCH);
   private static final int STUDENT_DIR = 1;
   private static final int STUDENT_ITEM = 2;
   static {
       uriMatcher.addURI(MyContentProviderMeta.AUTHORITY,"/student",STUDENT_DIR);
       uriMatcher.addURI(MyContentProviderMeta.AUTHORITY,"/student/#",STUDENT_ITEM);
   }
   public MyContentProvider() {
   }

   @Override
   public int delete(Uri uri, String selection, String[] selectionArgs) {
       SQLiteDatabase db = helper.getWritableDatabase();
       int count = db.delete(MyContentProviderMeta.TableSudentMeta.TABLE_NAME,
               selection,selectionArgs);
       return count;
```

```
    }

    @Override
    public String getType(Uri uri) {
        switch (uriMatcher.match(uri)){
            case STUDENT_DIR:
                return MyContentProviderMeta.TableSudentMeta.CONTENT_TYPE_DIR;
            case STUDENT_ITEM:
                return MyContentProviderMeta.TableSudentMeta.CONTENT_TYPE_ITEM;
            default:
                throw new IllegalArgumentException();
        }
    }
    @Override
    public Uri insert(Uri uri, ContentValues values) {
        SQLiteDatabase db = helper.getWritableDatabase();
        long rowid = db.insert(MyContentProviderMeta.TableSudentMeta.TABLE_NAME, null,
values);
        if (rowid > 0){
            Uri insertedUri = ContentUris.withAppendedId(
                    MyContentProviderMeta.TableSudentMeta.CONTENT_URI,
                    rowid);
            getContext().getContentResolver().notifyChange(insertedUri,null);
            return insertedUri;
        }else{
            throw new SQLException();
        }
    }

    @Override
    public boolean onCreate() {
        Log.i(TAG, "onCreate: ");
        helper = new MySQLiteOpenHelper(getContext(),
                MyContentProviderMeta.DATABASE_NAME);
        return true;
    }
    private static final HashMap<String,String> studentProjectionMap
            = new HashMap<>();
    static {
//        studentProjectionMap.put("id","id");
//        studentProjectionMap.put("city","city");
//        studentProjectionMap.put("name","name");
        studentProjectionMap.put(MyContentProviderMeta.TableSudentMeta.NAME,
                MyContentProviderMeta.TableSudentMeta.NAME);
        studentProjectionMap.put(MyContentProviderMeta.TableSudentMeta.ID,
                MyContentProviderMeta.TableSudentMeta.ID);
        studentProjectionMap.put(MyContentProviderMeta.TableSudentMeta.CITY,
                MyContentProviderMeta.TableSudentMeta.CITY);
    }
```

```
    @Override
    public Cursor query(Uri uri, String[] projection, String selection,
                     String[] selectionArgs, String sortOrder) {
        SQLiteDatabase db = helper.getReadableDatabase();
        SQLiteQueryBuilder qb = new SQLiteQueryBuilder();
        qb.setProjectionMap(studentProjectionMap);
        qb.setTables(MyContentProviderMeta.TableSudentMeta.TABLE_NAME);
        switch (uriMatcher.match(uri)){
            case STUDENT_DIR:
                break;
            case STUDENT_ITEM:
                break;
        }
        String orderBy;
        if (TextUtils.isEmpty(sortOrder)){
            orderBy = MyContentProviderMeta.TableSudentMeta.ORDERBY;
       }else{
           orderBy = sortOrder;
       }
       Cursor cursor = qb.query(db,projection,selection,selectionArgs,null,null,orderBy);
       cursor.setNotificationUri(getContext().getContentResolver(),uri);
       return cursor;
    }

    @Override
    public int update(Uri uri, ContentValues values, String selection,
                 String[] selectionArgs) {
       SQLiteDatabase db = helper.getWritableDatabase();
       int count = db.update(MyContentProviderMeta.TableSudentMeta.TABLE_NAME,
             values,selection,selectionArgs);
       return count;
    }
}
```

8. 编辑 MainActivity.java

MainActivity.java 内容如下：

```
package cuc.contentprovider1;
import android.content.ContentValues;
import android.database.Cursor;
import android.databinding.DataBindingUtil;
import android.os.Bundle;
import android.support.v7.app.AppCompatActivity;
import android.util.Log;
import android.view.View;
import android.widget.Toast;
import cuc.contentprovider1.databinding.ActivityMain1Binding;
```

```
public class MainActivity extends AppCompatActivity implements View.OnClickListener {
    private static final String TAG = "MainActivity";
   private ActivityMain1Binding binding;
    @Override
    protected void onCreate(Bundle savedInstanceState) {
        super.onCreate(savedInstanceState);
        binding = DataBindingUtil.setContentView(this,R.layout.activity_main1);
        binding.insertBtn.setOnClickListener(this);
        binding.deleteByIdBtn.setOnClickListener(this);
        binding.updateByIdBtn.setOnClickListener(this);
        binding.queryByIdBtn.setOnClickListener(this);
        binding.queryAllBtn.setOnClickListener(this);
    }
    @Override
    public void onClick(View view) {
        Cursor cursor;
        ContentValues values;
        Student student;
        switch (view.getId()){
            case R.id.insertBtn:
                values = new ContentValues();
                values.put(MyContentProviderMeta.TableSudentMeta.NAME,
                        binding.nameEdt.getText().toString());
                values.put(MyContentProviderMeta.TableSudentMeta.CITY,
                        binding.cityEdt.getText().toString());
                values.put(MyContentProviderMeta.TableSudentMeta.ID,
                        Integer.parseInt(binding.idEdt.getText().toString()));
                getContentResolver().insert(
                        MyContentProviderMeta.TableSudentMeta.CONTENT_URI,
                        values);
                Toast.makeText(MainActivity.this, "insert finished", Toast.LENGTH_SHORT).show();
                break;
            case R.id.deleteByIdBtn:
                    cursor = getContentResolver().query(
                        MyContentProviderMeta.TableSudentMeta.CONTENT_URI,
                         new String[]{"id"},"id == ?",new String[]{binding.idEdt.getText().
toString()},null);
                if (cursor.getCount() == 0){
                    Toast.makeText(MainActivity.this, "no records to deleteById", Toast.LENGTH_
SHORT).show();
                    return;
                }
                getContentResolver().delete(MyContentProviderMeta.TableSudentMeta.CONTENT_URI,
                        "id == ?",new String[]{binding.idEdt.getText().toString()});
                Toast.makeText(MainActivity.this, "deleteById finished", Toast.LENGTH_SHORT).
show();
                break;
            case R.id.updateByIdBtn:
```

```
                cursor = getContentResolver().query(
                        MyContentProviderMeta.TableSudentMeta.CONTENT_URI,
                         new String[]{"id"},"id == ?",new String[]{binding.idEdt.getText().
toString()},null

                );
                if (cursor.getCount() == 0){
                    Toast.makeText(MainActivity.this, "no records to updateById", Toast.LENGTH_
SHORT).show();
                    return;
                }
                values = new ContentValues();
                values.put(MyContentProviderMeta.TableSudentMeta.NAME,
                        binding.nameEdt.getText().toString());
                values.put(MyContentProviderMeta.TableSudentMeta.CITY,
                        binding.cityEdt.getText().toString());
                values.put(MyContentProviderMeta.TableSudentMeta.ID,
                        Integer.parseInt(binding.idEdt.getText().toString()));
                getContentResolver().update(MyContentProviderMeta.TableSudentMeta.CONTENT_URI,
                        values,"id == ?",new String[]{binding.idEdt.getText().toString()});
                Toast.makeText(MainActivity.this, "updateById finished", Toast.LENGTH_SHORT).
show();
                break;
            case R.id.queryByIdBtn:
                cursor = getContentResolver().query(
                        MyContentProviderMeta.TableSudentMeta.CONTENT_URI,
                        new String[]{"name","id","city"},"id == ?",new String[]{binding.idEdt.
getText().toString()},null

                );
                if (cursor.getCount() == 0){
                    Log.i(TAG, "queryById: no records");
                    return;
                }
                while(cursor.moveToNext()){
                    String name = cursor.getString(cursor.getColumnIndex("name"));
                    String city = cursor.getString(cursor.getColumnIndex("city"));
                    int id = cursor.getInt(cursor.getColumnIndex("id"));
                    student = new Student(name,id,city);
                    Log.i(TAG, "queryById: " + student);
                }
                break;
            case R.id.queryAllBtn:
                cursor = getContentResolver().query(
                        MyContentProviderMeta.TableSudentMeta.CONTENT_URI,
                        new String[]{"name","id","city"},null,null,"id asc"
                );
                if (cursor.getCount() == 0){
```

```
                Log.i(TAG, "queryAll: no records");
                return;
            }
            while(cursor.moveToNext()){
                String name = cursor.getString(cursor.getColumnIndex("name"));
                String city = cursor.getString(cursor.getColumnIndex("city"));
                int id = cursor.getInt(cursor.getColumnIndex("id"));
                student = new Student(name, id, city);
                Log.i(TAG, "queryAll: " + student);
            }
            break;
        }
    }
}
```

9. 编译运行

查询的数据显示在日志中,可以考虑使用 RecyclerView 显示在手机上。

15.7 本章主要参考文献

1. https://developer.android.google.cn/reference/android/content/ContentProvider.html
2. https://developer.android.google.cn/reference/android/content/ContentResolver.html
3. https://developer.android.google.cn/reference/android/content/UriMatcher
4. https://developer.android.google.cn/guide/topics/providers/content-providers.html
5. https://developer.android.google.cn/guide/topics/manifest/provider-element.html
6. https://developer.android.google.cn/reference/android/content/ContentUris.html
7. https://developer.android.google.cn/reference/android/support/v4/content/FileProvider
8. https://developer.android.google.cn/training/secure-file-sharing/request-file#java

第16章 访问互联网

CHAPTER 16

HTTP是基于客户端/服务端(C/S)的架构模型,通过一个可靠的链接来交换信息,是一个无状态的请求/响应协议。一个HTTP客户端是一个应用程序(Web浏览器或其他任何客户端),通过连接到服务器达到向服务器发送一个或多个HTTP的请求的目的。一个HTTP服务器同样也是一个应用程序(如Apache Web服务器),通过接收客户端的请求并向客户端发送HTTP响应数据。HTTP使用统一资源标识符(Uniform Resource Identifier, URI)来传输数据和建立连接。一旦建立连接后,数据消息就通过类似Internet邮件所使用的格式RFC5322和多用途Internet邮件扩展(MIME)RFC2045来传送。

HTTP是无连接的,每次连接只处理一个请求,服务器处理完客户的请求,并收到客户的应答后,就断开连接。HTTP是无状态的协议,对于事务处理没有记忆能力,如果后续处理需要前面的信息,则它必须重传。HTTP是媒体独立的,只要客户端和服务器知道如何处理数据内容,任何类型的数据都可以通过HTTP发送。HTTP 1.0定义了三种请求方法:GET、POST和HEAD方法,HTTP 1.1新增了五种请求方法:OPTIONS、PUT、DELETE、TRACE和CONNECT方法。表16-1列出了HTTP的请求方法。

表16-1 HTTP的请求方法

方法	描述
GET	请求指定的页面信息,并返回实体主体
HEAD	类似于GET请求,只不过返回的响应中没有具体的内容,用于获取报头
POST	向指定资源提交数据进行处理请求(例如提交表单或者上传文件)。数据被包含在请求体中,POST请求可能会导致新的资源建立和/或已有资源的修改
PUT	从客户端向服务器传送的数据取代指定的文档内容
DELETE	请求服务器删除指定的页面
CONNECT	HTTP/1.1协议中预留给能够将连接改为管道方式的代理服务器
OPTIONS	允许客户端查看服务器的性能
TRACE	回显服务器收到的请求,主要用于测试或诊断

GET请求可被缓存,保留在浏览器历史记录中,可被收藏为书签。GET方法把数据添加到URL,URL的长度是受限制的,因此GET请求有长度限制;GET请求应当用于取回数据;编码类型为application/x-www-form-urlencoded。GET请求的URL格式类似于:

```
URL ? name1 = value1&name2 = value2&..&nameN = valueN
```

GET 请求中，查询字符串(名称/值对)附在 URL 本身的后面，复制在 HTTP 请求的请求行中。查询字符串和 URL 之间用？隔开，查询字符串的形式为"名称＝值"(name＝value)对，相邻的"名称＝值"对之间用 & 隔开。

POST 请求不会被缓存，不会保留在浏览器历史记录中，不能被收藏为书签，编码类型为 application/x-www-form-urlencoded 或者 multipart/form-data。在 POST 请求中，查询字符串(名称/值对)是 http 请求的主体中发送的，通常，当填写一个在线表单并提交它时，这些填入的数据将以 POST 请求的方式发送给服务器。

客户端发送一个 http 请求到服务器的请求消息包括以下格式：请求行(request line)、请求头部(header)、空行和请求数据四个部分组成，图 16-1 给出了请求报文的一般格式。

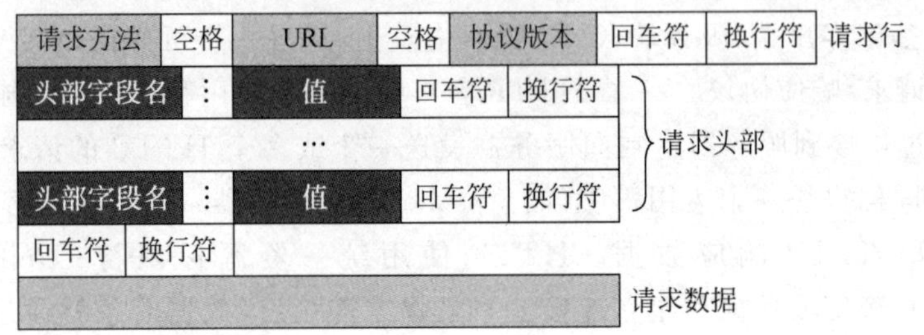

图 16-1 Http 请求报文的一般格式

一个 HTTP GET 请求的实例：

```
GET /books/?name = Professional % 20Ajax HTTP/1.1
Host: www.baidu.com
User - Agent: Mozilla/5.0 (Windows; U; Windows NT 5.1; en - US; rv:1.7.6)
Gecko/20050225 Firefox/1.0.1
Connection: Keep - Alive
```

一个 HTTP POST 请求的实例：

```
POST / HTTP/1.1
Host: www.baidu.com
User - Agent: Mozilla/5.0 (Windows; U; Windows NT 5.1; en - US; rv:1.7.6)
Gecko/20050225 Firefox/1.0.1
Content - Type: application/x - www - form - urlencoded
Content - Length: 40
Connection: Keep - Alive

name = Professional % 20Ajax&publisher = Wiley
```

第一行(请求行)包含 HTTP 请求方法、相对于主机的路径、HTTP 版本。第二行开始到空行是请求的头部，在最后一个头部之后必须有一个空行，即使不存在请求主体，这个空行也是必需的。

为了将文本 Professional Ajax 作为 URL 的参数，需要编码处理其内容，将空格替换成%20，称为 URL 编码(URL encoding)，常用于 HTTP 的许多地方(JavaScript 提供了内建的函数来处理 URL 编码和解码)。

对比上面的 GET 请求和 POST 请求的实例，可以发现两者的一些区别。首先，请求行开始处的 GET 改为了 POST，以表示不同的请求方式。Host 和 User-Agent 两者都有。POST 请求的后面有两个新行：Content-Type 和 Content-Length，空行后面的请求主体以简单的“名称＝值”对的形式给出的，其中 name 是 Professional Ajax，publisher 是 Wiley。如有必要，客户程序还可以选择发送其他的请求头，大多数请求头并不是必需的，但对于 POST 请求来说 Content-Length 必须出现。一些最常见的 HTTP 请求头部，如表 16-2 所示。

表 16-2 HTTP 请求头部

HTTP 请求头	描 述
Accept	指定客户端希望接收的内容类型，*/* 表示任何类型，type/* 表示该类型下的所有子类型
Accept-Charset	客户端可以接收的字符编码集
Accept-Encoding	客户端能够解码的压缩方法，例如 compress，gzip，deflate。如果 Accept-Encoding 的值是空，那么只有 identity 是会被接受的类型；如果 request 中没有 Accept-Encoding 头，那么服务器会假设所有的 Encoding 都是可以被接受的。Servlet 能够向支持 gzip 的浏览器返回经 gzip 编码的 HTML 页面
Accept-Language	客户端所希望的语言种类，当服务器能够提供多种语言版本时要用到。语言跟字符集的区别：中文是语言，中文有多种字符集，比如 big5，gb2312，gbk，等等
Accept-Ranges	可以请求网页实体的一个或者多个子范围字段
Authorization	授权信息，当客户端接收到来自 Web 服务器的 WWW-Authenticate 响应头部信息时，客户端使用 Authorization 头部来回答自己的身份验证信息给 Web 服务器
Cache-Control	指定请求和响应遵循的缓存机制
Connection	表示是否需要持久连接。close 告诉 Web 服务器或者代理服务器，在完成本次请求的响应后，断开连接，不要等待本次连接的后续请求了；keep-alive＝xx 告诉 Web 服务器或者代理服务器，在完成本次请求的响应后，保持连接，等待本次连接的后续请求。 如果 Servlet 看到这里的值为 Keep-Alive，或者看到请求使用的是 HTTP 1.1（HTTP 1.1 默认进行持久连接），它就可以利用持久连接的优点，当页面包含多个元素时（例如 Applet，图片），可以显著地减少下载所需要的时间。要实现这一点，Servlet 需要在应答中发送一个 Content-Length 头，最简单的实现方法是：先把内容写入 ByteArrayOutputStream，然后在正式写出内容之前计算它的大小
Cookie	http 请求发送时，会把保存在该请求域名下的所有 Cookie 值一起发送给 Web 服务器
Content-Length	请求主体的内容长度（字节数）
Content-Type	在本次请求中，请求主体的内容（客户端发送的实体数据）是如何编码的，浏览器始终以 application/x-www-form-urlencoded 的格式编码来传送数据
Date	请求发送的日期和时间
From	发出请求的用户的 Email
Host	请求的服务器的域名和端口号
If-Modified-Since	如果请求的部分在指定时间之后被修改则请求成功，未被修改则返回 304 代码
If-Unmodified-Since	只在实体在指定时间之后未被修改才请求成功
Range	只请求实体的一部分，指定范围
Referer	先前网页的地址，当前请求网页紧随其后
User-Agent	当前客户端信息，该信息由使用的浏览器来定义，在每个请求中将自动发送

HTTP 响应也由四个部分组成，分别是：状态行、消息报头、空行和响应正文。图 16-2 是一个 HTTP 响应的实例。HTTP 响应的头信息如表 16-3 所示。

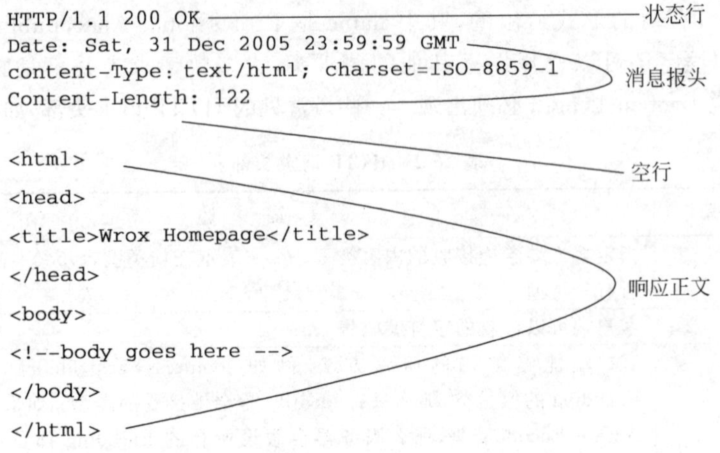

图 16-2　HTTP 响应报文的一般格式

表 16-3　HTTP 响应头信息

HTTP 响应头	说　明
Accept-Ranges	Web 服务器表明自己是否接受获取其某个实体的一部分(比如文件的一部分)的请求。bytes：表示接受，none：表示不接受
Age	从原始服务器到代理缓存形成的估算时间(单位秒，非负)
Allow	服务器支持哪些请求方法(如 GET、POST 等)
Cache-Control	服务器应遵循的缓存机制。public(Cached 内容可以回应任何用户)、private(缓存内容只能回应先前请求该内容的那个用户)、no-cache(可以缓存，但是只有在跟 Web 服务器验证了其有效后，才能返回给客户端)、max-age＝xxx(本响应包含的对象的过期时间)、no-store(不允许缓存)
Connection	是否需要持久连接。close(连接已经关闭，表明当前正在使用的 TCP 链接在请求处理完毕后会被断掉，以后 client 再进行新的请求时就必须创建新的 TCP 链接)、keep-alive(连接保持着，在等待本次连接的后续请求)。在 HTTP1.1 中，client 和 server 默认对方支持长链接的，如果 client 使用 HTTP1.1 协议，但又不希望使用长链接，则需要在 header 中指明 connection 的值为 close；如果 server 方也不想支持长链接，则在 response 中也需要明确说明 connection 的值为 close
Content-Encoding	Web 服务器表明自己使用了什么压缩方法(gzip，deflate)压缩响应中的对象。Servlet 应该通过查看 Accept-Encoding 请求头检查浏览器是否支持 gzip，为支持 gzip 的浏览器返回经 gzip 压缩的 HTML 页面，为其他浏览器返回普通页面
Content-Language	Web 服务器告诉浏览器自己响应的对象的语言
Content-Length	Web 服务器告诉浏览器自己响应的对象的长度。只有当浏览器使用持久 HTTP 连接时才需要这个数据。如果想要利用持久连接的优势，可以把输出文档写入 ByteArrayOutputStream，完成后查看其大小，然后把该值放入 Content-Length 头，最后通过 byteArrayStream.writeTo response.getOutputStream()发送内容
Content-Location	请求资源可替代的备用的另一地址
Content-MD5	返回资源的 MD5 校验值

续表

http 响应头	说　明
Content-Range	Web 服务器表明该响应包含的部分对象在整个对象中的字节位置
Content-Type	Web 服务器告诉浏览器自己响应的对象的类型。Servlet 默认为 text/plain，通常需要显式地指定为 text/html
Date	原始服务器消息发出的 GMT 时间，可以使用 setDateHeader 来设置这个头以避免转换时间格式的麻烦
ETag	请求变量的实体标签的当前值
Expired	Web 服务器表明该实体过期的日期和时间，对于过期了的对象，只有在跟 Web 服务器验证了其有效性后，才能用来响应客户请求
Last-Modified	请求资源的最后修改时间，比如文件的最后修改时间，动态页面的最后产生时间，等等
Location	Web 服务器告诉浏览器，试图访问的对象已经被移到别的位置了，到该头部指定的位置去取
Proxy-Authenticate	代理服务器响应浏览器，要求其提供代理身份验证信息
Server	Web 服务器表明自己是什么软件及版本等信息
Set-Cookie	设置 HTTP Cookie
WWW-Authenticate	表明客户端请求实体应该使用的授权方案

16.1 httpURLConnection

httpClient 类在 Android API 级别 22 中已经被废弃，Android 6.0(API 级别 23)已经移除了对 Apache HTTP 客户端的支持。java. net. httpURLConnection 是 java. net. URLConnection 的直接子类，比 httpClient 效率更高，因为它可以通过透明压缩和响应缓存减少网络使用，并可最大限度降低耗电量。javax. net. ssl. httpsURLConnection 是 java. net. httpURLConnection 的直接子类。

httpURLConneciton 和 httpsURLConnection 都是抽象类，自身不能被实例化，只能通过 URL. openConnection()方法创建 httpURLConnection 实例。任何网络连接都需要经过 socket 才能连接，httpURLConnection 并不是底层的连接，而是在底层连接上的一个请求。虽然底层的网络连接可以被多个 httpURLConnection 实例共享，但每一个 httpURLConnection 实例只能发送一个请求。请求结束之后，应该调用 httpURLConnection 实例的 InputStream 或 OutputStream 的 close()方法以释放请求的网络资源，不过这种方式对于持久化连接没用。对于持久化连接，必须使用 disconnect()方法关闭底层连接的 socket。

httpURLConnection 类的使用步骤如下：

第一步，创建连接，只能调用 URL. openConnection()并且将结果强制转成 httpURLConnection 类，从而获得一个新的 httpURLConnection 对象。

```
URL url = new URL("http://localhost:8080/TestHttpURLConnectionPro/index.jsp");
httpURLConnection httpURLConnection = (HttpURLConnection) url.openConnection();
```

第二步，设置连接参数，Request 的头部将会包含一些元数据例如凭证，首选内容类型和会话 Cookie。

```
httpURLConnection.setRequestMethod( "POST");                    // 提交模式
httpURLConnection.setRequestProperty( "Content-Type", "application/json;charset=UTF-8"
);                                                              //设置请求属性
httpURLConnection.setRequestProperty( "User-Agent", "Autoyol_gpsCenter" );
httpURLConnection.setConnectTimeout(100000);                    //连接超时 单位 ms
httpURLConnection.setReadTimeout(100000);                       //读取超时 单位 ms
httpURLConnection.setDoOutput( true);                           //是否输出参数
httpURLConnection.setDoInput( true);                            //是否读取参数
```

第三步，如果 HTTP 请求包含正文，实例必须被配置有 setDoOutput(true)，通过写入 getOutputStream()返回的流来传输数据。这一步是可选的，get 请求没有正文。

```
OutputStream outStream = httpUrlConnection.getOutputStream();
outStream.write(bytes);
outStream. flush();
outStream.close();
```

第四步，读取响应。响应标头通常包括元数据，例如响应主体(正文)的内容类型和长度，修改日期和会话 Cookie。响应主体可以从 getInputStream()返回的流中读取。如果响应没有主体，那么该方法将返回一个空的流。

```
InputStream in = conn.getInputStream();
 byte[] buffer = new byte[521];
 ByteArrayOutputStream baos = new ByteArrayOutputStream();
 for ( int len = 0; (len = in.read(buffer)) > 0;) {
     baos.write(buffer, 0, len);
 }
 String returnValue = new String(baos.toByteArray(), "utf-8" );
 reg= JSON. parseObject(returnValue, ReturnMessage.class );
 baos.flush();
 baos.close();
 in.close();
```

第五步，断开链接。一旦响应主体被读取，应该调用 disconnect()函数关闭 httpURLConnection。断开连接将会释放连接持有的资源，以便它们可以被关闭或重复使用。

http 请求是由 http 请求头和 http 请求主体两部分组成的。connect()方法会根据 httpURLConnection 对象的配置值生成 http 头部信息，因此必须先配置好 httpURLConnection，再调用 connect 方法。在 http 头后面紧跟着的是 http 请求的主体(正文)，http 请求主体的内容是通过 outputStream 流写入的，outputStream 不是一个网络流，往 outputStream 写入的东西不会立即发送到网络，而是存在于内存缓冲区中，待 outputStream 流关闭时，才生成 http 请求主体，至此，http 请求已经全部准备就绪。无论是 POST 请求还是 GET 请求，直到 HttpURLConnection 的 getInputStream()方法被调用的时候，http 请求才正式发送到服务器，然后返回一个输入流，用于读取服务器的返回信息。由于 http 请求(包括 http 请求头和请求主体)在 getInputStream 的时候已经发送出去了，因此在 getInputStream()调用之后对

connection 对象进行设置(对 http 头的信息进行修改)或者写入 outputStream(对 http 请求主体进行修改)都是没有意义的,即对 outputStream 的写操作必须要在 inputStream 的读操作之前完成。

httpURLConnection 是基于 HTTP 协议的,其底层通过 socket 通信实现。如果不设置超时(timeout),在网络异常的情况下,可能会导致程序僵死而不继续往下执行。在 JDK 1.5 及更高版本中,可以使用 httpURLConnection 的父类 URLConnection 的以下两个方法设置相应的超时时间:

```
httpURLConnection.setConnectTimeout(100000);        //设置连接主机超时(单位:ms)
httpURLConnection.setReadTimeout(100000);           //设置从主机读取数据超时(单位:ms):
```

16.2　Android 系统中 JSON 数据的解析

JSON(JavaScript Object Notation)是一种轻量级的数据交换格式,具有良好的可读性和便于快速编写的特性,可以在不同平台间进行数据交换。JSON 可以表示数组和复杂的对象,而不仅仅是键和值的简单列表。相对于 XML,JSON 数据的体积更小,解析速度更快,和 JavaScript 的交互更加方便。JSON 主要基于两种数据结构:JSON 对象和 JSON 数组。

JSON 对象是一个无序的"名称:值"映射的集合,"名称:值"映射也称为键值对(key:value)。一个对象以左花括号开始,以右花括号结束。每个 key 和它的 value 之间用冒号隔开,并列的键值对之间用逗号分隔。key 是唯一的字符串。不同的语言中,JSON 对象被理解为对象(object),记录(record),结构(struct)等。图 16-3 是 JSON 对象的语法结构。

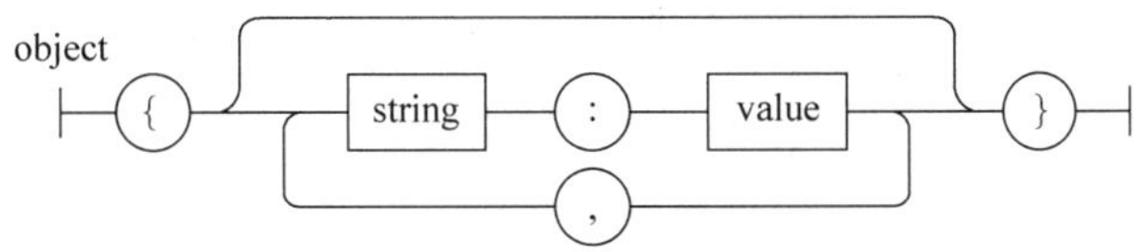

图 16-3　JSON 对象的语法结构

JSON 数组是值的有序列表,在大部分语言中,它被理解为数组(array)。图 16-4 是 JSON 数组的语法结构。

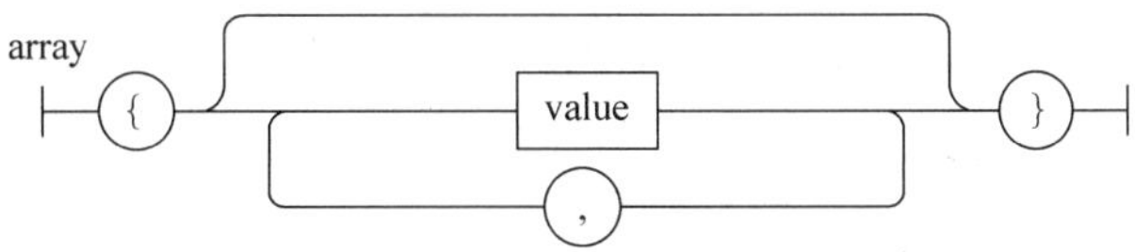

图 16-4　JSON 数组的语法结构

在 JSON 对象和 JSON 数组中提到的值可以是字符串、数字、一个 JSON 对象或者一个 JSON 数组、true、false、null,这些结构可以嵌套,如图 16-5 所示。字符串值是 0 个或者多个 Unicode 字符的序列,可以使用转义字符(\),如图 16-6 所示。字符串类似于 C 语言或者 Java 语言的字符串。数(number)语法结构如图 16-7 所示,类似于 C 语言或者 Java 语言的数,但不使用八进制和十六进制。

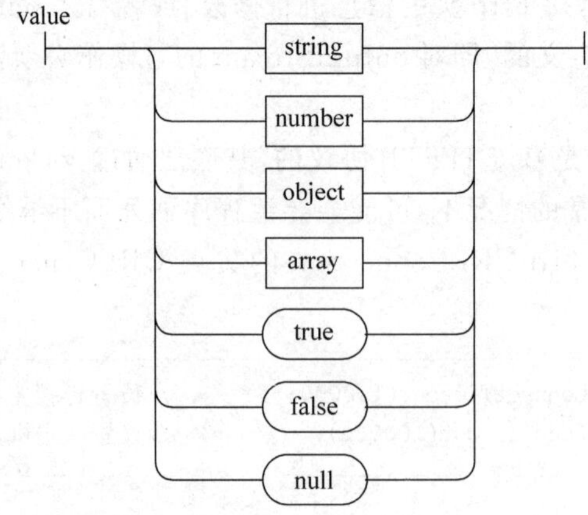

图 16-5 值的语法结构

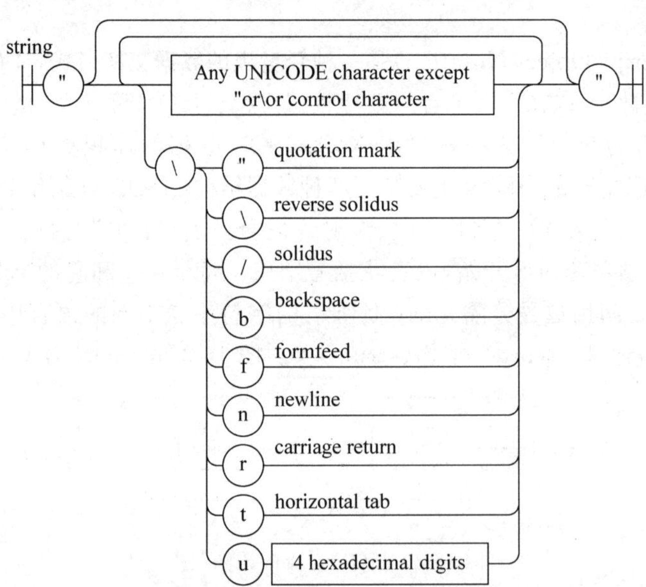

图 16-6 字符串的语法结构

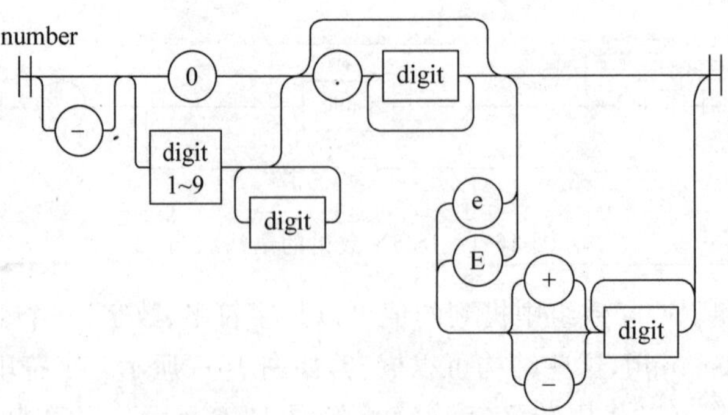

图 16-7 数的语法结构

JSON 语法规则总结如下：并列的数据之间用逗号分隔；映射的名称和值用冒号隔开；并列数据的集合(数组)用方括号括起来；映射的集合(对象)用花括号括起来。

JSON 数组表示有序数据的集合，数组的数据没有“名称”(name)，只有先后顺序。而 JSON 对象表示无序数据的集合，对象的数据有“名称”(name)，没有顺序。如果“键值对”的键的名称是任意字符串时，使用 JSON 对象。如果“键值对”的键的名称是顺序整数时，使用 JSON 数组。

16.2.1 org.json 基本用法

org.json 是 Java 常用的 JSON 解析工具，主要提供 JSONObject 和 JSONArray 类，和其他解析工具相比要轻量得多。org.json.JSONObject 是 org.json 包中有关 JSON 定义的基本单元，是键值对的集合，生成 JSON 对象时采用 put 方法添加或者替换值，解析时采用 getXXX(String name)、optXXX(String name)或者 optXXX(String name, XXX fallback)获取键对应的值，其中 XXX 可以是 Boolean、Double、Int、Long、String、JSONArray 或者 JSONObject。通过 toString()方法输出标准的 JSON 字符串，字符串的最外层被大括号包裹，其中的 Key 和 Value 被冒号分隔，例如{"name": "Zhangsan","age": 21}。org.json.JSONArray 代表一组有序的值，生成 JSON 对象时采用 put()方法添加或者替换值，解析时采用 getXXX(int index)、optXXX(int index)或者 optXXX(int index, XXX fallback)获取键对应的值，其中 XXX 可以是 Boolean、Double、Int、Long、String、JSONArray 或者 JSONObject。通过 toString()方法输出标准的 JSON 字符串，字符串的最外层被中括号包裹，值以逗号分隔，例如[value1,value2,value3]。getXXX 系列方法将要获取的键值转换为指定的类型，如果无法转换或没有值，则抛出 JSONException，optXXX 系列方法也是将要获取的键值转换为指定的类型，如果无法转换或没有值，则返回用户或系统提供的默认值。Value 的类型包括：JSONObject、JSONArray、Boolean、Number、String 或者默认值 JSONObject.NULL 对象。

org.json.JSONStringer 是 JSON 文本构建类，这个类可以帮助快速创建 JSON 文本。引用这个类可以自动严格按照 JSON 语法规则创建 JSON 文本，没有多余的空格键，所以生成的文本可以直接被储存或者传输，每一个 JSONStringer 的实例都能创建一个 JSON 文本。JSONStringer 的实例提供了 key()方法给键值对提供 key，value()方法给 JSON 对象的键值对提供值，或者给 JSON 数组提供值。JSON 对象的键值对必须先调用 key()方法添加 key、再调用 value()方法添加值，即 key()方法必须在 value()方法之前调用。object()方法用于开始一个 JSON 对象，endObject()方法用来结束一个 JSON 对象，object()方法和 endObject()方法必须成对出现。同样，array()方法用于开始一个 JSONArray 对象，endArray()方法用来结束一个 JSONArray 对象，array()方法和 endArray()方法必须成对出现。toString()方法返回一个 JSON 文本的字符串。

一个利用 JSONStringer 创建 JSON 对象的实例：

```
String jsonString = new JSONStringer()
        .object()
                .key("JSON").value("Hello, World!")
        .endObject()
        .toString();
```

一个利用 JSONStringer 创建 JSON 数组的实例：

```
String jsonString1 = new JSONStringer()
        .array()
                .value("zhangsan").value("lisi").value("wangwu")
        .endArray()
        .toString();
```

16.2.2 Google Gson 基本用法

Java 对象和 JSON 之间的相互转换，一般用得比较多的两个类库是 Jackson 和 Gson。Gson(又称 Google Gson)是 Google 公司发布的一个开放源代码的 Java 库，主要用途为序列化 Java 对象为 JSON 字符串，或反序列化 Json 字符串成 Java 对象，它提供一种简单的机制，来实现 Java 对象和 JSON 之间的互相转换，从而不需要手动编写代码进行解析。

com.google.gson.Gson 的 fromJson()方法用来把 JSON 字符串转换为 Java 对象和 ArrayList<T>。Gson 的 toJson()方法实现了序列化，用于把 Java 对象和 ArrayList<T> 转换成 JSON 字符串。

Gson 的创建方式一：直接 new Gson 对象。

```
Gson gson = new Gson();
```

Gson 的创建方式二：使用 GsonBuilder。

使用 new Gson()，此时会创建一个带有默认配置选项的 Gson 实例，如果不想使用默认配置，那么就可以使用 GsonBuilder。例如：

```
Gson gson = new GsonBuilder()
        .excludeFieldsWithoutExposeAnnotation()
//不对没有用@Expose 注解的属性进行操作
        .enableComplexMapKeySerialization()
//当 Map 的 key 为复杂对象时，需要开启该方法
        .serializeNulls()                    //当字段值为空时，依然对该字段进行转换
        .setDateFormat("yyyy-MM-dd HH:mm:ss:SSS")    //时间转化为特定格式
        .setPrettyPrinting()                 //对结果进行格式化，增加换行
        .disableHtmlEscaping()               //防止特殊字符出现乱码
        .registerTypeAdapter(User.class,new UserAdapter())
/*为某特定对象设置固定的序列或反序列方式，自定义 Adapter 需实现 JsonSerializer 或者
JsonDeserializer 接口*/
        .create();
```

把 Java 对象转换为 JSON 字符串，例如：

```
//user 是 User 类型的对象
String jsonString = gson.toJson(user,User.class);
```

把 Java 数组转换为 JSON 字符串，例如：

```
//List<User> userList;
String jsonString = gson.toJson(userList);
```

把 JSON 字符串转换为 Java 对象,例如

```
User user = gson.fromJson(jsonString,User.class);
```

把 JSON 字符串转换为 Java 数组,例如

```
Type type =  new TypeToken<List<User>>(){}.getType();
    List<User> personList = gson.fromJson(jsonString, type);
```

16.3 XML 文件解析

XML(eXtensible Markup Language,可扩展标记语言)是由 World Wide Web 联盟(W3C)定义的一种标记语言,它定义了一套基本规则,以人类和机器都可读的格式对文档进行编码,以便传输和存储数据,其焦点是数据的内容。开发者可以自行定义一个文档来约束一个 XML 文档的数据内容和格式,这称为 XML 约束。专门开发用来编写约束文档的语言称为 Schema 语言。Schema 语言的主要用途是规定 XML 文档的结构。XML Schema 是对 XML 文档类型的描述,主要表现为对文档结构和内容的约束,即超出 XML 自身的基本语法约束部分。Schema 定义了哪些元素可以被包含在别的元素里面、一个特定的元素有哪些合法的属性、元素的顺序、元素的内容和属性的数据类型、唯一性等。Schema 类似于语言中的语法,定义了语言中的词汇和有效的句子应该是什么形式。最常用的两种架构(Schema)语言是 Document Type Definition 与 XML Schema(注意,字符 S 是大写字母)。

Document Type Definition (DTD) 是一套为了进行程序间的数据交换而建立的关于标记符的语法规则。DTD 为 SGML 家族的标记语言(SGML,XML,HTML)定义文档类型,定义页面的元素、元素的属性及元素和属性间的关系。每一个元素都有一个用名字标识的类型,也称为它的通用标识符,并且还可以有一个属性说明集,每个属性说明都有一个名字和一个值作用是定义 XML 的合法构建模块,使用合法的元素和属性的列表定义文档的逻辑结构和合法的构建块。DTD 可以在 XML 文档中声明为内联的,或者是外部引用。XML 文件使用 DOCTYPE 声明语句来指明它所遵循的 DTD 文件,DOCTYPE 声明语句有两种形式:

当引用的文件在本地时,格式为:<!DOCTYPE 文档根结点 SYSTEM"DTD 文件的 URL">,例如<!DOCTYPE 书架 SYSTEM"book.dtd">。当引用的文件是一个公共文件时,格式为:<!DOCTYPE 文档根结点 PUBLIC"DTD 名称""DTD 文件的 URL">。

开发者定义了 DTD 后,就可以根据 DTD 建立文档实例。DTD 能够让 XML 语法分析程序确认页面标记使用的合法性。可通过比较实例文档和 DTD 来检查文档是否符合规范,元素和标签使用是否正确。DTD 不具强制性。对于简单的应用程序来说,开发商不需建立自己的 DTD,可以使用预先定义的公共 DTD 或不使用。即使某个文档已经有 DTD,

只要文档组织是良好的，语法分析程序也不必对照DTD来检验文档的合法性。

XML Schema是一种比DTD更新一些的Schema语言，现在已是W3C组织的标准。XML Schema的出现是为了克服DTD的局限性，它正在逐步取代DTD。XML Schema也是一种用于定义和描述XML文档结构与内容的模式语言，其主要的作用是用来约束XML文件，并验证XML文件有效性。一个XML Schema文档通常称之为模式文档(约束文档)，遵循这个文档书写的XML文件称之为实例文档。

XML Schema符合XML语法结构，它的扩展名通常为.xsd。和XML文件一样，一个XML Schema文档也必须有一个根结点，但这个根结点的名称为Schema。DOM、SAX等XML API很容易解析出XML Schema文档中的内容。XML Schema比DTD支持更多的数据类型，并支持用户自定义新的数据类型，Schema在这方面比DTD强大。DTD只能指定元素含有文本，不能定义元素文本的具体类型，如字符型、整型、日期型、自定义类型等。XML Schema文档结构性强，各元素之间的嵌套关系非常直观。DTD文档的结构是"平铺型"的，如果定义复杂的XML文档，很难把握各元素之间的嵌套关系。

XML Schema对名称空间支持得非常好。编写了一个XML Schema约束文档后，通常需要把这个文件中声明的元素绑定到一个URI(Uniform Resource Identifier，名称空间)上。在XML Schema中，每个约束模式文档都可以被赋予一个唯一的名称空间，以后XML文件就可以通过这个URI来告诉解析引擎，XML文档中编写的元素来自哪里，被谁约束。

在XML文件中书写标签时，可以通过名称空间(xmlns)来声明当前编写的标签来自哪个Schema约束文档。为了在一个XML文档中声明它所遵循的Schema文件的具体位置，通常需要在XML文档中的根结点中使用schemaLocation属性来指定，schemaLocation属性有两个值：第一个值是需要使用的命名空间，第二个值是供命名空间使用的XML Schema的位置，两者之间用空格分隔。

XML文档既要格式良好，又要有效。这意味着它包含架构文档的引用，元素和属性遵守架构文档定义的语法规则。

1. 文件头

XML文件头由XML声明组成，XML声明如果出现，必须出现在文档的第一行。XML声明后还可能紧跟着DTD文件类型声明。

```
<?xml version = "1.0" encoding = "utf - 8"?>
```

其中，"<?"代表一条指令的开始，"? >"代表一条指令的结束；xml代表此文件是XML文件；version="1.0"代表此文件用的是XML1.0标准；encoding="utf-8"代表此文件所用的是UTF-8字符编码。

2. 文件体

文件体中包含的是XML文件的内容，XML元素是XML文件内容的基本单元。从语法讲，一个元素包含一个起始标记、一个结束标记以及标记之间的数据内容。XML元素的格式如下：

```
<标记名称 属性名1 = "属性值1" 属性名2 = "属性值2" …>内容</标记名称>
```

XML 文档必须满足下面这些条件才能被认为是格式良好的：①每个开始标签必须有一个对应的结束标签，空元素必须以“/>”结束。②一个文档只可以有一个根元素。③元素可以嵌套。在使用嵌套结构时，标记之间不能交叉。④元素的所有属性的取值都要有引号。⑤一个元素不能有两个或更多的同名的属性，但是可以嵌套多个同名的元素。⑥XML 是严格区分大小写的，因此< FirstName >和</firstName >不能配对。⑦注释不能放入标签中（它们包含在<!--和-->标记中）。

实体引用是指分析文档时会被字符数据取代的元素。在一个甚至多个 XML 文档中频繁使用的数据，可以预先定义一个这条数据的“别名”，即一个 ENTITY。定义好的 ENTITY 在文档中通过“& 实体名；”来使用。XML 定义了两种类型的 ENTITY，一种在 XML 文档中使用，另一种作为参数在 DTD 文件中使用。对于重复信息或者可以存储在分离的文件中的较大的文本块，ENTITY 特别有用。

ENTITY 的定义语法：

```
<!DOCTYPE 文件名 [
    <!ENTITY 实体名 "实体内容">
]>
```

XML 中有 5 个预定义的实体引用，用于 XML 文档中的特殊字符，如表 16-4 所示。

表 16-4 XML 文件中 5 个预定义的实体引用

实体引用	字符	描述
<	<	小于号
>	>	大于号
&	&	与
'	'	单引号
"	"	双引号

16.3.1 面向文档的对象式解析

常用的 XML 解析技术主要包括：面向文档的对象式解析和面向事件的流式解析。对象式解析（Document Object Model，DOM）是用与平台和语言无关的方式对 XML 文档进行建模的 W3C 标准，它一次性将内容全部加载在内存中，它将一个 XML 文档看作一棵结点树，每个结点代表一个 XML 文档中的元素，生成一个树状层次结构，来组织结点和信息片段。可以在任何时候在树中导航，获取和操作任意部分的数据。DOM 的优点在于可以随机访问，但由于 DOM 在使用数据前需要完整的遍历 XML 文档，尤其是对于大型文档，在内存中构建树形结构，因此需要消耗大量的内存。DOM 解析过程如下：

第一步，创建 DocumentBuilder 对象，并调用它的 parse()方法解析 XML 文档，生成 Document 对象。例如：

```
DocumentBuilderFactory factory = DocumentBuilderFactory.newInstance();
DocumentBuilder documentBuilder = factory.newDocumentBuilder();
InputStream inputStream = this.getResources().openRawResource(R.raw.person);
Document document =  documentBuilder.parse(inputStream);
```

第二步，对 Document 对象的结点信息进行必要的处理。在 DOM(文档对象模型)中，每个部分都是结点。文档本身是文档结点，它是文档数据的根，Document 接口代表整个 XML 文档。①所有 XML 元素是 Element 结点，它可以拥有的类型为元素结点、文本结点、注释结点的子结点。元素也可以拥有属性，可用于获取属性参数。②所有 XML 属性是 Attribute 结点，XML 属性始终属于 XML 元素。③XML 元素内的文本是文本结点。④注释是注释结点。

NodeList 提供了有顺序的结点集合的抽象。NodeList 定义如下：

```
public interface NodeList {
    Node item(int var1);                //返回第 var1 个 Node
    int getLength();                    //返回 Node 的数量
}
```

Document 接口的 getElementsByTagName()方法可返回带有指定标签名的对象的集合。它返回元素的顺序是它们在文档中的顺序。如果把特殊字符"*"传递给 getElementsByTagName()方法，它将返回文档中所有元素的列表，元素排列的顺序就是它们在文档中的顺序。

16.3.2 面向事件的流式解析

流式解析是解析器顺序读取 XML 文档、捕获各种事件，如元素开始和元素结束等。流式解析又分为推式解析(SAX：Simple API for XML)和拉式解析(StAX：Streaming API for XML)两种解析方式。SAX 使用流式处理的方式，它并不记录所读内容的相关信息。推式解析中，解析器控制着读循环，对文档进行顺序扫描，在读取 XML 文档的过程中，每发现一个结点(例如 document 开始与结束、element 开始与结束)就报告一个事件，事件推给事件处理器，通过回调事件处理程序中相应的方法来进行处理，直到文档结束，在文档结束之前解析器不会把控制权返回给应用程序。推式解析是基于事件驱动的，事件的检测由解析器完成，事件的处理方法需要应用程序完成。使用 org. xmlpull. v1. XmlPullParser 的步骤如下：

第一步，新建 org. xml. sax. helpers. DefaultHandler 的子类，并重写 ContentHandler 接口中的 startDocument()、endDocument()、startElement()、endElement()和 characters()方法。DefaultHandler 是 SAX 事件处理程序的默认基类，实现了 EntityResolver、DTDHandler、ContentHandler 和 ErrorHandler 接口。如果需要为外部实体实现定制的处理，就必须实现 EntityResolver 接口，大部分 SAX 应用不需要实现 EntityResolver 接口。DTDHandler 接收基本的 DTD 有关的事件，仅仅包含 XML 处理器必须报告的事件：notation 声明以及没有解析的实体声明。notation 元素用于描述 XML 文档中非 XML 数据的格式。如果 SAX 应用需要实现定制的错误处理，就必须实现 ErrorHandler 接口。ContentHandler 接收文档的逻辑内容的通知。

```
import android.util.Log;
import org.xml.sax.Attributes;
import org.xml.sax.SAXException;
```

```
import org.xml.sax.helpers.DefaultHandler;
import java.util.ArrayList;
import java.util.List;

public class MyDefaultHandler extends DefaultHandler {
    public MyDefaultHandler() {
        super();
    }
//当开始读取文档标签时调用该方法
    @Override
    public void startDocument() throws SAXException {
        super.startDocument();
    }
    //当结束读取文档时调用该方法
    @Override
    public void endDocument() throws SAXException {
        super.endDocument();
    }
//当读取元素的开始标签时调用该方法
    @Override
    public void startElement (String uri, String localName, String qName, Attributes
attributes) throws SAXException {
        super.startElement(uri, localName, qName, attributes);
    }
//当读取元素的结束标签时调用该方法
    @Override
    public void endElement(String uri, String localName, String qName) throws SAXException {
        super.endElement(uri, localName, qName);
    }
//字符数据的接收通知
    @Override
    public void characters(char[] ch, int start, int length) throws SAXException {
      super.characters(ch, start, length);
    }
}
```

第二步，新建一个 javax.xml.parsers.SAXParser 解析器，然后调用 SAXParser 的 parse(InputStream is, DefaultHandler dh)方法进行解析。

```
try {
SAXParserFactory factory = SAXParserFactory.newInstance();
    SAXParser parser =  factory.newSAXParser();
        MyDefaultHandler myDefaultHandler = new MyDefaultHandler();
        InputStream inputStream = this.getResources().openRawResource(R.raw.person);
        parser.parse(inputStream,myDefaultHandler);
        inputStream.close();
    } catch (ParserConfigurationException e) {
        e.printStackTrace();
```

```
        } catch (SAXException e) {
            e.printStackTrace();
        } catch (IOException e) {
            e.printStackTrace();
        }
    }
```

16.3.3 XMLPullParser 类

拉式解析和推式解析类似，都是基于事件的模式，都提供了类似的事件，如开始元素和结束元素。不同的是，SAX 的事件驱动是回调相应方法，需要提供回调的方法，而后在 SAX 内部自动调用相应的方法。在 PULL 解析过程中，需要应用程序不断地请求事件、根据返回的数字（事件类型）做相应的操作，而不是像推式解析器那样由处理器触发特定事件的方法。Pull 解析器读取到 XML 的声明返回 START_DOCUMENT；文档结束返回 END_DOCUMENT；开始标签返回 START_TAG；结束标签返回 END_TAG；文本返回 TEXT。所有的事件必须应用自己去请求。XMLPullParser 是一种拉式解析器，常用的 XML Pull 的接口和类：

org.xmlpull.v1.XMLPullParserFactory：XML 解析器工厂类。一般都是用该工厂创建 XML Pull 解析器实例。

org.xmlpull.v1.XMLPullParser：该接口定义了 org.XMLpull.v1 API 中需要的解析功能。

org.xmlpull.v1.XMLSerializer：该接口定义了序列化 XML 的需要的方法。

org.xmlpull.v1.XMLPullParserException：在解析过程中遇到非正常情况就会抛出该异常，抛出单一的 XMLPull 解析器相关的错误。

XMLPullParser 中提供的获取事件的方法有 next()、nextToken()和 nextTag()。可能获取的事件类型如表 16-5 所示。

表 16-5 XMLPullParser 的事件类型

事件名称	描述
DOCDECL	读到 document type declaration 时（就是< DOCTYPE），就报告该事件。在 DOCTYPE 中没有解析的文本可以通过 getText()获得
PROCESSING_INSTRUCTION	处理指令
CDSECT	读到 CDATA Section 就报告该事件，仅仅 nextToken()可以获取该事件。next()把各种文本事件合并成单一的 TEXT 事件
COMMENT	读到注释时就报告该事件。仅仅 nextToken()可以获取该事件。next()跳过注释
IGNORABLE_WHITESPACE	可忽略的空白。在没用 dtd 约束文档时，IGNORABLE_WHITESPACE 只会出现在根元素外面；对于有 dtd 约束的文档，空白由 dtd 约束文档定义。（dtd 约束文档就是在 DOCTYPE 中指定的那个文件，它规定了可以在 XML 出现什么标签、以及标签可以出现在哪等）

续表

事件名称	描　述
START_DOCUMENT	XML 刚开始时，报告该事件
END_DOCUMENT	XML 结束时，就报告该事件。next()和 nextToken()都能获取该事件，随后调用 next()或者 nextToken()都将导致抛出异常
START_TAG	读到开始标签时，就报告该事件。例如：读到< div >
END_TAG	读到结束标签时，就报告该事件。例如：读到</div >。next()和 nextToken()都能获取该事件。开始标签的名字可以通过 getName()获取，名称空间和前缀可以分别通过 getNamespace()和 getPrefix()获取
TEXT	读到文本内容时，就报告该事件。如：读到< div >内容</div >标签间的内容时。可以通过 getText()读取文本内容。当处于验证模式时，next()将不报告可忽略的白空格，使用 nextToken()代替
ENTITY_REF	刚读到实体引用。仅仅 nextToken()可以获取该事件。通过 getName()可以获得实体名，通过 getText()可以获得替换文本。

1. int next()

此方法主要用于请求下一个比较高层的事件，包括：START_TAG，TEXT，END_TAG，END_DOCUMENT。元素内容将被合并，对于整个元素内容(注释和处理指令将被忽略，实体引用必须被扩展，否则如果实体引用不能被扩展，例外必须被抛出)，只有一个 TEXT 事件被返回，即把各种文本事件合并成单一的 TEXT 事件。如果元素内容是空的(内容是"")，将没有 TEXT 事件被报告。空元素(例如< tag/>)将被报告成 2 次分离的事件：START_TAG 和 END_TAG，它必须和< tag ></tag >形式的空元素保持分析的等价性。

2. int nextToken()

该方法能够返回所有的事件。与 next()相比，它将公开一些附加的事件类型(COMMENT，CDSECT，DOCDECL，ENTITY_REF，PROCESSING_INSTRUCTION，或者 IGNORABLE_WHITESPACE)，如果输入中有这些事件的话。

3. int nextTag()

调用 next()，如果是 START_TAG 或者 END_TAG，就返回事件，否则将抛出异常，它将跳过事件标签前面的空格。它相当于：

```
int eventType = next();
  if(eventType == TEXT && isWhitespace()) {        // skip whitespace
    eventType = next();
  }
  if (eventType != START_TAG && eventType != END_TAG) {
    throw new XmlPullParserException("expect start or end tag", this, null);
  }
  return eventType;
```

对于一般的解析，只需要用到 XMLPullParser 的几个获取值的 API 就够了。

4. String getName()

如果当前的事件不是 START_TAG、END_TAG 和 ENTITY_REF 将返回 null。对于 START_TAG 和 END_TAG 事件，当名称空间使能时，返回的是现在元素的本地名字(local name)；当名称空间处理被禁止时，返回的是现在元素的原始名字。对于 ENTITY_REF 事件，返回的是实体的名字。

5. String getText()

把现在事件的文本内容作为字符串返回，返回的值取决于现在的事件类型，例如：对于文本事件，返回的是元素内容(这是使用 next()时的典型情况)。CDSECT 事件，返回其内部的内容。COMMENT 事件，返回注释内容。如果现在的事件类型是 ENTITY_REF，返回的是实体引用替换后的文本(或者不可得到时，返回 null)，而 getTextCharacters()返回的是包含实体名的实际输入缓冲区，这是 getText()和 getTextCharacters()返回不同的值的唯一情形。其他的事件一般都返回 null。

6. String nextText()

该方法的正常执行是有条件的：需要当前的事件为 START_TAG，如果下一个元素是 TEXT，则元素内容被返回；或者下一个事件是 END_TAG，则返回空的字符串；否则将抛出异常。当调用这个函数成功后，解析器将定位到 END_TAG。这个函数的动机是保证解析空的元素和有非空内容的元素的一致性，例如对以下输入都可以使用相同的代码解析：< tag >foo </tag >和< tag ></tag >(等价于< tag/>)。

使用 XMLPullParser 的步骤

第一步，创建一个 XMLPullParser 解析器，并设置它的输入流为需要解析的 XML 文件。XML 文件可以保存在 app\src\main\res\xml、app\src\main\res\raw 或者 app\src\main\assets 文件夹。XML 文件保存位置不同，获取输入流的方式也有所不同。假如有 XML 文件 person. xml 保存在 app \ src \ main \ res \ xml，可以使用以下代码获取 XMLPullParser

```
XMLPullParser parser = this.getResources().getXml(R.xml.person);
```

对于 res/raw 和 assets 文件夹，相同点是两个目录下的文件在打包后会原封不动地保存在 apk 包中，不会被编译成二进制文件。不同点是：res/raw 中的文件会被映射到 R. java 文件中，访问的时候直接使用资源 ID 即 R. id. filename，因此 res/raw 不可以有目录结构；assets 文件夹下的文件不会被映射到 R. java 中，访问的时候需要 AssetManager 类，因而 assets 可以有目录结构，也就是 assets 目录下可以再建立文件夹。对于 res/raw 下的文件资源，通过以下方式获取输入流

```
InputStream inputStream = getResources().openRawResource(R.id.person);
```

对于 assets 下的文件资源，通过以下方式获取输入流：

```
AssetManager assetManager = getAssets();
InputStream inputStream = assetManager.open("person.xml");
```

获取 XMLPullParser 可以通过以下方式：

```
XMLPullParserFactory factory = XmlPullParserFactory.newInstance();
XMLPullParser parser = factory.newPullParser();
```

或者

```
XMLPullParser parser = Xml.newPullParser();
```

获取 XMLPullParser 后，就可以调用 setInput()设置解析器的输入流

```
parser.setInput(inputStream,"utf-8");
```

第二步，调用 getEventType()获取当前的事件，在 eventType 不是 END_DOCUMENT 的情况下，循环调用 next()、nextTag 或者 nextToken，并根据 eventType 做出相应的处理。

```
try {
    eventType = parser.getEventType();
    while(eventType!= XmlPullParser.END_DOCUMENT){
        switch (eventType){
          case XmlPullParser.START_DOCUMENT:
                    break;
          case XmlPullParser.END_TAG:
                    break;
          case XmlPullParser.START_TAG:
                    break;
          case XmlPullParser.TEXT:
                    break;
        }
        eventType = parser.next();
    }
} catch (XmlPullParserException e) {
    e.printStackTrace();
} catch (IOException e) {
    e.printStackTrace();
}
```

推式解析器和拉式解析器的区别如下。

推式解析器和拉式解析器的核心区别在于是由解析器还是应用程序控制读循环(读入文件的循环)。推式解析器是一种主动类型的解析器，它在解析 xml 的过程中，会将解析到的事件主动的推送给事件处理程序，整个解析过程都是全自动的(即必须从 XML 开始到 XML 结束)，比如，当解析到< div >时，它就会主动发出 startTag 的事件。拉式解析器与其相对，整个解析过程是被动的，应用程序想解析到哪就解析到哪。比如，应用程序想要< div >这种开始标签时，必须不断地调用 next()来找到该事件。同推式解析相比，拉式解析拉式解析允许过滤 XML 文件和跳过一些事件，代码更简单。

16.4 httpURLConnection 实验

1. 新建工程 httpurlconnection1

具体步骤：运行 Android Studio，单击 Start a new Android Studio project，在 Create New Project 页中，Application name 为 httpUrlConnection1，Company domain 为 cuc，Project location 为 C:\android\myproject\httpUrlConnection1，单击 Next 按钮，选中 Phone and Tablet，minimum sdk 设置为 API 19：Android 4.4(minSdkVersion)，单击 Next 按钮，选择 Empty Activity，再单击 Next 按钮，勾选 Generate Layout File 选项，勾选 Backwards Compatibility(AppCompat)选项，单击 Finish 按钮。

2. 新建布局文件 activity_main1.xml

具体步骤：在工程结构视图中，右击 layout，依次选择 New|Layout resource file，File name 输入框并输入 activity_main1，Root element 输入框输入 LinearLayout，Source set 为 main，单击 OK 按钮。layout/activity_main1.xml 内容如下：

```
<?xml version = "1.0" encoding = "utf - 8"?>
<LinearLayout xmlns:android = "http://schemas.android.com/apk/res/android"
    android:orientation = "vertical" android:layout_width = "match_parent"
    android:layout_height = "match_parent">
    <EditText
        android:id = "@ + id/textUriEdt"
        android:layout_width = "match_parent"
        android:layout_height = "wrap_content"
        android:hint = "net address"
        android:inputType = "textUri"
        android:text = "http://10.0.2.2:8080/RELEASE - NOTES.txt"/>
    <Button
        android:layout_width = "match_parent"
        android:layout_height = "wrap_content"
        android:id = "@ + id/downFileBtn"
        android:text = "downFile" />
    <Button
        android:layout_width = "match_parent"
        android:layout_height = "wrap_content"
        android:id = "@ + id/ConnectToServerSocketBtn"
        android:text = "ConnectToServerSocket" />
</LinearLayout>
```

Android 模拟器或者手机真机通过 USB 连接访问电脑本地 Web 服务，模拟器可以通过把服务器 ip 设置为 10.0.2.2 访问本地 Web 服务的，手机真机通过将服务器 IP 设置为 10.0.2.2 访问本地计算机 Web 服务是行不通的，这时需要手机真机和计算机在同一局域网内。计算机通过宽带网线或者 WiFi 连接路由器，手机开启 WiFi 功能、通过 WiFi 通过连接该路由器，即可让手机和计算机处在同一个局域网。手机真机访问时，将主机的 IP 地址更改为主机的实际的 IP 地址。在 Windows 操作系统中，运行 ipconfig 就可以查看主机的 IP 地址。

3. 新建一个 Java 类 DownloadFileRunnable，实现 Runnable 接口

具体步骤：在工程结构视图中 main 源集 Java 文件夹里，右击 cuc.httpurlconnection1 包，选择 New|Java Class，Name 输入框中输入 DownloadFileRunnable，Interface(s)输入框中输入 java.lang.Runnable，单击 OK 按钮。编辑 DownloadFileRunnable.java，文件内容如下：

```
package cuc.httpurlconnection1;
import android.os.Bundle;
import android.os.Environment;
import android.os.Handler;
import android.os.Message;
import java.io.File;
import java.io.FileOutputStream;
import java.io.InputStream;
import java.io.OutputStream;
import java.net.HttpURLConnection;
import java.net.URL;

public class DownloadFileRunnable implements Runnable {
   private String textUrl;
   private String fileName;
   private Handler handler;

   public DownloadFileRunnable(String textUrl, String fileName, Handler handler) {
       this.textUrl = textUrl;
       this.fileName = fileName;
       this.handler = handler;
   }
   @Override
   public void run() {
       try {
           URL url = new URL(textUrl);
           HttpURLConnection connection = (HttpURLConnection) url.openConnection();
           InputStream inputStream = connection.getInputStream();
           File sdcard = Environment.getExternalStorageDirectory();
           File file = new File(sdcard, fileName);
           file.createNewFile();
           OutputStream outputStream = new FileOutputStream(file);
           byte[] buffer = new byte[4 * 1024];
           int count;
           while ((count = inputStream.read(buffer))!= -1){
               outputStream.write(buffer,0,count);
           }
           outputStream.flush();
           outputStream.close();
           inputStream.close();
           Message msg = handler.obtainMessage();
           Bundle bundle = new Bundle();
           bundle.putString("downloadState","downloading finished");
           msg.setData(bundle);
```

```
                msg.sendToTarget();
            } catch (java.io.IOException e) {
                e.printStackTrace();
            }
        }
    }
```

4. 在 AndroidManifest.xml 文件添加权限

由于在 DownloadFileRunnable.java 中需要访问网络和写 SD 卡，所以在 AndroidManifest.xml 文件中添加如下权限，以便访问网络和写 SD 卡。

```
<uses-permission android:name="android.permission.INTERNET"/>
<uses-permission android:name="android.permission.WRITE_EXTERNAL_STORAGE"/>
```

添加权限后，AndroidManifest.xml 内容如下：

```
<?xml version="1.0" encoding="utf-8"?>
<manifest xmlns:android="http://schemas.android.com/apk/res/android"
    package="cuc.httpurlconnection1">
    <uses-permission android:name="android.permission.INTERNET"/>
    <uses-permission android:name="android.permission.WRITE_EXTERNAL_STORAGE"/>
    <application
        android:allowBackup="true"
        android:icon="@mipmap/ic_launcher"
        android:label="@string/app_name"
        android:roundIcon="@mipmap/ic_launcher_round"
        android:supportsRtl="true"
        android:theme="@style/AppTheme">
        <activity android:name=".MainActivity">
            <intent-filter>
                <action android:name="android.intent.action.MAIN"/>
                 <category android:name="android.intent.category.LAUNCHER"/>
            </intent-filter>
        </activity>
    </application>
</manifest>
```

5. 编辑 MainActivity.java

编辑后，MainActivity.java 内容如下：

```
package cuc.httpurlconnection1;
import android.os.Bundle;
import android.os.Handler;
import android.os.Message;
import android.support.v7.app.AppCompatActivity;
import android.text.TextUtils;
import android.view.View;
```

```
import android.widget.Button;
import android.widget.EditText;
import android.widget.Toast;

public class MainActivity extends AppCompatActivity {
   private EditText textUriEdt;
   private Button downFileBtn;
   private Button connectToServerSocketBtn;
   private void assignViews() {
       textUriEdt = (EditText) findViewById(R.id.textUriEdt);
       downFileBtn = (Button) findViewById(R.id.downFileBtn);
       connectToServerSocketBtn = (Button) findViewById(R.id.connectToServerSocketBtn);
   }

   @Override
   protected void onCreate(Bundle savedInstanceState) {
       super.onCreate(savedInstanceState);
       setContentView(R.layout.activity_main1);
       assignViews();
       final Handler handler = new Handler(){
           @Override
           public void handleMessage(Message msg) {
               super.handleMessage(msg);
               Bundle bundle = msg.getData();
               String info = bundle.getString("downloadState");
               Toast.makeText(MainActivity.this, info, Toast.LENGTH_LONG).show();
           }
       };
       downFileBtn.setOnClickListener(new View.OnClickListener() {
           @Override
           public void onClick(View v) {
               String textUrl;
               if (TextUtils.isEmpty(textUriEdt.getText().toString())){
                   textUrl = "http://10.0.2.2:8080/RELEASE - NOTES.txt";
               }else{
                   textUrl = textUriEdt.getText().toString();
               }
                   new Thread(new DownloadFileRunnable(textUrl,"RELEASE - NOTES.txt",
handler)).start();
           }
       });
   }
}
```

6. 下载 Tomcat，解压缩，并运行

具体步骤为：从 http://tomcat.apache.org/download-90.cgi # 9.0.0.M22 下载 apache-tomcat-9.0.0.M22-windows-x64.zip，并解压到某个文件夹如 C:\android，进入 C:\android\apache-tomcat-9.0.0.M22\bin 文件夹，双击 startup.bat 运行。不过在运行

startup.bat 之前，需要设置 Windows 系统的环境变量：设置 JAVA_HOME 为 JDK 的安装文件夹，例如 C:\android\Java\jdk1.8.0_91，具体设置过程参阅第 1 章。

7. 编译运行

创建模拟器，编译运行，单击 DOWNFILE 按钮，下载的文件(文件名 RELEASE-NOTES.txt)存储在模拟器的 sdcard 的根文件夹下。

上面这个例子是关于 httpURLConnection，下面几步是关于 Socket 编程。Android 模拟器到主机的连接是单向连接，Android 模拟器用作 Socket 客户端，PC 用作 Socket 服务器，为了在 PC 上运行 Java 程序，在 Android Studio 工程中新建一个 Java Library 模块。

8. 新建一个类，实现了 Runnable 接口

在包 cuc.httpurlconnection1 下，新建一个名为 SocketRunnable 的 Java class，实现了 Runnable 接口。SocketRunnable.java 内容如下：

```
package cuc.httpurlconnection1;
import android.content.Context;
import android.util.Log;
import java.io.IOException;
import java.io.InputStream;
import java.net.Socket;

public class SocketRunnable implements Runnable {
    private Context context;
    private static final String TAG = "SocketRunnable";

    @Override
    public void run() {
        try {
            Log.i(TAG, "run: before connect");
            Socket socket = new Socket("10.0.2.2",4567);
            Log.i(TAG, "run: after socket");
            InputStream inputStream = socket.getInputStream();
            int count = 0;
            byte[] buffer = new byte[1024];
            while((count = inputStream.read(buffer))!= -1){
                Log.i(TAG, "run: " + (new String(buffer,0,count)));
            }
            inputStream.close();
          socket.close();
          Log.i(TAG, "run: disconnect");
        } catch (IOException e) {
          e.printStackTrace();
        }
    }
}
```

9. 编辑 MainActivity.java

在类 MainActivity 的 onCreate 方法里添加以下代码，实现对 ConnectToServerSocketBtn 的监听。

```
ConnectToServerSocketBtn.setOnClickListener(new View.OnClickListener() {
    @Override
    public void onClick(View v) {
        new Thread(new SocketRunnable()).start();
    }
});
```

10. 新建一个 Java Library 模块

具体步骤：依次单击 Android Studio 菜单 File|New|New Module|Java Library，单击 Next 按钮，弹出 Create New Module 界面，Library name 输入框输入 LibPcAsServer，单击 Java package name 右面的 Edit 按钮，输入 cuc.httpurlconnection1，Java class name 输入框输入 PcAsServer，单击 Finish 按钮。编辑 PcAsServer.java，内容如下：

```
package cuc.httpurlconnection;

import java.io.DataInputStream;
import java.io.IOException;
import java.io.OutputStream;
import java.net.ServerSocket;
import java.net.Socket;

public class PcAsServer {
    public static void main(String[] args) throws Exception{
        System.out.println("PcAsServer: main is running ");
        DataInputStream dataInputStream = new DataInputStream(System.in);
        ServerSocket serverSocket = new ServerSocket(4567);
        System.out.println("server is ready!");
        while (true ){
            try {
                Socket socket = serverSocket.accept();
                System.out.println("socket connected");
                OutputStream outputStream = socket.getOutputStream();
                outputStream.write("Hello,Cuc!".getBytes());
                outputStream.flush();
                outputStream.close();
            }catch (IOException e) {
                e.printStackTrace();
            }
        }
    }
}
```

11. 先运行 ServerSocket 程序，再运行客户端程序

具体步骤：在工程结构视图中，选中 LibPcAsServer 的模块里的 PcAsServer，右击 PcAsServer，选择 Run 'LibPcAsServer' | Edit Configuration，如图 16-8 所示，在 Edit Configuration 界面中，Use classpath of module 从列表框中选择 LibPcAsServer，单击 Apply 按钮，单击 Run 按钮，运行 PcAsServer。

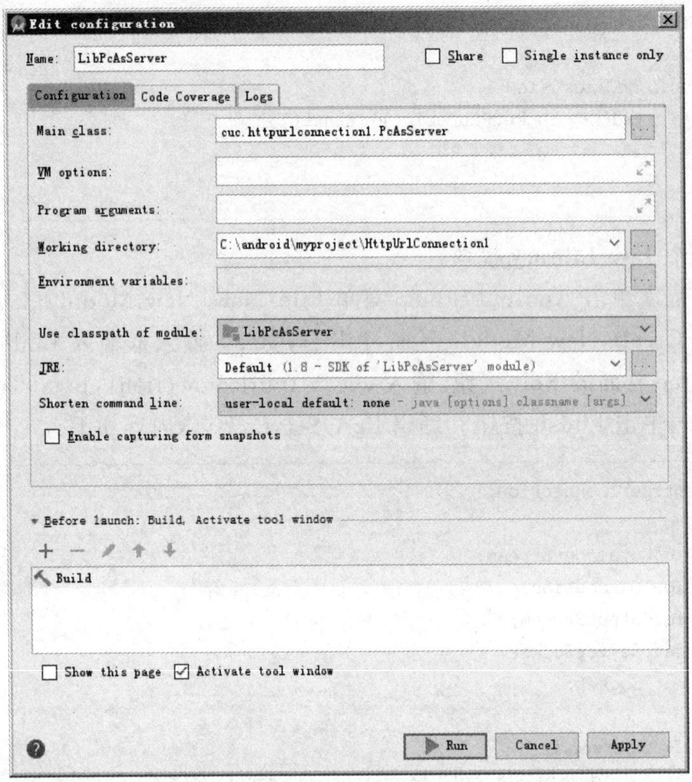

图 16-8 Android run 菜单的 Edit configuration 界面

单击 Android Studio 界面上的菜单 Run|Run|App，这时 App 模块可以在 Android 模拟器上运行，单击 Android 模拟器上的 ConnectToServerSocket 按钮，与 ServerSocket 建立连接，从日志中可以看到消息提示。

16.5 Json 解析实验

1. 新建工程 Json1

具体步骤：运行 Android Studio，单击 Start a new Android Studio project，在 Create New Project 页中，Application name 为 Json1，Company domain 为 cuc，Project location 为 C:\android\myproject\ Json1，单击 Next 按钮，选中 Phone and Tablet，minimum sdk 设置为 API 19：Android 4.4(minSdkVersion)，单击 Next 按钮，选择 Empty Activity，再单击 Next 按钮，勾选 Generate Layout File 选项，勾选 Backwards Compatibility(AppCompat)选项，单击 Finish 按钮。

2. 添加 Gson 库依赖

由于这个工程中，要使用 Google Gson 来生成和解析 JSON，所以需要添加 Gson 库。添加 Gson 库的具体步骤如下：依次单击 Android Studio 菜单 File|Project Structure|App，再单击右边的 Dependencies 选项，单击 Dependencies 选项卡右上角的加号，弹出的界面，如图 16-9 所示，单击“1 Library dependency”选项，弹出 Choose Library Dependency 界面，在

搜索框中输入 gson，并单击搜索图标进行搜索，在搜索结果中选择 com. google. code. gson：gson：2.8.1，再单击界面中的 OK 按钮，Gson 库就添加进 App 模块。再单击 OK 按钮，关闭 Project Structure 窗口。

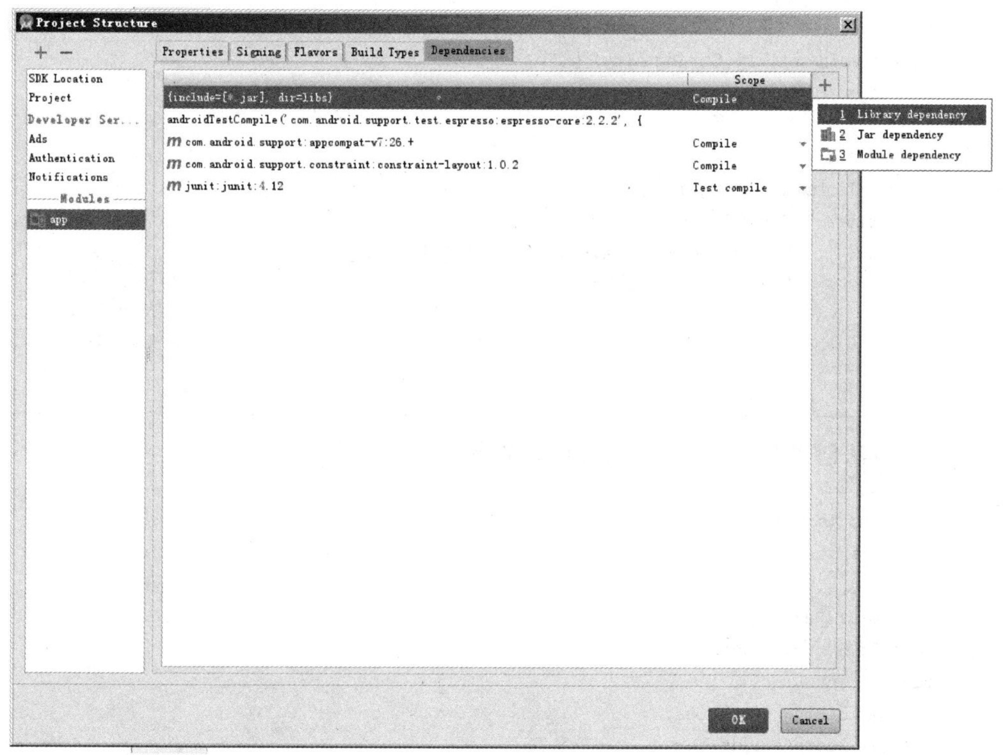

图 16-9 Project Structure 的 App 模块的 Dependencies 界面

3. 新建两个 Java 类

具体步骤：在工程结构视图中 main 源集 Java 文件夹里，右击 cuc. json1 包，依次选择 New|Java Class，Name 输入框中输入 Address，单击 OK 按钮，系统生成 Address. java 文件。利用菜单 Code | Generate，生成 Getter and Setter 方法，以及 Constructor。最后 Address. java 文件内容如下：

```
package cuc.json1;
public class Address {
    private String province;
    private String city;
    public Address(String province, String city) {
        this.province = province;
        this.city = city;
    }
    public String getProvince() {
         return province;
    }
    public void setProvince(String province) {
```

```
        this.province = province;
    }
    public String getCity() {
        return city;
    }
    public void setCity(String city) {
        this.city = city;
    }
    @Override
    public String toString() {
        return "Address{" +
                "province='" + province + '\'' +
                ", city='" + city + '\'' +
                '}';
    }
}
```

同样的方法创建 Person.java,文件内容如下:

```
package cuc.json1;
import java.util.Arrays;

public class Person {
    private String name;
    private int age;
    private String[] hobbies;
    private Address address;
    public Person(String name, int age, String[] hobbies, Address address) {
        this.name = name;
        this.age = age;
        this.hobbies = hobbies;
        this.address = address;
    }

    public Person(int age, Address address) {
        this.age = age;
        this.address = address;
    }
    public String getName() {
        return name;
    }
    public void setName(String name) {
        this.name = name;
    }

    public int getAge() {
        return age;
    }
    public void setAge(int age) {
```

```
        this.age = age;
    }

    public String[] getHobbies() {
        return hobbies;
    }
    public void setHobbies(String[] hobbies) {
        this.hobbies = hobbies;
    }
    public Address getAddress() {
        return address;
    }
    public void setAddress(Address address) {
        this.address = address;
    }
    @Override
    public String toString() {
         return "Person{" +
                "name='" + name + '\'' +
                ", age=" + age +
                ", hobbies=" + Arrays.toString(hobbies) +
                ", address=" + address +
                '}';
    }
}
```

4. 利用 Gson 进行 JavaBean 与 Json 的相互转换

具体步骤：在 MainActivity 类中添加方法 googleJavaBeanToJson()，该方法把一个 Person 对象转化为 Json 字符串，返回值是 Json 字符串。在 MainActivity 类中添加方法 googleJsonToJavaBean(String jsonString) 把 Json 字符串转化为 Person 对象。最后在 MainActivity 类的 doMyJob ()方法中调用 googleJsonToJavaBean(googleJavaBeanToJson())，编辑后 MainActivity.java 内容如下：

```
package cuc.json1;

import android.os.Bundle;
import android.support.v7.app.AppCompatActivity;
import android.util.Log;
import com.google.gson.Gson;
import com.google.gson.reflect.TypeToken;
import java.lang.reflect.Type;
import java.util.ArrayList;
import java.util.List;

public class MainActivity extends AppCompatActivity {
    private static final String TAG = "MainActivity";
    String googleJavaBeanToJson(){
```

```
        Person person = new Person("zhangsan",21,new String[]{"reading","swimming"},
                new Address("Hubei","wuhan"));
        Log.i(TAG, "googleJavaBeanToJson: JavaBean -> " + person);
        Gson gson = new Gson();
        String jsonString = gson.toJson(person,Person.class);
        Log.i(TAG, "googleJavaBeanToJson: Json -> " + jsonString);
        return jsonString;
    }
    void googleJsonToJavaBean(String jsonString){
        Gson gson = new Gson();
        Person person = gson.fromJson(jsonString,Person.class);
        Log.i(TAG, "googleJsonToJavaBean: JavaBean -> " + person);
    }
    void doMyJob(){
        googleJsonToJavaBean(googleJavaBeanToJson());
    }
    @Override
    protected void onCreate(Bundle savedInstanceState) {
        super.onCreate(savedInstanceState);
        setContentView(R.layout.activity_main);
        doMyJob();
    }
}
```

编译运行,日志输出如下:

```
I/MainActivity: googleJavaBeanToJson: JavaBean -> Person{name = 'zhangsan', age = 21, hobbies
= [reading, swimming], address = Address{province = 'Hubei', city = 'wuhan'}}
I/MainActivity: googleJavaBeanToJson: Json -> {"address":{"city":"wuhan","province":
"Hubei"},"name":"zhangsan","hobbies":["reading","swimming"],"age":21}
I/MainActivity: googleJsonToJavaBean: JavaBean -> Person{name = 'zhangsan', age = 21, hobbies
= [reading, swimming], address = Address{province = 'Hubei', city = 'wuhan'}}
```

5. 利用 Gson 进行 ArrayList<T>与 JSON 的相互转换

具体步骤:在 MainActivity 类中添加方法 googleArrayListnToJson(),该方法把一个 ArrayList<Person>转化为 Json 字符串,返回值是 Json 字符串。

在 MainActivity 类中添加方法 googleJsonToArrayList(String jsonString)把 JSON 字符串转化为 List<Person>。这两个方法的定义如下:

```
String googleArrayListnToJson(){
    List<Person> personList = new ArrayList<>();
    Person person = new Person("zhangsan",21,new String[]{"reading","swimming"},
            new Address("Hubei","wuhan"));
    personList.add(person);
    Log.i(TAG, "googleArrayListnToJson: javabean -> " + person);
    person = new Person("lisi",22,new String[]{"reading","swimming","dancing"},
            new Address("Hunan","changsha"));
```

```
        personList.add(person);
        Log.i(TAG, "googleArrayListnToJson: javabean->" + person);
        Gson gson = new Gson();
        String jsonString = gson.toJson(personList);
        Log.i(TAG, " googleArrayListnToJson: json->" + jsonString);
        return jsonString;
    }
    void googleJsonToArrayList(String jsonString){
        Gson gson = new Gson();
        Type type = new TypeToken<List<Person>>(){}.getType();
        List<Person> personList = gson.fromJson(jsonString,type);
        for (int i = 0;i<personList.size();i++){
            Log.i(TAG, " googleJsonToArrayList: javabean->" + personList.get(i));
        }
    }
```

最后编辑 MainActivity 类的 doMyJob()方法，doMyJob()方法的内容如下：

```
    void doMyJob(){
        googleJsonToArrayList(googleArrayListnToJson());
    }
```

编译运行，日志输出如下：

```
I/MainActivity: googleArrayListnToJson: javabean -> Person{name = 'zhangsan', age = 21,
hobbies = [reading, swimming], address = Address{province = 'Hubei', city = 'wuhan'}}
I/MainActivity: googleArrayListnToJson: javabean -> Person{name = 'lisi', age = 22, hobbies
 = [reading, swimming, dancing], address = Address{province = 'Hunan', city = 'changsha'}}
I/MainActivity: googleArrayListnToJson: json -> [{"address":{"city":"wuhan","province":
"Hubei"},"name":"zhangsan","hobbies":["reading","swimming"],"age":21},{"address":
{"city":"changsha","province":"Hunan"},"name":"lsi","hobbies":["reading","swimming",
"dancing"],"age":22}]
I/MainActivity: googleJsonToArrayList: javabean -> Person{name = 'zhangsan', age = 21,
hobbies = [reading, swimming], address = Address{province = 'Hubei', city = 'wuhan'}}
I/MainActivity: googleJsonToArrayList: javabean -> Person{name = 'lisi', age = 22, hobbies =
[reading, swimming, dancing], address = Address{province = 'Hunan', city = 'changsha'}}
```

6. 利用 Gson 进行 List 与 JSON 的转换的另一实例

在 MainActivity 类中添加方法 googleList2Json()，方法的定义如下：

```
    void googleList2Json(){
        Gson gson = new Gson();
        List<String> stringList = Arrays.asList("1","2","3","4","5");
        String jsonString = gson.toJson(stringList);
        Log.i(TAG, "gsonListToJson: 1 --> " + jsonString);
        Type type = new TypeToken<ArrayList<String>>(){}.getType();
        ArrayList<String> arrayList = gson.fromJson(jsonString,type);
        Log.i(TAG, "googleList2Json: 2 --> " + arrayList);
```

```
        Map<String,Object> map = new HashMap<>();
        map.put("name","zhangsan");
        map.put("age",25);
        Log.i(TAG, "googleList2Json:3 --> " + gson.toJson(map));
    }
```

最后编辑 MainActivity 类的 doMyJob()方法，doMyJob()方法的内容如下：

```
void doMyJob(){
    googleArrayList2Json();
}
```

编译运行，日志输出如下：

```
I/MainActivity: gsonListToJson: json->["1","2","3","4","5"]
I/MainActivity: googleList2Json: javabean->[1, 2, 3, 4, 5]
I/MainActivity: googleList2Json:json->{"age":25,"name":"zhangsan"}
```

7. 利用 org.json 包进行 JavaBean 与 JSON 的相互转换

具体步骤：在 MainActivity 类中添加方法 orgJavaBeanToJson()，该方法把一个 Person 对象转化为 JSON 字符串，返回值是 JSON 字符串。在 MainActivity 类中添加方法 orgJsonToJavaBean(String jsonString)把 JSON 字符串转化为 Person 对象。这两个方法的定义如下：

```
private final String NAME = "name";
    private final String AGE = "age";
    private final String PROVINCE = "province";
    private final String CITY = "city";
    private final String ADDRESS = "address";
    private final String HOBBIES = "hobbies";
    String orgJavaBeanToJson(){
        Person person = new Person("zhangsan",21,new String[]{"reading","swimming"},
                new Address("Hubei","wuhan"));
        Log.i(TAG, "orgJavaBeanToJson: javabean->" + person);
        JSONObject personObject = new JSONObject();
        JSONArray hobbiesJSONArray = new JSONArray();
        hobbiesJSONArray.put("reading");
        hobbiesJSONArray.put("swimming");
        try {
            JSONObject addressObject = new JSONObject();
            addressObject.put(PROVINCE,"Hubei");
            addressObject.put(CITY,"Wuhan");
            personObject.put(NAME,"zhangsan");
            personObject.put(AGE,21);
            personObject.put(ADDRESS,addressObject);
            personObject.put(HOBBIES,hobbiesJSONArray);
```

```
            Log.i(TAG, "orgJavaBeanToJson: json -> " + personObject.toString());
        } catch (JSONException e) {
            e.printStackTrace();
        }
        return personObject.toString();
    }
    void orgJsonToJavaBean(String jsonString){
        try {
         JSONObject personObject = new JSONObject(jsonString);
         JSONArray hobbiesJSONArray = personObject.getJSONArray(HOBBIES);
         JSONObject addressObject = personObject.getJSONObject(ADDRESS);
         Person person = new Person(personObject.getString(NAME),personObject.getInt(AGE),
             new String[]{hobbiesJSONArray.getString(0),hobbiesJSONArray.getString(1)},
             new Address(addressObject.getString(PROVINCE),addressObject.getString(CITY)));
             Log.i(TAG, "orgJsonToJavaBean: javaBean -> " + person);
         } catch (JSONException e) {
           e.printStackTrace();
         }
     }
```

最后编辑 MainActivity 类的 doMyJob()方法，doMyJob()方法的内容如下：

```
void doMyJob(){
    orgJsonToJavaBean(orgJavaBeanToJson());
}
```

编译运行，日志输出如下：

```
I/MainActivity: orgJavaBeanToJson: javabean -> Person{name = 'zhangsan', age = 21, hobbies =
[reading, swimming], address = Address{province = 'Hubei', city = 'wuhan'}}
I/MainActivity: orgJavaBeanToJson: json -> {"address":{"province":"Hubei","city":
"Wuhan"},"hobbies":["reading","swimming"],"age":21,"name":"zhangsan"}
I/MainActivity: orgJsonToJavaBean: javaBean -> Person{name = 'zhangsan', age = 21, hobbies =
[reading, swimming], address = Address{province = 'Hubei', city = 'Wuhan'}}
```

8. 利用 org.json 包进行 ArrayList<T>与 Json 的相互转换

具体步骤：在 MainActivity 类中添加方法 orgArrayListToJson()，该方法把一个 ArrayList<Person>转化为 Json 字符串，返回值是 Json 字符串。

在 MainActivity 类中添加方法 orgJsonToArrayList(String jsonString)把 Json 字符串转化为 List<Person>。这两个方法的定义如下：

```
String orgArrayListToJson(){
        List<Person> personList = new ArrayList<>();
        Person person = new Person("zhangsan",21,new String[]{"reading","swimming"},
                new Address("Hubei","wuhan"));
        personList.add(person);
        Log.i(TAG, "orgArrayListToJson: javaBean -> " + person);
```

```
        person = new Person("lisi",22,new String[]{"reading","dancing","playing football"},
                new Address("Hunan","changsha"));
        personList.add(person);
        Log.i(TAG, "orgArrayListToJson: javaBean -> " + person);
        JSONArray personArray = new JSONArray();
        Person personi = null;
        JSONObject personObject = null;
        for(int i = 0;i < personList.size();i++) {
            personi = personList.get(i);
            personObject = new JSONObject();
            JSONObject addressObject = new JSONObject();
            try {
                addressObject.put(PROVINCE, personi.getAddress().getProvince());
                addressObject.put(CITY, personi.getAddress().getCity());
                JSONArray hobbiesJSONArray = new JSONArray();
                for (int j = 0; j < personi.getHobbies().length; j++) {
                    hobbiesJSONArray.put(personi.getHobbies()[j]);
                }
                personObject.put(NAME, personi.getName());
                personObject.put(AGE, personi.getAge());
                personObject.put(ADDRESS, addressObject);
                personObject.put(HOBBIES, hobbiesJSONArray);
                personArray.put(personObject);
            } catch (JSONException e) {
                e.printStackTrace();
            }
        }
        Log.i(TAG, "orgArrayListToJson: json -> " + personArray.toString());
        return personArray.toString();
    }
    void orgJsonToArrayList(String jsonString){
        List<Person> personList = new ArrayList<>();
        Person person = null;
        try {
            JSONArray personArray = new JSONArray(jsonString);
            for(int i = 0;i < personArray.length();i++){
                JSONObject personObject = personArray.getJSONObject(i);
                JSONArray hobbiesJSONArray = personObject.getJSONArray(HOBBIES);
                String[] hobbies = new String[hobbiesJSONArray.length()];
                for(int j = 0;j < hobbiesJSONArray.length();j++){
                    hobbies[j] = hobbiesJSONArray.getString(j);
                }
                JSONObject addressObject = personObject.getJSONObject(ADDRESS);
                Address address = new Address(addressObject.getString(PROVINCE),addressObject.
getString(CITY));
                person = new Person(personObject.getString(NAME),personObject.getInt(AGE),
hobbies,address);
                personList.add(person);
                Log.i(TAG, "orgJsonToArrayList: JavaBean -> " + person);
```

```
            }
        } catch (JSONException e) {
            e.printStackTrace();
        }
    }
```

最后编辑 MainActivity 类的 doMyJob()方法，doMyJob()方法的内容如下：

```
void doMyJob(){
        orgJsonToArrayList(orgArrayListToJson());
    }
```

编译运行，日志输出如下：

```
I/MainActivity: orgArrayListToJson: javaBean -> Person{name = 'zhangsan', age = 21, hobbies = [reading, swimming], address = Address{province = 'Hubei', city = 'wuhan'}}
I/MainActivity: orgArrayListToJson: javaBean -> Person{name = 'lisi', age = 22, hobbies = [reading, dancing, playing football], address = Address{province = 'Hunan', city = 'changsha'}}
I/MainActivity: orgArrayListToJson: json -> [{"address":{"province":"Hubei","city":"wuhan"},"hobbies":["reading","swimming"],"age":21,"name":"zhangsan"},{"address":{"province":"Hunan","city":"changsha"},"hobbies":["reading","dancing","playing football"],"age":22,"name":"lisi"}]
I/MainActivity: orgJsonToArrayList: JavaBean -> Person{name = 'zhangsan', age = 21, hobbies = [reading, swimming], address = Address{province = 'Hubei', city = 'wuhan'}}
I/MainActivity: orgJsonToArrayList: JavaBean -> Person{name = 'lisi', age = 22, hobbies = [reading, dancing, playing football], address = Address{province = 'Hunan', city = 'changsha'}}
```

9. 利用 org.json 包的 JSONStringer 进行 JavaBean 与 Json 的相互转换

具体步骤：在 MainActivity 类中添加方法 orgJSONStringerJavaBeanToJson()，该方法利用 JsonStringer 把一个 Person 对象转化为 Json 字符串，返回值是 Json 字符串。方法的定义如下：

```
String orgJSONStringerJavaBeanToJson(){
        Person person = new Person("zhangsan",21,new String[]{"reading","swimming"},
                new Address("Hubei","wuhan"));
        Log.i(TAG, "orgJSONStringerJavaBeanToJson: javaBean -> " + person);
        JSONStringer jsonStringer = new JSONStringer();
        try {
            jsonStringer.object();
            jsonStringer.key(NAME).value("zhangsan");
            jsonStringer.key(AGE).value(21);

jsonStringer.key(HOBBIES).array().value("reading").value("swimming").endArray();
jsonStringer.key(ADDRESS).object().key(PROVINCE).value("Hubei").key(CITY).value("Wuhan").endObject();
            jsonStringer.endObject();
        } catch (JSONException e) {
```

```
        e.printStackTrace();
    }
    Log.i(TAG, "orgJSONStringerJavaBeanToJson: json -> " + jsonStringer.toString());
    return jsonStringer.toString();
}
```

最后编辑 MainActivity 类的 doMyJob()方法，doMyJob()方法的内容如下：

```
void doMyJob(){
    orgJsonToJavaBean(orgJSONStringerJavaBeanToJson());
}
```

编译运行，日志输出如下：

```
I/MainActivity: orgJSONStringerJavaBeanToJson: javaBean -> Person{name = 'zhangsan', age =
21, hobbies = [reading, swimming], address = Address{province = 'Hubei', city = 'wuhan'}}
I/MainActivity: orgJSONStringerJavaBeanToJson: json -> {"name":"zhangsan","age":21,
"hobbies":["reading","swimming"],"address":{"province":"Hubei","city":"Wuhan"}}
I/MainActivity: orgJsonToJavaBean: javaBean -> Person{name = 'zhangsan', age = 21, hobbies =
[reading, swimming], address = Address{province = 'Hubei', city = 'Wuhan'}}
```

16.6 XML 解析实验

1. 新建工程 XML1

具体步骤：运行 Android Studio，单击 Start a new Android Studio project，在 Create New Project 页中，Application name 为 XML1，Company domain 为 cuc，Project location 为 C:\android\myproject\XML1，单击 Next 按钮，选中 Phone and Tablet，minimum sdk 设置为 API 19：Android 4.4(minSdkVersion)，单击 Next 按钮，选择 Empty Activity，再单击 Next 按钮，勾选 Generate Layout File 选项，勾选 Backwards Compatibility(AppCompat)选项，单击 Finish 按钮。

2. 编辑布局文件 activity_main.xml

在布局文件里，添加一个按钮。activity_main.xml 内容如下：

```
<?xml version = "1.0" encoding = "utf - 8"?>
<android.support.constraint.ConstraintLayout xmlns:android = "http://schemas.android.com/
apk/res/android"
    xmlns:app = "http://schemas.android.com/apk/res - auto"
    xmlns:tools = "http://schemas.android.com/tools"
    android:layout_width = "match_parent"
    android:layout_height = "match_parent"
    tools:context = "cuc.xml1.MainActivity">
    <Button
        android:layout_width = "wrap_content"
```

```
        android:layout_height = "wrap_content"
        android:id = "@ + id/parseXmlBtn"
        android:text = "parseXml"/>
</android.support.constraint.ConstraintLayout>
```

3. 新建一个类 Person

具体步骤：在工程结构视图中 main 源集 Java 文件夹里，右击 cuc.xml1 包，选择 New|Java Class，Name 输入框中输入 Person，单击 OK 按钮，系统生成 Person.java 文件。利用菜单 Code|Generate，生成 Getter and Setter 方法，以及 Constructor。最后 Person.java 文件内容如下：

```
package cuc.xml1;

public class Person {
    private int id;
    private String name;
    private int age;
    public Person(int id) {
        this.id = id;
    }
    public Integer getId() {
        return id;
    }
    public void setId(Integer id) {
        this.id = id;
    }
    public String getName() {
        return name;
    }
    public void setName(String name) {
        this.name = name;
    }
    public int getAge() {
        return age;
    }
    public void setAge(int age) {
        this.age = age;
    }
    @Override
    public String toString() {
        return "Person{" +
                "id = " + id +
                ", name = '" + name + '\'' +
                ", age = " + age +
                '}';
    }
}
```

4. 新建一个 XML 文件 person. xml

具体步骤：如果在 res 文件夹里没有 XML 文件夹，就在 res 文件夹里新建 XML 文件夹。

具体步骤为：在工程结构视图中，右击 res 文件夹，选择 New | Android Resource Directory，从 Resource type 右边的下拉列表框中，选择 XML 选项，单击 OK 按钮，这时系统已经创建了 XML 资源文件夹。

如果 res 文件夹下没有 raw 文件夹，在工程结构视图中，右击 res 文件夹，选择 New | Android Resource Directory，从 Resource type 右边的下拉列表框中，选择 raw 选项，单击 OK 按钮，这时系统已经创建了 raw 资源文件夹。

再在 src/main 文件夹里新建一个 assets folder。具体步骤为：在工程结构视图中，右击 app，在弹出菜单中依次单击 New | Folder | Assets Folder，单击 Finish 按钮，这时系统就在 src/main 文件夹里新建一个 assets 文件夹。

然后在 XML 文件夹里新建文件 person. xml。文件 person. xml 内容如下：

```
<?xml version = "1.0" encoding = "utf - 8"?>
<persons xmlns:android = "http://schemas.android.com/apk/res/android">
    <person id = "5" name = "the 5 person">
        <name> zhangsan </name>
        <age> 21 </age>
    </person>
    <person id = "6" name = "the 6 person">
        <name> lisi </name>
        <age> 21 </age>
    </person>
</persons>
```

再把 XML 文件夹里的文件 person. xml 分别拷贝到 raw 文件夹和 assets 文件夹。

5. 使用 DOM 解析 XML

当单击 parseXmlBtn 按钮时，调用 domParse 对 res/raw 文件夹里的 person. xml 进行解析。使用 W3C 标准的 Document Object Model 解析 XML 文件，首先是根据指定的 XML 文件创建一个 Document 对象，然后调用 document. getElementsByTagName("person")，获得名字是 person 的所有元素，然后循环处理每一个 person 元素，检查 person 元素的子结点类型是元素结点的(name 和 age)，再对 name 和 age 这两个元素结点进行处理。

编辑 MainActivity. java，文件内容如下：

```
package cuc.xml1;
import android.os.Bundle;
import android.support.v7.app.AppCompatActivity;
import android.util.Log;
import android.view.View;
import android.widget.Button;
import org.w3c.dom.Document;
import org.w3c.dom.Node;
import org.w3c.dom.NodeList;
import org.w3c.dom.Element;
```

```
import org.xml.sax.SAXException;
import java.io.IOException;
import java.io.InputStream;
import java.util.ArrayList;
import java.util.List;
import javax.xml.parsers.DocumentBuilder;
import javax.xml.parsers.DocumentBuilderFactory;
import javax.xml.parsers.ParserConfigurationException;

public class MainActivity extends AppCompatActivity {
  String[] NodeType = new String[]{
        null,
        "ELEMENT_NODE", //1
        "ATTRIBUTE_NODE", //2
        "TEXT_NODE",//3
        "CDATA_SECTION_NODE",//4
        "ENTITY_REFERENCE_NODE",//5
        "ENTITY_NODE",//6
        "PROCESSING_INSTRUCTION_NODE",//7
        "COMMENT_NODE",//8
        "DOCUMENT_NODE",//9
        "DOCUMENT_TYPE_NODE",//10
        "DOCUMENT_FRAGMENT_NODE",//11
        "NOTATION_NODE "//12
  };

  private static final String TAG = "MainActivity";
  private Button parseXmlBtn;
  private void assignViews() {
     parseXmlBtn = (Button) findViewById(R.id.parseXmlBtn);
  }
  void domParse(){
     try {
        DocumentBuilderFactory factory = DocumentBuilderFactory.newInstance();
        DocumentBuilder documentBuilder = factory.newDocumentBuilder();
        InputStream inputStream = this.getResources().openRawResource(R.raw.person);
        Document document =  documentBuilder.parse(inputStream);
        List<Person> personList = new ArrayList<>();
        NodeList nodeList = document.getElementsByTagName("person");
        int countPerson = nodeList.getLength();
        Log.i(TAG, "domParse: countPerson:" + countPerson);
        for(int i = 0;i < countPerson;i++){
           Element personElement = (Element)nodeList.item(i);
           Person currentPerson = new Person(Integer.parseInt(personElement.getAttribute
("id")));
           NodeList childNodeList = personElement.getChildNodes();
           int countChild = childNodeList.getLength();
           Log.i(TAG, "domParse: countChild:" + countChild);
           for (int j = 0;j < countChild;j++){
```

```
                Node childNode = childNodeList.item(j);
                Log.i(TAG, "domParse: node " + j + ":" + NodeType[childNode.getNodeType()] +
                        ",NodeName:" + childNode.getNodeName());
                if (childNode.getNodeType() == Node.ELEMENT_NODE){
                    Element childElement = (Element)childNode;
                    if (childElement.getNodeName().equals("name")){
currentPerson.setName(childElement.getFirstChild().getNodeValue());

                    }else if (childElement.getNodeName().equals("age")){
currentPerson.setAge(Integer.parseInt(childElement.getFirstChild().getNodeValue()));
                    }
                }
            }
            personList.add(currentPerson);
        }
        for (int k = 0;k < personList.size();k++){
            Log.i(TAG, "domParse: " + personList.get(k));
        }

    } catch (ParserConfigurationException e) {
        e.printStackTrace();
    } catch (SAXException e) {
        e.printStackTrace();
    } catch (IOException e) {
        e.printStackTrace();
    }
}
@Override
protected void onCreate(Bundle savedInstanceState) {
    super.onCreate(savedInstanceState);
    setContentView(R.layout.activity_main);
    assignViews();
    parseXmlBtn.setOnClickListener(new View.OnClickListener() {
        @Override
        public void onClick(View view) {
            domParse();
        }
    });
}
}
```

上面代码中的 NodeType 字符串数组是根据 Node 类中的常量定义的。org.w3c.dom.Node 类中定义中以下常数：

```
short ELEMENT_NODE = 1;
short ATTRIBUTE_NODE = 2;
short TEXT_NODE = 3;
short CDATA_SECTION_NODE = 4;
```

```
    short ENTITY_REFERENCE_NODE = 5;
    short ENTITY_NODE = 6;
    short PROCESSING_INSTRUCTION_NODE = 7;
    short COMMENT_NODE = 8;
    short DOCUMENT_NODE = 9;
    short DOCUMENT_TYPE_NODE = 10;
    short DOCUMENT_FRAGMENT_NODE = 11;
    short NOTATION_NODE = 12;
```

编译运行，log 日志输出如下：

```
cuc.xml1 I/MainActivity: domParse: countPerson:2
cuc.xml1 I/MainActivity: domParse: countChild:5
cuc.xml1 I/MainActivity: domParse: node 0:TEXT_NODE,NodeName:#text
cuc.xml1 I/MainActivity: domParse: node 1:ELEMENT_NODE,NodeName:name
cuc.xml1 I/MainActivity: domParse: node 2:TEXT_NODE,NodeName:#text
cuc.xml1 I/MainActivity: domParse: node 3:ELEMENT_NODE,NodeName:age
cuc.xml1 I/MainActivity: domParse: node 4:TEXT_NODE,NodeName:#text
cuc.xml1 I/MainActivity: domParse: countChild:5
cuc.xml1 I/MainActivity: domParse: node 0:TEXT_NODE,NodeName:#text
cuc.xml1 I/MainActivity: domParse: node 1:ELEMENT_NODE,NodeName:name
cuc.xml1 I/MainActivity: domParse: node 2:TEXT_NODE,NodeName:#text
cuc.xml1 I/MainActivity: domParse: node 3:ELEMENT_NODE,NodeName:age
cuc.xml1 I/MainActivity: domParse: node 4:TEXT_NODE,NodeName:#text
cuc.xml1 I/MainActivity: domParse: Person{id=5, name='zhangsan', age=21}
cuc.xml1 I/MainActivity: domParse: Person{id=6, name='lisi', age=21}
```

6. 使用 SAXParser 解析 XML

首先在 cuc.xml1 包里，新建 MyDefaultHandler 类继承自 DefaultHandler，并重写 startDocument()、endDocument()、startElement()、endElement()和 characters()方法。MyDefaultHandler.java 内容如下：

```
package cuc.xml1;
import android.util.Log;
import org.xml.sax.Attributes;
import org.xml.sax.SAXException;
import org.xml.sax.helpers.DefaultHandler;
import java.util.ArrayList;
import java.util.List;

public class MyDefaultHandler extends DefaultHandler {
    private static final String TAG = "MyDefaultHandler";
    private Person currentPerson;
    private List<Person> personList;
    private String tagName = null;

    public MyDefaultHandler() {
        super();
```

```
    }
    public List<Person> getPersonList(){
        return personList;
    }
    @Override
    public void startDocument() throws SAXException {
        super.startDocument();
        Log.i(TAG, "startDocument: ");
    }
    @Override
    public void endDocument() throws SAXException {
        super.endDocument();
        Log.i(TAG, "endDocument: ");
    }
    @Override
    public void startElement(String uri, String localName, String qName, Attributes
attributes) throws SAXException {
        super.startElement(uri, localName, qName, attributes);
        if(localName.equals("persons")) {
            personList = new ArrayList<>();
        }else if(localName.equals("person")){
            currentPerson = new Person(Integer.parseInt(attributes.getValue("id")));
        }
        Log.i(TAG, "startElement: " + localName);
        tagName = localName;
    }
    @Override
    public void endElement(String uri, String localName, String qName) throws SAXException {
        super.endElement(uri, localName, qName);
        if (localName.equals("person")){
            personList.add(currentPerson);
            currentPerson = null;
        }
        Log.i(TAG, "endElement: " + localName);
        tagName = null;
    }

    @Override
    public void characters(char[] ch, int start, int length) throws SAXException {
        if(tagName!= null){
            String text = new String(ch,start,length);
            if(tagName.equals("name")){
                currentPerson.setName(text);
            }else if(tagName.equals("age")){
                currentPerson.setAge(Integer.parseInt(text));
            }
        }
    }
}
```

然后给类 MainActivity 添加方法 saxparse()，此方法使用 SAX 接口解析 XML 文件。saxparse()方法定义如下：

```
void saxparse() {
    SAXParserFactory factory = SAXParserFactory.newInstance();
    SAXParser parser = null;
    try {
        parser = factory.newSAXParser();
        MyDefaultHandler myDefaultHandler = new MyDefaultHandler();
        InputStream inputStream = this.getResources().openRawResource(R.raw.person);
        parser.parse(inputStream,myDefaultHandler);
        inputStream.close();
        List<Person> personList = myDefaultHandler.getPersonList();
        for (int i = 0;i<personList.size();i++){
            Log.i(TAG, "saxparser: " + personList.get(i));
        }
    } catch (ParserConfigurationException e) {
        e.printStackTrace();
    } catch (SAXException e) {
        e.printStackTrace();
    } catch (IOException e) {
        e.printStackTrace();
    }
}
```

最后在 parseXmlBtn 按钮的监听函数 OnClick 中把调用函数 domParse()改为调用 saxparse()，即原来使用文档对象模型解析 person.xml，现在采用 Simple API for XML 解析，person.xml 存储在 app/src/main/res/raw 文件夹里，上述的编辑完成后，编译运行，运行后单击界面中的 parseXml 按钮，就可以调用 saxparse()对 person.xml 进行解析了。

7. 采用 XMLPullParser 解析字符串

XMLPullParser 属于拉式解析。拉式解析的过程主要分成两步：第一步，创建一个首先创建一个 XMLPullParser 解析器，并设置它的输入流；第二步，应用程序根据自己的需要调用解析器的 next()、nextTag()或者 nextToken()获取下一个解析事件或者调用 getEventType()获取现在的解析事件的类型。

首先在类 MainActivity 中定义一个字符串数组 XMLPullParserEventType 的定义，因为 next()返回的事件类型是个整数，在日志输出中输出代表事件类型的字符串更容易理解。然后在类 MainActivity 添加一个方法 parseString()，该方法定义了解析字符串 "<foo>Hello，the World<hi> Hello，CUC</hi></foo>"的方法。

在类 MainActivity 中添加的代码如下：

```
String[] XmlPullParserEventType = new String[]{
        "START_DOCUMENT",//0
        "END_DOCUMENT",//1
        "START_TAG",//2
        "END_TAG",//3
```

```
            "TEXT",//4
            "CDSECT",//5
            "ENTITY_REF",//6
            "IGNORABLE_WHITESPACE",//7
            "PROCESSING_INSTRUCTION",//8
            "COMMENT",//9
            "DOCDECL",//10
    };

    void parseString() {
        XmlPullParserFactory factory = null;
        try {
            factory = XmlPullParserFactory.newInstance();
            factory.setNamespaceAware(true);
            XmlPullParser parser = factory.newPullParser();
            StringReader stringReader = new StringReader("<foo>Hello,the World<hi>Hello,CUC
    </hi></foo>");
            parser.setInput(stringReader);
            int eventType = parser.getEventType();
            while (eventType!= XmlPullParser.END_DOCUMENT){
                Log.i(TAG, "parseString: eventType:" + XmlPullParserEventType[eventType]
                 + ((eventType == XmlPullParser.START_TAG)||(eventType == XmlPullParser.END_
    TAG)?(",name:" + parser.getName()):"" )
                 + ((eventType == XmlPullParser.TEXT)?(":" + parser.getText()):"")
                );
                eventType = parser.next();
            }
        } catch (XmlPullParserException e) {
            e.printStackTrace();
        } catch (IOException e) {
            e.printStackTrace();
        }
        Log.i(TAG, "parseString: END_DOCUMENT");
    }
```

上面代码中的字符串数组 XMLPullParserEventType 是根据 XMLPullParser 接口中的整型常数定义的。下面列举了 XMLPullParser 接口中定义的一些整型常数：

```
int START_DOCUMENT = 0;
int END_DOCUMENT = 1;
int START_TAG = 2;
int END_TAG = 3;
int TEXT = 4;
int CDSECT = 5;
int ENTITY_REF = 6;
int IGNORABLE_WHITESPACE = 7;
int PROCESSING_INSTRUCTION = 8;
int COMMENT = 9;
```

```
int DOCDECL = 10;
```

最后在 parseXmlBtn 按钮的监听函数 OnClick 中把调用函数 saxparse()改为调用 parseString()，这里解析的是一个字符串，上述的编辑完成后，编译运行，运行后单击界面中的 parseXML 按钮，就可以调用 parseString()对字符串进行解析了，log 输出如下：

```
cuc.xml1 I/MainActivity: parseString: eventType:START_DOCUMENT
cuc.xml1 I/MainActivity: parseString: eventType:START_TAG,name:foo
cuc.xml1 I/MainActivity: parseString: eventType:TEXT:Hello,the World
cuc.xml1 I/MainActivity: parseString: eventType:START_TAG,name:hi
cuc.xml1 I/MainActivity: parseString: eventType:TEXT:Hello,CUC
cuc.xml1 I/MainActivity: parseString: eventType:END_TAG,name:hi
cuc.xml1 I/MainActivity: parseString: eventType:END_TAG,name:foo
cuc.xml1 I/MainActivity: parseString: END_DOCUMENT
```

8. 采用 XMLPullParser 解析 person. xml

首先在类 MainActivity 中添加 3 个方法：getParserForXML()、getParserForRaw()和 getParserForAssets()用于获取 XMLPullParser，以便解析不同文件夹下的 person. xml；再定义一个方法 pullParse()，该方法定义了利用刚刚获取的 XMLPullParser 解析 person. xml 的具体过程。

```
private XmlPullParser getParserForXml(){
    return this.getResources().getXml(R.xml.person);
}
private XmlPullParser getParserForRaw() {
    XmlPullParser parser = null;
    try {
        XmlPullParserFactory factory  =  XmlPullParserFactory.newInstance();
        parser = factory.newPullParser();
        InputStream inputStream = this.getResources().openRawResource(R.raw.person);
        parser.setInput(inputStream,"utf-8");
    } catch (XmlPullParserException e) {
        e.printStackTrace();
    }
    /* * 4.xml 文件放在 SD 卡，path 路径根据实际项目修改，此次获取 SDcard 根目录：
 * String path  =  Environment.getExternalStorageDirectory().toString();
 * File xmlFlie  =  new File(path + fileName);
 * InputStream inputStream  =  new FileInputStream(xmlFlie);
 * */
    return parser;
}
private XmlPullParser getParserForAssets(){

    XmlPullParser parser = null;
    try {
        XmlPullParserFactory factory  =  XmlPullParserFactory.newInstance();
        factory.setNamespaceAware(true);
```

```
            parser = factory.newPullParser();
            AssetManager assetManager = getAssets();
            InputStream inputStream = assetManager.open("person.xml");
            parser.setInput(inputStream,"utf-8");
        } catch (XmlPullParserException e) {
            e.printStackTrace();
        } catch (IOException e) {
            e.printStackTrace();
        }
        return parser;
    }
    void pullParse(XmlPullParser parser){
        int eventType = -1;
        String text;
        List<Person> personList = null;
        Person currentPerson = null;
        try {
            eventType = parser.getEventType();
            while(eventType!= XmlPullParser.END_DOCUMENT){
                switch (eventType){
                    case XmlPullParser.END_DOCUMENT:
                        break;
                    case XmlPullParser.END_TAG:
                        Log.i(TAG, "parseXml1: END_TAG:" + parser.getName());
                        if (parser.getName().equals("person")){
                            personList.add(currentPerson);
                            currentPerson = null;
                        }else if (parser.getName().equals("persons")){
                            for (int i = 0;i<personList.size();i++){
                                Log.i(TAG, "parseXml1: " + personList.get(i));
                            }
                        }
                        break;
                    case XmlPullParser.START_TAG:
                        Log.i(TAG, "parseXml1: START_TAG:" + parser.getName());
                        if (parser.getName().equals("persons")){
                            personList = new ArrayList<>();
                        }else if (parser.getName().equals("person")){
                            String idStr = parser.getAttributeValue(null,"id");
                            String nameAttr = parser.getAttributeValue(1);
                            Log.i(TAG, "parseXml1:id: " + idStr + " ,name:" + nameAttr);
                            currentPerson = new Person(Integer.parseInt(idStr));

                        }else if (parser.getName().equals("name")){
                            text = parser.nextText();
                            Log.i(TAG, "parseXml1: " + text);
                            currentPerson.setName(text);
                        }else if (parser.getName().equals("age")){
                            text = parser.nextText();
```

```
                        Log.i(TAG, "parseXml1: " + text);
                        currentPerson.setAge(Integer.parseInt(text));
                    }

            }
            eventType = parser.next();
        }
    } catch (XmlPullParserException e) {
        e.printStackTrace();
    } catch (IOException e) {
        e.printStackTrace();
    }
}
```

这里根据 XML 文件的存储位置定义了不同的获取 XMLPullParser 对象的方法：方法 getParserForXML()获取的解析器，用于解析 app\src\main\res\xml 文件夹下 person.xml；方法 getParserForRaw()获取的解析器，用于解析 app\src\main\res\raw 文件夹下 person.xml；方法 getParserForAssets()获取的解析器，用于解析 app\src\main\assets 文件夹下 person.xml。

最后在 parseXMLBtn 按钮的监听函数 OnClick 中把调用函数 parseString ()改为调用 pullParse (getParserForXml ( ))、pullParse (getParserForRaw ( )) 或者 pullParse (getParserForAssets()) 就可以进行对 person.xml 的解析了。上述的编辑完成后，编译运行，运行后单击界面中 parseXML 的按钮，就可以调用 saxparse()对 person.xml 进行解析，日志输出如下：

```
cuc.xml1 I/MainActivity: parseXml1: START_TAG:persons
cuc.xml1 I/MainActivity: parseXml1: START_TAG:person
cuc.xml1 I/MainActivity: parseXml1:id: 5 ,name:the 5 person
cuc.xml1 I/MainActivity: parseXml1: START_TAG:name
cuc.xml1 I/MainActivity: parseXml1: zhangsan
cuc.xml1 I/MainActivity: parseXml1: START_TAG:age
cuc.xml1 I/MainActivity: parseXml1: 21
cuc.xml1 I/MainActivity: parseXml1: END_TAG:person
cuc.xml1 I/MainActivity: parseXml1: START_TAG:person
cuc.xml1 I/MainActivity: parseXml1:id: 6 ,name:the 6 person
cuc.xml1 I/MainActivity: parseXml1: START_TAG:name
cuc.xml1 I/MainActivity: parseXml1: lisi
cuc.xml1 I/MainActivity: parseXml1: START_TAG:age
cuc.xml1 I/MainActivity: parseXml1: 21
cuc.xml1 I/MainActivity: parseXml1: END_TAG:person
cuc.xml1 I/MainActivity: parseXml1: END_TAG:persons
cuc.xml1 I/MainActivity: parseXml1: Person{id = 5, name = 'zhangsan', age = 21}
cuc.xml1 I/MainActivity: parseXml1: Person{id = 6, name = 'lisi', age = 21}
```

16.7 本章主要参考文献

1. https://developer.android.google.cn/reference/javax/net/ssl/HttpsURLConnection.html
2. https://github.com/google/gson/blob/master/UserGuide.md
3. https://developer.android.google.cn/reference/org/xmlpull/v1/XmlPullParser.html
4. https://developer.android.google.cn/reference/javax/xml/parsers/DocumentBuilder.html
5. https://developer.android.google.cn/reference/javax/xml/parsers/DocumentBuilderFactory.html
6. https://developer.android.google.cn/reference/javax/xml/parsers/SAXParserFactory.html
7. https://developer.android.google.cn/reference/javax/xml/parsers/SAXParser.html

图书资源支持

感谢您一直以来对清华版图书的支持和爱护。为了配合本书的使用，本书提供配套的资源，有需求的读者请扫描下方的"清华电子"微信公众号二维码，在图书专区下载，也可以拨打电话或发送电子邮件咨询。

如果您在使用本书的过程中遇到了什么问题，或者有相关图书出版计划，也请您发邮件告诉我们，以便我们更好地为您服务。

我们的联系方式：

地　　址：北京市海淀区双清路学研大厦 A 座 701

邮　　编：100084

电　　话：010－62770175－4608

资源下载：http://www.tup.com.cn

客服邮箱：tupjsj@vip.163.com

QQ：2301891038（请写明您的单位和姓名）

教学交流、课程交流

清华电子

扫一扫，获取最新目录

用微信扫一扫右边的二维码，即可关注清华大学出版社公众号"清华电子"。